Urban High-Resolution Remote Sensing

Urban High-Resolution Remote Sensing

Algorithms and Modeling

Guoqing Zhou

CRC Press is an imprint of the
Taylor & Francis Group, an informa business

First edition published 2021
by CRC Press
6000 Broken Sound Parkway NW, Suite 300, Boca Raton, FL 33487-2742

and by CRC Press
2 Park Square, Milton Park, Abingdon, Oxon, OX14 4RN

CRC Press is an imprint of Taylor & Francis Group, LLC

Library of Congress Cataloging-in-Publication Data

ISBN: 978-0-367-85750-9 (hbk)
ISBN: 978-1-003-08243-9 (ebk)

Typeset in Times
by SPi Global, India

Contents

SECTION IV Advanced Algorithms Urban Remote Sensing Application

Foreword

We humans have drawn maps in sand, on clay, parchment, and paper, and now on computer screens and as holograms. More impressively, we have the spatio-temporal information in various types of databases available for complex computer analysis. This is important because our increasing population puts ever greater demands on our natural and societal resources. Better data and more complex analyses are needed to help meet those demands.

As our population has grown, our demographics have shifted increasingly toward greater concentrations of people. The World Bank estimated that the global population distribution shifted from being mostly rural to mostly urban in 2007. We have added an ever-expanding number of population centers and have grown many of them outward, upward, and downward to accommodate our population, industry, and commerce.

What is it that changes? Of course, the number of people changes. Whether there is growth or decline, the urban features change to accommodate it. How we use the land changes. Transportation networks grow or evolve to accommodate new needs. Communications networks adapt to new technologies. Buildings become taller or alter shape to adapt to physical conditions and changing architectural capabilities and aesthetics. Obtaining, maintaining, interpreting, and using information about these changes is a never-ending task.

In this book, *Urban High-Resolution Remote Sensing: Algorithms and Modeling*, Guoqing Zhou presents remote sensing techniques to help ensure the quality of data and effectiveness of analysis to address some of the issues that have emerged in our growing urban infrastructure. He provides tested algorithms and experimentally proven mathematical models that give the student in remote sensing an excellent base of expertise on which to build. The book is also a useful reference for the seasoned remote sensing scientist and geospatial expert. It gives users of geospatial information in urban areas ideas of some things that can be done with remote sensing to help address questions about the urban physical condition, as well as mathematical tools that help meet some of their analytical needs.

Dr. Zhou has invested his career in advancing our knowledge of high-resolution remote sensing, developing new methods and applications, and passing that knowledge on to his students and other scientists around the world. He has been selfless in sharing his knowledge with all of us.

John A. Kelmelis, PhD, GISP
Board of Directors, GIS Certification Institute
Chief Scientist for Geography, US Geological Survey (retired)
Senior Counselor for Earth Science, US Department of State (retired)
Professor, the Pennsylvania State University (retired)

Preface

With increasing interest in and requests for applications of remote sensing in urban areas, the development of urban remote sensing algorithms and methodologies driven by applications from different fields has become imperative. In addition to the traditional fields of application such as land use and cover and environmental pollution monitoring, remote sensing has been widely applied in other fields such as micro-climate analysis, cell phone planning, and socioeconomic analysis.

The objective of this book is to provide readers with algorithms and methods with an emphasis on urban three-dimensional modeling, urban orthophotomap generation, and urban feature extraction from high-resolution urban remotely sensed imagery. On the other hand, it has been observed that none of the eight published books relevant to urban remote sensing has specially presented algorithms and methods relevant to urban 3D modeling, orthophotomap generation, and feature extractions. Therefore, another objective of this book is to compensate for the gap.

When we talk about urban remote sensing, we should NOT forget its roots. In the early nineteenth century, balloon-based aerial photography was applied in city monitoring and observations. Since that era, three major tasks in urban remote sensing are urban digital surface model (DSM) creation, including digital terrain models (DTM), orthophotomap generation, and feature extractions. The book therefore presents novel developments of methods and algorithms surrounding this root through dealing with various data resources, such as UAV-based aerial image video, LiDAR (light detection and ranging) data, high-resolution satellite imagery, and so on. Readers should be very interested in many novel algorithms and methods, such as high building orthorectification with self-constraints, deep learning neural networks for urban classification, co-location decision trees for urban classification. The arrangement of the book is below:

- Section I is an introduction to urban remote sensing fundamentals.
- Section II emphasizes the algorithms and methods of urban feature extractions, including DSM extraction, power line extraction, tree extraction, and car/vehicle extraction on the freeway using advanced methods such as co-location decision trees, and deep learning neural networks.
- Section III focuses on the algorithms and methods for urban orthophotomap generation, especially urban true orthorectification.
- Section IV presents novel algorithms and methods for case applications, including soil surface moisture (SSM) retrieval and asphalt road aging monitoring.

I believe several features of the book will enhance its value as a textbook or reference book. First, the algorithms and methods described in this book are almost all for high-resolution imagery, that is, the ground sampling distance (GSD) is less than 1.0 meter. Second, the algorithms and methods presented in this book focus on the three traditional urban remote sensing products, namely, the 3D urban model, orthophotomap, and feature extraction. Third, both linear and sporadic reading are appropriate, in that sporadic readers can completely understand the essence of any chapter without needing to read all the previous chapters. Readers may therefore read the chapters relevant to his/her background. However, the chapters have a logical order, which is given above especially for various interested readers.

This book is based on my early research work and articles published in journals. These unconnected subjects have become consolidated in lecture notes for a one-semester, three-credit graduate course. This book can therefore be used as a textbook in universities for upper-level undergraduate and graduate students, and may also serve as a reference book for researchers in academia, and in governmental and industrial sectors. Prerequisites for upper-level undergraduate and/or graduate courses are "Introduction to Remote Sensing", or "Remote Sensing and Digital Image Processing", or "Digital Image Processing with Remote Sensing Perspective".

Guoqing Zhou

Acknowledgments

This book is based on my current research work and the early articles published in journals. The early work dates back to 1996 when I worked in the Technical University of Berlin (TU-Berlin), Germany as the Alexander von Humboldt (AvH) Fellow. Therefore, I give particular thanks to Prof. Joerg Albertz for his bringing me into the field of aerial image flow processing. Many thanks are also due to my coauthors, who were basically my graduate students and my supervisors while pursuing my PhD and post-doctoral research, for their valuable inputs and impetus to my research. In addition, I would like to thank my graduate students and my research assistant, who have been involved in various aspects of this book, such as image provision, and format editions.

Finally, I should very much like to thank my wife, Michelle M. Xie, and two sons, Deqi Zhou and Mitchell Zhou. I thank them for those occasions when I dedicated myself to work on the manuscript rather than spending more time with them.

Guoqing Zhou

Acknowledgments

[illegible]

A Note on the Author

Guoqing Zhou received a PhD degree from Wuhan University, Wuhan, China, in 1994. He was a Visiting Scholar with the Department of Computer Science and Technology, Tsinghua University, Beijing, China, and a Post-Doctoral Researcher with the Institute of Information Science, Beijing Jiaotong University, Beijing. He continued his research as an Alexander von Humboldt Fellow with the Technical University of Berlin, Berlin, Germany from 1996 to 1998, and was a Post-Doctoral Researcher with The Ohio State University, Columbus, OH, USA from 1998 to 2000. He was made an Assistant Professor, an Associate Professor, and a Full Professor of Old Dominion University, Norfolk, VA, USA in 2000, 2005, and 2010, respectively. Dr. Zhou received a second PhD degree from Virginia Tech, Blacksburg, Virginia, USA in 2011. He has authored 6 books and refereed more than 400 publications.

About the Author

[illegible]

List of Abbreviations

ADMS	attribute data management system
AFOV	angular field of view
AIMS	Airborne Integrated Mapping System
AL	Attribute Linkage
API	Application Programming Interface
APS	Atmospheric Phase Screen
ARAN	Automatic Road Analyzer
ASD	Analytical Spectral Devices
ATI	approximation of thermal inertia
AVI	automatic vehicle identification
BM	Bench Mark
BP	Backpropagation Processes
BR	Boundary Representation
CART	Classification and Regression Trees
CC	Correlation Coefficient
CCD	Charge Coupled Device
CED	Configurational Edge Density
CI	Clumpiness Index
CL	cultivated land
CL-DT	co-location-based decision tree
CNNs	Convolutional Neural Networks
CSG	Constructive Solid Geometry
CWSI	crop water stress index
DBD	Digital Building Data
DBH	Diameter at Breast Height
DBM	Digital Building Model
DEM	Digital Elevation Model
DGPS	differential GPS
DINSAR	Differential Interferometric Synthetic Aperture Radar
DISP	Declassified Intelligence Satellite Photography
DLR	German Aerospace Research Establishment
DLT	Direct Linear Transformation
DMI	distance measuring instrument
DN	digital number
DNNs	deep neural networks
DOQ	Digital Orthophoto Quadrangle
DOQs	Digital Orthophoto Quarter-Quadrangles
DPU	Digital Processing Unit
DR	Density Ratio
DRO	Duda Road Operators
D-S	Dempster-Shafer
DSM	digital surface model
DT	decision tree
DTM	digital terrain model
DOQ	Digital Orthophoto Quadrangle
EC	Exposed Carbonatite
ED	Edge Density

EMPs	Extended Morphological Profiles
EMVU	Extended Maximum Variance Unfolding
EOPs	Exterior Orientation Parameters
EOS	Earth Observing System
EPIs	Epipolar Plane Images
ESRI	Environmental Science Research Institute
FHWA	Federal Highway Administration
FI	Facet Identifier
FN	False Negative
FOV	Field of View
FP	False Positive
FR	Full Range
FSF	First scale Spatial Features
GCPs	Ground Control Points
GE	Google Earth
GIS	Geographic Information System
GLCM	Gray-Level Co-Occurrence Matrix
GPS	Global Positioning System
G-roof	ghost-roof
GRS	Geodetic Reference System
GSD	ground sample distance
GUI	Graphic User Interface
HCNNs	Hierarchical CNNs
HMSFs	Hierarchical Multiscale Spatial Features
HRSC	High-Resolution Stereo Camera
HSF	Hierarchical Spatial Features
HST	HyperSpecTIR
ID	Identification
IDW	Inverse Distance Weight
INS	Inertial Navigation System
IOP	Interior Orientation Parameter
IP	Iterative Photogrammetric
IR	Infrared
IRMSS	Infrared Multispectral Scanner
IRT	Iterative Ray-Tracing
ISA	Impervious surface area
JDK	Java Development Kit
JPL	Jet Propulsion Laboratory
JSDK	Java Servlet Development Kit
JSP	Java Server Pages
KC	Kappa Coefficients
KRD	Karst Rocky Desertification
LAI	Leaf Area Index
LI	Line Identifier
LiDAR	Light Detection And Ranging
LMLFs	large-scale man-made linear features
LOS	Level of Service
LST	Land Surface Temperature
LUCC	Land Use and Cover Change
LULC	Land use and land cover
MAS	MODIS Airborne Simulator

MAV Miniature Air Vehicle
MCL MVU-based co-locations
MDL Minimum Description Length
MLP Multilayer Perceptron
MNDV Mean NDVI
MOMS Modular Opto-electronic Multispectral Scanner
MSL MSL Mean Sea Level
MSNN Morphological Shared-Weight Neural Networks
MSS Multispectral Modular Spacecraft
MST Minimum Spanning Tree
MT-INSAR Multi-temporal INSAR
MTMP mean temperature
MUD MVU unfolded distance
MVU Maximum Variance Unfolding
MVUP Maximum Variance Unfolding Projections
NAD83 North American Datum of 1983
NCRST National Center for Remote Sensing in Transportation
NDOP National Digital Orthophoto Program
NDVI Normalized Difference Vegetation Index
NGS National Geodetic Survey
NIR Near Infrared
NSDI National Spatial Data Infrastructure
NSF National Science Foundation
NSFC National Natural Science Foundation of China
NTOI Near True Orthoimage
NUSDI National Urban Spatial Data Infrastructure
OA Over Accuracy
ODBC Open Database Connectivity
ODOT Ohio Department of Transportation
OLs Orientation Lines
OODM Object-Orientated Database Management
OSF Original Spectral Features
PAR Photosynthetically Active Radiation
PCA Principle Component Analysis
PCI Pavement Condition Index
PCNNs pixel-level CNNs
PDEN Population Density
PESNVS Probe Eye Scanner/Normal Color Video System
PHI Polyhedron Identifier
PI Point Identifier
PID Point Identification
PLA percentage of landscape area
PR precipitation radar
PS Permanent Scatterer
PSC Persistent Scatterers Candidate
PSI Persistent Scatterers Interferometry
PURB Percentage of Urban Surfaces
QOL quality of life
QPS Quasi-PS
RA Relative Accuracy
RD Relative Difference

RDBMS	Relational Database Management System
RMS	Root Mean Square
RMSE	Root Mean Square Error
RMVU	Relaxed MVU
ROI	Region of Interest
RK	Rocky karstification
ROSAN	Road Surface Analyzer
RRS	RRSR-relationship
RS	Remote Sensing
RT	Ray-Tracing
RTD	Real-Time Display
RTK	Real-Time Kinematic
SB	Santa Barbara
SBAS	Small Baseline Subset
SCSG	Spatial CSG
SDI	Shannon's diversity index
SDP	Semidefinite Programming
SHEI	Shannon's Evenness Index
SN	Skid Number
SR	State Route
SRG	Seed Region Grow
SRTM	Shuttle Radar Topography Mission
SSM	Surface Soil Moisture
STAISA	Spatio-Temporal Technique of Aerial Image Sequence Analysis
TB	Transportation Research Board
TD	Tract Density
TEX	texture
3D	Three-Dimensional
TID	Texture Image Data
TIN	Triangulated Irregular Network
TIR	Temperature Thermal Infrared
TM	Thematic Mapper
TP	True Positive
TSI	Tract Shape Index
TSX	TerraSAR-X
UAV	Unmanned Aerial Vehicle
UCSB	University of California, Santa Barbara
UHI	Urban Heat Island
URLs	Uniform Resource Locators
USGS	US Geological Survey
UTM	Universal Transverse Mercator
VC	Vegetation Coverage
VDOT	Virginia Department of Transportation
VHR	Very High Resolution
VNIR	Visible and Near Infrared
VR	Virtual Reality
VRML	Virtual Reality Modeling Language
WFI	Wide Field Imager
WT	Water
WTUSM	Wuhan Technical University of Surveying and Mapping
WWW	World Wide Web

Section I

Introduction

1 Introduction

1.1 WHAT IS URBAN REMOTE SENSING?

In order to help completely understand the definition of urban remote sensing, the prerequisite is becoming familiar with the definition of remote sensing. Many scholars, such as Zhou (2003), Zhou et al. (2004), Lillesand and Kiefer (2000), and Schowengerdt (1997), have given definitions regarding "What is Remote Sensing?" in their books. A relatively complete definition is from Colwell (1997), whose definition is:

> Remote Sensing is an art, science and technology of obtaining reliable information about physical objects and the environment through recording, measuring and interpreting imagery and digital representation of energy patterns derived from non-contact sensor system.
>
> **(Colwell 1997)**

The characteristics of remote sensing, according to Lillesand and Kiefer (2000), are:

1. Acquiring information about objects or phenomena using sensors/detectors from a distance, rather than *in situ*, such as from spaceborne and/or airborne, and/or ground-borne platforms;
2. Imaging the objects at spectral ranges exceeding the visible range of electromagnetic energy that our eyes sense;
3. Observing perspectives extending from regional to global scale; and
4. Reserving the imaged objects or phenomena for several decades, and beyond.

Therefore, remote sensing has been recognized as a very valuable tool at various levels in academia, governmental agencies, industrial sectors, and national security communities to observe, analyze, characterize, and recognize objects of interest (Sherbinin et al. 2002).

Strictly speaking, there is no special term for urban remote sensing. For convenience, one usually calls it "urban remote sensing." However, a more likely term should be application of remote sensing in urban areas, just like remote sensing applied in, for instance, the environment and oceans is called "environmental remote sensing" or "remote sensing for environments", and "oceanic remote sensing". Another reason is probably because urban systems are very complicated, with many associated special properties, for instance, natural environments (e.g., mountains, land, lakes), artificial constructions (e.g., houses, bridges, TV towers), social factors (e.g., population migration), and socioeconomic factors (e.g., night life). Such a complicated system results in the fact that although remote sensing has increasingly been applied in city areas, many new problems have emerged that are continuously challenging urban remote sensing's theory and technology. As a result, more and more researchers are interested in how urban remote sensing can solve existing and emerging problems.

Urban remote sensing has been considered as an important branch of remote sensing science and technology, associated with urban science, and is consistently taken as cross-disciplinary in international society and communities. Therefore, urban remote sensing has become one of the most valuable tools for monitoring and analyzing urban changes on landscapes, urbanization processes and their evolution, and furthermore to extract information about the ecological environment response and socioeconomic factors both explicitly and implicitly from multiple sources and at multiple scales, using active/passive remotely sensed image data. With increasing requirements for various applications, data acquisition, processing and analysis, urban remote sensing's theory, methods, and algorithms have developed rapidly in recent years (Du et al. 2013; Du 2018).

1.2 SIMPLE OVERVIEW OF THE HISTORY OF URBAN REMOTE SENSING

The early development of urban remote sensing can be traced back to 1979, when the International Symposia on Remote Sensing of Urban Areas, sponsored by the International Society for Photogrammetry and Remote Sensing (ISPRS) was held. Before 1979, many scholars, such as Anderson (1977), Lillestrand (1972), Lodwik (1979), and, especially, Lo (1971), pioneered urban remote sensing through uses of Landsat image data for mapping land use and land cover (LULC) in China, and estimated the population using aerial photography. Professor Lo pioneered the use of large format camera photography, Landsat MSS, Landsat TM, SPOT, DMSP-OLS, and Shuttle Imaging Radar images for investigation of the bio- and socioeconomic systems of the city of Hong Kong and settlements in China and the United States.

With the great advances in remote sensing applied in urban areas in the 1980s through the 1990s, the forums on urban remote sensing began to appear in 1995 within the meeting on remote sensing and urban analysis as part of the GISDATA Program, which was sponsored by the European Science Foundation (ESF). Afterwards, the Institute of Electrical and Electronic Engineers and Geoscience and Remote Sensing Society/International Society for Photogrammetry and Remote Sensing (IEEE GRSS/ISPRS) Joint Workshop on Remote Sensing and Data Fusion over Urban Areas was held in 2001 in Rome, Italy. Since 2005, the two forums have co-located to form a joint event that was officially named the Joint Urban Remote Sensing Event (URSE) in 2007.

After the joint workshop in Tempe (2005), the 4th GRSS/ISPRS Joint Workshop on Remote Sensing and Data Fusion over Urban Areas (URBAN) and the 6th International Symposium on Remote Sensing of Urban Areas (URS) were jointly held in Paris (2007). The meeting was named the Joint Urban Remote Sensing Event (JURSE). The last JURSEs were held in Shanghai (2009), Munich (2011), Sao Paulo (2013), Lausanne (2015), Dubai (2017), and Vannes (2019), France. http://jurse2019.org/previous-conferences/

The Association of American Geographers (AAG) annual meeting in the United States began to include special sessions on remote sensing and geographic information systems (GIS) for urban analysis in 2000. Following almost 20 years of development, the remote sensing and GIS for urban analysis special session sponsored by the American Association of Geodetic Surveying (AAGS) has become a major urban remote sensing forum in the United States. While urban remote sensing is rapidly emerging as a major field of study, receiving more attention than ever, all of these activities have facilitated the advances of urban remote sensing. The topics include (1) data requirements for urban areas; (2) digital image processing methodologies and algorithms for urban feature extraction; (3) parameters deriving from remotely sensed data for urban socioeconomic indicators; (4) urban and landscape modeling; (5) urban change monitoring, and so on.

1.3 JOURNALS RELEVANT TO URBAN REMOTE SENSING

Approximately 30 journals relevant to remote sensing have been published in English globally. In addition to those journals, there exist other journals in different languages, such as Germany, French, Japanese, and Chinese. The investigation and analysis of these journals presented below are based on English as the language of publication.

- ***Environment of Remote Sensing:*** https://www.journals.elsevier.com/remote-sensing-of-environment The journal publishes papers relevant to scientific and technical results on theory, experiments, and applications of remote sensing of the Earth's resources and environment, especially, with new sections on terrestrial, atmospheric, and oceanic sensing, and is thoroughly interdisciplinary.
- ***IEEE Transactions on Geoscience and Remote Sensing:*** https://ieeexplore.ieee.org/xpl/RecentIssue.jsp?punumber=36 The journal publishes papers on the theory, concepts, and techniques of science and engineering as applied to sensing the land, oceans, atmosphere, and

space, as well as the processing, interpretation, and dissemination of information, with technical papers disclosing new and significant research.

- ***IEEE Geoscience and Remote Sensing Letters:*** https://ieeexplore.ieee.org/xpl/RecentIssue.jsp?punumber=8859 The journal publishes short papers (maximum length five pages) with the focus on new ideas and formative concepts in remote sensing as well as important new and timely results and concepts.
- ***IEEE Journal of Selected Features in Applied Earth Observations and Remote Sensing:*** https://ieeexplore.ieee.org/xpl/RecentIssue.jsp?punumber=4609443 The journal publishes papers covering Earth observations and remote sensing, and current issues and techniques in applied remote and *in situ* sensing, their integration, and applied modeling and information creation for understanding the Earth, oceans, and atmosphere.
- ***International Journal of Remote Sensing:*** https://www.tandfonline.com/loi/tres20 The journal publishes papers with the focus on the science and technology of remote sensing and the applications of remotely sensed data in all major disciplines, including data collection, analysis, interpretation, and display; surveying from space, air, and water platforms; sensors; image processing; use of remotely sensed data; and economic surveys and cost–benefit analyses.
- ***Remote Sensing Letters:*** https://www.tandfonline.com/loi/trsl20 The journal publishes papers for rapid communication on the theory, science, and technology of remote sensing and novel applications of remotely sensed data. The journal's focus includes remote sensing of the atmosphere, biosphere, cryosphere, and the terrestrial Earth, as well as human modifications to the Earth system.
- ***Journal of Applied Remote Sensing:*** https://www.spiedigitallibrary.org/journals/journal-of-applied-remote-sensing?SSO=1 The journal publishes papers relevant to the past, current, and future experimental, research, and operational atmospheric and environmental remote sensing programs and experiments, as well as the planning, implementation, strategic partnerships, policies, and measures of success leading to the optimal utilization of remote sensing data.
- ***Remote Sensing Open Access Journal:*** https://www.mdpi.com/journal/remotesensing The journal publishes research papers, reviews, letters, and communications covering all aspects of remote sensing science, from sensor design, validation/calibration, to its applications in all aspects, such as geosciences, environmental sciences, ecology, and so on.
- ***European Journal of Remote Sensing:*** https://www.tandfonline.com/loi/tejr20 The journal publishes papers related to the use of remote sensing technologies and research on all applications of active or passive remote sensing technologies related to terrestrial, oceanic, and atmospheric environments.
- ***Canadian Journal of Remote Sensing:*** https://www.tandfonline.com/loi/ujrs20 The journal publishes papers relevant to sensor and algorithm development, image processing techniques, and advances focused on a wide range of remote sensing applications including, but not restricted to, forestry and agriculture; ecology; hydrology and water resources; oceans and ice; geology; urban; atmosphere; and environmental science. Papers can be published in English or French.
- ***International Journal of Applied Earth Observation and Geoinformation (JAG):*** https://www.journals.elsevier.com/international-journal-of-applied-earth-observation-and-geoinformation. The journal publishes original papers that apply Earth observation data for investigation and management of natural resources and the environment. Earth observation data are normally those acquired from remote sensing platforms such as satellites and aircraft. Natural resources include forests, agricultural land, soils, water resources, and mineral deposits. Environmental issues include biodiversity, land degradation, industrial pollution, and natural hazards such as earthquakes, floods, and landslides. The focus includes all major themes in geoinformation, like capturing, databasing, visualization, and interpretation of data, but also issues of data quality and spatial uncertainty.

- ***Journal of the Indian Society of Remote Sensing:*** https://www.springer.com/earth +sciences +and+geography/journal/12524 The journal publishes original research contributions in all the related fields of remote sensing and its applications, and on the advancement, dissemination, and application of the knowledge of remote sensing technology, which includes aspects such as photo interpretation, photogrammetry, aerial photography, image processing, and other related technologies in the field of surveying, planning, and management of natural resources and other areas of application.
- ***ISPRS Journal of Photogrammetry & Remote Sensing:*** https://www.journals.elsevier.com/isprs-journal-of-photogrammetry-and-remote-sensing The journal focuses on photogrammetric theory, technology, remote sensing theory and applications, spatial information systems, computer vision, and related fields.
- ***ISPRS International Journal of Geo-Information:*** https://www.mdpi.com/journal/ijgi The journal is an open access journal on geoinformation that publishes with a focus on the features of photogrammetric theory, technology, and remote sensing theory and applications.
- ***Photogrammetric Engineering and Remote Sensing:*** https://www.asprs.org/asprs-publications/pers The journal is an official journal of ASPRS for imaging and geospatial information science and technology, and publishes documents, reports, codes, and informational papers about the industries relating to geospatial sciences, remote sensing, photogrammetry, and other imaging sciences.
- ***Photogrammetric Record:*** https://onlinelibrary.wiley.com/journal/14779730 The journal publishes original, independently and rapidly refereed papers that reflect modern advancements in photogrammetry, 3D imaging, computer vision, and other related fields. All aspects of the measurement workflow are relevant, from sensor characterization and modeling, data acquisition, processing algorithms and product generation, to novel applications.
- ***Photogrammetrie Fernerkundung Geoinformation (PFG), Journal of Photogrammetry, Remote Sensing and Geoinformation Processing:*** https://www.springer.com/earth+ sciences+and +geography/geography/journal/41064 The journal is the official journal of PFG and publishes papers on the progress and application of photogrammetric methods, remote sensing technology, and the intricately connected field of geoinformation science, especially highlighting new developments and applications of these technologies in practice, and new methodologies in data acquisition, new approaches to optimized processing, and interpretation of all types of data which were acquired by photogrammetric methods, remote sensing, image processing, and the computer-aided interpretation of such data in general.
- ***International Journal of Digital Earth:*** https://www.tandfonline.com/loi/tjde20 The journal is the official journal of ISDE and publishes papers with a focus on knowledge-based solutions to build and advance applications to improve human conditions, protect ecological services, and support future sustainable development for environmental, social, and economic conditions.
- ***GeoInformatica:*** https://link.springer.com/journal/10707 The journal publishes articles with a special emphasis on GIS and computer information technology applied in geospatial science and technology. The journal covers spatial modeling and databases; human–computer interfaces for GIS; digital cartography; space imagery; parallelism, distribution, and communication through GIS; spatio-temporal reasoning, and more. GeoInformatica presents the most innovative research results in the application of computer science to GIS.
- ***Surveying and Land Information System Journal (SaLIS):*** https://www.aagsmo.org/salis-journal/ The journal is the official publication of the AAGS, the Geographic and Land Information Society (GLIS), and the National Society of Professional Surveyors (NSPS), who are all member organizations of the American Congress on Surveying and Mapping (ACSM). It publishes research papers, technical papers, and technical notes on surveying and land information theory, technology, and systems. Every four years, the journal publishes the US Report to the International Federation of Surveyors (FIG).

- ***GIScience & Remote Sensing (Mapping Sciences and Remote Sensing, 1984–2003):*** https://www.tandfonline.com/loi/tgrs20 The journal publishes papers relevant to GIS, remote sensing of the environment (including digital image processing), geocomputation, spatial data mining, and geographic environmental modeling. Papers reflecting both basic and applied research are published.
- ***International Journal of Geographical Information Science (International Journal of Geographical Information Systems, 1987–1996):*** https://www.tandfonline.com/toc/tgis20/current The journal publishes original research in fundamental and computational geographic information science, including applying geographical information science to monitoring, prediction, and decision making, as well as natural resources, social systems, computer science, cartography, surveying, geography, and engineering.
- ***Cartographica: The International Journal for Geographic Information and Geovisualization:*** https://utpjournals.press/loi/cart The journal publishes a wide range of cartographic studies including the production, design, use, and cognitive understanding of maps, the history of maps, and GIS.
- ***Cartography and Geographical Information Science:*** https://www.tandfonline.com/toc/tcag20/current The journal publishes a wide range of cartographic studies including the major features such as mapping design, map theory, map technology, data structure, and models.
- ***Journal of Spatial Science (formerly, Cartography (1900–2003)):*** https://www.tandfonline.com/loi/tjss20 The journal publishes original research contributing to the theory and practice of the spatial sciences, and papers that describe aspects of professional practice and implementation of techniques related to cartography, geodesy, GIS, hydrography, photogrammetry, remote sensing, or surveying.
- ***Survey Review:*** https://www.tandfonline.com/toc/ysre20/current The journal publishes papers with emphasis on research, theory, and practice of positioning and measurement, engineering surveying, cadaster and land management, and spatial information management.
- ***IEEE Sensors Journal:*** https://ieee-sensors.org/sensors-journal/ The journal publishes papers on the theory, design, fabrication, manufacturing, and applications of devices for sensing and transducing physical, chemical, and biological phenomena, with emphasis on the electronics and physics aspects of sensors and integrated sensors and actuators.
- ***Sensors:*** https://www.mdpi.com/journal/sensors The journal publishes papers on the science and technology of sensors and biosensors, reviews (including comprehensive reviews on the complete sensors products), regular research papers, and short notes.

Others

1. *International Journal of Imaging Systems and Technology*
2. *Journal of Electronic Imaging*
3. *Journal of Imaging Science and Technology*
4. *IEEE Transactions on Imaging*

1.4 ABOUT THIS BOOK

The development of urban remote sensing methodology and modeling, driven by wide applications in many different fields of science, has become of interest to many academic, governmental, and industrial sectors, such as micro-climate, cell phone planning, and photorealistic visualization, in addition to the classical fields of application in, for instance, land use and cover monitoring, and environmental analysis. Meanwhile, the advanced sensor technologies and image processing methodologies, such as deep learning and data mining, facilitate the wide application of remote sensing in urban areas.

The objective of this book is to present methodologies and algorithms, with an emphasis on urban three-dimensional (3D) model generation, urban orthophotomap generation, and feature extraction from high-resolution urban remotely sensed imagery, to upper-level students, graduate students, scientists, and engineers in academia, and governmental and industrial sectors.

When we talk about urban remote sensing, we should NOT forget the roots of urban remote sensing. In the nineteenth century, balloon-based aerial photography was applied in city monitoring and observations. During that era, three major tasks in urban remote sensing were urban digital surface models (DSM), orthophotomaps, and feature extractions or classification by artificial interpretation of photography (no computers in that era). For this reason, this book presents complete coverage of three topics, dealing with urban 3D model generation, orthophotomaps, and feature extraction using the most current emergence of the advanced algorithms and methodologies on the basis of various data resources, including traditional aerial photography, aerial image video, UAV-based video, LiDAR data, and so on. Therefore, the subject areas in this book are categorized below, which reflect my belief that the roots of urban remote sensing are in its methodologies and algorithms.

- Section I is an introduction including urban remote sensing terminology, contents, its relationship with urban study, and advances in recent years.
- Section II emphasizes the algorithms and methodologies of urban 3D building extractions from various imagery resources, including airplane-based video, LiDAR data, high-resolution satellite imagery, and traditional aerial imagery.
- Section III focuses on the various methodologies for true orthorectification for urban orthophotomap generation.
- Section IV presents the most current and novel algorithms and methodologies for urban remote sensing applications, such as co-location decision trees and deep learning neural networks.

I believe that several of the book's features will enhance its value. First, the algorithms and methodologies described in this book are for handling high-resolution imagery, namely, the ground sampling distance (GSD) is less than 1.0 meter. Second, the algorithms and methodologies presented in this book follow the root of urban remote sensing, that is, 3D urban models, orthophotomaps, and feature extraction. Third, linear readers will probably notice that various algorithms and methodologies from diverse referenced literature are presented to handle the same topic, for example, orthorectification. This is deliberate to help the sporadic readers who want to understand the essence of a topic but who have not read all the previous chapters. Such an arrangement of chapters allows readers to pick up apparently different streams of thought and tie them together.

It is my hope that this book can be read in varying depths by readers with remote sensing backgrounds including image processing, but this book is not intended to be read from cover to cover. Of course, the chapters have a logical order. In addition, some chapters in the book, – for instance, on co-location decision trees for urban image classification, deep learning neural networks for image processing – are close to the frontiers of theoretical research, while applications-oriented chapters of the book – for instance, on urban remote sensing big data – provide a general understanding of the methods. Therefore, the contents in this book should appeal to a large number of scientists, engineers, and graduate students.

This book is based on my early research work and articles published in journals that were consolidated as lecture notes for a one-semester, three-credit graduate course. Therefore, this book can be used as a textbook in universities and also serve as a reference book for researchers in academia, and in governmental and industrial sectors. This book is ideal for upper-level undergraduate and graduate students, and can also serve as a reference for researchers or those individuals interested in urban remote sensing in academia, and governmental and industrial sectors.

Prerequisites courses for this book as textbooks for upper-level undergraduates are Introduction to Remote Sensing, or Remote Sensing and Digital Image Processing, or Digital Image Processing with Remote Sensing Perspective.

The aim is to present as many different methods as possible developed in different fields of urban remote sensing to a broader readership. However, it is, in fact, impossible to present all the existing knowledge and methods in urban remote sensing.

REFERENCES

Anderson, J. R. (1977). Land use and land cover changes: A framework for monitoring. *Journal of Research by the Geological Survey*, 5, 143–153.

Colwell, R. N. (1997). History and place of photographic interpretation, in *Manual of Photographic Interpretation* (2nd edition) (ed W. R. Phillipson), ASPRS, Bethesda, pp. 33–48.

Du, P. (2018). Progress and trends of urban remote sensing: Reading guidance for the special issue on urban remote sensing. *Geography and Geo-Information Science*, 34, 3, 1–4.

Du, P. J., K. Tan, J. Xie. (2013). *Urban Remote Sensing Methods and Practices*. Academic Press, ISBN: 978-7-03-038049-4. (Chinese)

Lillesand, T. M., R. W. Kiefer. (2000). *Remote Sensing and Image Interpretation*, 4th edition. John Wiley & Sons. ISBN: 0-471-25515-7.

Lillestrand, R. L. (1972). Techniques for change detection. *IEEE Transactions on Computers*, 21, 654–659.

Lo, C. P. (1971). *Aerial photographic analysis of the urban environment: a study of the three-dimensional aspects of land use in the city centres of Glasgow and Hong Kong*, PhD dissertation, University of Glasgow, United Kingdom.

Lodwik, G. D. (1979). Measuring ecological changes in multitemporal Landsat data using principal components. *Proceedings of the 13th International Symposium on Remote Sensing of Environment* (Ann Arbor, MI: Environmental Research Institute of Michigan), pp. 1–11.

Schowengerdt, R. A. (1997). *Remote Sensing: Models and Methods for Image Processing*, 2nd edition, Academic Press. ISBN: 0-12-628981-6.

Sherbinin, A., D. Balk, K. Yager, M. Jaiteh, F. Pozzi, C. Giri, A. Wannebo. (2002). *A CIESIN Thematic Guide to Social Science Applications of Remote Sensing*. Available https://sedac.ciesin.columbia.edu/binaries/web/sedac/thematic-guides/ciesin_ssars_tg.pdf (accessed August 20, 2019).

Zhou, G. (2003). *Real-time Information Technology for Future Intelligent Earth Observing Satellite System*, Hierophantes Press, ISBN: 0-9727940-0-X.

Zhou, G., O. Baysal, J. Kaye. (2004). Concept design of future intelligent earth observing satellites. *International Journal of Remote Sensing*, 25, 14, 2667–2685.

2 Urban Remote Sensing and Urban Studies

2.1 CHARACTERISTICS OF URBAN REMOTE SENSING

In accordance with analysis from the United Nations Population Fund (UNFPA), more than half of the world's population now lives in towns and cities, and it is estimated that this number will swell to about 5 billion by 2030 (https://www.unfpa.org/urbanization). Since a city is a complex system composed of many of natural and artificial elements with a functional integrated ecosystem and socioeconomic system, rapid urbanization not only brings huge social, economic, and environmental transformations, resulting in well-being, resource efficiency, and economic growth, but also brings huge new environmental, ecosystem, and socioeconomic problems (Bloom et al. 2008). Therefore, urban remote sensing is typically different from the remote sensing applied to natural resources, environments, ecosystems, and so on. The major characteristics include (Zou 1990; Wang 1987):

1. *Impervious area.* The impervious area is usually identified as coverage by artificial materials, such as cement, asphalt, gravel, tiles, and metal. Their spectral characteristics are quite different from those of natural materials, such as plants, water, soil, and rocks. Many efforts on the spectral characteristics and patterns of the impervious areas in city have been made in the last decades, including image interpretation, image processing, and quantitative analysis of artificial materials, and many findings and accomplishments have been achieved (Weng 2012; Herold et al. 2004). With the emergence of new materials associated with the appearance of environmental changes in cities, such as air pollution, further studies of the spectral analysis of the impervious areas in city still need to be made, since often different objects may be imaged with the same spectrum in practice, such as an asphalt roof and an asphalt road. On the other hand, the same object may occur in different spectrums in practice, such as a tree in different seasons (Zhou and Zhang 2016). Therefore, highly accurate image interpretation and spectral analysis of the impervious area are still needed.
2. *"Heterogeneous" area.* The area is usually identified according to the type of texture features (e.g., tile, cement, asphalt, plants, water, etc.) with a narrow and small geometric shape (e.g., size, height, direction, etc.). Furthermore, the size of different textures may be smaller than the ground spatial resolution (also called "ground sample distance (GSD)") of the remotely sensed image pixels, which results in a number of mixed pixels happening (Figure 2.1). This fact largely decreases consistency of spectral properties for those similar features, and the difference in spectral characteristics for those heterogeneous features (Batty 2008). As a result, the probability of image misclassification and object recognition are significantly increased. It has been demonstrated for many studies that when the size of a pixel is close to the urban basic unit (elements), such as housing, factories, and roads, the spectral response variance of the target pixel and neighboring pixels will significantly increase (Alberti et al. 2004). Therefore, a high-resolution image is more suitable for analysis and processing in a city area than a low-resolution one (Du 2018). On the other hand, if a heterogeneous region is larger than several pixels cover, the outline of various features and their types is obvious and its boundary is clear (Stefanov et al. 2001). This phenomenon is largely helpful to avoid the misclassification of the artificial objects from the natural

FIGURE 2.1 Urban imagery.

objects, due to its blurring of the boundaries of natural objects. Consequently, the accuracy and reliability of interpretation, classification, and mapping for the heterogeneous region larger than several pixels can be improved (Herold et al. 2004).

3. *Linear structure elements:* Many city basic units (elements), for instance, power poles, street lamp posts, and power lines, have a linear structure in vertical profile, and occupy less than one pixel to a few pixels when imaged from a nadir-looking onboard spaceborne and airborne sensor, although aerial photographs and high-resolution satellite imagery can achieve higher than 1.0 m GSD. The oblique imaging of linear structure objects is probably suitable in urban remote sensing, and has an irreplaceable role. Thereby, many efforts in mobile mapping have been proposed internationally (Yang 2002).
4. *Air pollution:* Air pollution in a city, such as dust, soot, and smog in the air, can be identified. Their impacts on radiation transmission should be fully estimated. This is because a lot of exhaust gases and soot are discharged into the air in a city, resulting in increasing turbidity of the atmosphere, thereby attenuating solar radiation and thus reducing the radiance of the features. It has been demonstrated that the city's solar radiation can be reduced by approximately 15%–20% compared with the suburbs due to atmospheric pollution (Wang 1987; Yu and Wu 2006).
5. *Artificial ecosystem:* The city has become a typical artificial ecosystem due to many artificial buildings and construction. This implies that it is not rhythmic and characteristic over time like a natural ecosystem (e.g., a biological cycle) (Seto and Kaufmann 2003). It is reluctantly accepted that the changes in urban elements and their spectral responses have become subject to economic and social changes (Stefanov and Netzband 2005). Therefore, the choice of phase in urban remote sensing should consider these economic and social factors and their impacts.

2.2 STUDIES ON URBAN REMOTE SENSING

2.2.1 Architecture of Urban Remote Sensing

Du et al. (2018) gave an architecture of urban remote sensing (Figure 2.2). As observed from Figure 2.2, urban remote sensing is not only comprehensively capable of monitoring the urban environmental changes, urban land use and land cover changes, urban functional distribution patterns and urbanization evolution, but also of providing urban environmental assessment, landscape planning, and urbanization prediction, further extracting a city's natural, human, economic, and social information.

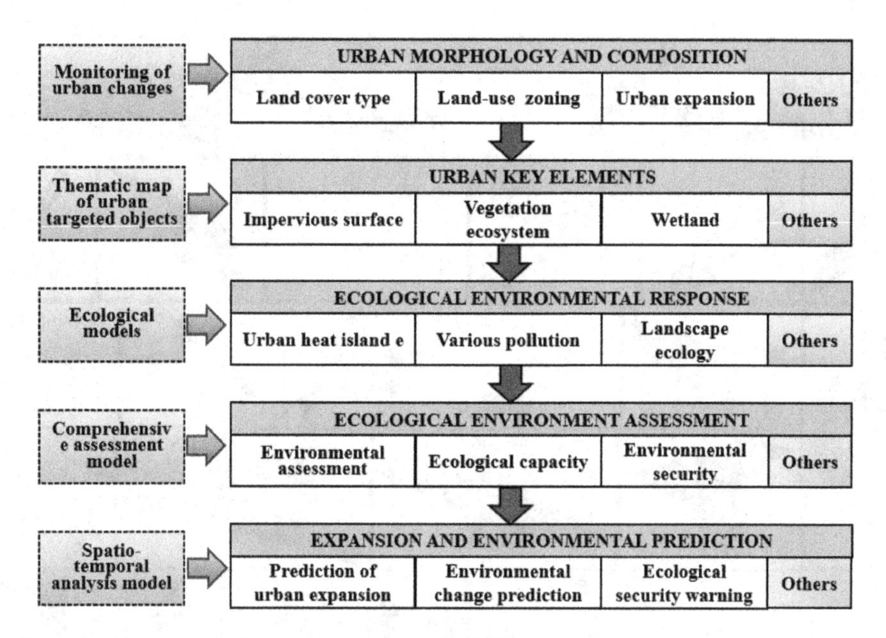

FIGURE 2.2 Architecture of urban remote sensing (Du 2018).

2.2.2 Topics of Studies on Urban Remote Sensing

Urban remote sensing is required to provide the continuous understanding of the city's current situation, changes, and development. The topics of urban remote sensing have therefore been very wide, including urban investigations of land use and land cover, ecological environment, municipal construction, transportation, water conservancy, agriculture, forestry, tourism, and so on. A few examples are listed below.

(1) Sensor and imaging modes

Sensor and imaging modes for urban studies should meet certain conditions in terms of spatial, spectral, radiometric, and temporal characteristics (Jensen and Cowen 1999). Nowadays, a wide range of spatial, spectral, and temporal parameters, and/or their different combinations are offered to meet the needs of different users for various applications (Figure 2.3). For example, some users may be interested in high-frequency, high-repetition revisits with a relatively low ground resolution (or GSD) (e.g., meteorology); some users may desire as high a GSD as possible with low-frequency coverage (e.g., terrestrial mapping); and some other users may need simultaneously both high GSD and high-frequency coverage, plus rapid image delivery (e.g., military surveillance) (Zhou 2003). These applications have basically been met with the rapid development of information technology in the most recent years.

With an increasing requirement for rapid response to natural disasters, such as earthquakes, landslides, and debris flows in urban areas, the sensors associated with imaging modes are required to be capable of acquiring comprehensive, near real-time images, with full coverage of the disaster area, in particular under fog, cloud, rain, or snow in an inclement climate. For this reason, spacecraft need to become more autonomous. Therefore, Earth observations may involve flying several small, low-cost spacecraft in formation to achieve the correlated instrument measurements formerly possible only by flying many instruments on a single large platform.

An example is a method for the guidance and control of the Earth Observing-1 (EO-l) spacecraft to fly in formation with the Landsat-7 spacecraft using an autonomous closed loop three-axis navigation control, GPS, and Cross link navigation support Figure 2.4.

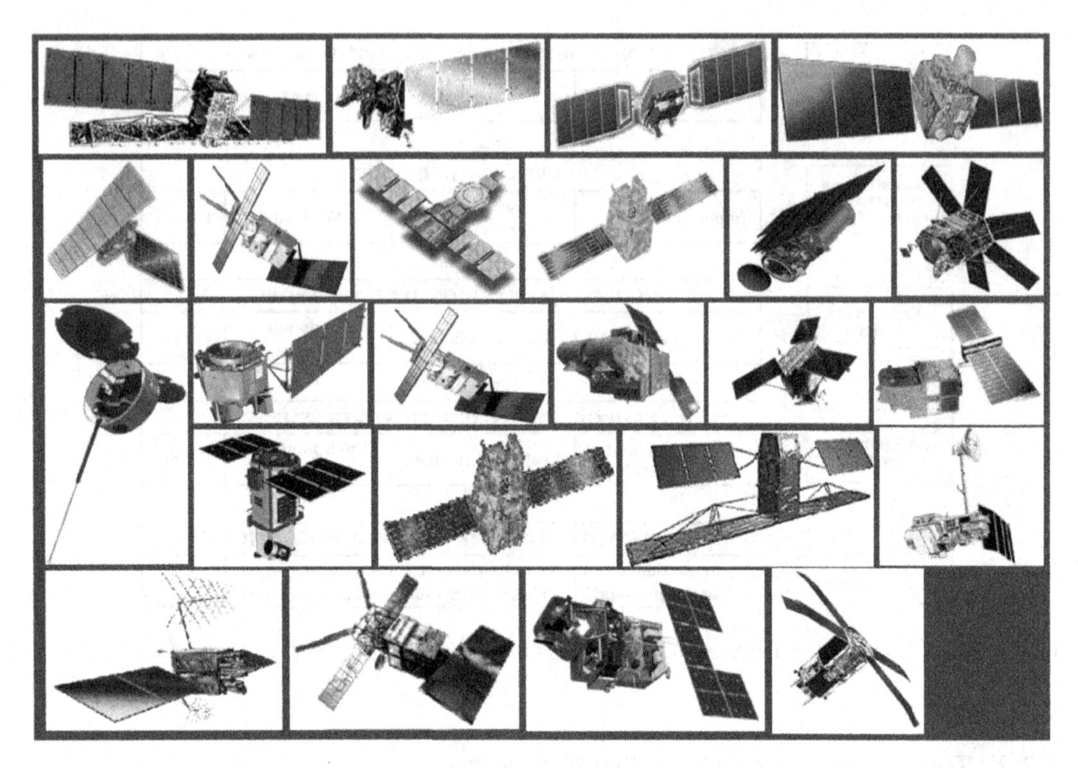

FIGURE 2.3 Various spaceborne sensors for imaging.

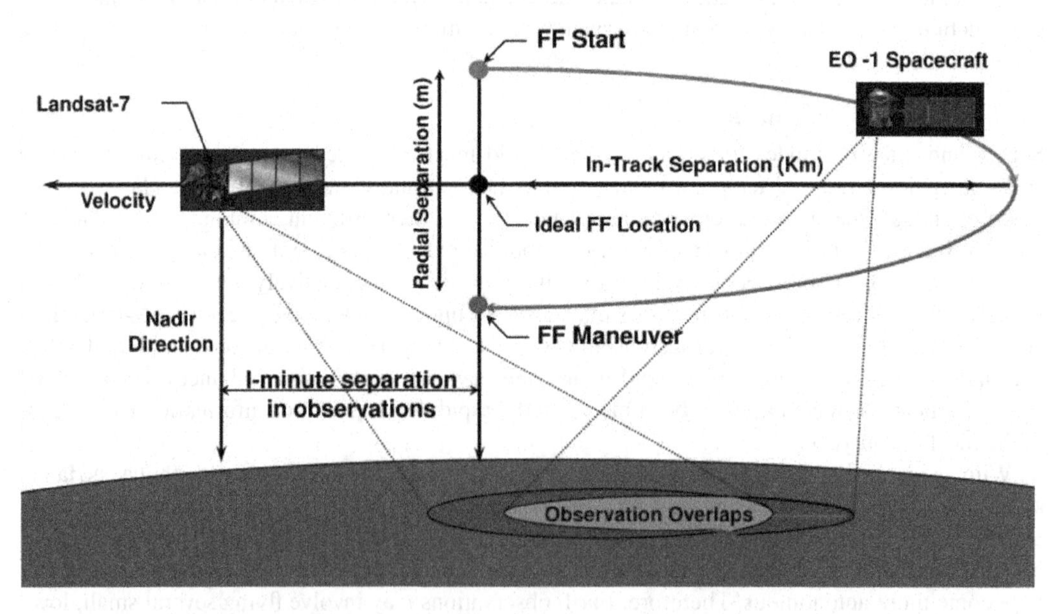

FIGURE 2.4 Formation flying in Earth, libration, and distant retrograde orbits (by David Folta courtesy of NASA-Goddard Space Flight Center on Advanced Topics in Astrodynamics, Barcelona, Spain, July 5–10, 2004).

(2) Algorithms and methodologies for urban remote sensing applications

The urban environment is characterized by the presence of heterogeneous surface covers with large interpixel and intrapixel spectral variations, thus challenging the applicability and robustness of conventional image processing algorithms and techniques (Du 2018; Yang 2011a). The current advances in algorithms and methodologies for urban remote sensing applications cover many disciplines, such as land use and land cover classification, urban change detection, urban environment and quantitative parameter retrieval, urban risk assessment, hydrological/geological environmental assessment, disaster management algorithms, and public safety-related applications. For example, methods and algorithms for urban vegetation analysis mainly include surface radiant energy budget, surface moisture and temperature changes, vegetation cover and land degradation, vegetation biomass and net primary production volume (NPP), vegetation structure and ecological parameters, surface ecological processes, land use and agricultural vegetation extraction, land surface vegetation ecological processes, and so on (Du et al. 2013b).

To improve urban mapping performance, many advanced techniques, algorithms, and methods were present in the last century in per-pixel and sub-pixel image processing, the spectral mixture analysis (SMA) technique, multiple endmember spectral mixture analysis (MESMA), object-based image analysis (OBIA), the image fusion technique, and advanced pan-sharpening algorithms.

In addition, multisource data fusion, object-oriented image analysis, and so on have also become popular methods in urban remote sensing research. With the rapid development of artificial intelligence theory and methods, urban remote sensing will apply more intelligent and automated analysis methods to achieve rapid and intelligent processing of massive remote sensing data, and better meet the needs of smart city construction in the era of artificial intelligence, fusion of GIS and images, and multisource remote sensing data fusion (Du et al. 2013a).

In the most recent years, some advanced methods and algorithms are continuously being developed, especially in anomaly detection and association rules; meanwhile, different types of models, such as artificial neural networks, decision trees, support vector machines, Bayesian networks, and genetic algorithms, as well as training models, such as federated learning, machine learning (Jean et al. 2016), artificial intelligence methods (e.g., Stefanov et al. 2001), support vector machines, supervised learning, unsupervised learning, reinforcement learning, feature learning, sparse dictionary learning (e.g., Yang 2011b), or a fuzzy classifier (e.g., Shalan et al. 2003) are increasingly used in urban remote sensing processing.

(3) Algorithms and methodologies for three classical urban products

The three most classical products in urban remote sensing are the digital terrain model (DTM) (Figure 2.5), digital orthophotomap (DOM) (Figure 2.6), digital aerial image (i.e., monoscopic mapping) (Figure 2.7), all of which have been the most important urban geospatial data infrastructure. The additional products derived through further processing include digital elevation models (DEM), digital landscape models, digital building models (DBM) (Figure 2.8), digital thematic maps, and various specialized geographic information data. With the development of smart cities, urban remote sensing is required to be able to produce different types and various scales of thematic maps or image maps to meet the needs of various users.

- *Digital elevation models (DEM):* A digital elevation model is a three-dimensional (3D) representation of the terrain elevations referenced to a vertical datum. DEM is typically created through stereo photogrammetry, which is traditionally carried out from a stereo pair of aerial images, before the emerging light detection and ranging (LiDAR) system. The LiDAR system uses the laser round-trip time difference between laser emitter and

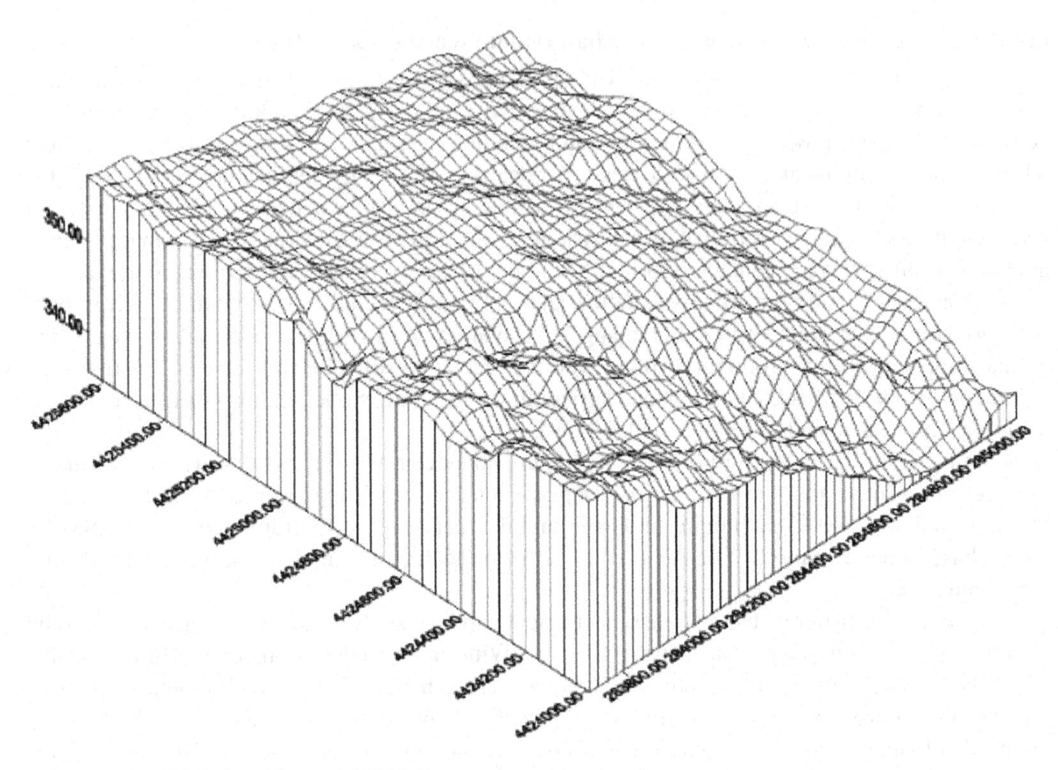

FIGURE 2.5 Digital terrain model (DTM).

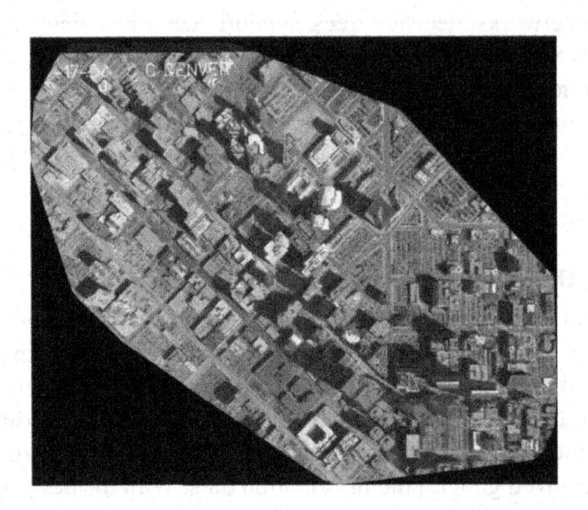

FIGURE 2.6 Aerial imagery.

target surface to measure ranges (variable distances) and calculate the 3D ground coordinates through other ancillary data, such as GPS and IMU data plus a series of coordinate transformations.

- *Digital Terrain Model (DTM):* DTM is a bare-earth surface DEM, in which terrain data has been further enhanced. Usually, DTM will NOT contain the non-ground points, such as bridges, so a smooth digital elevation model (DEM) can be obtained. In other words, the

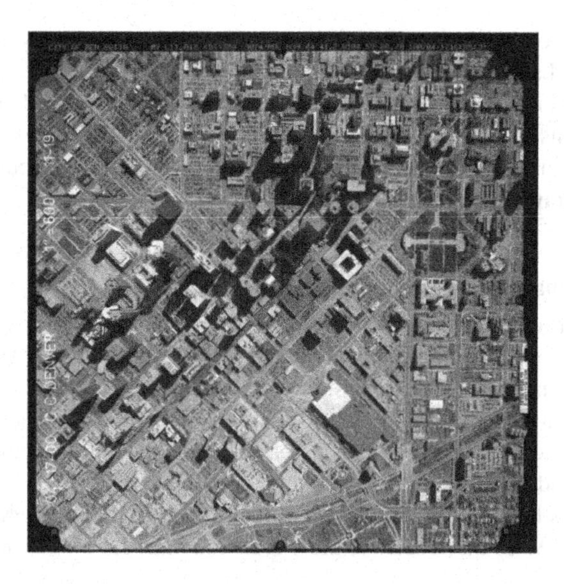

FIGURE 2.7 Urban orthophotomap.

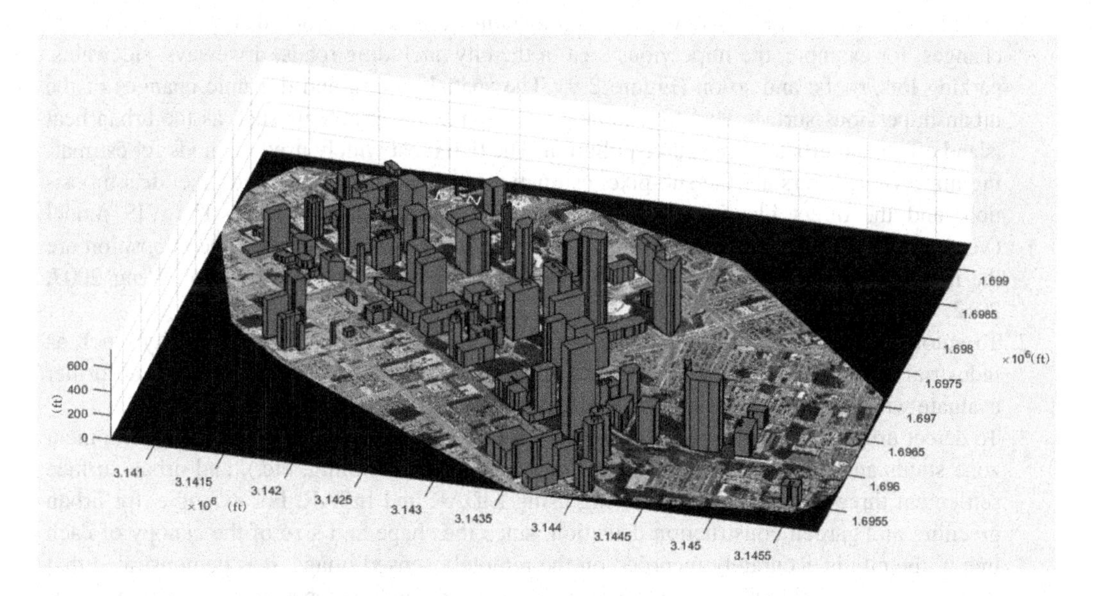

FIGURE 2.8 Urban building model (DBM) superimposed on orthophotomap.

artificial buildings (houses and TV towers) and natural objects (trees and other types of vegetation) are not included in a DTM (Figure 2.5).

- *Digital Surface Model (DSM):* A DSM contains elevations of the Earth's surface, with the elevations representing the first reflected surface detected by the sensor. These first returns may be reflected by bare ground or by surface features, such as trees and artificial buildings (Maune et al. 2001).
- *Digital orthophotomap (DOM):* An orthophotomap (DOM), also called an orthoimagemap, is produced by photogrammetric orthorectification (Figure 2.7). The orthorectification method is able to orthorectify the objects imaged through perspective projection into their correct

and upright positions and remove sufficient radiometric differences and various distortion. Therefore, a DOM serves to measure the geographical location, scope, shape, structure, distribution, and spatial relationships and remaining characteristics of the buildings, roads, vegetation, water systems, mountains, and so on of the city's objects. Therefore, an orthoimagemap is much more intuitive, accurate, profound, informative, rapid, and current than the conventional line map drawn with lines and symbols.

(4) Urban landscape changes investigations

One of the most important studies in urban remote sensing is to investigate the urban landscape changes, such as urban expansion, and land use and land cover changes. The methodologies and algorithms include the uses of various classifiers, such as support vector machines, rotating forests, spatial-spectral feature synthesis, object-oriented classification method, and multi-classifier integration, as well as a combination of various prior knowledge and semantic information (Sohn and Dowman 2007; Soudani et al. 2006; Goldblatt et al. 2018). It has been demonstrated that these methods and algorithms have had a good effect in land use classification when using medium and high-resolution images. With the results of the classification of land use/land cover, it is possible (Du 2018):

1. To extract the patterns of land use, dynamically analyze the urban land use evolution process, including its expansion, its development modeling, urban trajectory theory, and structural changes, for example, the impervious area in the city, including roads, driveways, sidewalks, parking lots, roofs, and so on (Figure 2.9). The spatial pattern and dynamic changes of the urban impervious surface lead to a series of environmental problems such as the urban heat island effect and urban point source pollution. The two most widely used methods for estimating impervious layers are the sub-pixel estimation method based on mixed pixel decomposition, and the direct identification method based on hard classification. The VIS model (Vegetation-Impervious surface-Soil) and the combination of mixed pixel decomposition are the most widely used methods for extracting impervious surface information (Weng 2007, 2012; Wu and Murray 2003).
2. To automatically calculate the area and proportion of various land uses in a city (such as industrial land, residential land, transportation, greening, garbage dumping, etc.) and further evaluate whether it is reasonable to use the proportion of different types of land use.
3. To detect urban area structures. This type of research includes investigation of development zone status and change, target area detection (e.g. street, building, etc.), and urban surface settlement three-dimensional monitoring using LiDAR and InSAR. For example, for urban greening and garden construction detection, since the shape and size of the canopy of each tree in the city is accurately recorded on the remotely sensed image, it is demonstrated that researchers can distinguish the types and tree species using hyperspectral images, and to calculate the area of various types of green spaces, which can be used for analysis of gardening, the scale, layout, service radius, tourist capacity, and so on in city gardens.
4. To detect urban building types and analyze density. It has been demonstrated that urban remote sensing is able to quickly and accurately obtain the types of urban buildings and their distribution, the actual construction area of urban districts and streets, the density of urban buildings and the proportion of urban land occupied. These parameters are important, and profoundly reflect the level of urban construction. In particular, the density of urban residential houses is inextricably related to the density of urban population distribution. Therefore, using housing density as a variable, it can be used for population census and demographic research.
5. To detect urban traffic. Investigation of the status on urban roads, highways, railways, waterways, transportation facilities (e.g., stations, terminals, parking lots, airports, etc.), and port site selection, construction and/or reconstruction, road line selection, railway line selection

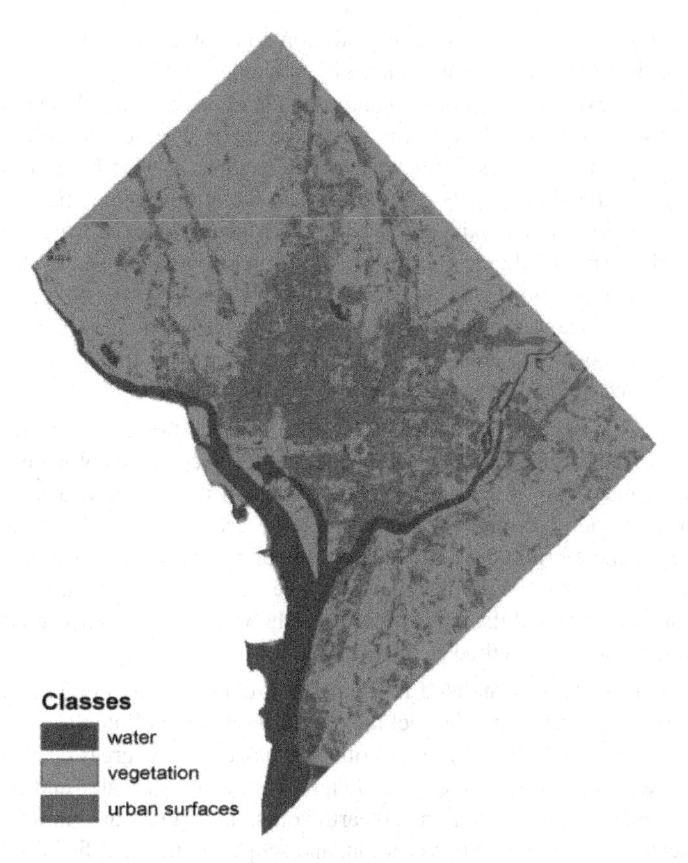

FIGURE 2.9 Urban land cover classification of Washington DC derived from Terra ASTER VNIR.

and so on can be assisted through urban remotely sensed images. In addition, remotely sensed images can help to understand the instantaneous movements of various vehicles, and traffic distribution, flow, and flow direction on various roads in the city and provide basic data for urban road system planning.

(5) Natural environment investigation

Investigation of urban environments using remote sensing is an important topic in both remote sensing and urban studies, since urban environmental changes significantly impact hydrology, biodiversity, ecosystems, and climate at local, regional, and global scales (Grimm et al. 2008). Du (2018) thought that urban environmental pollution monitoring and environmental quality assessment usually apply environmental quality indicators such as chlorophyll-a, suspended solids concentration, and particle pollutants NO_2, O_3, CO, TN, TP, while surface emission estimation usually applies indicators such as nitrogen oxides, volatile organic compounds, CO, aerosol pollution sources, and so on (Wang et al. 2011b; Zheng et al. 2016; Van et al. 2016; Guo et al. 2017). These indicators can be retrieved from the remotely sensed images (Xian 2015; Shen et al. 2017; Wen et al. 2018). The major topics in urban environmental investigation mainly include:

Atmospheric environment: Monitoring city atmospheric pollution resources allows identification of the number, height, and distribution of chimneys from high-resolution remotely sensed images. The air pollution concentration can directly be calculated through the correlation between the number of chimneys and coal combustion, fuel quantity, and soot. Also, the numbers and types of motor vehicles can be recognized from high-resolution images, and the pollution concentration from motor vehicles can directly be derived and calculated through analyzing the correlation between numbers of vehicles and exhaust emissions. In addition, the image acquired by a few special sensors such as gas filter analyzer, infrared interferometer, Fourier transform interferometer, visible light radiation polarimeter, and laser radar can also be used in accordance with the physics and chemistry of pollutants in the atmosphere. The indicators in the atmosphere include sulfur oxides (SO_2, etc.), nitrogen oxides (NO, etc.), carbon oxides (CO_2), hydrocarbons (alkanes, etc.), photochemical oxidants, suspended particulates (shape, size, composition), and other pollution substances such as the presence of concentration and concentration changes can be retrieved from images (Du 2018). All these investigations above can be used for dynamic monitoring of atmospheric pollution from remotely sensed images (Wang 1987).

Water environment: Because the differences in the composition and concentration of pollutants dissolved or suspended in water cause a difference in the color, density, transparency, and temperature of each water body, they will inevitably lead to changes in the reflected light energy of the water body (Figure 2.10), with such differences in hue, grayscale, image morphology, and shadow features appearing on the remotely sensed images (Carlson 2004). With these features, the source of pollution, the extent of contamination, the area, and the concentration of the water can be retrieved through quantitative remote sensing methods (Wang 1987).

Waste environment: Waste generated in industrial production, civil engineering, and construction projects, unreasonable stacking and disposal will pollute the surrounding environment (Wang 1987). This type of pollution has caused a great deal of attention in many developing countries since it is a significant cause of water and food pollution, and civil engineering construction and earthworks, their sites, surrounding environmental characteristics, the nature of pollution, and scope can be identified on the images.

Thermal environment: Infrared remote sensing is capable of effectively recording the infrared radiation from the ground object, resulting in the hue on the image representing the

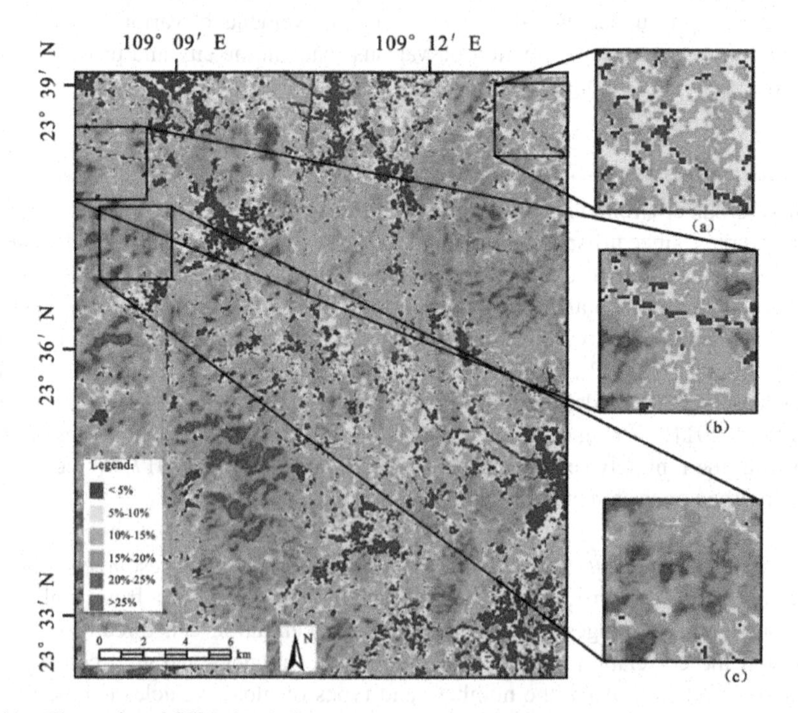

FIGURE 2.10 The retrieved SSMs level diagram from Landsat TM satellite imagery.

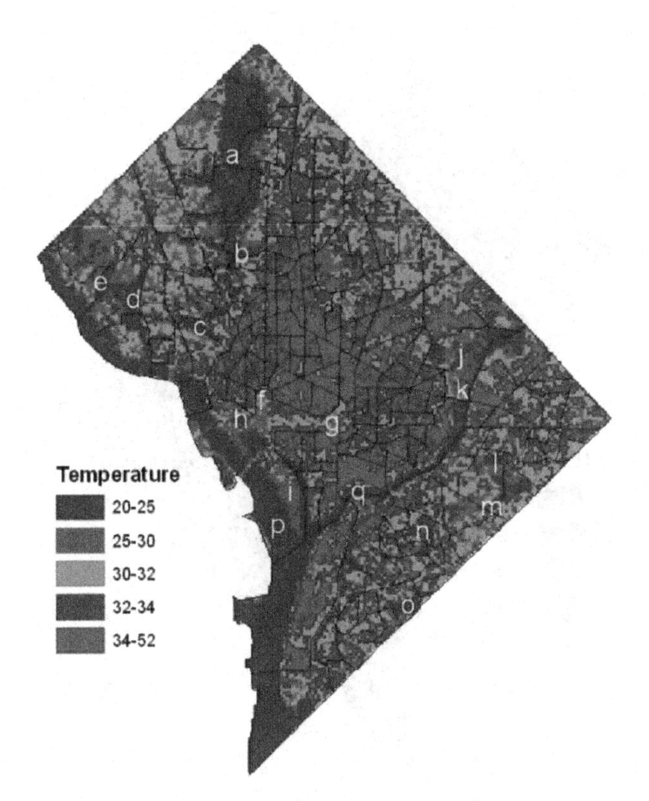

FIGURE 2.11 Surface temperature (in °C) of Washington DC superimposed by census tracts. Note that the lighter the tone, the higher the temperature. a – Rock Creek Park, b – Piney Branch Park, c – Dumbarton Oaks-Montrose Park, d – Glover-Archbold Park, e – Battery Kemble Park, f – White House, g – U. S. Capitol, h – Potomac Park (west), i – Potomac Park (east), j – National Arboretum, k – Langston Golf Course, l – Fort Dupont Park, m – Fort Davis Park, n – Fort Station Park, o – Oxon Run Park, p – Potomac River, q – Anacostia River.

actual temperature information at each point on the ground (commonly known as thermal images) (Arthur-Hartranft et al. 2003). Therefore, investigations into heat pollution, such as the urban heat island effect, can be carried out to accurately find the location of the heat loss point and its precise temperature and influence from the urban infrared remotely images (Figure 2.11).

Ecological environment: Many investigators analyzed the ecological burden capacity of urban agglomerations through a combination of remotely sensed images and ecological models (Du et al. 2013). A few studies applied urban patterns, extended indicators, and models to conduct research on urbanization and ecological environment response (Langemeyer et al. 2018). On the other hand, urban/urban surrounding ecology, including modeling of landscape ecology, ecological security evaluation, and ecological processes, is an imperative factor for natural–human association environment investigation. Many examples for comprehensive assessments of urban ecological security (Chen et al. 2016), human settlements (Yang and Zhang 2016), ecological risk (Li et al. 2017), and ecological burden capacity are usually investigated through combining the derived information from remotely sensed images, field observation, and various ecological environment data, and thematic models in the current years and beyond (Figure 2.12).

Socioeconomic environment investigation: Urban remote sensing has widely been used for social sciences, including socioeconomic analysis, population statistics, human rights analysis, health investigation, children's crime, public safety and emergency management, and typical social activities, for example, the Olympic Games, World Expo (Du et al., 2013a, 2013b; Du 2018), O-D estimation (Zhou and Wei 2008). Yang (2002) and Yang and Lo (2003)

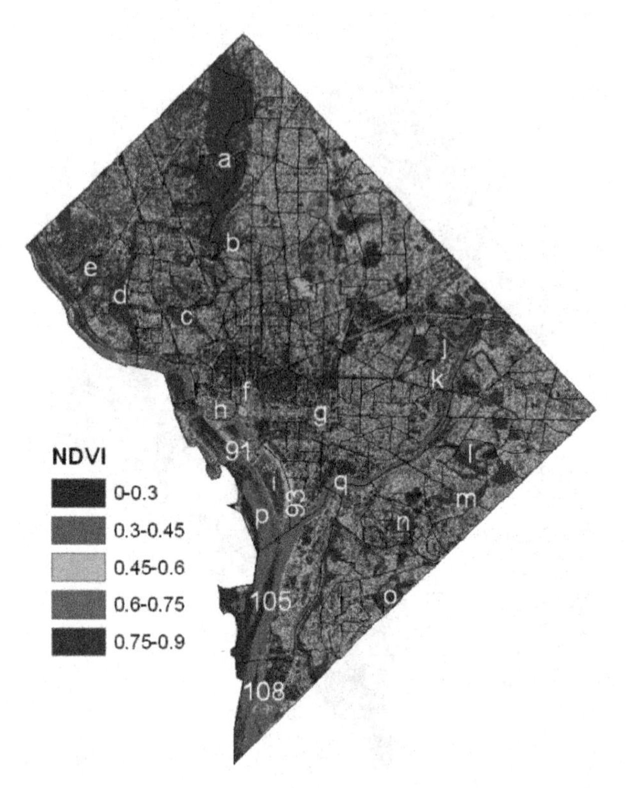

FIGURE 2.12 NDVI of Washington DC superimposed by census tracts. Note that the darker the tone, the higher the NDVI (indicating a greater amount or cover of vegetation). a – Rock Creek Park, b – Piney Branch Park, c – Dumbarton Oaks-Montrose Park, d – Glover-Archbold Park, e – Battery Kemble Park, f – White House, g – U. S. Capitol, h – Potomac Park (west), i – Potomac Park (east), j – National Arboretum, k – Langston Golf Course, l – Fort Dupont Park, m – Fort Davis Park, n – Fort Station Park, o – Oxon Run Park, p – Potomac River, q – Anacostia River. Numbers are the census tract IDs.

introduced the two major types of socioeconomic analyses: (a) linking socioeconomic data to land use/cover change data derived from remote sensing to identify the driving factors of landscape changes (e.g., Lo and Yang 2002; Seto and Kaufmann 2003), and (b) developing urban indicators of urban socioeconomic status, such as a pluralistic approach to defining and measuring urban sprawl, by combined use of remote sensing and census or field-survey data (e.g., Lo and Faber 1997; Yu and Wu 2006), from which developed urban socioeconomic indicators.

An example is the US Department of Transportation (US DOT), which initiated the use of commercial remote sensing and spatial information technology application in the transportation program in 1999 in collaboration with the National Aeronautics and Space Administration (NASA), in accordance with Section 5113 of the Transportation Equity Act for the 21st Century (DOT-NASA 2002, 2003). The collaborative program with NASA is administered by the US DOT Research and Special Programs Administration (RSPA). The program was intended to focus on unique and cost-effective applications of remote sensing and spatial information technologies for delivering smarter, more efficient and responsive transportation services with enhanced safety and security (DOT-NASA 2002, 2003). The five (originally four) application areas within the program have been (DOT-NASA 2002, Figure 2.13):

1. Environmental assessment, integration, and streamlining for faster decision making at reduced costs;
2. Transportation infrastructure management for improving maintenance service efficiency;

3. Traffic surveillance, monitoring and management for monitoring and managing traffic and freight flow;
4. Safety hazards and disaster assessment for unplanned events and security of critical transportation lifeline systems; and
5. Highway and runway pavement construction, quality control, and maintenance.

The above major priority areas of the collaborative program were deployed through national consortia, each of which consists of teams from leading institutions, industries and service providers. The major administrations from the US DOT are:

- Bureau of Transportation Statistics;
- Federal Aviation Administration;
- Federal Highway Administration;
- Federal Motor Carrier Safety Administration;
- Federal Railroad Administration;
- Federal Transit Administration;
- Maritime Administration;
- National Highway Traffic Safety Administration; and
- Research and Special Programs Administration.

The research centers of NASA consist of:

- Ames Research Center;
- Dryden Flight Research Center;
- Glenn Research Center at Lewis Field;
- Jet Propulsion Laboratory;
- Johnson Space Center;
- Kennedy Space Center;
- Langley Research Center;
- George C. Marshall Space Flight Center;
- Goddard Space Flight Center; and
- John C. Stennis Space Center.

FIGURE 2.13 Collaborative program between US DOT and NASA on remote sensing and geospatial information technologies application to transportation (courtesy of DOT, 2003).

(6) Natural resources investigation

Geological and geographical conditions investigation: In addition, the remotely sensed images have been widely applied in investigation of urban geoscience, such as geomorphology, quaternary geology, engineering geology, disaster geology, environmental geology, geological structure, active fault movement, earthquake, and regional geological stability evaluation. For example, Zhou et al. (2019) propose the application of remotely sensed multispectral imagery to create a 1:100,000 geological map. First, the analysis of the spectral characteristics of six types of lithologies from the USGS spectral library and from actual measurements by Field Spec®4, ASD Inc. was conducted (Figure 2.14). Using the analyzed spectral characteristics, the carbonate rock was separated from the other rocks using band ratios of multispectral imagery, and then the study area was divided into carbonate and non-carbonate areas (Figure 2.15). In the non-carbonate area, five types of rocks, namely, shale, metamorphic, sandstone, acid igneous, and basalt, were classified using their spectral characteristics (Figure 2.16) and the training data sets obtained from the previous 1:200,000 geographic map (Figure 2.17).

Mineral resources investigation: In addition to its application in investigation of metals and non-metallic minerals, remote sensing has also been very effective in the investigation of building materials resources that are closely related to urban construction, for example, ancient rivers and alluvial fans, where building materials and mineral resources such as sand and gravel can be found.

Water resources investigation: Remote sensing data can clearly show the current status of water conservancy projects, and can be used for river channel evolution, lake utilization, and floods.

Coastal zones in coastal cities investigation: Remote sensing technology can be used to investigate coastal zone types, coastal changes, tidal flat resources, and port location argumentation (Bartlett et al. 2000; Zhou and Xie 2009).

Forestry resources: It is widely recognized that forest remote sensing has provided more strong techniques and methods for extracting metrics of interest with equal or better accuracy than ground-based forest inventory methods. The use of LiDAR technology, which uses laser-pulse time-of-flight data to measure distances and corresponding back-scatter intensities, is expanding due to its characteristics of penetration through a coniferous forest canopy to reach the ground. It has been

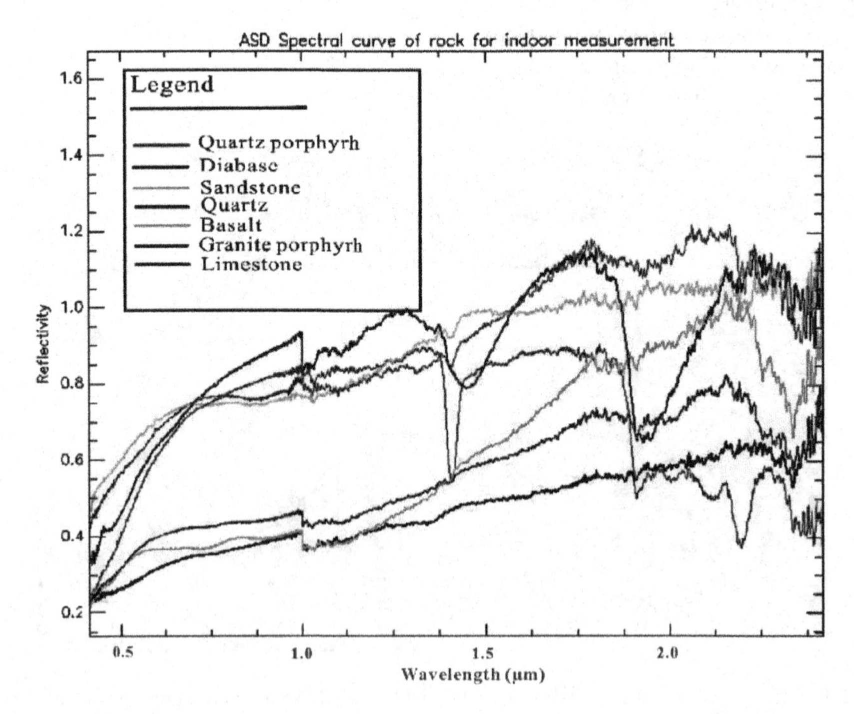

FIGURE 2.14 Spectral curves of laboratory measurements of seven rock samples (courtesy of Zhou et al. 2019).

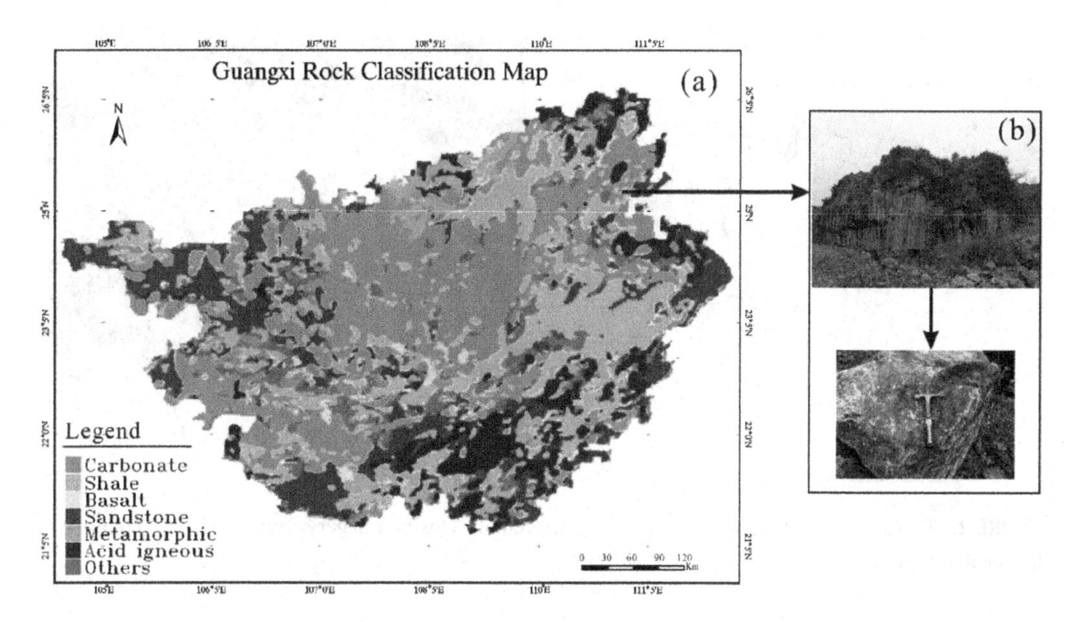

FIGURE 2.15 Classification results for six types of rocks using Landsat 8 imagery and their field verification (courtesy of Zhou et al. 2019).

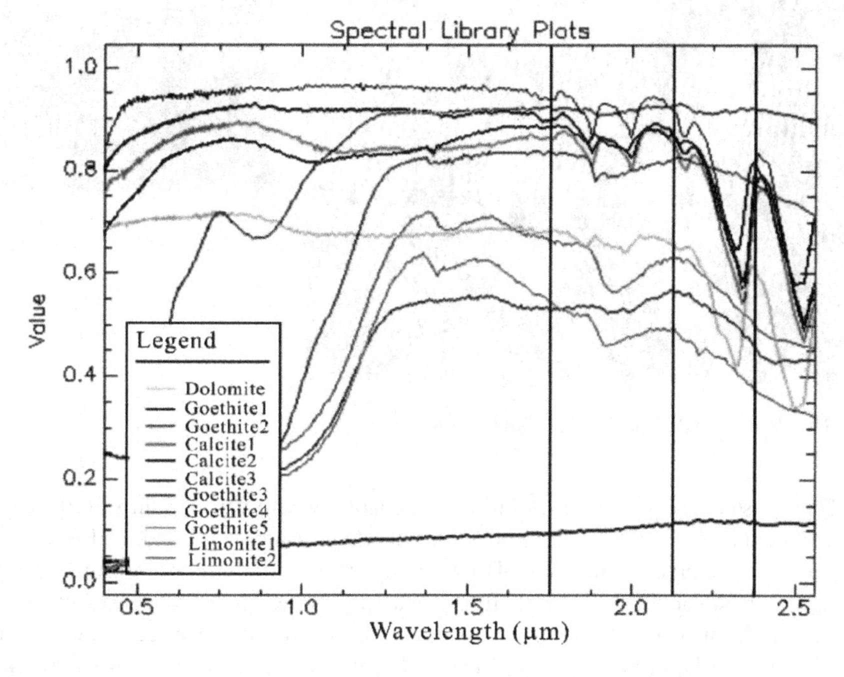

FIGURE 2.16 Spectral characteristics of the minerals in carbonate (from USGS_MIN spectral library retrieved from ENVI 5.0 software).

demonstrated that LiDAR can provide accurate estimations of many key forest characteristics, such as canopy height, topography, and vertical distribution; above-ground biomass, average basal area, average stem diameter, and canopy volume; and leaf area indices and canopy cover (Figure 2.18). The fusion of hyperspectral and LiDAR data for the classification of complex forest areas has been shown to be superior to results obtained using a single data source (Pirotti 2011).

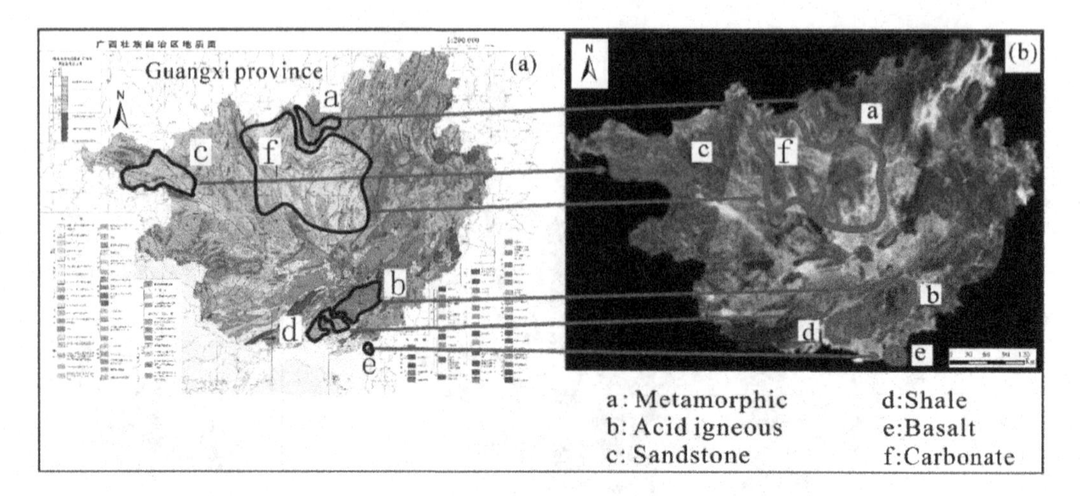

FIGURE 2.17 (a) 1:200,000 geographic map, and (b) remote sensing imagery from Landsat 8 (courtesy of Zhou et al. 2019).

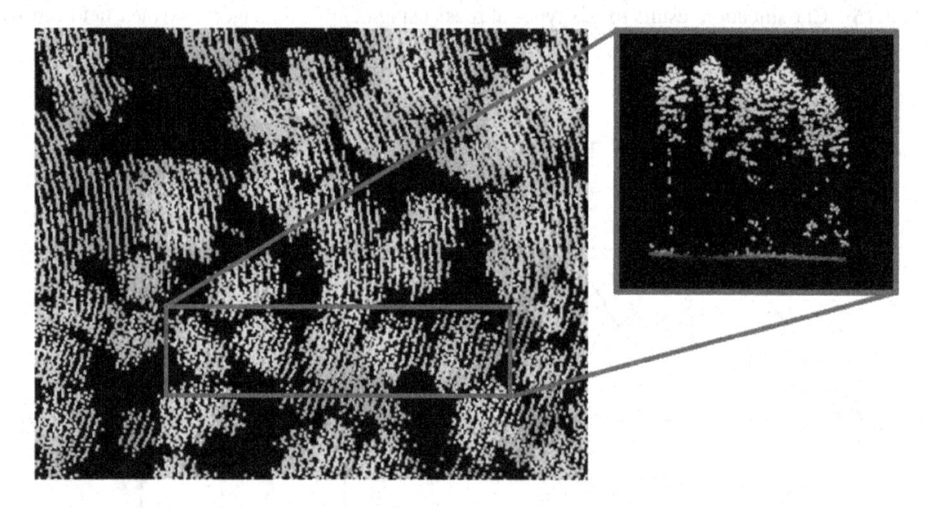

FIGURE 2.18 Point cloud visualization from LiDAR in forest area.

The LiDAR systems with full-waveform (FW) sensors, where waves return to the sensor after being back-scattered by objects on the Earth's surface (e.g., canopy, ground, and water) are sampled at regular time intervals. The FW LiDAR collects substantially more information relative to discrete-return LiDAR, and "shapes" the forest structure by using the properties of all elements that intersect the beam path, including forest cover, the canopy surface, the crown volume, the understory layer, and the ground surface. Many scholars have demonstrated that the FW LiDAR has provided improved information retrieval, such as height estimations (Duong et al. 2008), point density measurements (Chauve et al. 2008), and range determinations (Wagner et al. 2008) over discrete-return LiDAR methods.

Long-term observation of the targeted events (such as economical environments) using remotely sensed data can be used for comprehensive analysis of the dynamic changes in the urban environment and their driving factors. It also gives more opportunities to further analyze the dynamic changes of specific factors in urban environments through multitemporal change detection or post-processing.

The analyzed results can help understand the status, trends, and threats in urban areas so that appropriate management measures can be taken. The analyzed results can also be used for governance agencies to scientifically and rationally plan the urban environment, and take action over issues such as waste heat utilization, and/or disposal.

In summary, the bullet lists above of remotely sensed images applied in urban areas do not indeed include all applications. Since the urban remotely sensed image contains comprehensive information, various professionals can conduct their own research and applications in accordance with the requirements. Therefore, the content of urban remote sensing work is not limited to the above.

REFERENCES

Alberti, M., Weeks, R., Coe, S. (2004). Urban land-cover change analysis in Central Puget Sound. *Photogrammetric Engineering and Remote Sensing*, 70, 1043–1052.

Arthur-Hartranft, S.T., Carlson, T.N., Clarke, K.C. (2003). Satellite and ground-based microclimate and hydrologic analyses coupled with a regional urban growth model. *Remote Sensing of Environment*, 86, 385–400.

Bartlett, J.G., Mageean, D.M., O'Connor, R.J. (2000). Residential expansion as a continental threat to US coastal ecosystems. *Population and Environment*, 21, 429–468.

Batty, M. (2008). The size, scale, and shape of cities. *Science*, 319, 769–771.

Bloom, D.E., Canning, D., Fink, G. (2008). Urbanization and the Wealth of Nations. *Science*, 319, 772–775.

Carlson, T.N. (2004). Analysis and prediction of surface runoff in an urbanizing watershed using satellite imagery. *Journal of the American Water Resources Association*, 40, 1087–1098.

Chauve, A., Vega, C., Bretar, F., Durrieu, S., Allouis, T., Pierrot-Deseilligny, M., Puech, W. (2008). Processing full-waveform LiDAR data in an alpine coniferous forest: assessing terrain and tree height quality. *International Journal of Remote Sensing*, 30, 5211–5228.

Chen, L., Zhou, W., Han, L. (2016). Developing key technologies for establishing ecological security patterns at the Beijing-Tianjin-Hebei urban megaregion. *Acta Ecologica Sinica*, 36(22), 7125–7129.

DOT-NASA (2002). *Commercial Remote Sensing and Spatial Information Technologies Application to Transportation, a partnership for advancing transportation practice, a collaborative research Program, Progress Report, January 2002. U.S. Department of Transportation, National Aeronautics and Space Administration.* http//www.ncgia.ucsb.edu/ncrst/synthesis/Brochure200201/brochure2001.pdf

DOT-NASA (2003). *Remote Sensing and Geospatial Information Technologies Application to Multimodal Transportation 2003*, www.ncgia.ucsb.edu/ncrst/.../SynthRep 2003/6pager-2003.pdf

Du, P. (2018). Progress and trends of urban remote sensing: Reading guidance for the special issue on urban remote sensing. *Geography and Geo-Information Science*, 34(3) (May), 1–4.

Du, P., Xia, J., Du, Q. (2013b). Evaluation of the spatio-temporal pattern of urban ecological security using remote sensing and GIS. *International Journal of Remote Sensing*, 34(3), 848–863.

Du, P.J., Tan, K., Xie, J. (2013a). *Urban Remote Sensing Methods and Practices*. Academic Press, ISBN: 978-7-03-038049-4, 2013. (Chinese)

Duong H., Pfeifer, N., Lindenbergh, R., Vosselman, G. (2008). Single and two epoch analysis of ICESat full-waveform data over forested areas. *International Journal of Remote Sensing* 29, 1453–1473.

Goldblatt, R., Stuhlmacher, M.F., Tellman, B. (2018). Using Landsat and nighttime lights for supervised pixel-based image classification of urban land cover. *Remote Sensing of Environment*, 205, 253–275.

Grimm, N., Foster, D., Groffman, P., Grove, M., Hopkinson, C., Nadelhoffer, K., Pataki, D., Debra, P. (2008). The changing landscape: ecosystem responses to urbanization and pollution across climatic and societal gradients. *Front Ecol Environ*, 6(5), 264–272, doi:10.1890/070147.

Guo, Y.X., Tang, Q.H., Gong, D.Y. (2017). Estimating ground level PM2. 5 concentrations in Beijing using a satellite based geographically and temporally weighted regression model. *Remote Sensing of Environment*, 198, 140–149.

Herold, M., Roberts, D.A., Gardner, M.E., Dennison, P.E. (2004). Spectrometry for urban area remote sensing-Development and analysis of a spectral library from 350 to 2400nm. *Remote Sensing of Environment*, 91, 304–319.

Herold, M., Scepan, J., Clarke, K.C. (2002). The use of remote sensing and landscape metrics to describe structures and changes in urban land uses. *Environment and Planning A*, 34, 1443–1458.

Jean, N., Burke, M., Xie, M. (2016). Combining satellite imagery and machine learning to predict poverty. *Science*, 353(6301), 790–794.

Jensen, J.R., Cowen, D.C. (1999). Remote sensing of urban suburban infrastructure and socio-economic attributes. *Photogrammetric Engineering and Remote Sensing*, 65, 611–622.
Langemeyer, J., Camps-Calvet, M., Calvet-Mir, L. (2018). Stewardship of urban ecosystem services: understanding the value(s) of urban gardens in Barcelona. *Landscape and Urban Planning*, 170, 79–89.
Li, H., Li, X., Tian, S. (2017). Temporal and spatial variation characteristics and mechanism of urban human settlements: A case study of Liaoning province. *Geographical Research*, 36(7), 1323–1338.
Lo, C.P., Faber, B.J. (1997). Integration of Landsat Thematic Mapper and census data for quality of life assessment. *Remote Sensing of Environment*, 62, 143–157.
Lo, C.P., Quattrochi, D.A. (2003). Land-use and land-cover change, urban heat island phenomenon, and health implications: A remote sensing approach. *Photogrammetric Engineering and Remote Sensing*, 69, 1053–1063.
Maune, D.F., Maitra, J.B., Mckay, E.J. (ed) (2001). *Digital Elevation Model Technologies and Applications: The Dem Users Manual*, American Society for Photogrammetry and Remote Sensing, Bethesda, MD, 539 p.
Pirotti, F. (2011). Analysis of full-waveform LiDAR data for forestry applications: a review of investigations and methods. *iForest - Biogeosciences and Forestry*, 4(3), 100–106. doi: https://doi.org/10.3832/ifor0562-004
Seto, K.C., Kaufmann, R.K. (2003). Modeling the drivers of urban land use change in the Pearl River Delta, China Integrating remote sensing with socioeconomic data. *Land Economics*, 79, 106–121.
Shalan, M.A., Arora, M.K., Ghosh, S.K. (2003). An evaluation of fuzzy classifications from IRS 1C LISS III imagery: A case study. *International of Remote Sensing*, 23(15), 3179–3186.
Shen, Q., Zhu, L., Cao, H. (2017). A review remote sensing monitoring and screening for urban black and odorous water body. *Chinese Journal of Applied Ecology*, 28(10), 3433–3439.
Small, C., Nicholls, R.J. (2003) A global analysis of human settlement in coastal zones. *Journal of Coastal Research*, 18, 584–599.
Sohn, G., I. Dowman (2007). Data fusion of high-resolution satellite imagery and LiDAR data for automatic building extraction. *ISPRS Journal of Photogrammetry and Remote Sensing*, 62(1), 43–63.
Soudani, K., Francois, C., Maire, G.L. (2006). Comparative analysis of IKONOS, SPOT, and ETM+ data for leaf area index estimation in temperate coniferous and deciduous forest stands. *Remote Sensing of Environment*, 102(1/2), 161–175.
Stefanov, W.L., Netzband, M. (2005). Assessment of ASTER land cover and MODIS NDVI data at multiple scales for ecological characterization of an urban center. *Remote Sensing of Environment*, 99, 31–43.
Stefanov, W.L., Ramsey, M.S., Christensen, P.R. (2001). Monitoring urban land cover change: An expert system approach to land cover classification of semiarid to arid urban centers. *Remote Sensing of Environment*, 77, 173–185.
Van, D.A., R.V. Martin, M. Brauer et al. (2016). Global estimates of fine particulate matter using combined geophysical-statistical method with information from satellites, models, and monitors. *Environmental Science and Technology*, 50(7), 3762–3772.
Wagner, W., Hollaus, M., Briese, C., Ducic, V. (2008). 3D vegetation mapping using small-footprint full-waveform airborne laser scanners. *International Journal of Remote Sensing*, 29, 1433–1452.
Wang, C. (1987). Characteristics and methods of urban remote sensing, *Urban Planning*, 3, 19–22.
Wang, Q., Heng, L., Li, S.S. (2011a). *Evaluation and Simulation of Water Environment Based on Satellite HJ-1*. Science Press, 2011.
Wang, Q., Q. Li, L. Chen. *Technology and Application of Atmospheric Environment Remote Sensing*. Science Press, 2011b.
Wen, S., Wang, Q., Li, Y. (2018). Remote sensing identification of urban black-odor water bodies based on high-resolution images: A case study in Nanjing. *Environmental Science*, 39(1), 57–67.
Weng, Q.H. (2007). *Remote Sensing of Impervious Surfaces*. Boca Raton, FL: CRC Press. ISBN-13: 978-1420043747, 494 pages.
Weng, Q.H. (2012). Remote sensing of impervious surfaces in the urban areas: Requirements, methods, and trends. *Remote Sensing of Environment*, 117(2), 34–49.
Wu, C.S., Murray, A.T. (2003). Estimating impervious surface distribution by spectral mixture analysis. *Remote Sensing of Environment*, 84(4), 493–505.
Xian, G.Z. (2015). *Remote Sensing Applications for the Urban Environment*. CRC Press, October 2015 ISBN: ISBN-13: 978-1420089844, 234 pages.
Yang, X. (2002) Satellite monitoring of urban spatial growth in the Atlanta metropolitan region. *Photogrammetrical Engineering and Remote Sensing*, 68, 725–734.

Yang, X. (2011a) Parameterizing support vector machines for land cover classification. *Photogrammetric Engineering and Remote Sensing*, 77, 27–37.

Yang, X. (2011b). *Urban Remote Sensing: Monitoring, Synthesis and Modeling in the Urban Environment*, Apr 25, 2011. CRC Press, ISBN: 978-0470749586, 408 pages.

Yang, X., Lo, C.P. (2003) Modelling urban growth and land scape changes in the Atlanta metropolitan area. *International Journal of geographical Information Science*, 17, 463–488.

Yang, X., Zhang, W. (2016). Combining natural and human elements to evaluate regional human settlements quality based on raster data: A case study in Beijing-Tianjin-Hebei region. *Acta Geographica Sinica*, 71(12), 2141–2154.

Yu, D.L., Wu, C.S. (2006). Incorporating remote sensing information in modeling house values: A regression tree approach. *Photogrammetric Engineering and Remote Sensing*, 72, 129–138.

Zheng, Y.X., Zhang, Q., Liu, Y. (2016). Estimating ground-level PM2. 5 concentrations over three megalopolises in China using satellite-derived aerosol optical depth measurements. *Atmospheric Environment*, 124(B), 232–242.

Zhou, G. (2003). *Real-time Information Technology for Future Intelligent Earth Observing Satellite System*, Hierophantes Press, ISBN: 0-9727940-0-X, February 2003.

Zhou, G., Baysal, O., Kaye, J. (2004). Concept design of future intelligent earth observing satellites. *International Journal of Remote Sensing*, 25(14), 2667–2685.

Zhou, G., Wang, H. (2020). Impacts of urban land surface temperature on tract landscape pattern, physical and social variables. *International Journal of Remote Sensing*, 41(2), 683–703.

Zhou, G., Wang, H., Sun, Y., Shao, Y., Yue, T. (2019). Lithologic classification using multilevel spectral characteristics, *Journal of Applied Remote Sensing*, 13(1), 016513, doi:10.1117/1.JRS.13.016513.

Zhou, G., Wei, D. (2008). Traffic spatial measures and interpretation of road network using aerial remotely sensed data. *2008 IEEE International Geoscience and Remote Sensing (IGARSS 2008)*, Boston, MA, July 7–11.

Zhou, G., Xie, M. (2009). GIS-based three-dimensional morphologic analysis of Assateague Island National Seashore from LIDAR series datasets. *Journal of Coastal Research*, 25(2), 435–447.

Zhou, G., Zhang, R. (2016). Manifold learning co-location decision tree for remotely sensed imagery classification. *Remote Sensing*, 8, 855. doi:10.3390/rs8100855.

Zou, S. (1990). The characteristics of urban remote sensing and its technical requirement, *Remote Sensing for Land Resources*, 1990(2), 51–55.

3 Advances in Urban Remote Sensing

3.1 URBAN REMOTE SENSING BIG DATA

In recent years, with the development of information and communication technologies, especially the development of mobile communication technologies, the ability to acquire and transmit information has been greatly enhanced. Everyone can act as a sensor to collect data with individual tags and spatio-temporal semantic information at anytime, anywhere (Li et al. 2014a, 2014b). This has led to a sharp increase in the amount of urban remote sensing data, even a "big data" trend, that is so-called urban remote sensing big data, which are the data related to the city, mainly through interpretation, analysis, mining and other technologies, from a variety of remote sensing platforms to obtain useful multi-resolution, multi-source, multi-band mass data (Casu et al. 2017; Chi et al. 2016; Liu et al. 2018). Therefore, the calculation, storage, and management of urban remote sensing big data are a challenging task that we are facing.

The topic of urban remote sensing big data has attracted international scholars. In March 2012, the US government officially launched the "big data" project. NASA, USGS, and other space remote sensing research institutions conducted a series of studies on issues, including analysis, processing, storage, and sharing, relevant to remote sensing big data (Ma et al. 2015a, 2015b; Rathore et al. 2015). In 2017, ESA held the conference on "Big Data in Space," which brought together researchers and users of big data on topics of resources sharing, services, and infrastructure. In the same year, China delivered the national GEOSS data sharing platform at the GEO Week exhibition and participated in relevant projects of the Earth Observation Organization (GEO), which promoted the development of big data (Liu et al. 2018). On the other hand, many papers have been published by IEEE JSTARS, GeoInformatica of Springer, Environmental Remote Sensing, The International Journal of Digital Earth, and other international top journals.

From the perspective of application of remote sensing big data, NASA (2015) has built a supercomputing platform (NEX) to quickly calculate and analyze remotely sensed big data, and China Resource Satellite Application Center, National Basic Geographic Information Center, and other institutions have provided corresponding products and services for remotely sensed big data. In addition, products and services relating to remotely sensed big data have been launched by Google, Microsoft, Amazon, Baidu, and other companies. This means that the commercial application of remotely sensed big data has been integrated into human lives, and it inversely promotes the deep research and further development of remotely sensed big data in urban areas.

This chapter gives a whole picture on the storage management, processing, and applications of remote sensing big data in urban areas, and possible future developments.

3.2 CHARACTERISTICS OF URBAN REMOTE SENSING BIG DATA

Urban remote sensing big data typically comes from a variety of remote sensing platforms, including radar, infrared, microwave, panchromatic, multispectral, hyperspectral and other data, including environment, economy, traffic, population, and so on. The characteristics include its external characteristics (e.g., massive, diversity, high resolution, true record, high value, Chang updating, uncertainty) and internal characteristics (e.g., high dimensionality, multiscale, real-time) are analyzed in Table 3.1 (Chi et al. 2016; Liu 2015; Ma et al. 2015a, 2015b; Song et al. 2014; Wang et al. 2014, 2018).

TABLE 3.1
Characteristics of Urban Remote Sensing Big Data

	Characteristics	Implications	Challenges	Solutions
External Characteristics	Massive	With increasing variety of sensors, the volume of data is increasing significantly.	Data processing and storage are facing challenges.	Enlarge memory, cloud computing.
	Diversity	Various platforms and various sensors resulting in different types of data.	Data processing models and methods are inconsistent.	A compatible processing model for multiple data.
	High resolution	The temporal, spatial, and spectral resolutions are largely high.	Significant increase in data volume.	Increase memory, storage.
	True record	Data truly record the geometric, natural physical properties of the objects on the Earth's surface in urban areas.	Cannot falsify the data.	Keep them true.
	High value	Data are highly valuable for accurately recording information such as the geography, social economy, ecology, environment, and so on.	The data analysis and mining are complicated.	Further deep mining algorithm.
	Change updating	Data updating is fast, especially in developing countries, resulting in real-time processing, such as traffic jams, fire emergency, and so on.	Dynamic, real-time processing with high-performance requirements.	Real-time information technology, and dynamic processing.
	Uncertainty	Influences of external factors such as platforms, sensors, and so on make the acquired data uncertain, for example inconsistent, incomplete, and so on.	Data uncertainty and error propagation.	Tolerable and reconfigurable method.
Internal Characteristics	High dimensionality	The diversity of data results in its higher dimensionality.	The high dimensionality leads to complicated data structure.	Data dimension reduction using AI algorithms.
	Multiscale	Various applications make data obtained using different observation sensors, different time, spectral and spatial resolutions.	Large amount of data and complex processing.	Strong, deep excavation technology, high accommodation.
	Real-time	The real-time information technology makes the data applied in real-time.	Data volume is huge and some data may never be used, so data processing in real-time is required.	Real-time data processing and depth mining information technology.

3.3 CHALLENGES IN URBAN REMOTELY SENSED BIG DATA PROCESSING

3.3.1 Storage and Management of Urban Remote Sensing Big Data

There is much research related to urban remote sensing big data management. Traditionally, it includes database management, file-based management, and hybrid management based on files and databases (Lv et al. 2011). For the management of a massive remote sensing image data set, three storage architectures are adopted: centralized storage architecture, network storage architecture, and distributed file system storage architecture (Liu 2015).

1. *Centralized storage architecture*. Taking the central server as the core, and using the file form to store, manage, and maintain the image data resources with a complex structure, the scalability of such storage architecture is not high enough.
2. *Network storage architecture*. This kind of architecture realizes the connection and sharing of data resources by combining the performance advantages of network and I/O, and stores and processes data on different nodes, which enhances the availability of the system.
3. *Distributed file system storage architecture*. It is mainly used in large-scale cluster systems, using a master node to manage resources, multiple storage nodes to share storage tasks, and through the means of backup to increase the scalability of the system.

Urban remote sensing big data management needs to provide interface language to manage and call data efficiently, so as to better meet the needs of further application. The storage technologies majorly include the distributed file method (Aji et al. 2013), distributed database method (Aljarrah et al. 2015), access interface (Bauermarschallinger et al. 2014), and query language (Cheng 2014).

Distributed file method. GFS (Google File System), HDFS (Hadoop distribute file system), NFS (Network file system), DFS (Distributed File System), and so on are common. The distributed file method adopts a master/slave structure to store and manage metadata through the memory of the master node, to further optimize the storage performance of big data sets (Nie et al. 2018). As shown in Figure 3.1, the

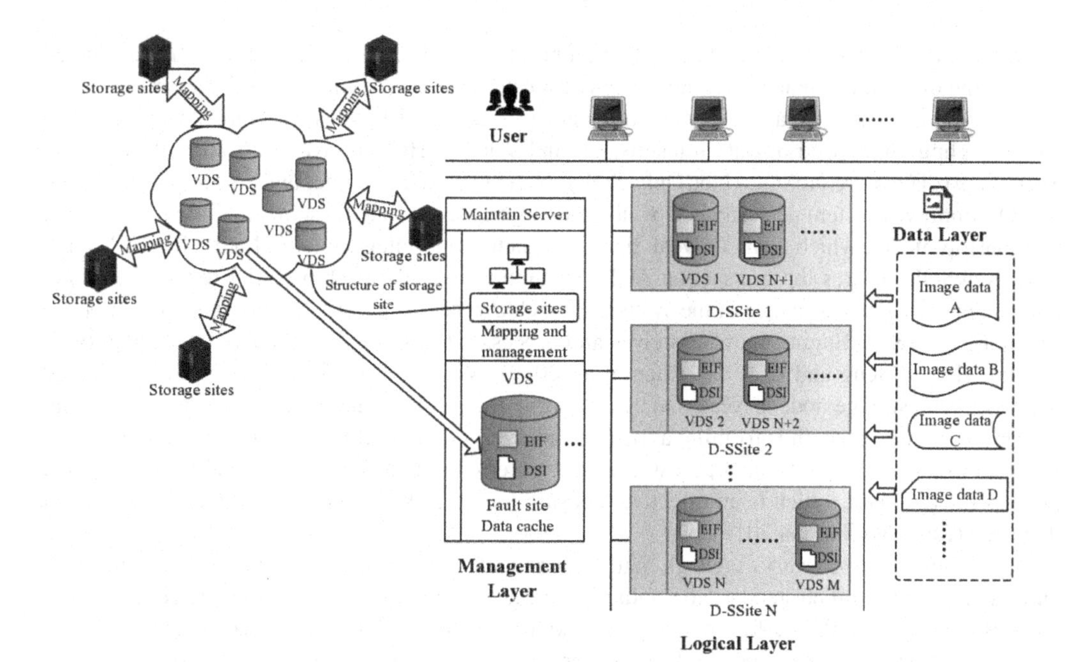

FIGURE 3.1 A distributed remote sensing image data storage and management system based on RSC-DOM architecture.

distributed file system includes maintenance server (MServer) and distribute-storage site (D-SSite). Each D-SSite includes multiple virtual disk spaces (VDS), and each VDS is composed of an embedded index file (EIF) and data storage infrastructure (DSI).

- *Maintenance Server*
  It is the central control station of the whole distributed storage system. The maintenance server is mainly responsible for the management and monitoring of the D-SSite, mapping between the storage site and the VDS, task allocation and scheduling, data temporary storage, and service of the failed site (Lai et al. 2013).

- *Distribute-Storage Site*
  It is indeed a storage site in the distributed storage system. Each D-SSite contains multiple VDS, backs up the data, and ensures that the data will not be lost due to failure. Moreover, the D-SSite migrates the VDS through the maintenance server to expand and change the storage site.

- *Virtual Disk Spaces*
  A VDS is a storage unit of the logical layer in the distributed storage system. Each VDS is composed of an EIF and DSI.

- *Embedded Index File*
  It is an index file in the VDS, and supports SQL retrieval. It stores and indexes metadata in the VDS, and supports multiple EIF operations at the same time, and it can improve the parallel retrieval ability of the system.

- *Data Storage Infrastructure*
  It is the internal data file in the VDS. It is constructed according to the organization structure of the direct addressing model in the image data, so as to realize the rapid retrieval and management of the data.

Distributed databases. As the data collection point for different storage nodes in the distributed system, the distributed database mainly assigns data storage tasks to each partition according to the workload performance, so as to support the query processing of various data models and improve the throughput and speed of data transmission and storage (Hercezelaya et al. 2019; Muzammal et al. 2019). There are NoSQL (Not Only SQL), NewSQL, SQL Engine, and so on. Among them, NoSQL improves system performance by abandoning transaction ACID semantics and the complex relationship model, which leads to weak data consistency and integrity (e.g., Etcd, HBase, Mongo DB); NewSQL ensures the transaction ACID semantics and relationship model without reducing the scalability and performance of the system (e.g., CockroachDB, TiDB, Spanner). As an extension of NoSQL, the SQL Engine mainly receives and processes tasks based on the NoSQL system (e.g., Hadapt, Hive) (Meng and Ci 2013; Ruan et al. 2019). As shown in Figure 3.2, in the distributed database, each storage node is related to each other in the form of a network, and they are in a relationship of mutual trust, thus forming a trust boundary to judge whether there is a malicious attack. External users need to have data access control to break through the trust boundary to query and process internal data, which is an effective measure to ensure that the data in the database are not damaged (Rani and Sharma 2019).

Access interface and query language. With different interface and query languages in different systems, there is not a unified standard to standardize the access interface and query language. For example, HBase used the API interface to query and modify the database; the Hive used HiveQL interface for data queries; and Sqoop used the data transfer language for queries (Ting and Cecho 2013).

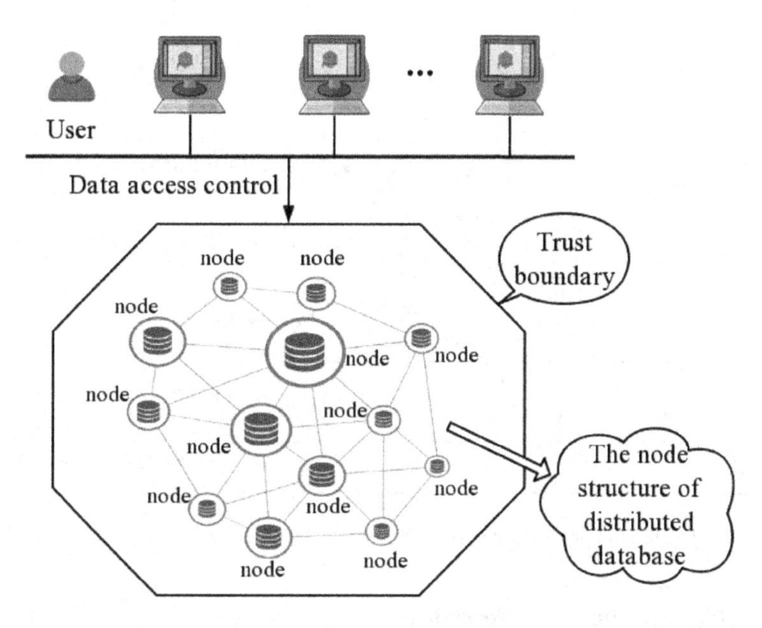

FIGURE 3.2 The structure of a distributed database.

3.3.2 Processing of Urban Remote Sensing Big Data

Urban remote sensing big data processing is facing many challenges, such as intensive data, multiple tasks, high load, high access, high throughput, efficient computing, and so on. In addition, there are many kinds of urban features, complex relationships, and wide application fields, which lead to the increasing amount of urban remote sensing data, and the processing of urban remote sensing big data is more complex and difficult. Therefore, it is an urgent need to develop a big data processing system for urban remote sensing with high availability, scalability, and reliability. Urban remote sensing big data processing mainly includes the following:

(1) GPU-based computing system for urban remote sensing big data processing

There are many proposals on GPU (Graphic Process Unit)-based remote sensing big data computing. The urban remote sensing big data processing system based on GPU is mainly categorized two types as follow:

A. GPU-based rapid big data processing

Generally, the programming algorithm of a GPU acceleration processing system is closely related to the hardware structure of GPU, which makes the programming algorithm of the system unable to be used repeatedly, limits the expansion of the programming function library of the system, and reduces the efficiency of data acceleration processing in the system. In order to make up for this deficiency, the GPU accelerated processing system shown in Figure 3.3 improves the original system. The HPGFS file system is used to transform the data format to provide a unified I/O interface to read and write remote sensing image files. The interface layer, including the GPU parallel programming model and function library, is added between the system layer and algorithm layer of the system, so as to improve the performance of reading, writing, storing, and computing of remote sensing data (Ma et al. 2016).

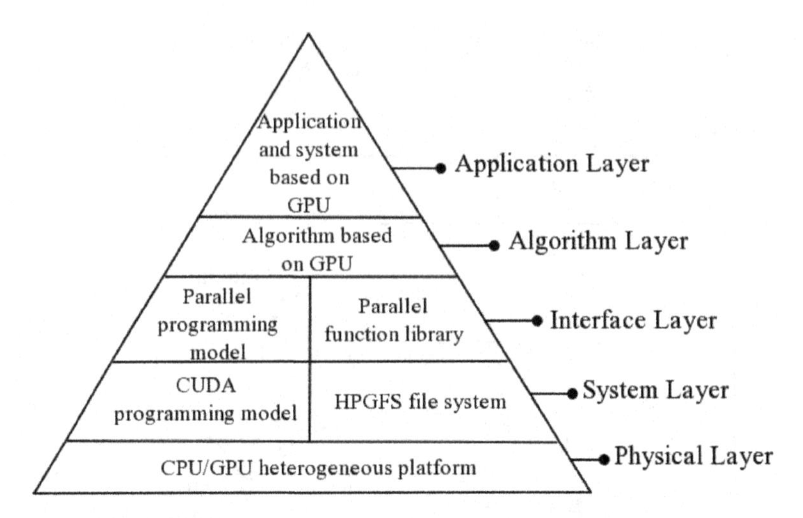

FIGURE 3.3 GPU accelerated processing system.

B. GPU cluster-based big data processing

Computing using the GPU cluster method has matured. According to this method, each node is equipped with one or more GPU accelerators to carry out high parallel cluster computing. The processing speed is continuously being improved. Out of these methods, the CPU/GPU heterogeneous cluster method is widely applied. This method allocates computing tasks according to the computing performance characteristics of CPU and GPU. The CPU is for the logic calculations, while the GPU is for the calculations with parallel arithmetic processing (Dai and Yang 2011; Lei et al. 2014). For the hardware in the GPU cluster method, the GPU is considered as peripheral, and uses a high-speed PCI bus to connect internal nodes to carry out the communication between them (Lv 2015) (Figure 3.4).

Compared with the traditional cluster method, the GPU cluster method has advantages in floating-point computing, general computing, parallel computing, and others. For example, Fan et al. (2004) applied the GPU cluster method in high-performance computing to speed data processing, including cellular automata, finite difference, finite element method, and other numerical methods through the GPU cluster; Fang et al. (2015) designed a parallel computing system using MPI, OpenMP and CUDA. This system provided a real-time service through rapid data processing.

In practice, the processing of urban remote sensing big data, either GPU-based rapid big data processing or GPU-based accelerating big data processing, is only for common operations of remotely sensed big data, such as image registration, geometric correction, image compression, and so on. This implies that the current GPU-based system is still not enough to analyze and mine the data in depth, and further research and deep learning are needed to improve the more complicated processing.

(2) Grid computing system for urban remote sensing big data processing

Grid computing-based big data processing brings together computers which are distributed in different locations with various system structures using the network. Each computer is regarded as a computing node. Through such a large number of computing nodes, a virtual supercomputer structure is built to achieve high-performance processing of urban remote sensing big data, and to solve all types of remote sensing application problems encountered in the city (Bauermarschallinger et al. 2014; Wang 2018). Such a data processing method is called grid computing. Grid computing-based urban remote sensing big data processing can not only make full use of all types of computing resources, but also provide a huge and reliable computing environment to meet the needs of various urban applications.

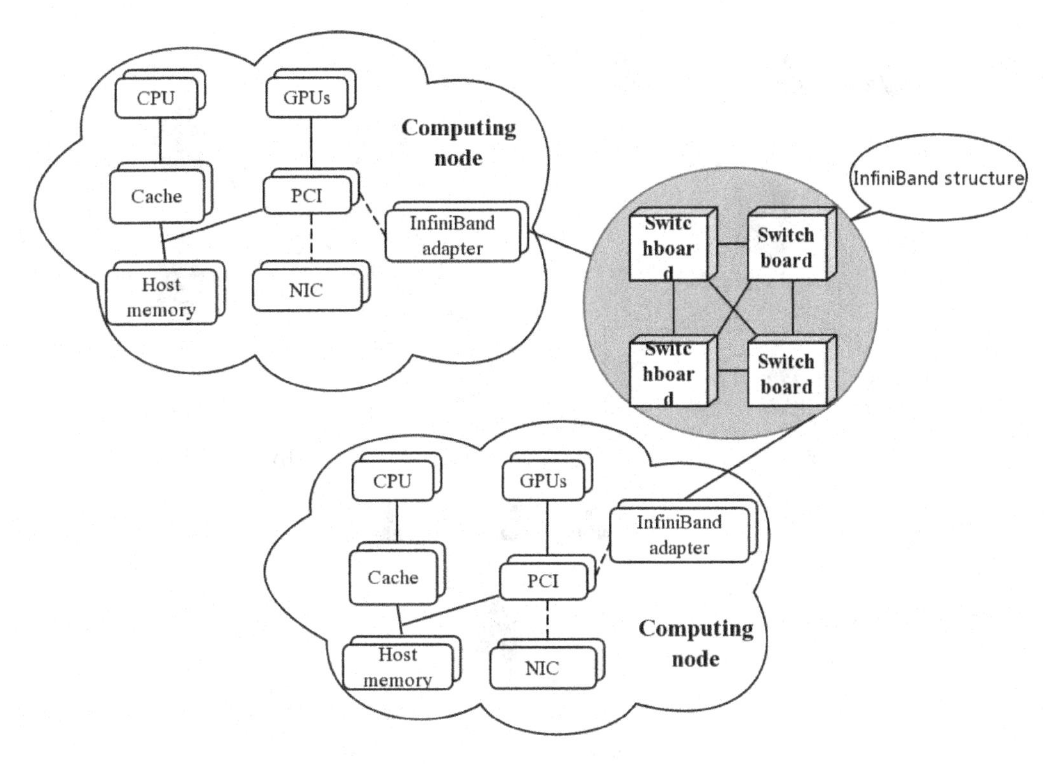

FIGURE 3.4 Hardware architecture of GPU cluster mode.

The main characteristics of grid computing are (Hua and Zhang 2015):

- *Sharing and self similarity.* In the grid structure, resources are stored in different nodes, and they share resource information through network connection. Because of this, there is a self similarity of "part reflects the whole, the whole expresses the part."
- *Heterogeneous.* With the diversity of the resource acquisition and operating systems, the resource structure of different nodes in the grid structure is diverse and inconsistent, thus they are heterogeneous.
- *Dynamic.* With the continuous change of application requirements, the resources in the grid structure are frequently faced with increasing, deleting, modifying, and other operations, which means the whole grid system is in a dynamic state.
- *Autonomy and management multiplicity.* In order to better share and manage the resources of each node in the grid structure, the nodes located in different locations have the management control and allocation policy for their own resources, which also means that there are multiple management nodes for system resources.

Grid computing mainly involves the following key technologies (Wang 2018):

1. *High-performance task scheduling technology.* Due to the large scale, various types, different structures, and dynamic changes of computing resources involved in grid computing, how to efficiently schedule computing resources in a grid system has become a key technology in grid computing. As shown in Figure 3.5, task scheduling mainly includes three grid scheduling methods (i.e., centralized, distributed, hierarchical) (Liu 2017; Shen et al. 2007).

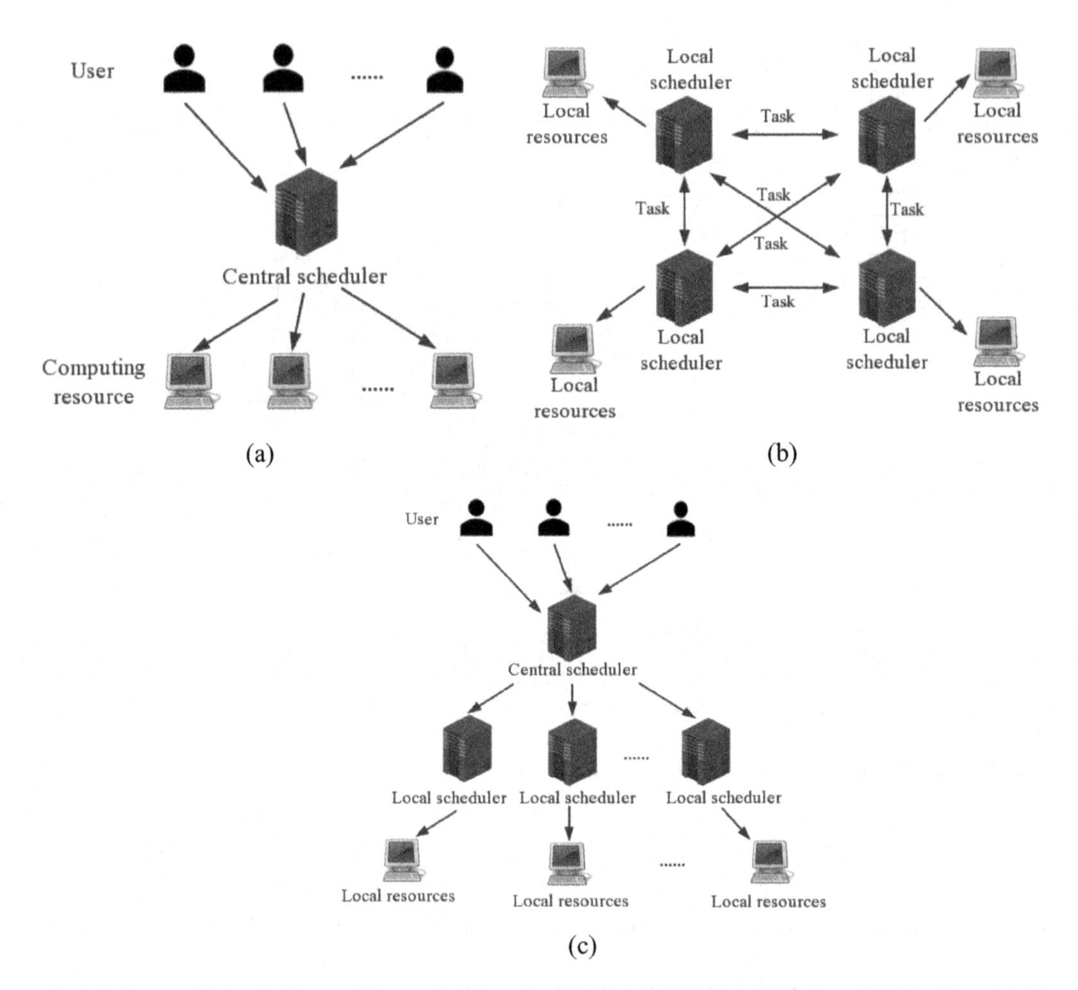

FIGURE 3.5 Grid scheduling structure. (a) Centralized grid scheduling architecture. (b) Distributed grid scheduling architecture. (c) Hierarchical grid scheduling architecture.

From Figure 3.5a, it can be seen that the centralized grid scheduling is mainly used to allocate tasks sent by user entities to all computing resources in the grid through the central scheduler, so as to complete the scheduling of huge computing tasks. Such structure is conducive to management, but its security guarantee is not strong enough. From the results shown in Figure 3.5b, the distributed grid scheduling is mainly used to allocate large and complex computing tasks to each local scheduler interacting with each other, and then use each local scheduler to separately call resources for task processing. Compared with the centralized structure, this kind of structure is more secure and reliable, but the management is relatively lacking, and there is no unified scheduling decision. The hierarchical grid scheduling shown in Figure 3.5c is a comprehensive embodiment of the above two. It has the advantages of unified and autonomous task allocation by the central scheduler and resource invocation by the local scheduler.

2. *Resource management technology.* The allocation of such large-scale processing tasks as urban remote sensing big data is a problem that grid computing must face, pay attention to, and solve. Only with efficient resource management technology can tasks be better allocated and scheduled.
3. *Network security technology.* Because grid computing involves a variety of dynamic heterogeneous computing resources, and the communication protocols and organizations among the computing resources are different, the security assurance of the network communication environment is one of the key technologies in grid computing.

Grid computing related to remotely sensed big data processing includes Hawick et al. (2003), who used a grid computing platform to design corresponding distributed architecture to process massive data; Shen et al. (2004) have designed a grid GIS structure to realize corresponding data processing based on grid computing and GIS technology; and the wide area grid project proposed by the French space agency mainly integrates distributed resources through the InterGrid integrated grid, so as to achieve efficient processing of remote sensing image data (Gorgan et al. 2009; Hassan and Abdullah 2011).

(3) Cloud computing system for urban remote sensing big data processing

After the cluster computing and grid computing appeared, a new form of data computing, namely *cloud computing*, has appeared in the field of big data. The characteristics of the cloud computing system are (Mell and Grance 2011, Zhang et al. 2010):

- *High network access.* Based on network services, in cloud computing, users can access the corresponding resources, functions, and services provided by the cloud through the network, and support the access of mobile phones, laptops, palmtops, workstations, and other large or small platforms.
- *High scalability.* According to the demand of users, the resources in the cloud are released in the way of internal contraction or external expansion, so as to save the cost of system operation and the use of resources.
- *Demand service.* Users can obtain data, computing, storage, and other resources directly according to their own needs without negotiating with service providers. Meanwhile, users can also publish resources and services to the cloud, which greatly reduces the cost of system operation, management, and maintenance, as well as the corresponding business risks.
- *Measure service.* The system displays and explains the progress, speed, and quantity of the use of resources, so that the cloud can reasonably adjust the allocation of resources, and users can more clearly understand the use of resources.

The cloud computing system effectively places all types of data resources, computing resources, storage resources, and other resources in the "cloud" node, to process big and various data, and to provide "cloud" users with sharing and payment services in areas such as data, computing, storage, and others. The four types of cloud services provided are:

1. *IaaS (Infrastructure as a Service).* This type of service mainly provides users with some basic virtual technology services, including computing, storage, network communication, and server storage. Users can use these services through the network with payment for a variety of applications. Typical services vendors include Go Grid, Amazon EC2, Drop Box, and Akamai (Li 2013; Wang et al. 2013; Youssef 2012).
2. *PaaS (Platform as a Service).* This type of service provides users with a platform with a programming environment, development tools, and other combinations operated in the cloud node (Geetha and Robin 2017). The users can operate, develop, publish, distribute, and share software resources through the platform service. Typical platform service vendors include Google application engine, Alibaba cloud, Baidu cloud, and Microsoft Azure (Buyya et al. 2014; Espadas et al. 2013).
3. *SaaS (Software as a Service).* In this type of service, users can directly access and use the software resources placed in the cloud node through the network in accordance with their own needs. The users do not need to install the software or obtain software licenses, and just pay pocket fees. This form of service greatly saves resources. The typical service vendors include YouTube, Zoho, Google Apps, and Salesforce.com (Li et al. 2015; Rao et al. 2012; Yang et al. 2015).

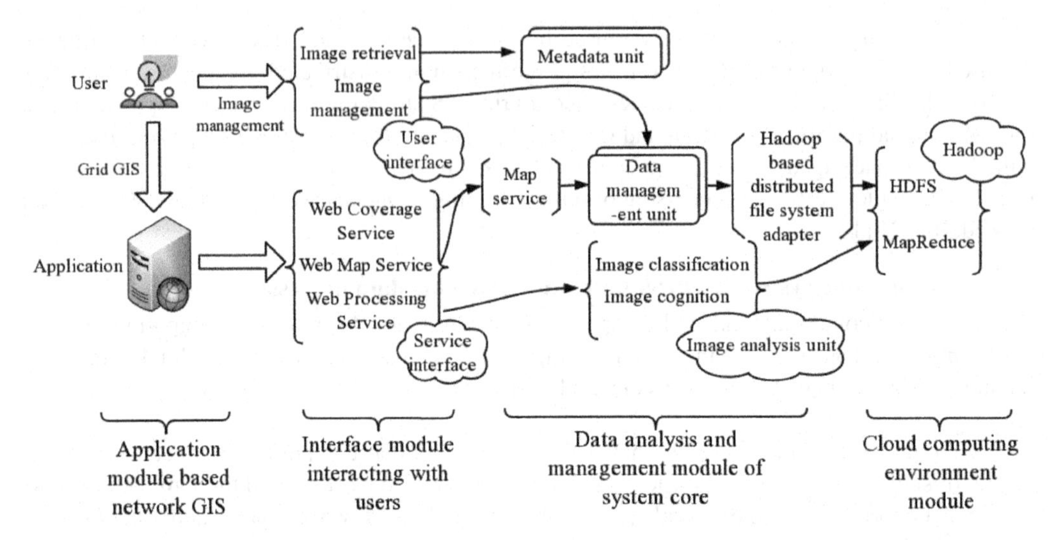

FIGURE 3.6 Remote sensing data processing system based on Hadoop cloud computing platform.

4. *DaaS (Data as a Service).* This type of service can provide users with a comprehensive and rich visualization through a browser with large, multi-dimensional data. Typical service vendors are Google Earth, Google Map, and Yahoo Map (Sénica et al. 2011; Zhu et al. 2016).

For urban remote sensing big data, a cloud computing system based on Hadoop architecture is proposed, as shown in Figure 3.6. It includes (Lin et al. 2013):

1. An application module based on network GIS;
2. An interface module interacting with users;
3. A data analysis and management module of system core; and
4. A cloud computing environment module as system infrastructure.

Apart from the cloud computing system based on Hadoop described above, the other cloud computing-based urban remote sensing big data processing systems include Hadoop-GIS, Google Earth Engine, OpenRS Cloud, and G-Cloud.

- *Hadoop-GIS.* A spatial database system based on MapReduce realizes parallel processing of large-scale data by block processing, and expands a new spatial query language based on the HiveQL interface query language of Hive, so as to query various types of spatial data (e.g., points, lines, faces) and relationships (e.g., including, separated, nearest) (Aji et al. 2013).
- *Google Earth Engine.* This is a browser platform based on cloud computing, with rich data resources, multi-functional data analysis, and processing algorithms and software development tools to support parallel processing, monitoring, visual analysis, and other operations of large geospatial data sets, and is a good tool for graphics processing, professional mapping, terrain modeling, and machine learning (Brooke et al. 2020; Tamiminia et al. 2020).
- *OpenRS Cloud.* As an open-source cloud platform, it mainly deals with the intensive remote sensing image in a fast and parallel way. Based on the Hypertable and Hadoop environment, it uses an object-oriented and attribute-oriented plug-in system (including object recognition, registration, interface query, and other functions) to realize the abstraction and independence of programming, as well as the expansibility of application functions. At the same time, it also provides three services for users: a web site system, independent desktop processing system, and web service based on SOAP Protocol, so that users can query, obtain, manage data, and develop algorithms (Guo et al. 2010).

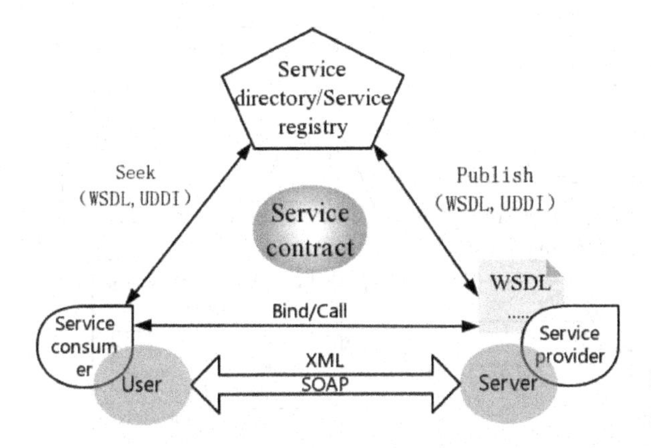

FIGURE 3.7 Basic architecture of SOA (Endrei et al. 2004).

- *G-Cloud.* Under the Internet environment, the virtual service infrastructure including rich information and technology is used to provide the corresponding virtual operating system and rental service platform for the public sector (e.g., education, medical care, government, and enterprises), aiming to reduce the energy consumption and cost of the system by simplifying resource deployment, and realize a more "low-carbon" and "green" data processing platform (Chandra and Bhadoria 2012; Jeong et al. 2014).

(4) Cloud grid-based urban remote sensing big data processing

The purpose of a grid computing system is to gather those idle computing resources together through a network to achieve the effect of resource sharing, while the purpose of the cloud computing system is to place various resources in the individual cloud node to provide services for users with computing, storage, and professional applications under payment. The former focuses on resource sharing and parallel computing, while the latter focuses on application services on the basis of resource sharing.

How to combine the advantages of both the grid computing system and the cloud computing system to build a high-performance and high-service big data processing system is very interesting and attractive to users and scholars. The cloud grid is the result of integrating grid computing and cloud computing through service-oriented architecture (SOA) (Rings and Grabowski 2012).

The basic architecture of SOA is shown in Figure 3.7. As observed from Figure 3.7, it can be found that under the constraint of service contract, the vendors register and publish their services through a service center, while users find their needed services through the service center, and the service center feeds back the results to the vendors through called-on services (Endrei et al. 2004). A typical cloud grid-based service system for remote sensing big data application was reported by Zeng (2012). This system is based on a SOA structure, dynamically allocates computing resources using P2P technology according to demand in the cloud grid environment, assigns the service resources using a GL-RSLM scheduling model, and establishes SOA in the cloud grid environment for disaster monitoring.

3.4 CHALLENGES OF URBAN REMOTE SENSING BIG DATA

Various resolutions, diversities, and multi-levels of urban remote sensing big data increase the difficulty of data processing, including computing, storage, analysis, mining, and others. The mismatch between the technology improvement and data increments results in the challenge. Summarily, it includes (see Figure 3.8):

1. Data processing;
2. Storage and management;

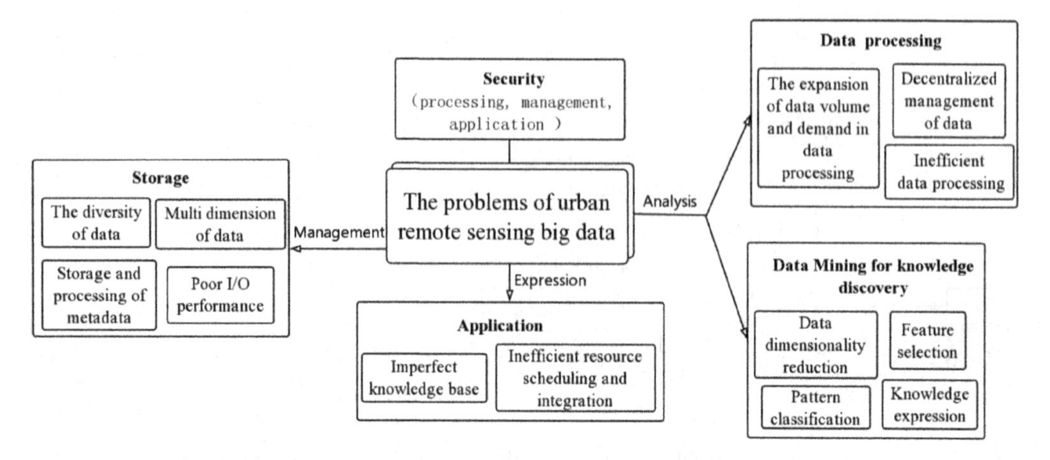

FIGURE 3.8 The problems of urban remote sensing big data.

3. Data mining for knowledge discovery;
4. Security; and
5. Applications.

(1) Data processing vs data increasing

Due to dramatically increasing remote sensing data, associated with high dimensionality, diversity, and real-time, the following problems have been exposed:

a. *The expansion of big data volume and the growth of demand in data processing*. The remotely sensed data collection platform is growing in quantity and in quality, which leads to the continuous expansion in data volume. The current big data analysis and processing methods are not enough to solve the problem of urban remote sensing big data. How to build a more suitable analysis model for big data is a challenging task.
b. *Decentralized management of data*. The difference between various types of sensors results in the inconsistency of data resolution, data format, and so on. The data processing methods and semantic expression technologies should not be uniform. This means that highly diverse and decentralized data lead to decentralized management. Therefore, how to establish a unified and standardized semantic expression model and data processing method to solve the problem of decentralized management for urban remote sensing big data is challenging.
c. *Low data processing efficiency.* The original data unavoidably contains errors, ambiguity, incompleteness, inconsistency, and uncertainty, which aggravates the difficulty of data processing and analysis. Although there are many kinds of high performance computing methods, such as cluster computing, grid computing, cloud computing, and so on, none of the unified computing modes can efficiently process the high-dimensional data. In other words, the current processing system is still inefficient.

(2) Storage of big data

The storage and management of big data are prerequisites in the process of data mining and applications. The problems of big data storage and management include:

a. *The diversity of data*. The data obtained from various platforms, including spaceborne, airborne, and personal cell phones, are widely varied and distributed. This challenging fact makes the storage system structure a higher requirement.

b. *Multi-dimensionality of data.* Urban remote sensing big data is multi-dimensional in terms of, for instance, time, space, and spectrum, which results in difficulty in data storage management and processing. How to properly reduce complex multi-dimensional data into one-dimensional data for storage is a challenging issue.
c. *Storage and processing of metadata.* As the core of big data, metadata is an important criterion in the storage performance of the remote sensing big data. How to establish a unified standard and a unified storage management method to efficiently manage metadata is a challenging topic in urban remote sensing big data.
d. *Poor I/O performance.* The storage of big data needs not only enough space, but also high-performance I/O response. The current storage system relies on the performance of physical equipment, which greatly reduces the efficiency of data storage, and cannot meet the requirements of users for high performance of I/O response. In addition, if the I/O mode is in an unstable state, the processing of I/O is extremely difficult.

The future storage management will be an integration of both storage and computing, which can largely improve the metadata processing, I/O performance, data reading, and processing, in order to solve the low effectiveness of I/O performance and data utilization, which arise under the present conditions of separation of storage and computing. Just like the "human brain plan" proposed by the European Union in 2013, the future big data storage will use computers to simulate the operation mechanisms of the human brain, with human brain cells gathering to organize through neurons, and build a "cell body" structure of a human brain for data storage and management. This cell body of the human brain takes cells as the smallest storage unit, and realizes the evolution from geospatial data cells to data clusters, to data blocks, and then carry out expression of geospatial space (see Figure 3.9). This cell body of the human brain facilitates the rapid data processing and wide application of urban remote sensing big data.

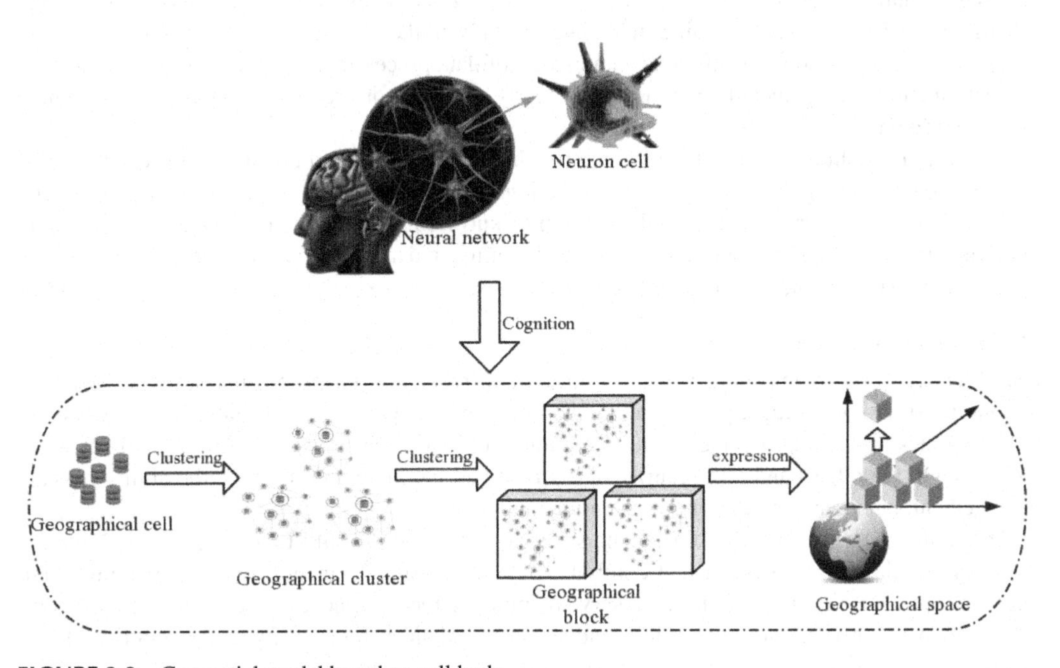

FIGURE 3.9 Geospatial model based on cell body.

(3) Machine learning and data mining vs big data

Machine learning is a powerful tool for analyzing data and acquiring knowledge. The current machine learning method is only suitable for the analysis of small amounts of data, and not suitable for the deep learning analysis of big data (Aljarrah et al. 2015; Leuenberger and Kanevski 2015; Lary et al. 2016; Yu et al. 2014). The current research on machine learning is as follows:

a. *Data dimensionality reduction.* Urban remote sensing big data is high-dimensional, including rich attribute information. Therefore, during applications in practice, the big data is often dimensionally reduced to obtain the essential information hidden in big data, and to better mine data knowledge. Meanwhile, the essential characteristics of the original data are required to remain. This increases the difficulty of data learning.
b. *Feature selection.* The urban remote sensing big data have spatial, temporal, spectral, and other characteristics. They require selecting an appropriate feature to carry out in-depth machine learning in accordance with the various application requirements. For this reason, it is most difficult to determine which feature is most appropriate.
c. *Pattern classification.* According to the prior knowledge, it is an effective method to import the training set acquired by training (learning) into the designed classification pattern according to the attributes of the training set for matching and classification, which is to realize information recovery and data deep interpretation analysis. Therefore, pattern classification is a problem that must be considered in the process of effective use of data information, and research and learning.
d. *Knowledge expression.* Layers generally acquire the information knowledge after the machine learning and deep mining of big data. Thus, semantic expression of knowledge must be researched. Therefore, the standardized expression of semantic knowledge is the core of knowledge expression and application. The realization to be achieved is very challenging.

The mining technology of urban remote sensing big data aims to explore laws and reveal the deep information hidden in the big data. The challenges during big data mining are the huge volume of big data, the fast update cycle, the diversity of big data, and inconsistent format of various data. These facts significantly impact development of big data mining and analysis. Future development in data mining should be information mining, including not only in the accuracy and recognition ability of data processing, but also in the efficiency and speed of data processing, especially in comprehensive data/information mining, including urban disaster early warning, construction planning, environmental safety, and so on.

In addition, machine learning and artificial intelligence are both used for mining the deep knowledge of remote sensing data to execute the functions of data interpretation, data analysis and data classification, and usher/guide the applications. For knowledge acquisition, knowledge expression of urban remote sensing big data, future machine learning, and artificial intelligence methods should be intelligent, simple, and fast for complicated analysis.

(4) Security of big data

Data security is becoming more and more important, requiring the proper measures to be taken in data transmission, storage, processing, and so on using a dynamic controllable security system that integrates processing, management, operation, and maintenance using hardware and software facilities. On the other hand, the security of big data can facilitate and improve the ability of data management, emergency processing, prevention, and others.

The future measures to be taken for the security of big data should be to establish a dynamic and controllable security system that combines data processing, management, and operation and maintenance. For end-users, it is necessary to improve the data access security, and carry out special identity verification, encryption, and so on to enhance the security and reliability of the system.

(5) Application problems of urban remote sensing big data

The urban remote sensing big data have widely been applied in fields such as agriculture, industry, water conservancy, climate change, disaster emergency, human geographical environment, and social activities. However, these applications have exposed two problems:

a. The applications are limited due to the imperfect knowledge base, which includes intelligent algorithms similar to deep learning, such as target detection, classification, parameter extraction, and parameter retrieval.
b. The low efficiency and high running loads during allocating and scheduling the computing, data, and storage resources are exposed. In addition, since the structure and function of each application are varied, how to integrate these resources is also one of the greatest challenges.

3.5 INTELLIGENT EARTH OBSERVING SATELLITE SYSTEM FOR URBAN REMOTE SENSING

3.5.1 Change in Users' Need from Image-Based Product to Image-Based Information/Knowledge

Although many remote sensing systems offer a wide range of spatial, spectral, and temporal parameters, users have still not satisfied most provisions from the Earth observation program. With the development of information technology, users' needs have migrated from traditional image-based data to advanced image-based information/knowledge, such as production estimation of rice, flood coverage area, on-board forest fire detection, and so on (see Figure 3.10). To this end, many innovative Earth observing systems have been emerging in recent years. Zhou has provided a detailed

Various Users		Illustration
Mobile user	A real-time user, e.g., a mobile GIS user, requires a real-time downlink for geo-referenced satellite imagery with a portable receiver, small antenna and laptop computer.	
Real-time user	A mobile user, e.g., a search-and-rescue pilot, requires a real-time downlink for geo-referenced panchromatic or multispectral imagery in a helicopter.	
Lay user	A lay user, e.g., a farmer, requires geo-referenced, multispectral imagery at a frequency of 1-3 days for investigation of his harvest.	
Professional user	A professional user, e.g., a mineralogist, requires hyperspectral imagery for distinguishing different minerals.	
Professional user	A topographic cartographer, e.g., a photogrammetrist, requires panchromatic images for stereo mapping.	

FIGURE 3.10 Some examples of future direct end-users in the land surface remote sensing (Courtesy of Zhou et al. 2004).

review of a future Earth observation system (Zhou 2003; Zhou et al. 2004), which is called the Intelligent Earth Observing Satellite (IEOS).

More and more users want the imagery provider to provide the value-added content they need, but these users are not concerned with the technical complexities of image processing. Therefore, timely, reliable, and accurate information, with the capability of direct downlink of various bands of satellite data/information and operation as simple as selecting a TV channel, is highly preferred. It is apparent that no single satellite can meet all of the requirements presented by users above.

3.5.2 Intelligent Earth Observing Satellite System for Urban Remote Sensing

Zhou et al. (2004) presented in early 2000 an envisioned future intelligent Earth observing satellite system (FIEOS), which is a space-based architecture for the dynamic and comprehensive on-board integration of Earth observing sensors, data processors, and communication systems. The architecture and implementation strategies suggest a seamless integration of diverse components into a smart, adaptable, and robust Earth observation satellite system. It is intended to enable simultaneous, global measurements and timely analyses of the Earth's environment for a variety of users. Especially, common users would directly access data in a manner similar to selecting a TV channel.

The IEOS is imagined as a two-layer satellite network. This satellite network is enough to reach all functions required by users (see Figure 3.11). In contrast, a satellite network with more than two layers will add the load of data communication of cross-links. The first layer, which consists of many Earth observing satellites (EOSs) viewing the entire Earth, is distributed in low orbits ranging from 300 km to beyond. Each EOS is small, lightweight, and inexpensive relative to current satellites. These satellites are divided into groups called satellite groups. Each EOS is equipped with a different sensor for collection of different data and an on-board data processor that enables it to act autonomously, reacting to significant measurement events on and above the Earth. They collaboratively work together to conduct the range of functions currently performed by a few large satellites today. There is a lead satellite in each group, called group-lead; the other satellites are called member-satellites. The group-lead is responsible for management of the member-satellites and communication with other group-leads in the network (constellation) in addition to communication with

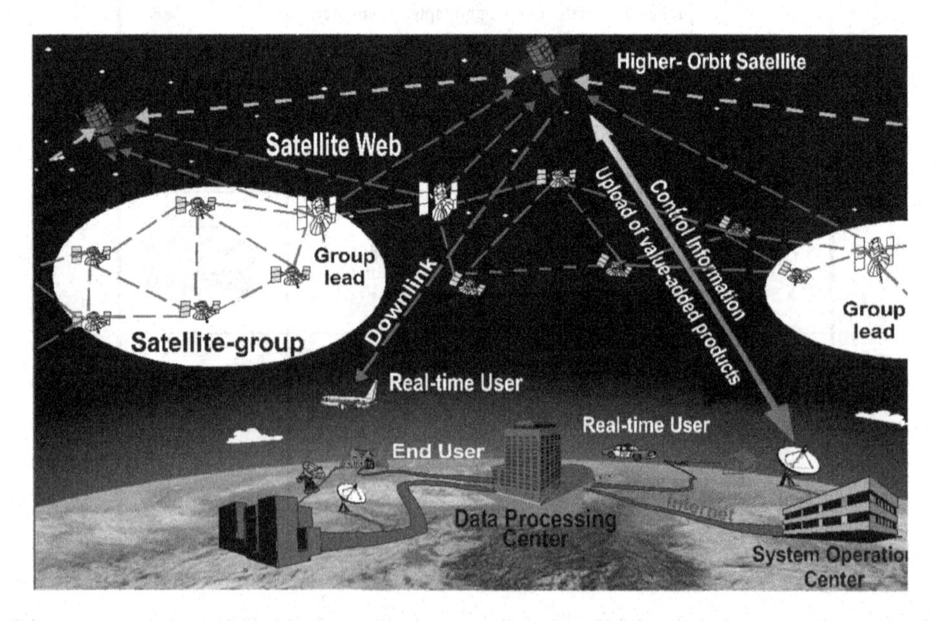

FIGURE 3.11 The architecture of Intelligent Earth Observing Satellite (IEOS) (Courtesy of Zhou et al. 2004).

the geostationary satellites. This mode of operation is similar to an intranet. The group-lead looks like a local server, and the member-satellites look like the computer terminals. The local server (group-lead) is responsible for Internet (external) communication in addition to management of the intranet (local) network. This design can reduce the communication load and ensure effectiveness of management and coverage of data collection.

The second layer is composed of geostationary satellites because not all EOSs are in view of or in communication with worldwide users. The second layer satellite network is responsible for communication with end-users (e.g., data downlink) and ground control stations, and ground data processing centers, in addition to further processing of data from group-lead satellites.

All of the satellites are networked together into an organic measurement system with high-speed optical and radio frequency links. User requests are routed to specific instruments maximizing the transfer of data to archive facilities on the ground and on the satellite. Thus, all group-leads must establish and maintain a high-speed data cross-link with one another in addition to an up-link with one or more geostationary satellites, which in turn maintain high-speed data cross-links and down-links with end-users and ground control stations and processing centers.

REFERENCES

Aji, A., Wang, F., Vo, H., Lee, R., Liu, Q., Zhang, X., & Saltz, J. H., Hadoop GIS: A high performance spatial data warehousing system over MapReduce, *The 39th International Conference on Very Large Data Bases,* Trento, Italy: Very Large Database Endowment, 2013, pp. 1009–1020.

Aljarrah, O. Y., Yoo, P. D., Muhaidat, S., Karagiannidis, G. K., & Taha, K., Efficient machine learning for big data: A review, *Big Data Research*, 2015, vol. 2, no.3, pp. 87–93.

Bauermarschallinger, B., Sabel, D., & Wagner, W., Optimization of global grids for high-resolution remote sensing data, *Computers & Geosciences*, 2014, vol. 72, pp. 84–93.

Brooke, S. A., Darcy, M., Mason, P. J., & Whittaker, A. C., Rapid multispectral data sampling using Google Earth Engine, *Computers & Geosciences*, 2020, vol. 135, pp. 104366.

Buyya, R., Calheiros, R. N., Son, J., Dastjerdi, A. V., & Yoon, Y., Software - Defined cloud computing: Architectural elements and open challenges, *2014 International Conference on Advances in Computing, Communications and Informatics (ICACCI)*, New Delhi, 2014, pp. 1–12.

Casu, F., Manunta, M., Agram, P. S., Crippen, R. E., Big remotely sensed data: Tools, applications and experiences, *Remote Sensing of Environment*, 2017, vol. 202, pp. 1–2.

Chandra, D. G., & Bhadoria, R. S., Role of G-Cloud in citizen centric governance, *Grid Computing*, 2012, pp. 44–48.

Cheng, F. C., Research on Distributed Hybrid Computing Technologies for Massive Remote Sensing Data, Chengdu University of Technology, 2014. (Chinese)

Chi, M., Plaza, A., Benediktsson, J. A., Sun, Z., Shen, J., & Zhu, Y., Big Data for remote sensing: Challenges and opportunities, *Proceedings of the IEEE*, 2016, vol. 104, no.11, pp. 2207–2219.

Dai, C. G., & Yang, J. Y., Research on orthorectification of remote sensing images using GPU-CPU cooperative processing, *2011 International Symposium on Image and Data Fusion*, Teng-chong, Yunnan: IEEE, 2011, pp. 1–4.

Endrei, M., Ang, J., Arsanjani, A., Ang, A., & Chua, S., Patterns: Service - Oriented Architecture and Web Services, http://publib-b.boulder.ibm.com/Redbooks.nsf/RedbookAbstracts/sg246303.html? Open [EB/OL], 2004-7-6.

Espadas, J., Molina, A., Jimenez, G., Molina, M., Ramirez, R., & Concha, D., A tenant - based resource allocation model for scaling Software-as-a-Service applications over cloud computing infrastructures, *Future Generation Computer Systems*, 2013, vol. 29, no.1, pp. 273–286.

Fan, J., Yan, J., Ma, Y., Wang, L., Big data integration in remote sensing across a distributed metadata-based spatial infrastructure, *Remote Sensing*, 2018, vol. 10, p.7, doi:10.3390/rs10010007.

Fan, Z., Qiu, F., Kaufman, A. E., & Yoakumstover, S., GPU Cluster for High Performance Computing, *Conference on High Performance Computing (Supercomputing)*, 2004, pp. 47–47.

Fang, L. Y., Wang, M., Li, D. R., & Pan, J., MOC-based parallel preprocessing of ZY-3 satellite images, *IEEE Geoscience and Remote Sensing Letters*, 2015, vol. 12, no.2, pp. 419–423.

Geetha, P., & Robin, C. R. R., A comparative-study of load-cloud balancing algorithms in cloud environments, *2017 International Conference on Energy, Communication, Data Analytics and Soft Computing (ICECDS)*, Chennai, 2017, pp. 806–810.

Gorgan, D., Stefanut, T., & Bacu, V., Grid based training environment for earth observation, *Proceedings of the 4th International Conference on Advances in Grid and Pervasive Computing*, Berlin, Germany: Springer, 2009, vol. 5529, pp. 98–109.

Guo, W., Gong, J., Jiang, W., Liu, Y., & She, B., OpenRS-Cloud:A remote sensing image processing platform based on cloud computing environment, *Science China-technological Sciences*, 2010, vol. 53, no.1, pp. 221–230.

Hassan, M. I., & Abdullah, A., A semantic service discovery framework for inter grid, *Proceedings of the Second International Conference on Software Engineering and Computer Systems*, Berlin, Germany: Springer, 2011, pp. 448–462.

Hawick, K. A., Coddington, P. D., & James, H. A., Distributed frameworks and parallel algorithms for processing large-scale geographic data, *IEEE International Conference on High Performance Computing Data and Analytics*, 2003, vol. 29, no.10, pp. 1297–1333.

Hercezelaya, J., Porcel, C., Bernabemoreno, J., Tejedalorente, A., & Herreraviedma, E., Web platform for learning distributed databases' queries processing, *Procedia Computer Science*, 2019, pp. 827–834.

Hua, Y. C., & Zhang, G. K., Research and analysis of grid computing model based on big data, *Information Technology and Informatization*, 2015, no.10, pp. 230–231. (Chinese)

Jeong, H., Jeong, Y., & Park, J. H., G-cloud monitor: A cloud monitoring system for factory automation for sustainable green computing, *Sustainability*, 2014, vol. 6, no.12, pp. 8510–8521.

Lai, J. B., Luo, X. L., Yu, T., & Jia, P. Y., Remote sensing data organization model based on cloud computing, *Computer Science*, 2013, vol. 40, no.7, pp. 80–83+115. (Chinese)

Lary, D. J., Alavi, A. H., Gandomi, A. H., & Walker, A. L., Machine learning in geosciences and remote sensing, *Geoscience Frontiers*, 2016, vol. 7, no.1, pp. 3–10.

Lei, Z., Wang, M., Li, D. R., & Lei, T. L., Stream model-based orthorectification in a GPU cluster environment, *IEEE Geoscience and Remote Sensing Letters*, 2014, vol. 11, no.12, pp. 2115–2119.

Leuenberger, L., & Kanevski, M., Extreme learning machines for spatial environmental data, *Computers & Geosciences*, 2015, vol. 83, pp. 64–73.

Li, C. L., Research and prototype implementation of remote sensing business process based on cloud platform, China University of Geosciences (Beijing), 2013. (Chinese)

Li, D. R., Yao, Y., & Shao, Z. F., Big data in smart cities, *China Construction Information*, 2014b, vol. 39, no.6, pp. 631–640.

Li, D. R., Zhang, L. P., & Xia, G. S., Automatic analysis and data mining of remote sensing big data, *Acta Geodaetica et Cartographica Sinica*, 2014a, vol. 43, no.12, pp. 1211–1216.(Chinese).

Li, Z. J., Li, X. J., Liu, T., Xie, J. W., & Yang, S., Remote Sensing Cloud Computing: Current Research and Prospect, *Journal of Equipment Academy*, 2015, vol. 26, no.5, pp. 95–100. (Chinese)

Lin, F. C., Chung, L. K., Ku, W. Y., Chu, L. R., & Chou, T. Y., The framework of cloud computing platform for massive remote sensing images, *Proceedings of the IEEE 27th International Conference on Advanced Information Networking and Applications,* Barcelona: IEEE, 2013, pp. 621–628.

Liu, P., A survey of remote-sensing big data, *Frontiers in Environmental Science*, 2015, no.3, pp. 45. doi:10.3389/fenvs.2015.00045.

Liu, P., Di, L., Du, Q., & Wang, L., Remote sensing big data: Theory, methods and applications, *Remote Sensing*, 2018, vol. 10, no.5, pp. 711.

Liu, Y., Research on Task Scheduling Algorithms in Grid Computing Environment, *Xi'an University of Electronic Science and Technology*, 2017. (Chinese)

Lv, X. F., Cheng, C. Q., Gong, J. Y., & Guan, L., Review of data storage and management technologies for massive remote sensing data, *Scientia Sinica (Technologica)*, 2011, vol. 41, no.12, pp. 1561–1573. (Chinese)

Lv, X. W., Research on GPU Parallel Computing and Application for HPC Cloud, *Nanjing University of Aeronautics and Astronautics*, 2015. (Chinese)

Ma, Y., Chen, L. J., Liu, P., & Lu, K., Parallel programing templates for remote sensing image processing on GPU architectures: Design and implementation, *Computing*, 2016, vol. 98, no.1, pp. 7–33.

Ma, Y., Wang, L., Liu, P., Ranjan, R., Towards building a data-intensive index for big data computing – A case study of remote sensing data processing, *Information Sciences*, 2015a, vol. 319, pp. 171–188.

Ma, Y., Wu, H., Wang, L., Huang, B., Ranjan, R., Zomaya, A., Jie, W., Remote sensing big data computing: Challenges and opportunities, *Future Generation Computer Systems*, 2015b, vol. 51, pp. 47–60.

Mell, P., & Grance, P., The NIST Definition of Cloud Computing, *National Institute of Standards and Technology*, 2011.

Meng, X. F., & Ci, X., Big data management: Concept, technology and challenge, *Journal of Computer Research and Development*, 2013, vol. 50, no.1, pp. 146–169. (Chinese)

Muzammal, M., Qu, Q., & Nasrulin, B., Renovating blockchain with distributed databases: An open source system, *Future Generation Computer Systems*, 2019, vol. 90, pp. 105–117.

NASA. NASA NEX [OL]. [2015-02-28]. http://aws.amazon.com/cn/nasa/nex/ [2015-02-28].

Nie, P., Chen, G. S., & Jing, W. P., A distributed storage method for remote sensing images, *Engineering of Surveying and Mapping*, 2018, vol. 27, no.11, pp. 40–45. (Chinese)

Rani, K., & Sharma, C., Tampering detection of distributed databases using blockchain technology, *International Conference on Contemporary Computing*, 2019, pp. 1–4.

Rao, B. B. P., Saluia, P., Sharma, N., Mittal, A., & Sharma, S. V., Cloud computing for Internet of Things & sensing based applications, *2012 Sixth International Conference on Sensing Technology (ICST)*, 2012, pp. 374–380.

Rathore, M. M. U. et al., Real-time big data analytical architecture for remote sensing application, *IEEE Journal of Selected Topics in Applied Earth Observations and Remote Sensing*, 2015, vol. 8, no.10, pp. 4610–4621.

Rings, T., & Grabowski, J., *Pragmatic Integration of Cloud and Grid Computing Infrastructures*, *Proceedings of the 2012 IEEE 5th International Conference on Cloud Computing,* Honolulu, Hawaii: IEEE, 2012, pp. 710–717.

Ruan, P., Chen, G., Dinh, T. T., Lin, Q., Loghin, D., Ooi, B. C., & Zhang, M., Blockchains and Distributed Databases: A Twin Study, *arXiv: Databases*, 2019.

Sénica, N., Teixeira, C., & Pinto, J. S., *Cloud computing: A platform of services for services*, *Proceedings of International Conference on ENTERprise Information Systems,* Berlin Heidelberg: Springer, 2011, pp. 91–100.

Shen, Z., Luo, J., Zhou, C., Cai, S., Zheng, J., Chen, Q., Ming, D., & Sun, Q., Architecture design of grid GIS and its applications on image processing based on LAN, *Information Sciences*, 2004, vol. 166, no.1, pp. 1–17.

Shen, Z. F., Luo, J. C., Huang, G. Y., Ming, D. P., Ma, W. F., & Sheng, H., Distributed computing model for processing remotely sensed images based on grid computing, *Information Sciences*, 2007, vol. 177, no.2, pp. 504–518.

Song, W. J., Liu, P., Wang, L. Z., & Lv, K., Intelligent processing of remote sensing big data: Current situation and challenges, *Journal of Engineering Studies*, 2014, vol. 603, pp. 259–265. (Chinese)

Tamiminia, H., Salehi, B., Mahdianpari, M., Quackenbush, L., Adeli, S., & Brisco, B., Google Earth Engine for geo-big data applications: A meta-analysis and systematic review, *ISPRS Journal of Photogrammetry and Remote Sensing*, 2020, vol. 164, pp. 152–170.

Ting, K., & Cecho, J. J., *Apache Sqoop Cookbook*, O'Reilly Media, 2013.

Wang, J., Grid computing technology, *TV Guide China*, 2018, no.1, pp. 250. (Chinese)

Wang, L., Ma, Y., Yan, J., Chang, V., & Zomaya, A. Y., pipsCloud: High performance cloud computing for remote sensing big data management and processing, *Future Generation Computer Systems*, 2018, vol. 78, pp. 353–368.

Wang, L., Zhong, H., Ranjan, R., Zomaya, A., & Liu, P., Estimating the statistical characteristics of remote sensing big data in the wavelet transform domain, *IEEE Transactions on Emerging Topics in Computing*, 2014, vol. 2, no.3, pp. 324–337.

Wang, P. Y., Wang, J. Q., Chen, Y., & Ni, G. Y., Rapid processing of remote sensing images based on cloud computing, *Future Generation Computer Systems*, 2013, vol. 29, no.8, pp. 1963–1968.

Yang, J., Zhang, L., & Wang, X. A., *On Cloud Computing Middleware Architecture*, *2015 10th International Conference on P2P, Parallel, Grid, Cloud and Internet Computing (3PGCIC)*, Krakow, 2015, pp. 832–835.

Youssef, A. E., Exploring cloud computing services and applications, *Journal of Emerging Trends in Computing and Information Sciences*, 2012, pp. 838–847.

Yu, B., Li, S. Z., Xu, S. X., & Ji, R. R., Deep Learning: A key of stepping into the era of big data, *Journal of Engineering Studies*, 2014, vol. 6, no.3, pp. 233–243. (Chinese)

Zeng, Z., Research on resource and service provisioning with high efficiently for massive high resolution remote sensing image processing under cloud environment, *Zhejiang University*, 2012. (Chinese)

Zhang, Q., Cheng, L., & Boutaba, R., Cloud computing: State-of-the-art and research challenges, *Journal of Internet Services and Applications*, 2010, vol. 1, no.1, pp. 7–18.

Zhou, G., *Real-time Information Technology for Future Intelligent Earth Observing Satellite System*, Hierophantes Press, 2003, ISBN: 0-9727940-0-X.

Zhou, G., O. Baysal, J. Kaye, Concept design of future intelligent earth observing satellites, *International Journal of Remote Sensing*, vol. 25, no.14, pp. 2667–2685.

Zhu, J. Z., Shi, Q., Chen, F. E., Shi, X. D., Dong, Z. M., & Qin, Q. Q., Research status and development trends of remote sensing big data, *Journal of Image and Graphics*, 2016, vol. 21, no.11, pp. 1425–1439. (Chinese)

Section II

Information Extroduction

4 Urban 3D Surface Information Extraction from Aerial Image Sequences

4.1 INTRODUCTION

Urban three-dimensional (3D) modeling for various applications such as town planning, microclimate investigation, transmitter placement in telecommunication, noise simulation, and heat and exhaust spread in big cities has been an interesting research topic. In the past decades, Light Detection And Ranging (LiDAR) has been widely applied for extraction of urban buildings. A number of methods have been proposed and investigated, such as Baltsavias et al. (1995), Haala (1995), Haala et al. (1998), Eckstein and Muenkelt (1995), Hug (1997), Morgan and Tempfli (2000), and Morgan and Habib (2002). In general, these methods can be grouped into two categories (Yoon and Shan 2002): the classification approach and adjustment approach. The classification approach detects the ground points using certain operators designed based on mathematical morphology (Lindenberger 1993; Vosselman 2000), or terrain slope (Axelsson 1999), or local elevation difference (Wang et al. 2001). The refined classification approach uses the triangulated irregular network (TIN) data structure (Tao and Hu 2001) and iterative calculation (Sithole 2001) to consider the discontinuity in the LiDAR data or terrain surface. The adjustment approach essentially uses a mathematical function to approximate the ground surface, which is determined in an iterative least adjustment process while outliers of non-ground points are detected and eliminated (Kraus & Pefifer 2001). Despite plenty of efforts, difficulties for high accuracy and high reliability of extraction of urban buildings still remain (Vosselman and Mass 2001). It has been widely accepted by many scholars in photogrammetry, remote sensing, artificial intelligence, computer vision, and image processing communities that methods based on single terrain characteristics or criteria often fail to obtain satisfactory results in other terrain types.

On the other hand, the cost of using LiDAR for urban 3D DSM creation is large, so many small private companies cannot bear the cost. For this reason, UAV-based oblique photogrammetry is widely attracting many researchers. The basis of this method is from traditional photogrammetry, which generates the 3D urban model using stereo image matching. However, this method decreases rapidly for complex scenes in dense urban areas using large-scale imagery. The degradation in the performance of photogrammetric processes is mainly due to the failures of image matching, which are primarily caused by, for example, occlusions, depth discontinuities, shadows, poor or repeated textures, poor image quality, foreshortening and motion artifacts, and the lack of models for man-made objects (Schenk and Toth 1992; Zhou et al. 1999; Ackermann 1999; Förstner 1999). To offset the effect of these problems, the extraction of buildings and DEM generation in urban areas is currently still done by human-guided interactive operations, such as stereo compilation from a screen. The whole process is both labor-intensive and time-consuming.

This chapter proposes a spatio-temporal analysis technique to extract 3D surface information from aerial image sequences. The method was originally presented by Baker et al. (1986), Baker (1987), Baker and Bolles (1988, 1989), Bolles and Baker (1985a, 1985b), Bolles et al. (1987) for analysis of motion image sequences. Their initial experimental environment fully met four constraint conditions (see Section 4.2). Bolles, Baker, and Marimont extended the analysis to arbitrary motions using projective duality in space. Their generalizations are on (a) the generalization of their technique for varying viewing directions, including variation over time; (b) providing the three-dimensional connectivity information for building coherent spatial descriptions of observed objects; and (c) a sequential implementation, allowing initiation and refinement of scene feature estimates while the sensor is in motion (Baker and Bolles 1988, 1989). Zhou et al. (1999) experimented with the feasibility and applicability of this method for solving the problems of occlusions and depth discontinuities encountered in the traditional photogrammetry. Three test fields were built in Berlin, Germany. The experimental results show that the technique can solve some problems that the conventional photogrammetry is not able to solve.

4.2 PRINCIPLE OF TEMPORAL AND SPATIAL ANALYSIS

The method was based on the four conditions as follows:

Condition 1: The camera's movement is restricted along a straight linear path.
Condition 2: Image capture is rapid enough with respect to the camera's movement and scene scale to ensure that the data are temporally continuous.
Condition 3: The velocity of the camera's movement is a constant.
Condition 4: The camera's position and attitude at each imaging site are known.

Theoretically, if a camera mounted on an airborne platform flies along a straight-line path (flight course), and the camera's optical axis is orthogonal to the direction of motion, this operational condition meets the four constraint conditions above. The impact of the irregular real flight movements will be discussed in Section 4.4.

The meaning of Condition 2 is that the image sequences captured by the camera are so close that none of the image features moves more than a pixel or so. This sampling frequency guarantees continuity in the temporal domain that is similar to continuity in the spatial domain (Figure 4.1). Thus, an edge of an object in one image appears temporally adjacent to its occurrence in the following image. This temporal continuity makes it possible to construct cube data in which time is the third dimension and continuity is maintained over all three dimensions. This solid of data is referred to as spatio-temporal data (see Figure 4.2). Figure 4.3 shows three individual images used to form the solid of data. Typically, one hundred or more images are used, making the trajectory of a point, such as P, through a continuous path. With Condition 1, suppose that there is a simple motion in which a camera moves from left to right, with its optical axis orthogonal to its direction of motion. For this type of motion, the epipolar plane for a point P is the same for all camera positions. This plane is called as an epipolar plane of the point P for the entire motion. If the velocity of the camera is a constant (Condition 4), the trajectories in the epipolar plane imagines (EPIs) are straight lines. Figure 4.3 illustrates the construction of the EPIs.

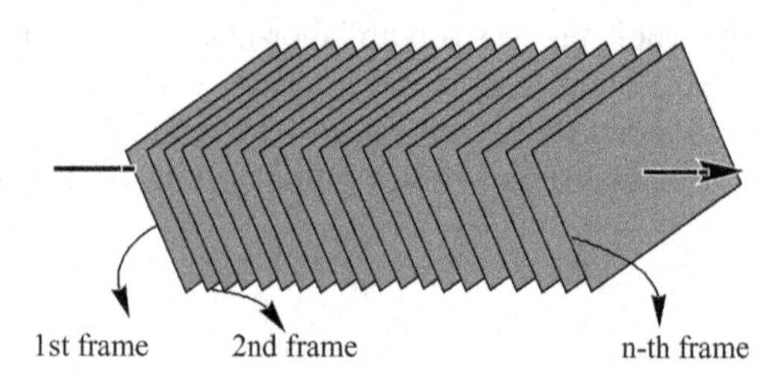

FIGURE 4.1 Construction of spatio-temporal data.

FIGURE 4.2 Spatio-temporal solid of data.

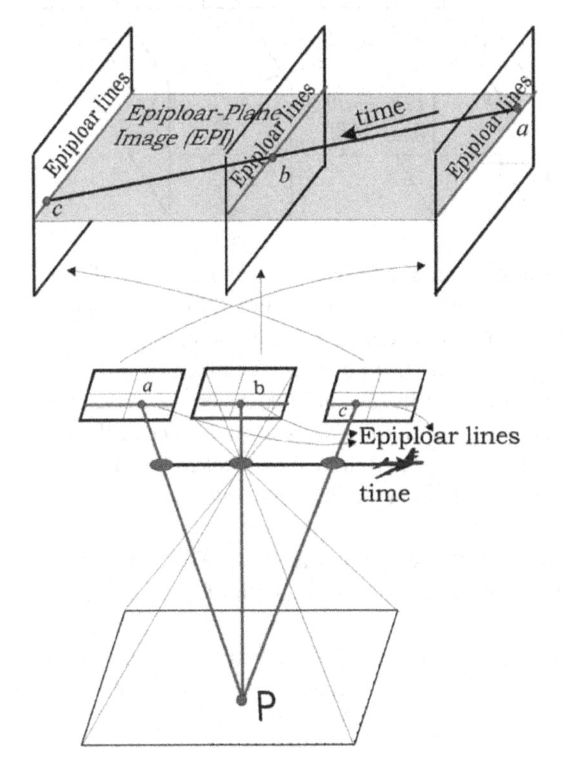

FIGURE 4.3 Construction of EPIs.

The projection of *P* onto the epipolar lines moves from the right to the left as the camera moves from the left to the right. The velocity of this movement along the epipolar line is a function of *P's* distance from the line through the lens centers. The closer it is, the faster it moves. Therefore, a vertical slice of the spatio-temporal solid of data contains all the epipolar lines associated with one epipolar plane. If we slice the spatio-temporal data along the temporal dimension, a new "image plane," which is called an EPI, can be formed (see Figure 4.3).

(x, y, z) coordinates can be computed from Condition 4. An equation for the trajectory of a scene point in the EPI constructed from a motion is analyzed, and how to compute the (x, y, z) coordinates

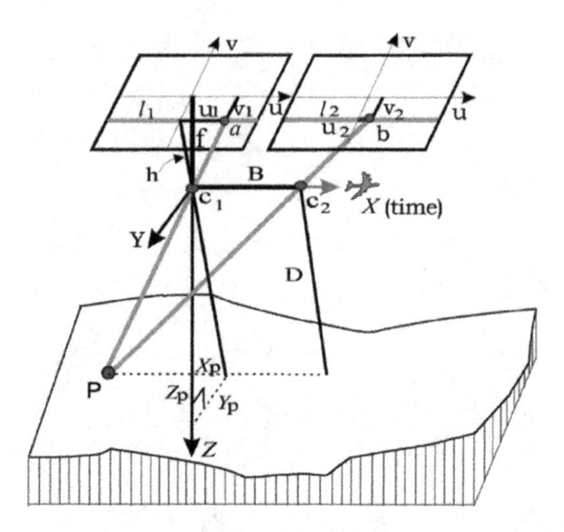

FIGURE 4.4 Trajectory in EPI.

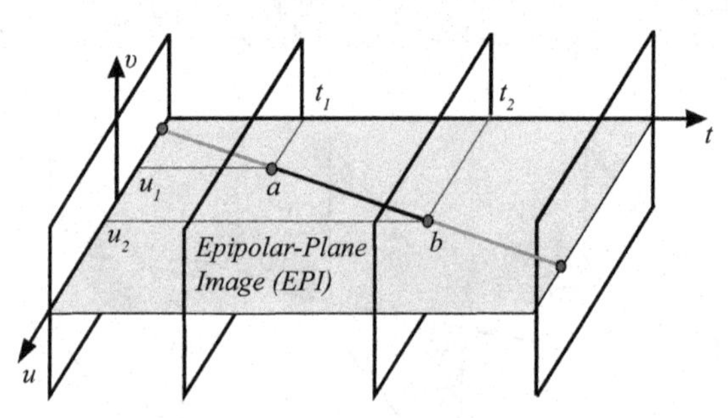

FIGURE 4.5 Geometric configuration for computing 3D coordinates.

of such a point is presented in Figure 4.4, which is a diagram of a trajectory in an EPI derived from the left-to-right motion illustrated in Figure 4.5. The image row at t_1 in Figure 4.4 corresponds to the epipolar line l_1 in Figure 4.5. Similarly, the image row at t_2 corresponds to the epipolar line l_2. (Remember that the EPI is constructed by extracting one line from each image taken by a camera as it moves along the straight line from c_1 and c_2. Since the images are taken very closely, there would be several images taken between c_1 and c_2. However, to simplify the diagram, none of them is shown.) The point (u_1, t_1) in the EPI corresponds to the point (u_1, v_1) in the image taken by camera at time t_1 and position c_1. Thus, as the camera moves from c_1 to c_2 in the time interval t_1 to t_2, the scene point moves in the EPI from (u_1, t_1) to (u_2, t_2).

The intent of the section above is to characterize the shape of this trajectory. The following is how to compute the three-dimensional coordinates when given the focal length of the camera and the camera's speed.

In the following analysis, a right-hand coordinate system is defined that is centered on the initial position of the camera. Given the speed of the camera, s, which is assumed to be a constant, the distance from c_1 to c_2 can be computed as follows (see Figure 4.5):

$$B = s\Delta t \tag{4.1}$$

where $\Delta t = (t_2 - t_1)$. From similar triangles in Figure 4.5, we have

$$\frac{u_1}{h} = \frac{x_p}{D} \tag{4.2}$$

$$\frac{u_2}{h} = \frac{B + x_p}{D} \tag{4.3}$$

where u_1 and u_2 have been converted from pixel values into distances on the image plane, h is the distance from the projected lens center to the epipolar line in the image plane, x_p is the x-coordinate of P in the scene coordinate system, and D is the distance from P to the line through the lens centers. Since h is the hypotenuse of a right triangle, it can be computed by:

$$h = \sqrt{f^2 + v_1^2} \tag{4.4}$$

where f is the focal length of the camera. From Equations 4.2 and 4.3, we have

$$\Delta u = \left(u_2 - u_1\right) = \frac{h\left(B + x_p\right)}{D} - \frac{hx_p}{D} = \frac{h}{D}B \tag{4.5}$$

Equation 4.5 shows that Δu is a linear function of B, while B is also a linear function of Δt. Thus, Δt is linearly related to Δu. That means that all trajectories in the EPIs are straight lines in the constrained straight-line motion.

With the analysis above, the (x, y, z) coordinates of P can be computed by u_1, u_2 and f, i.e.,

$$\left(x, y, z\right) = \left(\frac{D}{h}u_1, \frac{D}{h}v_1, \frac{D}{h}f\right) \tag{4.6}$$

Equation 4.7 is rewritten as

$$\left(x, y, z\right) = \left(mu_1, mv_1, mf\right) \tag{4.7}$$

where

$$m = \frac{D}{h} = \frac{B}{\Delta u} \tag{4.8}$$

which represents the slope of the trajectory computed in terms of the distance traveled by the camera (B as opposed to Δt) and the distance which the point moves along the epipolar line.

If the first camera position c_1, on an observed trajectory, is different from the camera position c_0, defining a global camera coordinate system, the x coordinate has to be adjusted by the distance traveled from c_0 to c_1. Thus,

$$\left(x, y, z\right) = \left(\left(t_1 - t_0\right)s + mu_1, mv_1, mf\right) \tag{4.9}$$

where t_0 is the time of the first image and s is the camera's speed. This correction is equivalent to computing the intercept x of the trajectory, which is the first camera's position. Therefore, the (x, y, z) coordinates of the points can be easily computed from the slopes and intercepts of the trajectories.

4.3 GEOMETRIC RECTIFICATION OF IMAGE SEQUENCES

The irregular motions of an airborne sensor mentioned above cause unpredictable changes to a camera's altitude (translation parameters) and attitude (rotation angles), resulting in the pixel point deviation from their ideal positions. Correction of the deviation should be carried out so that ground coordinates calculated by Equation 4.6 achieve a high accuracy. To this end, a group of definitions is first given for development of the rectification model.

- *Distorted image*. The original image is geometrically distorted due to the various error sources, such as atmospheric turbulence, flight vibration, and instrument defect, resulting in geometric deviation of the pixel points from their ideal positions. The image is called a distorted image.
- *Geometric rectification.* For a distorted image, correcting the deviation of the real pixel position to the ideal position using a mathematical model is called geometric rectification, i.e., the process of rectifying the geometric distortion is called geometric rectification.
- *Rectified image (normal image).* The image rectified using a mathematical model is called a rectified image or normal image.
- *Camera coordinate system (S_i-UVW):* In order to derive the rectification model, the camera coordinate system is defined. Each projective center of camera S_i ($i = 1, 2, \ldots, n$) is defined as the origin, and the W axis is defined as the connection from the principal point to camera center, the U axis is chosen along the direction of flight (for a stereo pair, air base direction is U axis direction), and V axis is orthogonal to U axis (see Figure 4.6).

Figure 4.6 illustrates the geometric configuration of distortion rectification. *Plane1′* is a distorted image and *plane1* is a rectified image (normal image). According to the principle of EPI analysis, all conjugate epipolar lines for any point, e.g., P in the image sequences, are coplanar when the camera's movement meets the four conditions (Zhou et al. 1999). In other words, the four constraint conditions in Section 4.2 ensure that all coordinates along the v direction in the image sequence planes are equal. These constraints construct a basis for development of a mathematical model and geometric rectification method. The geometric distortion can be decomposed into u and v components along the u and v axes in the image plane.

First, the rectification of the v component is considered, with the u component considered later. Assume that c_1 denotes the first projective center, B_1 is the length of the air base, the point p'_1 in the original image and the point p_1 in the normal image are the projections of the point P in the object space in Figure 4.6. The relationship between the original (distorted) image and the normal (rectified) image for the point P can be expressed by

$$\frac{U_{p_1}}{-f} = \frac{U_{p'_1}}{W_{p'_1}} \text{ and } \frac{V_{p_1}}{-f} = \frac{V_{p'_1}}{W_{p'_1}} \tag{4.10}$$

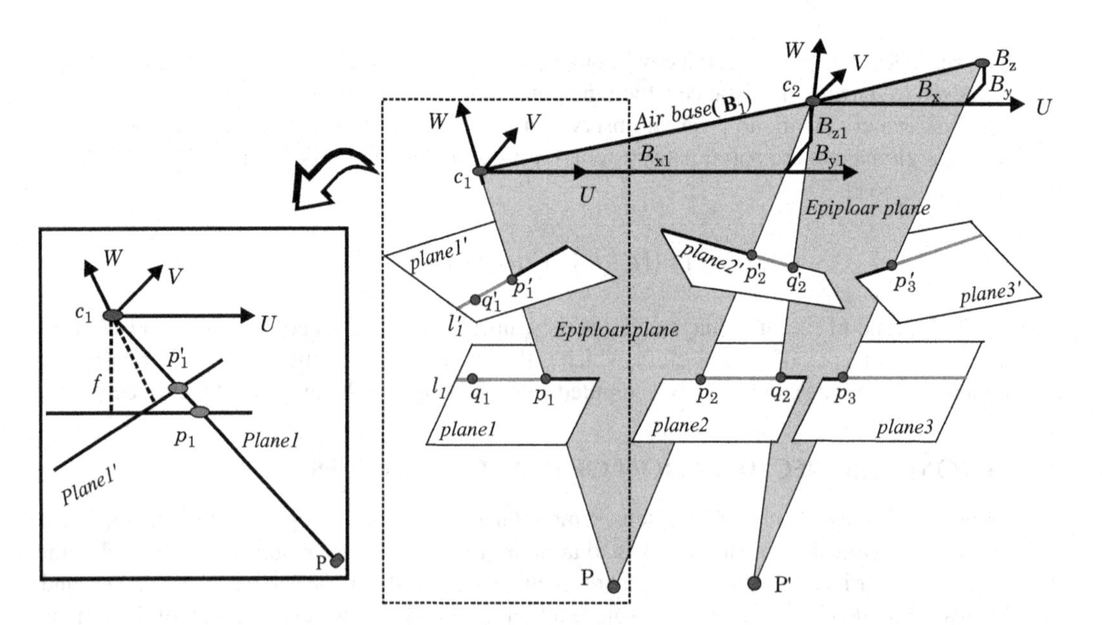

FIGURE 4.6 Geometry of image sequences and the relationship between the original and normal images.

where U_{p_1}, V_{p_1} and $U_{p'_1}, V_{p'_1}$ denote the coordinates of the point p_1 in the normal plane and the point p'_1 in the distorted plane with respect to the camera coordinate system, respectively. The coordinates for the point p'_1 are calculated by

$$\begin{pmatrix} U_{p'_1} \\ V_{p'_1} \\ W_{p'_1} \end{pmatrix} = \begin{pmatrix} m_{11} & m_{12} & m_{13} \\ m_{21} & m_{22} & m_{23} \\ m_{31} & m_{32} & m_{33} \end{pmatrix} \begin{pmatrix} u_{p'_1} \\ v_{p'_1} \\ w_{p'_1} \end{pmatrix} = R_1 \begin{pmatrix} u_{p'_1} \\ v_{p'_1} \\ w_{p'_1} \end{pmatrix} \tag{4.11}$$

where $m_{i,j}$ $(i = 1, 2, 3; j = 1, 2, 3)$ are the elements of the rotation matrix R_1 for the first image, f denotes the focal length, $u_{p'_1}, v_{p'_1}$ denote the coordinates of the point p'_1 in the image plane, *plane1′*. With Equations 4.10 and 4.11, we have

$$U_{p_1} = -f \frac{m_{11}u_{p'_1} + m_{12}v_{p'_1} - m_{13}f}{m_{31}u_{p'_1} + m_{32}v_{p'_1} - m_{33}f} = \frac{m'_{11}u_{p'_1} + m'_{12}v_{p'_1} + m'_{13}}{m'_{31}u_{p'_1} + m'_{32}v_{p'_1} + 1} \tag{4.12a}$$

$$V_{p_1} = -f \frac{m_{21}u_{p'_1} + m_{22}v_{p'_1} - m_{23}f}{m_{31}u_{p'_1} + m_{32}v_{p'_1} - m_{33}f} = \frac{m'_{21}u_{p'_1} + m'_{22}v_{p'_1} + m'_{23}}{m'_{31}u_{p'_1} + m'_{32}v_{p'_1} + 1} \tag{4.12b}$$

where $m'_{i,j}(i = 1,2,3; j = 1,2,3)$ are a new set of eight independent transformation parameters which are functions of the original unknowns ω, φ, and κ as well as the focal length of camera, f. Similarly, the relationship between the point p'_2 in the second original image plane, *plane2′*, and the point p_2 in the second normal image plane, *plane2*, is

$$U_{p_2} = \frac{n'_{11}u_{p'_2} + n'_{12}v_{p'_2} + n'_{13}}{n'_{31}u_{p'_2} + n'_{32}v_{p'_2} + 1} \tag{4.13a}$$

$$V_{p_2} = \frac{n'_{21}u_{p'_2} + n'_{22}v_{p'_2} + n'_{23}}{n'_{31}u_{p'_2} + n'_{32}v_{p'_2} + 1} \tag{4.13b}$$

where $n'_{i,j}(i = 1,2,3; j = 1,2,3)$ are also a set of eight independent transformation parameters which are functions of the rotation angles ω, φ, and κ for the second image as well as the focal length of camera, f. Similarly, we can get an equation describing the relationship between the Nth rectified image and the original images.

Because the orientation parameters of the camera are unknown, it is difficult to determine the relationship between the original image and the normal image directly. However, according to the characteristics of the epipolar plane and epipolar lines, all conjugate epipolar lines in the image sequences should lie in a common epipolar plane. This means that the v coordinates of all conjugate points in the normal image sequences are constrained to be equal. Thus, a relative relationship among the normal image sequences (if the original image sequences have been rectified) can be built by

$$\overbrace{\frac{m'_{21}u_{p'_1} + m'_{22}v_{p'_1} + m'_{23}}{m'_{31}u_{p'_1} + m'_{32}v_{p'_1} + 1}}^{\text{the first image}} = \overbrace{\frac{n'_{21}u_{p'_2} + n'_{22}v_{p'_2} + n'_{23}}{n'_{31}u_{p'_2} + n'_{32}v_{p'_2} + 1}}^{\text{the second image}} = \overbrace{\cdots}^{\text{many images}} = \overbrace{\frac{r'_{21}u_{p'_n} + r'_{22}v_{p'_n} + r'_{23}}{r'_{31}u_{p'_n} + r'_{32}v_{p'_n} + 1}}^{\text{the n-th image}} \tag{4.14}$$

where $r'_{i,j}(i = 1,2,3; j = 1,2,3)$ are eight independent transformation parameters for the Nth image. Equation 4.14 implies that all of the rectified conjugate points in the image sequences lie in a common plane, that is, they are coplanar. Thus, the trajectories in the "new" EPI (called rectified EPI) generated by the rectified image sequences meet the EPI requirement to be analyzed, that is, *all trajectories in the EPIs are straight lines when the camera moves along a constrained straight-line path and viewing direction is orthogonal to the direction of flight* (Zhou et al. 1999). In order to simplify the derivation, the first condition equation, describing the relationship between the first and the second image, is chosen for discussion of the computing process. It can be rewritten by

$$K_1 v_{p'_1} u_{p'_2} + K_2 v_{p'_1} v_{p'_2} - K_3 v_{p'_1} + K_4 u_{p'_2} + K_5 v_{p'_2} - K_6 v_{p'_1} + K_7 u_{p'_1} u_{p'_2} + K_8 u_{p'_1} v_{p'_2} - K_9 u_{p'_1} = 0 \quad (4.15)$$

where

$$K_1 = m'_{22} n'_{31} - m'_{32} n'_{21},\ K_2 = m'_{22} n'_{32} - m'_{32} n'_{22},\ K_3 = m'_{22} - m'_{32} n'_{23},$$
$$K_4 = m'_{23} n'_{31} - n'_{21},\ K_5 = m'_{23} n'_{32} - n'_{22}, K_6 = m'_{23} - n'_{23},$$
$$K_7 = m'_{21} n'_{31} - m'_{31} n'_{21},\ K_8 = m'_{21} n'_{32} - m'_{31} n'_{22}, K_9 = m'_{21} - m'_{31} n'_{23}$$

The Equation 4.13 is divided by K_5, we have

$$K_1^0 v_{p'_1} u_{p'_2} + K_2^0 v_{p'_1} v_{p'_2} - K_3^0 v_{p'_1} + K_4^0 u_{p'_2} + K_5^0 v_{p'_2} - K_6^0 v_{p'_1} + K_7^0 u_{p'_1} u_{p'_2} + K_8^0 u_{p'_1} v_{p'_2} - K_9^0 u_{p'_1} = 0 \quad (4.16)$$

where $K_i^0 = K_i / K_5$, and $K_5^0 = 1$. Equation 4.16 describes the relative relationship of two adjacent images. When using the traditional least square method, the eight parameters, K_1–K_8, can be obtained, and further the exterior orientation elements of two images can be obtained by K_1–K_8, although each stereo pair may be constructed by two neighboring images, and their geometric rectification is carried out separately. However, from the experimental results conducted in Zhou et al. (1999), if the rectification for an image sequence (e.g., 600 images) is operated, all conjugate epipolar lines still cannot be guaranteed to be in a common plane, that is, coplanar. This is because these stereo pairs constructed by two neighboring images are based on individual independent rectification coordinate systems. They have individual origin points, and individual u axes. Therefore, if the first image is taken as a fixed image, the other images are rectified relative to it. This method is capable of ensuring that all conjugate points/lines lie in a common plane. Moreover, the computational procedures become simple. So, the rotation angles of the first image are set $\omega = \phi = \kappa = 0$, that is, the R_1 is a unit matrix. The elements of the second rotation matrix can be computed by

$$n'_{11} = (K_3K_5 - K_6K_2 - B_ZK_1 - B_YK_4)/(B_X^2 + B_Y^2 + B_Z^2), n'_{12} = (B_Y + K_4)/B_X, n'_{13} = (B_Z + K_1)/B_X,$$
$$n'_{21} = (K_1K_6 - K_3K_4 - B_ZK_2 - B_YK_5)/(B_X^2 + B_Y^2 + B_Z^2), n'_{22} = (B_Y + K_5)/B_X, n'_{23} = (B_Z + K_2)/B_X,$$
$$n'_{31} = (K_2K_4 - K_1K_5 - B_ZK_3 - B_YK_6)/(B_X^2 + B_Y^2 + B_Z^2), n'_{32} = (B_Y + K_6)/B_X, n'_{33} = (B_Z + K_3)/B_X,$$

where Bx, By, and Bz are components of airbase, and K_5 is calculated by

$$K_5 = \pm\sqrt{2B_X^2 / \left(K_1^{0^2} + K_2^{0^2} + K_3^{0^2} + K_4^{0^2} + K_5^{0^2} + K_6^{0^2} - K_7^{0^2} - K_8^{0^2} - K_9^{0^2}\right)} \quad (4.17)$$

As observed from Equation 4.17, K_5 can be positive and negative, that is, have two values. This fact causes two sets of rotation angles for the second image. However, only one of the two solutions is the correct one. Thereby, it is necessary to determine which solution is correct.

In geometry, if the photogrammetric bundle of rays is rotated around the airbase (see Figure 4.7), obviously the V coordinates still meet the condition of epipolar plane constraint, but the photogrammetric principle requires that the first and second images of a stereo pair of images should capture the identical object of the Earth's surface. It means that the ranges of rotation angles ω and ϕ are limited

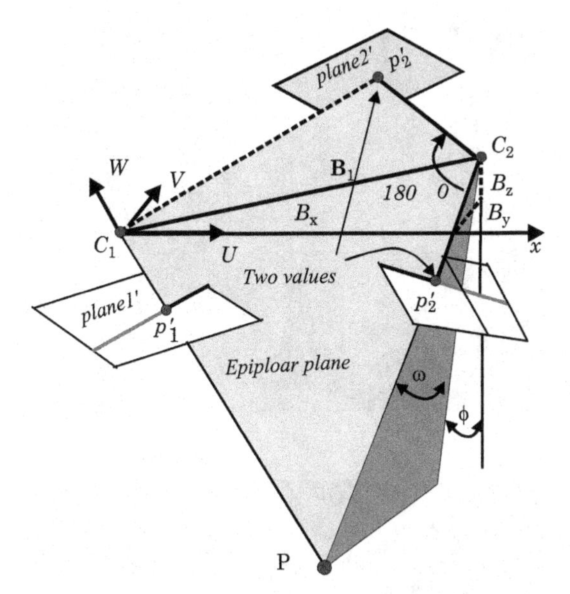

FIGURE 4.7 Two solutions of rotation angles for the second image.

$$-\pi/2 < \phi \text{ and } \omega > \pi/2 \tag{4.18}$$

With the condition restriction of Equation 4.18, three rotation angles can be determined uniquely.

In summary, the basic idea of rectifying the distorted image into the correct one is to use the epipolar constrain condition (i.e., coplanarity). The constraint, by which all conjugate epipolar lines for any point, such as *P*, in a sequence of images are coplanar, only rectifies image distortion along the *v* direction. Distortion along the *u* direction has not been considered above. However, this type of distortion will have little impact on the accuracy of ground point coordinates. The algorithm first requires the determination of more than eight conjugate points in the left and right images in order to establish the observation equations of Equation 4.16 for solving the implicit parameters K_1~K_8 using the least-squares method, and then resample the original images into rectified images, that is, EPIs.

4.4 EXPERIMENTS IN SPATIO-TEMPORAL ANALYSIS

Three test fields in the region of Berlin (Berlin city, Schönefeld, Werder) were used to test this technical application in aerial photogrammetry. In October 1995, the image sequences for three experimental fields were captured using a video camera mounted on a CESSNA 207 T flying platform. Details of the imaging parameters are listed in Table 4.1. The original data were recorded on video cassette, and the digital image sequences were obtained by resampling with a frequency of 10 frames/second. The first image of each test field is shown in Figures 4.8 through 4.10.

TABLE 4.1
Imaging Flight Parameters

Name	Parameters	Name	Parameters
Platform	CESSNA 207 T	**Flight velocity**	100 Knots
Flight height	about 800 m	**Camera type**	S-VHS, Panasonic Videorecorder
Focal Length	35 mm	**Scale**	1:2500

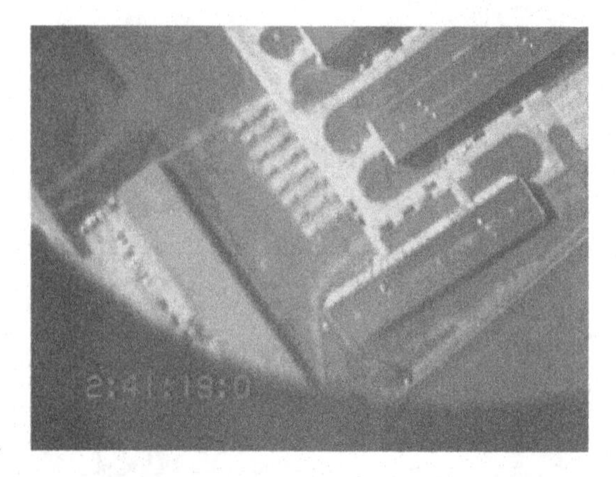

FIGURE 4.8 The first image in the Berlin test field.

FIGURE 4.9 The first image in the Schönefeld test field.

FIGURE 4.10 The first image in the Werder test field.

For further analysis, software was developed to implement the algorithms proposed above, which includes the following steps:

4.4.1 Image Data Processing

The purpose of image data preprocessing is to obtain the high accuracy of slope and intercept of straight line in the EPIs. The process has been given in Section 4.4. The key step is extraction of straight linear features from EPIs. The zero-crossing operator was used to detect the "edges" in

EPIs, and then image thinning, gap connection, short line (less than 5 pixels) eliminating for these detected "edges" are implemented to obtain line segments. A straight-line regression algorithm is finally used for fitting the detected line segments for extraction of straight lines, which are described by the slope and intercept. Figure 4.11 is one image after noise removal. The corresponding original image is depicted in Figure 4.8. The 540th EPI generated from a sequence of 620 images in the second test field is illustrated in Figure 4.12. The detected edges using the zero-crossing operator are displayed in Figure 4.13. Line segments are depicted in Figure 4.14. The regressed straight line is shown in Figure 4.15, in which the slope and intercept are given. The steps include partitioning the edges at sharp corners and applying a line regression algorithm to recursively partition the smooth segments into continuous straight-line segments.

FIGURE 4.11 The image of noise removal.

FIGURE 4.12 The 540th EPI.

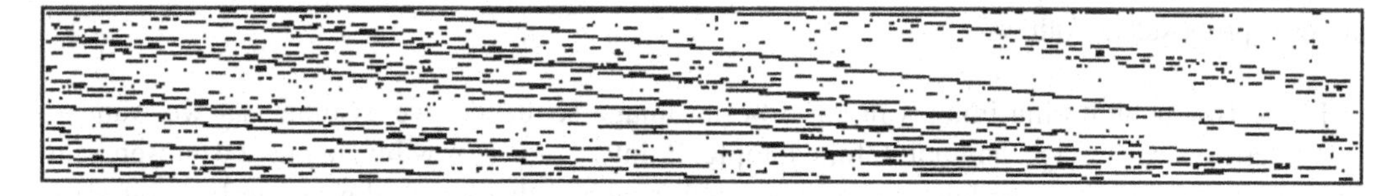

FIGURE 4.13 The detected edges for the 540th EPI.

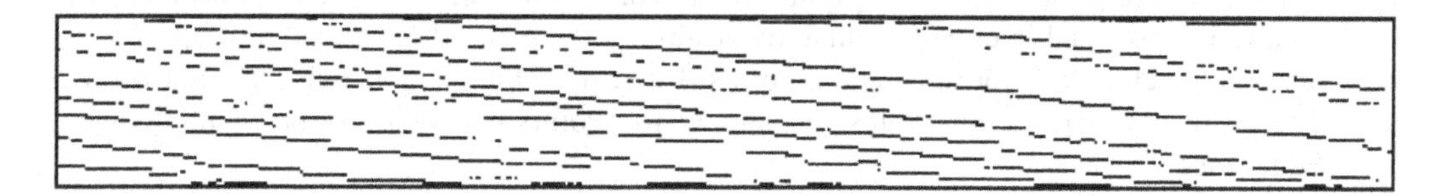

FIGURE 4.14 From edges to line segments.

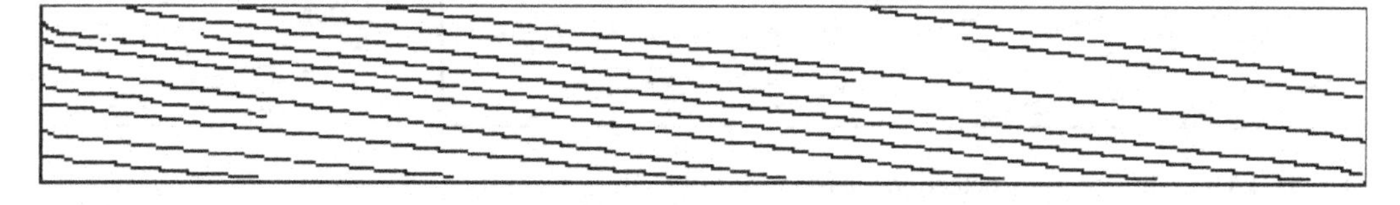

FIGURE 4.15 Line segment description with slope and interception in EPIs.

4.4.2 Sensitivity to Occlusion

In order to demonstrate the capability of the EPI technique for detecting occlusion, one of the EPIs in the second test field is chosen. In this EPI, the trajectory of a chimney is involved, and all linear features are extracted (see Figure 4.16). As observed in Figure 4.16, it can be found that the trajectories of the occluded objects are temporally broken by the trajectory of a high object. This means that the occlusion can be manifested in EPIs obviously, and detected via the fact that if an object temporally occludes another object, the trajectories for corresponding occluding and occluded features intersect in the EPI. Moreover, for each intersection, the trajectory of a low building is temporally broken by that of a high building, which is in fact the chimney in the second test field. Therefore, the approach presented in this chapter is more robust against occlusion and depth discontinuities than traditional photogrammetric stereo matching.

4.4.3 Rectification of Distorted Image Sequence Data

Section 4.3 has presented the model for rectifying the errors caused by a real operational process. Two essential steps are point extraction and geometric rectification. In fact, after the numerically distinctive points, associated with their descriptions, are extracted from the first frame, it is unnecessary to directly extract distinctive points from the other images using time-consuming operators, for example, the Förstner operator. In other words, the conjugate points in the other images can be tracked, since the overlap of neighboring images reaches over 90% (the resample rate of the image sequence is 10 frames per second). Thus, a method, called multi-criterion, is proposed for this purpose. The first criterion is the maximum parallax constraint for determining the search range. A horizontal parallax of about 10–15 pixels (thanks to the resampling rate of the image sequences) and a vertical parallax of about 3–5 pixels are recommended. The second criterion is a correlation coefficient maximum between window patches around the distinctive point pairs.

After over eight conjugate points have been obtained above, the rotation angles of the second image can be calculated via the parameters K_1–K_8, which express the relationship between the original image and the normal image. However, the computation process is time-consuming because of nonlinear processing, and computational errors are included because of unknown interior orientation parameters (f, xo, yo). For this reason, a linear rectification method without solving the rotation angles is proposed as follows:

1. For two neighboring images in an airborne image sequence, we assume that we want to rectify i^{th} row image, any point, for example, Point 1 in the ith row image may be chosen, and its v coordinate in the left image is denoted v_1, while the u_1 coordinate can be arbitrarily selected.
2. In terms of the principle of the epipolar plane constraint condition, the v coordinates are required to be equal in epipolar plane regardless of the u coordinates. Therefore, we may arbitrarily select a u coordinate for Points 3 and 4 in the right image, denoted u_3, u_4. The corresponding v coordinates, v_3, and v_4 for Points 3 and 4 can be computed via Equation 4.16.

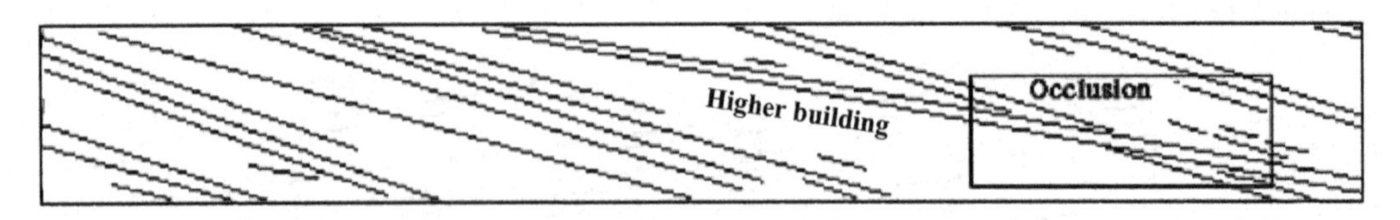

FIGURE 4.16 The occlusion detection via trajectories in EPI.

3. Using Point 3 or Point 4, another point in the left image plane, denoted Point 2, can be computed as follows: selecting any u coordinate for Point 2, signed u_2, in the left image (selected u_2 should be as far from u_1 as possible), the corresponding v coordinate of Point 2 will be calculated via Equation 4.16 again. Thus, a straight line formed by Points 1 and 2 in the left image plane and a straight line formed by Points 3 and 4 in the right image plane are conjugate epipolar lines. Points 1 and 3, and Points 2 and 4 are conjugate points, respectively.
4. The conjugate epipolar lines in the left and right images need to be resampled along the epipolar line to guarantee their coplanarity. To this end, the slope of the epipolar line and gray resampling should be carried out.
5. Repeat steps 1 to 4 to calculate other epipolar lines until the entire image is rectified. Finally, two normal images (a stereo pair of images) are produced.

Table 4.2 lists the comparison of computational times for two rectification algorithms. Method 1 rectifies the original images through the rotation angles computed by parameters K1 to K8, while Method 2 directly rectifies the original image through the proposed method above. As seen from Table 4.2, the computational time decreases by 20% over the three test fields. The rectified effectiveness of EPIs in the three test fields is depicted in Figures 4.17b through 4.19b. The corresponding original EPIs are displayed in Figures 4.17a through 4.19a. In terms of the EPI analysis principle, all trajectories in the EPIs should be straight lines if the real operation meets the four conditions. This criterion can be used to evaluate our algorithms by comparing the original EPIs with the rectified EPIs. We have noticed that:

- *The Berlin City Test Field:* Trajectory 2 first shows a wide white strip at the top, and then splits into two strips at 2/3 heights from the EPI's bottom (see Figure 4.17a). The rectified EPI shows that the correct trajectory is indeed a straight line (see Figure 4.17b). Additionally, Trajectory 1 slowly converges to a triangle in the original EPI, but it is an entire straight line in the rectified EPI (see Figure 4.17b).
- *The Schönefeld Test Field:* Trajectory 1 in Figure 4.18a is a segment in the original EPI. The rectified EPIs (see Figure 4.18b) show that its trajectory is a straight line, and is neither broken nor hidden. Additionally, Trajectory 2 (a wide white bar) in Figure 4.18a (the original EPI) is an irregular curve. The rectified trajectory in Figure 4.18b is a straight line.
- *The Werder Test Field:* It is not hard to find Trajectory 1, and it appears only in the middle of the original EPI in Figure 4.19a, while the rectified trajectory is a continuous straight line (see Figure 4.19b). Other similar broken and disappearing trajectories are restored in the rectified EPI versions of the scenes.

TABLE 4.2

Comparison of Computational Times for Two Rectification Algorithms. The Computational Time Consists of Extracting Interesting Points, Refining Conjugate Points, Calculating Parameters, K_1 to K_8, Rectifying Errors, and Resampling Gray

	Method 1 (Nonlinear Algorithm)	Method 2 (Linear Algorithm)	Decreasing Rate
Test field 1	17 minute 52 seconds	13 minute 33 seconds	24%
Test field 2	19 minute 32 seconds	15 minute 24 seconds	22%
Test field 3	21 minute 42 seconds	17 minute 25 seconds	20%

FIGURE 4.17 The original image (a) and rectified image (b) for the 400th frame at the Berlin test field.

FIGURE 4.18 The original image (a) and rectified image (b) for the 300th frame at the Schönefeld test field.

FIGURE 4.19 The original image (a) and rectified image (b) for the 220th frame at the Werder test field.

4.4.4 DEM Generation

After the 3D point coordinates (X, Y, Z) in three test fields are generated by the method presented above, these point clouds are gridded into the DEMs via a so-called image knowledge-based interpolation method. The final results of DEMs are shown in Figures 4.20b through 4.22b. In order to compare the DEM accuracy, the DEMs generated from the original image sequences without rectification are depicted in Figures 4.20a through 4.22a. These DEMs are called original DEMs. In addition, the traditional stereo pair of images are used to produce DEMs as well as using a VirtuoZo v3.5 Softcopy Photogrammetric Workstation (DPW) (Supresoft Inc.). The DEMs for the three test fields are shown in Figures 4.20c through 4.22c, for which a lot of manual compilation, which was highly labor-intensive and time-consuming, is needed.

By observing the three sets of DEMs, it can be found that the DEMs generated manually have the highest accuracy out of the three sets of DEMs. In order to quantify the improvement effect, X, Y, and Z coordinates for over 800 points in each test field are measured. The DEMs' variances before and after the rectifications are listed in Table 4.3. As observed from Table 4.3, accuracy of the rectified DEMs increases by up to 34% in X, 46% in Y, and 33% in the Z direction.

TABLE 4.3
DEM Accuracy Evaluation. The Variances of Before and after rectification are computed by $var = \sqrt{\sum_{i=1}^{N} (X_i - X_{ref})^2 / (N-1)}$, The increment rate is computed by *rate* = (*before* – *after*)/*before*%.

	X (m) Variances			Y(m) Variances			Z(m) Variances			Number of Points
	Before	After	Rate(%)	Before	After	Rate(%)	Before	After	Rate (%)	
Berlin City	1.63	1.21	33.3	1.81	1.01	44.9	2.67	2.06	30.4	810 pts
Schönefeld	2.87	2.03	32.9	3.50	2.31	44.2	4.21	3.02	31.7	1304 pts
Werder	5.11	3.21	36.4	6.23	3.32	49.1	8.51	4.79	39.3	2050 pts
Average			34.2			46.0			33.8	

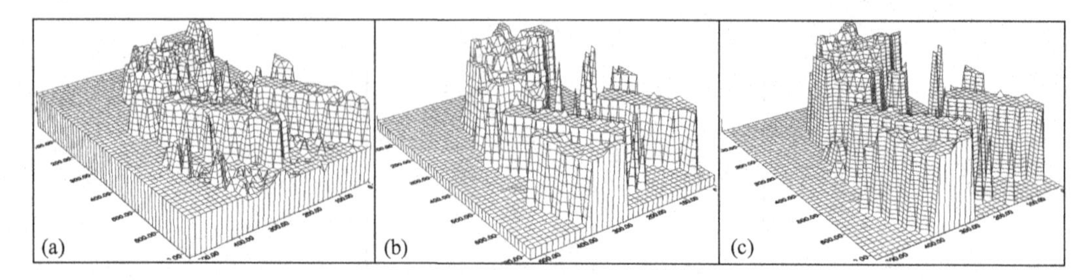

FIGURE 4.20 The DEM generated in the Berlin test field by (a) original image sequences, (b) rectified image sequences, and (c) VirtuoZo software.

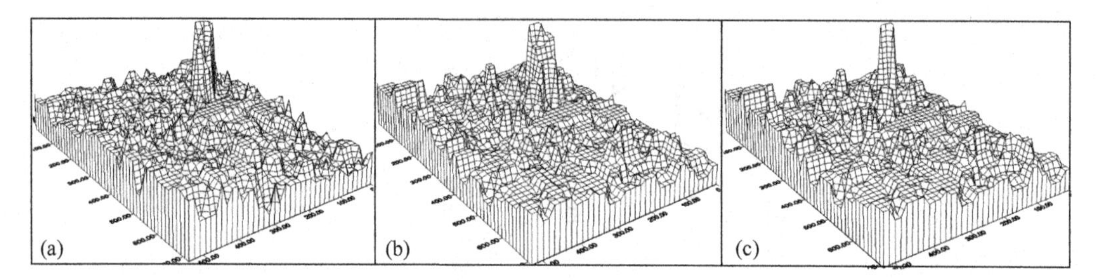

FIGURE 4.21 The DEM generated in the Schölenfeld test field by (a) original image sequences, (b) rectified image sequences, and (c) VirtuoZo software.

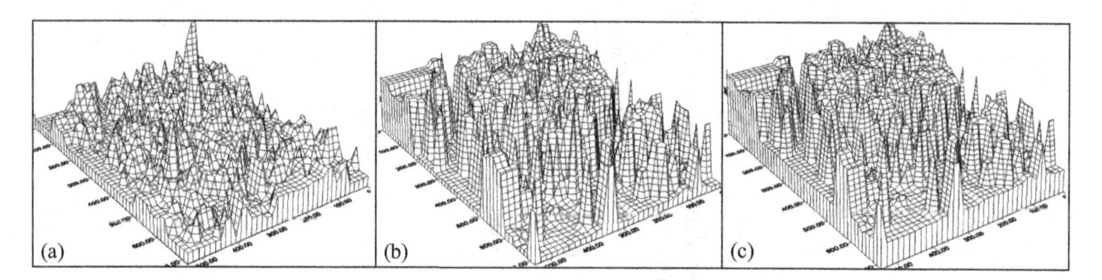

FIGURE 4.22 The DEM generated in the Werder test field by (a) original image sequences, (b) rectified image sequences, and (c) VirtuoZo software.

4.5 ERROR ANALYSIS

As mentioned above, the operational conditions of airborne sensors are much more complicated than supposed. In other words, the real operation of the camera shows considerable deviations from an ideal linear movement. Atmospheric turbulence, flight vibration, and other influences result in a rather irregular flight course, which results in the trajectories of the features in EPIs not being straight lines. On the other hand, the camera's viewing direction randomly varies due to changes of the flight attitude parameters, including roll, pitch, and yaw angles, which result in the aerial image sequences not exactly meeting the four constraints. These various error resources cause rather large errors in global coordinates. Several typical error paradigms include:

1. *Deviation from the ideal camera path on the horizontal plane.* When the camera path deviates from the ideal one on the horizontal plane, the ideal camera path is distorted (Figure 4.23) and the trajectories for the features in EPIs show a "zigzag" shape. Figure 4.24 is an EPI from the first experimental field. The edge-like features (a narrow white bar in Figure 4.24) show an obvious "zigzag" shape. If the deviation is restricted within a finite range and is a stochastic

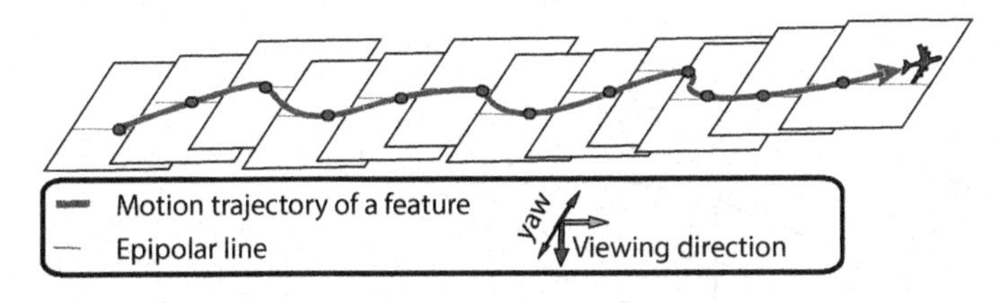

FIGURE 4.23 Trajectory deviation and yaw angle of the camera.

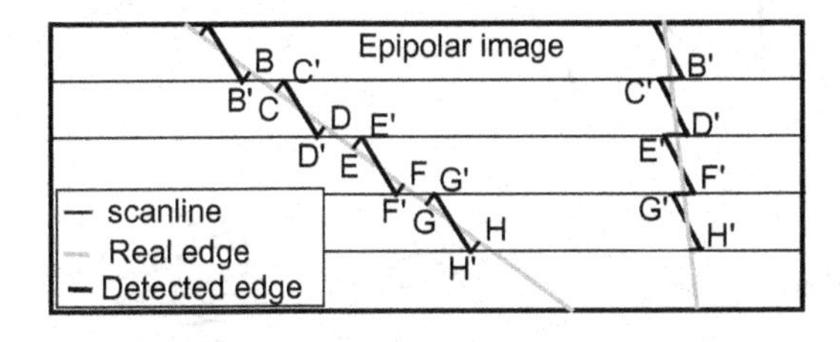

FIGURE 4.24 Error compensation algorithm.

FIGURE 4.25 The trajectories of features and deviations of the camera path in the horizontal plane.

process, the errors can be compensated by our program automatically because the trajectories in EPIs are generated by more than one hundred images and line segments that are described by applying a regression algorithm associated with a gross error eliminating technique. Figure 4.25 illustrates the compensation principle employed in our program.

2. *Deviation from the ideal camera path on the vertical plane.* When the optical axis of the camera (the camera viewing direction) is not orthogonal to the direction of motion (see Figure 4.26), the error can be modeled by Wang (1991):

$$\delta_r = -\frac{\frac{r^2}{f} \sin\varphi \sin\theta}{1 - \frac{r^2}{f} \sin\varphi \sin\theta} \tag{4.19}$$

where $\delta_r = r - r_0$ is a radius difference (see Figure 4.26). When the angle θ is very small, Equation 4.19 can be approximated

$$\delta_r = -\frac{r^2}{f} \sin\varphi \sin\theta \tag{4.20}$$

In Equation 4.20, when ϕ equals 0° or 180°, $\delta_r = r - r_0$ reaches the maximum value, that is, the deviation is biggest; when ϕ equals 90° or 270°, the deviation is smallest. Moreover, the positive and negative signs change randomly with angles varying from 0° to 360°. Thus, the deviation can still be considered as a stochastic process. This kind of error can also be compensated automatically.

If the camera's viewing direction deviates at a fixed angle relative to the direction of motion, which means the deviation is not a stochastic process, the trajectories of the features in EPIs are simple hyperbolas (Baker and Bolles 1988, 1989). If the angle of the camera's viewing

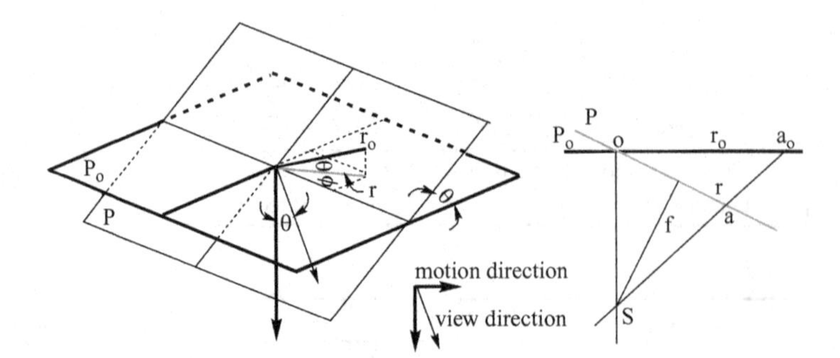

FIGURE 4.26 Camera path deviation in vertical plane.

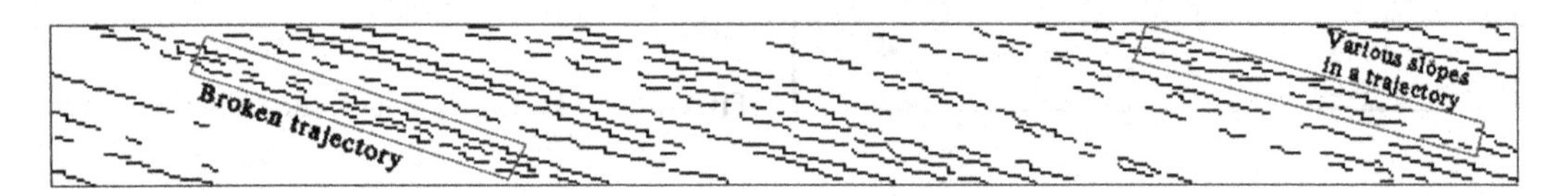

FIGURE 4.27 Trajectories and varying movement velocity, low frequency resampling.

direction relative to the direction of motion shows systemic errors, it is impossible to partition the scene into a fixed set of planes, which in turn means that it is not possible to construct EPIs for such a motion. In the two cases, it is very difficult to compensate the errors.

3. *The velocity of the camera's movement is not a constant.* The theoretic analysis (see *Appendix A*) shows that the trajectories of features in EPIs are composed of several line segments with various slopes rather than simple straight lines. If no-constant speed of the camera's movement happens occasionally in the whole movement process, this type of error can automatically be compensated by employing blunder elimination algorithm; if this case usually happens during the entire movement process, this type of error is difficulty to be compensated. On the other hand, if the error cannot be controlled effectively, it will reach a rather large (x, y, z) deviation because it can change the slope of the trajectory (see Appendix A). Figure 4.27 illustrates the true trajectories of the features in EPI from the third experimental field.
4. *The sampling frequency cannot guarantee a temporal continuity.* If the edge of an object in an image does not appear temporally adjacent to its following images, the trajectory of the feature consists of several line segments (broken lines) rather than an entire straight line. The algorithm proposed above can fill the gap automatically. Figure 4.27 illustrates the fact that the low sampling frequency cannot guarantee a continuity in spatial domain (several broken line segments).
5. *Edge location error.* The (x, y, z) errors caused by the edge location error can be computed in terms of error propagation laws. The mean error for (x, y, z) can be given by (from Equation 4.9)

$$\begin{aligned} \Phi_x &= m\Phi_{u1} = s \cdot slope \cdot \Phi_{u1} \; when\left(t = t_0\right) \\ \Phi_y &= s \cdot slope \cdot \Phi_{v1} \\ \Phi_z &= s \cdot slope \cdot f \end{aligned} \tag{4.21}$$

Equation (4.21) shows that the mean errors of the coordinates (x, y, z) are a linear function of u and v, and the scale factor is the velocity of the camera's movement.

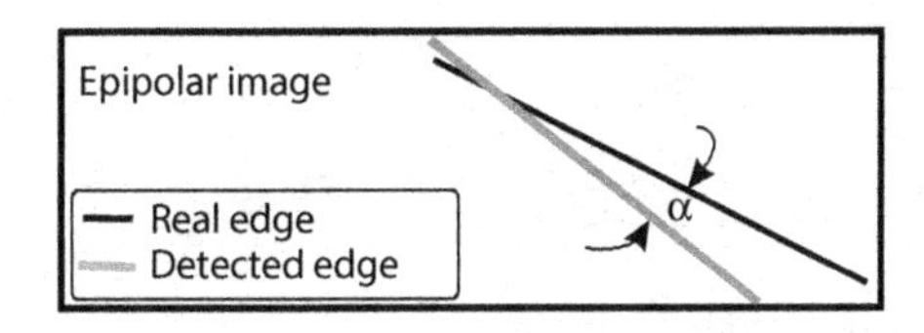

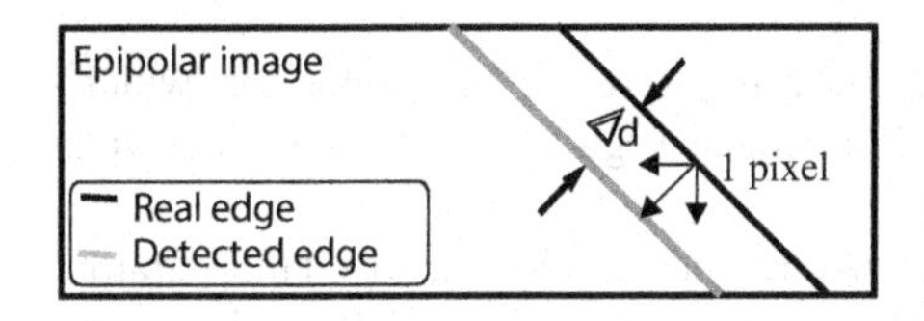

FIGURE 4.28 Detected and real edge. (a) Trajectory deviates from ideal direction. (b) Trajectory shifts along u and v axis.

If $s = 170(m/second)$, $\Phi_{u1} = \Phi_{v1} = 1pixel = 0.001m$, that is, the trajectory of a feature shifts 1 pixel along u and v directions in EPI respectively, *slope* = 1(45°) (see Figure 4.28). When the slope value of trajectory is maintained, we have

$$\Phi_x = 170 \cdot slope = 0.17(m) \tag{4.22a}$$

$$\Phi_y = 170 \cdot slope = 0.17(m) \tag{4.22b}$$

$$\Phi_z = 170 \cdot slope \cdot f \tag{4.22c}$$

Equation 4.22 shows that if the trajectory of a feature shifts 1 pixel along the axes u and v in EPI, the coordinates x and y in the world can reach a 0.17 m deviation, whereas the z-coordinate does not change.

In the same way, if the slope of the trajectory of a feature deviates 1 pixel (this is in fact quite possible), the error of coordinates (x, y, z) can reach a rather large deviation. Therefore, high-accuracy edge location is significant.

4.6 SYSTEM DEVELOPMENT

In order to extract the urban surface information from an image sequence, we initially developed a system called spatio-temporal technique of aerial image sequence analysis (STAISA) from 1997 through 1998 at the Technical University of Berlin, Germany. All codes were written in C language on a Silicon Graphics/Indigo WorkStation. In recent years, we have improved this system and have migrated the STAISA into our early stereovision software package as a subsystem. This subsystem consists of the following five modules (see Figure 4.29):

- ***Image preprocessing:*** This module contains (1) grain noise filtering, isolated point noise filtering, 3 × 3 mean filtering, N × N mean filtering, selective averaging filtering (adaptive smoothing filtering), N × N median filtering, N × N cross median filtering; (2) image enhancement, including linear transform, log transform, gray histogram transform, gray distribution normalization, histogram equalization, gray histogram computation, gray accumulated histogram computation, gray contour extraction; and (3) image operation, including size enlargement (integer factor), size scaling (integer factor), Zoom in/out with specified area/window, image translation, image rotation (90°, 180°, 270°, free), image skew, and so on.
- ***Extraction of edges/lines and distinctive points:*** This module mainly contains the algorithms for extracting two basic features: lines/edges and distinctive points.

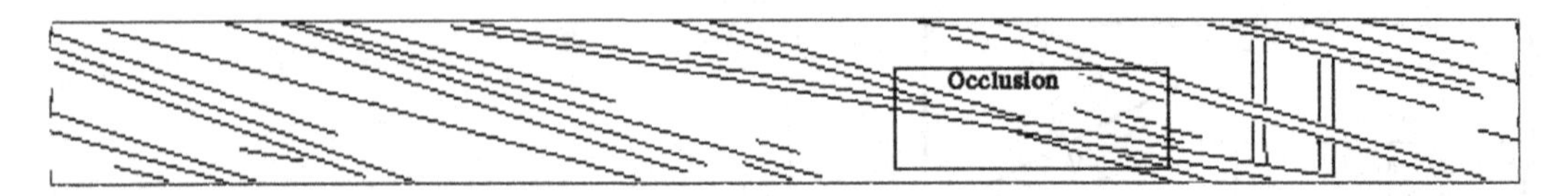

FIGURE 4.29 The relationship between trajectories and occlusions, depth discontinuities.

- o Edge detectors contain, for example, Laplacian operator, zero-crossing operator, which are used to detect "edge-like" features in the EPIs. Linear segment fits to the edges and description of the line segments are also involved.
- o Distinctive point operators include, for example, Wong and Ho (1986), Mikhail operator (1984), Zuniga-Haralick operator (1983), Kitschen-Rosenfeld operator (1982), Liu-Tsai operator (1990), and Förstner operator (1986). The Förstner operator used to be for distinctive point detection of aerial images in photogrammetry because it is associated with various descriptors such as coordinates, weights, interesting values, and corner or circular centers for interesting points.

4.7 CONCLUSIONS

Compared to traditional two-view stereo photogrammetric matching, the approach proposed above has some apparent advantages. Firstly, since the rapid image sampling gives minimal change from frame to frame, that is, not choosing quite disparate views (a large baseline length), the correspondence problem (stereo matching) is eliminated. Secondly, all trajectories in EPIs are straight lines, so only massive similar data is processed. Thereby, the approach is very simple and robust. Thirdly, this technique involves the processing of a very large number of images acquired by a moving camera, that is, a great deal of redundant information, which is unused in the conventional stereo photogrammetric approach. Fourthly, the approach is feature-based, but is not restricted to point features. Linear features that are perpendicular to the direction of motion can also be used. Fifthly, spatial structures in EPIs are much simpler than the original, which means they are easy to be interpreted and analyzed.

By observing the described EPI in Figure 4.29, which corresponds to the original EPI (see Figure 4.12), it can be found that the trajectories of the occluded objects are temporally broken by other trajectories, which correspond to the high objects. That means that the occlusion can be manifested in EPIs obviously, that is, the proposed method provides many chances to detect them. It can be theoretically proved (see Appendix B) that if an object temporally occludes another object, the trajectories respectively corresponding to occluding and occluded features intersect in the EPI. Moreover, for each intersection, the feature with the smaller slope occludes the one with the steeper slope. It means the higher building occludes the low one. Additionally, the trajectories of higher buildings, which cause depth discontinuities, can obviously be reflected in EPIs because of their steeper slopes. As mentioned above, the 3D coordinates are directly obtained from the trajectories of the highs buildings in EPIs without determining the values of horizontal parallaxes. Therefore, the approach is more robust against the occlusion and depth continuities than traditional multiple (two)-view matching.

Nevertheless, the technical **PRACTICAL** application in aerial photogrammetry is the first time so far, thus the realization of practical application still needs much elaboration, particularly since the operational conditions are very complicated, and the real camera's path deviates from an ideal one considerably. Rectifying the distorted image sequences into normal ones is imperative. Additionally, the bundle adjustment combining GPS + IMU (POS) and image data to obtain the orientation and attitude parameters of the camera is a fundament of the practical application.

In summary, the approach proposed above can become a useful tool in solving occlusion and depth discontinuities, with which stereo photogrammetrists have been wresting for decades.

APPENDIX A

We use a counterevidence method to prove that if the speed of the camera is non-constant, the trajectory of a feature in the EPI is not a straight line and instead consists of many line segments with various slopes.

Without loss generality, assume that the camera's movement speed falls into s1, s2, and s1 is unequal to s2.
Original proposition:

$$s_1 \neq s_2 \quad (\#)$$

From Equation (9), we have

$$z_1 = m_1 f_1 = s_1 slope_1 f_1 \quad (a)$$

$$z_2 = m_2 f_2 = s_2 slope_2 f_2 \quad (b)$$

where: $slope_1 = \Delta t_1/\Delta u_1$, $slope_2 = \Delta t_2/\Delta u_2$. Supposing that the slopes corresponding to the speeds s_1, s_2 are $slope_1$, $slope_2$ respectively, moreover the slopes are equal, *i.e.* $slope_1 = slope_2$, we have

$$z_1 = s_1 slopef \quad (c)$$

$$z_2 = s_2 slopef \quad (d)$$

For a feature in the scene, the coordinate z does not change with the camera's speed s_1, s_2, that is, $z_1 = z_2$. From Equations (c), (d), the following equation can be obtained

$$s_1 = s_2 \quad (e)$$

Equation (e) shows that the speeds s_1, s_2 should be equal. The conclusion contradicts the original proposition (Equation #).

Therefore, when the speed of the camera's movement is a non-constant, the trajectory of a feature is not a straight line. If the camera's movement accelerates or decelerates smoothly, the trajectory of a feature is a smooth curve; if the camera's movement accelerates or decelerates suddenly, the trajectory of a feature is a discontinuous curve.

APPENDIX B

Suppose that there are two feature points P_1, P_2 in the scene, which correspond to the heights z_1, z_2, and $z_1 > z_2$. If the flight speed is a constant, s, from Equation (9), we have

$$z_1 = m_1 f_1 = s slope_1 f_1 \quad (a)$$

$$z_2 = m_2 f_2 = s slope_2 f_2 \quad (b)$$

where: $slope_1 = \Delta t_1/\Delta u_1$, $slope_2 = \Delta t_2/\Delta u_2$. Since the focal length of the camera is a constant:

$$f_1 = f_2 \quad (c)$$

Substituting Equation (c) into Equations (a), (b), and considering $z_1 > z_2$, we get

$$slope_1 = slope_2 \tag{d}$$

Therefore, we can conclude that if the features in the scene have various heights (z coordinates), the trajectories of these features will intersect in the EPI or in infinity. If a feature is too much higher than another's, that is, $z_1 >> z_2$, in the scene, it means one feature is occluded by another, and the trajectories of the two features should intersect in the EPI.

REFERENCES

Ackermann, F., "Airborne Laser Scanning—Present Status and Future Expectations," *ISPRS Journal of Photogrammetry and Remote Sensing*, Vol. 54, pp. 64–67, 1999.

Axelsson, P., "Processing of Laser Scanner Data—Algorithms and Applications," *ISPRS Journal of Photogrammetry and Remote Sensing*, Vol. 54, No. 2–3, pp. 138–147, 1999.

Baker, H.H., "*Multiple-Image Computer Vision*," *41st Photogrammetric Week,* Stuttgart Germany, pp. 7–13, September 1987.

Baker, H.H., and Bolles, R.C., "*Generalizing Epipolar-Plane Image Analysis on the Spatiotemporal Surface*," *IEEE Computer Vision and Pattern Recognition*, Ann Arbor, MI, USA, pp. 2–9, June 1988.

Baker, H.H., and Bolles, R.C., "Generalizing Epipolar-Plane Image Analysis on the Spatiotemporal Surface," *International Journal of Computer Vision,* Vol. 3, No. 1, pp. 33–49, 1989.

Baker, H.H., Bolles, R.C., and Marimont, D.H., "*A New Technique for Obtaining Depth Information from a Moving Sensor*," *Proceeding of the ISPRS Commission II Symposium on Photogrammetric and Remote Sensing System for Data Processing and Analysis*, Baltimore, MD, pp. 120–129, May 1986.

Baltsavias, E., Mason, S., and Stallmann, D., "Use of DTMs/DSMs and Orthoimages to Support Building Extraction," *Automatic Extraction of Man-Made Objects from Aerial and Space Images*, pp. 199–210, Birkhäuser, Basel, 1995.

Bolles, R.C., and Baker, H.H., "*Epipolar-Plane Image Analysis: A Technique for Analyzing Motion Sequence*," *International Symposium on Robotics Research*, Gouvieux, France, pp. 41–48, 1985a.

Bolles, R.C., and Baker, H.H., "*Epipolar-Plane Image Analysis: A Technique for Analyzing Motion Sequence*," *IEEE Third Workshop on Computer Vision: Representation and Control*, Bellaire, MI, USA, pp. 168–176, 1985b.

Bolles, R. C., Baker, H.H., and Marimont, D.H., "Epipolar-Plane Image Analysis: An Approach to Determining Structure from Motion," *International Journal of Computer Vision*, Vol. 1, No. 1, pp. 7–55, 1987.

Eckstein, W., and Muenkelt, O., "Extracting Objects from Digital Terrain Models. In Remote Sensing and Reconstruction for Three-Dimensional Objects and Scenes," *International Society for Optics and Photonics*, Vol. 2572, pp. 43–51, August 1995.

Förstner, W., "A Feature Based Correspondence Algorithm for Image Matching," *International Archives of Photogrammetry*, Vol. 26, Part III, 1986.

Förstner, W., "3D-City Models: Automatic and Semiautomatic Acquisition Methods,". *Photogrammetric Week*, eds. D. Fritsch and R. Spiller, pp. 291–303, 1999.

Haala, N., "3D Building Reconstruction Using Linear Edge Segments," *Photogrammetric Week*, Vol. 95, pp. 19–28, Wichmann, Karlsruhe, 1995.

Haala, N., Brenner, C., and Anders, K.H., "3D Urban GIS from Laser Altimeter and 2D Map Data," *International Archives of Photogrammetry and Remote Sensing*, Vol. 32, pp. 339–346, 1998.

Hug, C., "Extracting Artificial Surface Objects from Airborne Laser Scanner Data," *Automatic Extraction of Man-Made Objects from Aerial and Space Images (II)*, pp. 203–212, Birkhäuser, Basel, 1997.

Kraus, K., and Pefifer, N., "Advanced DEM Generation from LiDAR Data," In: Hofton, M.A. (Ed.), *Proceedings of the ISPRS Workshop on Land Surface Mapping and Characterization using Laser Altimetry, Annapolis, Maryland. International Archives of the Photogrammetry, Remote Sensing and Spatial Information Sciences*, Vol. XXXIV, Part 3/W4 Commission III, 2001.

Kruck, E., "Advanced Combined Bundle Block Adjustment with Kinematics GPS Data," *International Archives of Photogrammetry and Remote Sensing*, Vol. XXXI, Part B3, pp. 294–398, 1996.

Larry, M., and Kanade, T., "Kalman Filter-based Algorithms for Estimating Depth from Image Sequences," *International Journal of Computer Vision*, Vol. 3, No. 1, pp. 209–236, 1989.

Lindenberger, J., "Laser-Profilmessungen zur topographischen Geländeaufnahme," *Deutsche Geodaetische Kommission*, Series C, No. 400, Munich, 1993.

Morgan, M., and Habib, A., "*Interpolation of LiDAR data and Automatic Building Extraction*," *ACSM-ASPRS Annual Conference Proceedings*, pp. 432–441, April 2002.

Morgan, M., and Tempfli, K., "Automatic Building Extraction from Airborne Laser Scanning Data," *International Archives of Photogrammetry and Remote Sensing*, Vol. 33, B3/2, Part 3, pp. 616–623, 2000.

Schade, H., "On the Use of Modern GPS Receiver and Software Technology for Photogrammetric Applications," *International Archives of Photogrammetry and Remote Sensing*, Vol. XXXI, Part B3, pp. 729–734, 1996.

Schenk, T., and Toth, C., "Conceptual Issues of Softcopy Photogrammetric Workstations," *Photogrammetric Engineering and Remote Sensing*, Vol. 58, pp. 101–110, 1992.

Sithole, G., "Filtering of Laser Altimetry Data Using a Slope Adaptive Filter," In: Hofton, M.A. (Ed.), *Proceedings of the ISPRS Workshop on Land Surface Mapping and Characterization Using Laser Altimetry, Annapolis, Maryland. The International Archives of the Photogrammetry, Remote Sensing and Spatial Information Sciences*, Vol. XXXIV, Part 3/W4 Commission III, pp. 203–210, 2001.

Song, P., Zhou G., and Cheng P., "Urban Surface Model Generation from Remotely Sensed Airborne Image Sequence Data," *International Journal of Remote Sensing*, Vol. 26, No. 1, pp.79–98, 2005.

Tao, C., and Hu, Y., "*A Review of Post-Processing Algorithms for Airborne LiDAR Data*," *CD-ROM Proceedings of ASPRS Annual Conference*, St. Louis, MO, USA, 23–27 April, 2001.

Vosselman, G., "Slope Based Filtering of Laser Altimetry Data," *International Archives of Photogrammetry and Remote Sensing*, Vol. 33, B3/2, Part 3, pp. 935–942, 2000.

Vosselman, G., and Mass, H., "*Adjustment and Filtering of Raw Laser Altimetry Data*," *Proceedings of the OEEPE Workshop on Airborne Laser Scanning and Interferometric SAR for Detailed Digital Elevation Models*, Stockholm, 1–3 March 2001.

Wang, Y., Mercer, B., Tao, V.C., Sharma, J., and Crawford, S., "*Automatic Generation of Bald Earth Digital Elevation Models from Digital Surface Models Created Using Airborne IFSAR*," *Proceedings of 2001 ASPRS Annual Conference*, St. Louis, MO, USA, 23–27 April 2001.

Wong, K.W., and Ho, W.H., "Close-Range Mapping with a Solid State Camera," *Photogrammetric Engineering and Remote Sensing*, Vol. 2, pp. 67–74, 1986.

Yoon, J.S., and Shan, J., "Urban DEM Generation from Raw Airborne LiDAR Data," *Proceedings of the Annual ASPRS Conference*, Washington, DC, 22–26 April, 2002.

Zhou, G., Albertz, J., and Gwinner, K., "Extracting 3D Information Using Spatio-Temporal Analysis of Aerial Image Sequences," *Photogrammetric Engineering & Remote Sensing*, Vol. 65, No. 7, pp. 823–832, 1999.

5 Urban 3D Surface Information Extraction from Linear Pushbroom Stereo Imagery

5.1 INTRODUCTION

The launch of the first satellite, SPOT-1, from the Kourou Launch Range in French Guiana on February 21, 1986 on board an Ariane launch vehicle began a new era in imaging systems, for it first employed a linear array sensor and pushbroom scanning techniques (Kiefer 1997). However, it has no along-track capability, which means that the stereo image pair is only constructed by cross-track images. In the 1990s, several American private companies launched their high-resolution satellites, such as EarlyBird (3 m resolution, launched in early 1998 and failed two-way communication) and QuickBird (1 m/4 m) from EarthWatch Inc. and OrbView-1 (1 m/2 m) from Orbital Science Corporation, Space Imaging/EOSAT, IKONOS (0.82 m). The imaging systems of all these satellites employed a linear array stereo imaging mode with pushbroom scanning techniques. Thereby, a stereo pair of images can be constructed by along-track images, which are acquired in most atmospheric conditions. Besides satellite imaging systems, some airborne systems of flying shuttles, such as Modular Opto-electronic Multispectral Scanner (MOMS) series products and High-Resolution Stereo Camera (HRSC) have mounted linear array sensors. The new generation of commercial high-resolution (especially up to 1 m) linear stereo satellite imaging systems is considered a revolution in 3D location and reconstruction by the photogrammetric and remote sensing communities (Fritz 1996). This is because the linear array imaging system has the following obvious advantages relative to existing frame imaging systems: (1) high-resolution (up to 1 m) imagery for long focal length used, such as 10 m in IKONOS; (2) enormous potential with their flexible pointing ability, such as along-track stereo and cross-track stereo capability; (3) high geometric fidelity for linear array sensors with the pushbroom imaging mode; (4) rapid conversion of image data into deliverable spatial information products; and (5) greater base–height ratio in comparison with aerial photographs.

Three-dimensional (3D) coordinate determination from the linear array sensor with pushbroom imaging mode is a fundamental task in photogrammetry, remote sensing, computer vision, and pattern recognition. Even though the SPOT-1 satellite also employed the pushbroom technique, the imaging system is a little different from the linear array stereo mode. Moreover, it has a lower resolution (10 m ground resolution in panchromatic channel). The absolute accuracy of computed 3D ground coordinates only reaches 15 m in planimetry and 8 m in height (Kiefer 1997). Thus, the SPOT satellite cannot meet the high accuracy location requirement. Even though aerial photography can provide high accuracy of 3D coordinates relative to SPOT-1, it cannot be used in areas where the airplane cannot reach, such as the North Pole, South Pole, and Atlantic surface. The new generation of high-resolution linear array stereo satellites, typically, IKONOS in the 1990s, has overcome these faults and provides the potential for reconstructing and locating ground objects with high accuracy. Furthermore, the reconstructed area and located object probably appear somewhere humans cannot reach.

Urban 3D building extraction from linear pushbroom stereo imagery is the other important issue (Gupta and Hartley 1997). Work should be undertaken to evaluate the attainable accuracy of 3D coordinates in the case of with and without GCPs and to hunt for the attainable highest accuracy of 3D coordinates with a lower cost.

The three-line stereo imaging technique in the shuttle was pioneered by Hofmann et al. (1982). It was implemented by the German Aerospace Research Establishment (DLR) as the MOMS system series first on board a space shuttle and then on the space station MIR (Kornus and Lehner 1998). Ebner et al. (1991) simulated the geometry and estimated the accuracy of MOMS-02/D2. Ebner and Strunz (1988) investigated the accuracy of obtained ground points using a DTM as control information. Ebner et al. (1992) finished their simulation study on influences of the precision of observed exterior orientation parameters, the type and density of GCPs, camera inclination across flight direction, and simultaneous adjustment of two crossing strips with different intersection angles on the theoretical accuracy of the point determination. The desired height accuracy of about 5 m can be achieved by the simultaneous adjustment of two (or more) crossing strips within the overlapping area. Fraser and Shao (1996) did similar work evaluating the accuracy of ground points when employing the Australian control field. Fritsch et al. (1998) reported their recent results of processing MOMS-2P data.

The very same concept was used by DLR to develop HRSC to map the Mars 3D terrain surface (Albertz et al. 1996; Wewel 1996). After the failure of its launch, HRSC has been used for mapping applications on the Earth. A modified version of three-line imaging is, for example, implemented in the IKONOS Space Imaging satellite.

McGlone (1998) introduced his experiments in the bundle adjustment of linear pushbroom sensor imagery incorporating object–space straight-line constraints. The results demonstrated that the flexibility of the interpolative model makes it better able to describe complex platform motion and to utilize the geometric strength given by the straight-line constraints by testing polynomial and interpolative platform models. Gupta and Hartley (1997) introduced a simplified model for ground point determination from the SPOT satellite. This model has advantages, such as computational simplicity and very accurate results compared with the full orbiting pushbroom model. However, SPOT-1 did not provide along-track stereoscopic ability.

In this section, a mathematical model for ground coordinate determination is developed to process data acquired by linear array stereo imaging systems. Two test fields are built up for checking our mathematical model and the attainable accuracy of new generation high-resolution satellite images. Based on simulated data sets from IKONOS (Space Imaging), HRSC, and MOMS-2P, various experiments for hunting for the attainable highest accuracy of 3D ground points for creations of digital surface model, and also for extraction of buildings are performed.

5.2 LINEAR ARRAY STEREO IMAGING PRINCIPLE WITH PUSHBROOM SCANNING TECHNIQUE

5.2.1 Basic Principle of Linear Array Stereo Imaging System

High-resolution airborne, shuttle, and satellite imaging employ the pushbroom imaging principle, where a single linear array is mounted in the focal plane, which is orthogonal to the direction of flight. In the pushbroom imaging mode, a continuous succession of the one-dimensional image is electronically sampled in such way that an entire linear array is read out during the integration time. When traveling at a certain speed over ground altitude ranging from about 2500 m to 680 km, it pitches forward at an angle (such as IKONOS, 45°) and begins collecting data. Then it pitches to nadir and collects data over the same ground area. After collection at nadir, it pitches aft at the same angle and collects the same area for the third time. Such a configuration provides a so-called *along-track stereo model.* Stereo pairs created from fore-, nadir-, and aft-looking ensure high-quality collections because images are acquired in almost the same ground and atmospheric conditions. The stereo imaging collections cover a wide swath. Stereo pairs have varied convergence angles in different imaging configurations.

5.2.2 Three Typical Linear Array Stereo Imaging Systems

Three typical linear array stereo imaging systems, MOMS-2P, HRSC, and IKONOS, which are representative of shuttle, airborne, and satellite imaging respectively, were implemented in the last century and different ground resolutions, 17.8 m, 10.5 cm, and 0.82 m respectively, would be introduced here.

1. *MOMS-2P*. The MOMS project started in the early 1970s in DLR. A three-line optical-electronic sensor, designated a, b and c, is placed in the focal planes, as shown in Figure 5.1. The sensors are oriented perpendicular to the direction of flight. During flight, the sensors continuously scan the terrain with a constant frequency, recording three images. As a consequence, three pushbroom image strips cover the terrain surface. By modifying the CCD integration frequency relative to the velocity of the camera platform, the image scale, and the size of the sensor elements, nearly square ground pixels are generated (see Figure 5.1). The MOMS-2P provides four data collection modes. Mode A uses the panchromatic (PAN) channels 5, 6, and 7 only, which provide three-fold along-track stereo scanning with different ground resolutions. The nadir-looking (HR channel) is 6 m, and the other is 18 m. Mode B delivers all four multispectral (MS) data channels simultaneously (1, 2, 3, and 4). Mode C offers high-resolution MS imaging (5A PAN, 2, 3, and 4). Mode D provides simultaneous stereo and multispectral capability. In general, each three-line has its own set of exterior orientation parameters. In contrast to conventional frame cameras, the image recording is performed in dynamic mode.

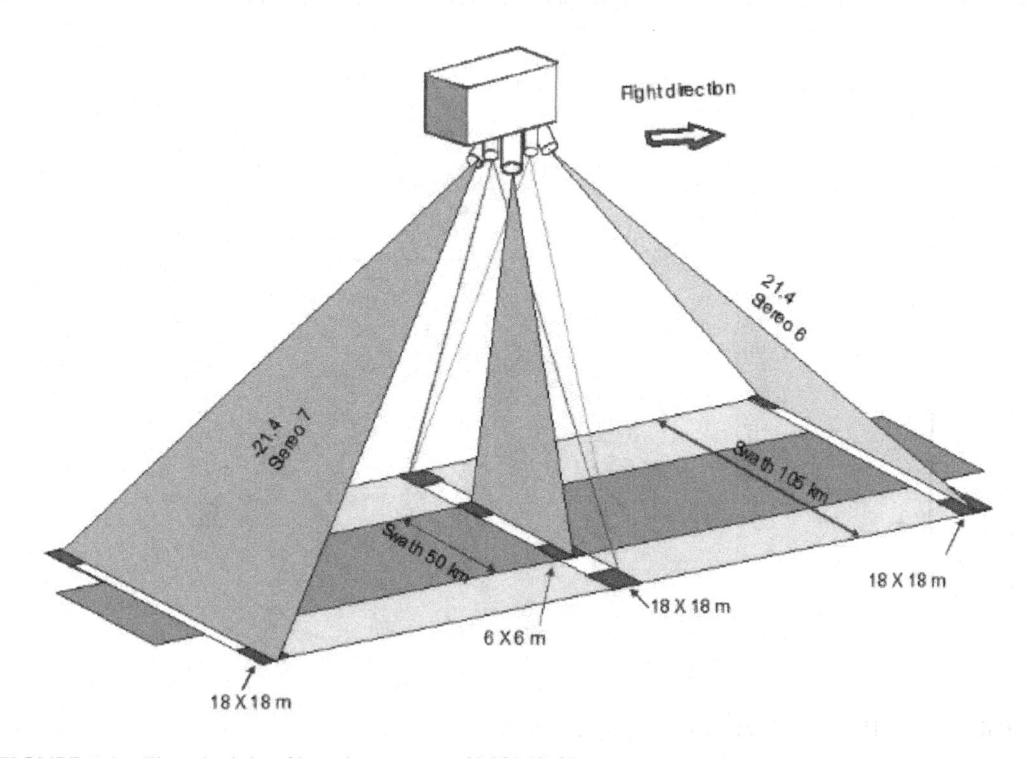

FIGURE 5.1 The principle of imaging system of MOMS-2P.

2. *The HRSC*. The HRSC is designed as a compact single-optic pushbroom instrument with nine sensors mounted in parallel in DLR, and provides multiple along-track stereo images as well as high-resolution hyperspectral information. Depending on the actual ground track velocity, the integration time of the CCD lines is adjusted in such a way that the center pixels are kept square. The ground pixel size varies depending on the current flying height. Nine superimposed image strips are acquired almost simultaneously by the forward motion of the airborne sensor over the terrain. Five of the nine CCD lines are arranged at specific viewing angles for providing stereo imaging and photometric viewing capability (see Figure 5.2). Four of the nine CCD lines are covered with different filters for the acquisition of hyperspectral images. Together, the three focal planes form one high-resolution panchromatic (clear) nadir channel, two panchromatic (clear) stereo channels with convergence angles of ± 19°, four color channels at ± 3° and ± 6°, and two additional channels at ± 13° for photometric purposes.

3. *IKONOS*. The geometric configuration of IKONOS is depicted in Figure 5.3. Special characteristics of IKONOS, in comparison to other similar systems, are:

   - A new optical system folds a very long focal length (10 m) into 2 m using a mirror system;
   - In addition to along-track stereo capability, it is still able to collect cross-track images at distance of 725 km on either side of the ground track;
   - At least 1 m ground sample distance (GSD) (Corbley 1996); and
   - GPS antennas and three digital star trackers are carried to maintain a precise camera position and attitude.

Even though three types of linear array stereo imaging principle are described above, the image acquisition principle outlined above is applicable to many other imaging situations.

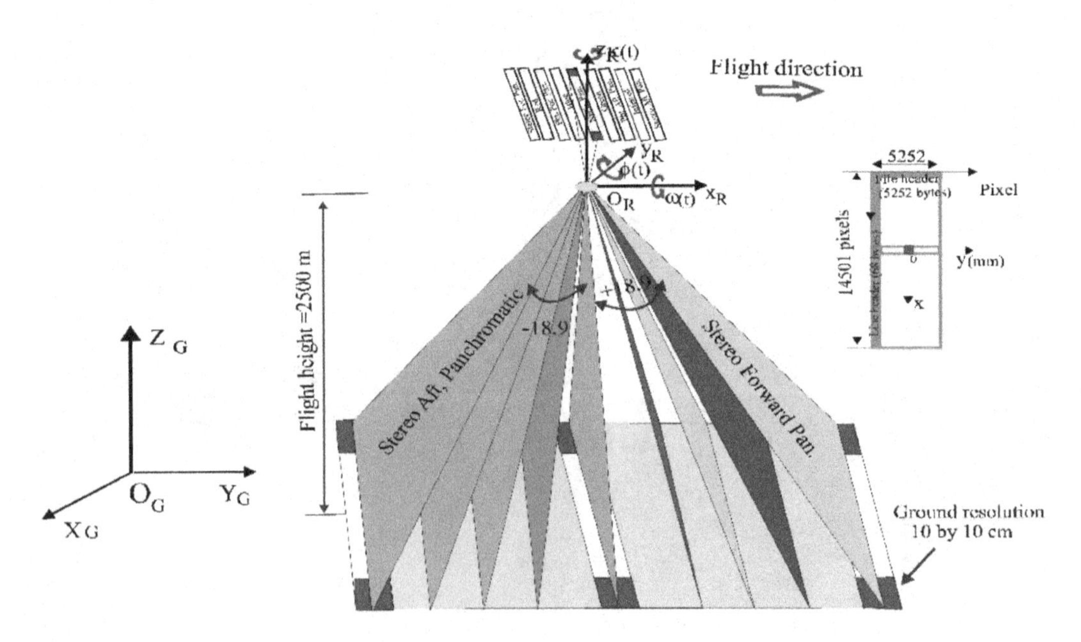

FIGURE 5.2 The principle of imaging system of HRSC.

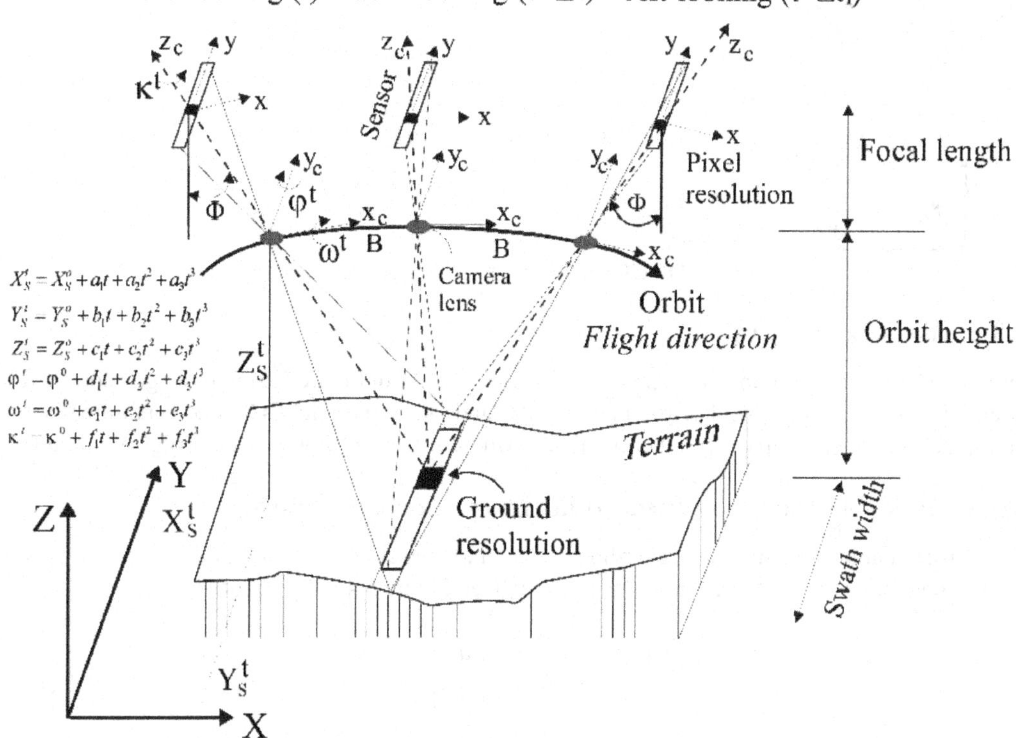

FIGURE 5.3 Geometric configuration of IKONOS.

5.3 MATHEMATICAL MODEL OF 3D GROUND COORDINATES FROM LINEAR ARRAY STEREO IMAGING SYSTEM

5.3.1 Coordinate Systems

Figure 5.3 shows the established image, camera, and georeference coordinate systems. These coordinate systems are right-handed. In fact, each of the fore-, nadir- and aft-looking arrays has its own image coordinate system (x_c, y_c and z_c). An image reference coordinate system (x_R, y_R and z_R) is defined to unify image coordinates from all three arrays (see Figure 5.4).

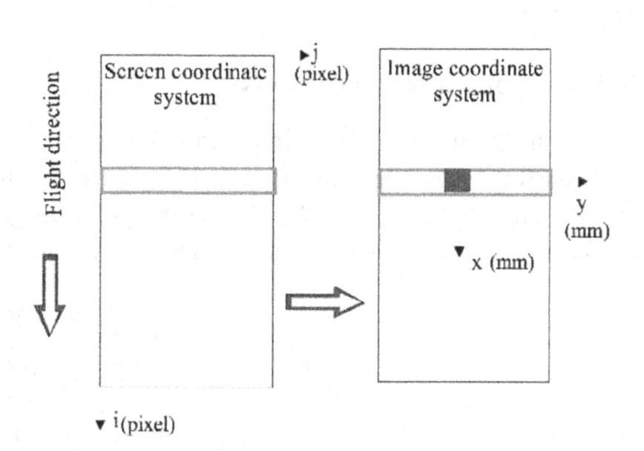

FIGURE 5.4 From image coordinate to image reference coordinate system.

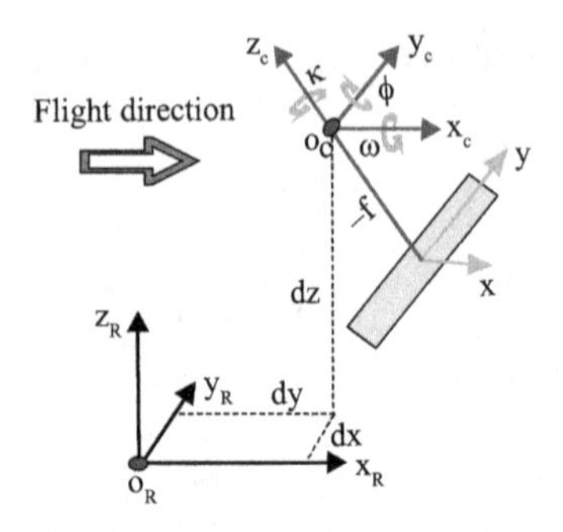

FIGURE 5.5 Screen and image coordinate system.

5.3.2 Interior Orientation

Interior orientation is to transform screen coordinates (i and j in Figure 5.5) into image coordinates (x and y), and to correct lens distortion (symmetric and tangential) and CCD line curvature distortion, in general, the principal offset that should be estimated from a laboratory or in-flight calibration.

5.3.3 Transformation from Image to Reference Coordinate System

The transformation from an image coordinate system to the reference coordinate system involves a translation (dx, dy, dz) and three rotations (ω, ϕ, κ) (see Figure 5.4). The equation is

$$\begin{pmatrix} x_R \\ y_R \\ z_R \end{pmatrix} = R_c^R \begin{pmatrix} x_c \\ y_c \\ z_c \end{pmatrix} + \begin{pmatrix} dx \\ dy \\ dz \end{pmatrix} \tag{5.1}$$

where

$$\begin{pmatrix} x_c \\ y_c \\ z_c \end{pmatrix} = \begin{pmatrix} x \\ y \\ -f \end{pmatrix},$$

$$R_c^R = \begin{pmatrix} \cos\phi\cos k & \cos\omega\sin k + \sin\omega\sin\phi\cos k & \sin\omega\sin k - \cos\omega\sin\phi\cos k \\ -\cos\phi\sin k & \cos\omega\cos k - \sin\omega\sin\phi\sin k & \sin\omega\cos k + \cos\omega\sin\phi\sin k \\ \sin\phi & -\sin\omega\cos\phi & \cos\omega\cos\phi \end{pmatrix}$$

R_c^R is an orthogonal matrix, that is, $\left(R_R^c\right)^T = \left(R_R^c\right)^{-1}$. The counterclockwise rotation angle is defined as positive.

5.3.4 Collinearity Equations

For any image point within a linear array image plane, its image reference coordinates are (x_R, y_R, z_R). The coordinates of the exposure center of the array in the ground coordinate system at the imaging epoch t are ($X_S(t)$, $Y_S(t)$, $Z_S(t)$). The corresponding ground point coordinates are (X_G, Y_G, Z_G). The collinearity condition states that all these three points must be on the same line:

$$\begin{aligned} x_R &= z_R \frac{r_{11}\left(X_G - X_S(t)\right) + r_{12}\left(Y_G - Y_S(t)\right) + r_{13}\left(Z_G - Z_S(t)\right)}{r_{31}\left(X_G - X_S(t)\right) + r_{32}\left(Y_G - Y_S(t)\right) + r_{33}\left(Z_G - Z_S(t)\right)} \\ y_R &= z_R \frac{r_{21}\left(X_G - X_S(t)\right) + r_{22}\left(Y_G - Y_S(t)\right) + r_{23}\left(Z_G - Z_S(t)\right)}{r_{31}\left(X_G - X_S(t)\right) + r_{32}\left(Y_G - Y_S(t)\right) + r_{33}\left(Z_G - Z_S(t)\right)} \end{aligned} \tag{5.2}$$

where R_G^R is a rotation matrix from the ground coordinate system to the reference coordinate system and is defined by

$$R_G^R = \begin{pmatrix} \cos\phi\cos k & \cos\omega\sin k + \sin\omega\sin\phi\cos k & \sin\omega\sin k - \cos\omega\sin\phi\cos k \\ -\cos\phi\sin k & \cos\omega\cos k - \sin\omega\sin\phi\sin k & \sin\omega\cos k + \cos\omega\sin\phi\sin k \\ \sin\phi & -\sin\omega\cos\phi & \cos\omega\cos\phi \end{pmatrix} \tag{5.3}$$

The rotation angles $\phi(t)$, $\omega(t)$, and $k(t)$ are defined for each linear array at the epoch t. Depending on types of observations, 3D coordinates may be treated as knowns and unknowns differently in various situations.

5.3.5 Navigation Data as Exterior Orientation Parameters

Current satellites carry navigation equipment of GPS receivers and digital star trackers that can provide positions (X_S, Y_S, Z_S) and attitudes (ω, ϕ, κ) of linear arrays if appropriate calibrations are performed. Since the navigation data have a lower data collection rate, they are not acquired for every image line. Those lines are called orientation lines (OLs). Navigation data at OLs can be used as their initial exterior orientation parameters (EOPs). The initial EOPs of the OLs are introduced at a certain time interval. Previous investigations based on simulated orbit data showed that a third-order polynomial function can exactly approximate EOP changes (Wu 1986; Ebner et al. 1992). The EOPs of lines between OLs are computed by a polynomial interpolation:

$$\begin{aligned} X_S(t) &= a_0 + a_1 t + a_2 t^2 + a_3 t^3 \\ Y_S(t) &= b_0 + b_1 t + b_2 t^2 + b_3 t^3 \\ Z_S(t) &= c_0 + c_1 t + c_2 t^2 + c_3 t^3 \\ \phi_S(t) &= d_0 + d_1 t + d_2 t^2 + d_3 t^3 \\ \omega_S(t) &= e_0 + e_1 t + e_2 t^2 + e_3 t^3 \\ \kappa_S(t) &= f_0 + f_1 t + f_2 t^2 + f_3 t^3 \end{aligned} \tag{5.4}$$

The parameter t may be beginning time at a certain epoch or image line number from a certain orbit position. All three images will be referenced to the same t. The unknown coefficients in Equation 5.4 can be determined either independently or by Equation 5.2.

5.3.6 Observation Equations

Observations include GCPs, checkpoints/unknown ground points, and navigation data.

- ***GCPs.*** Ground coordinates and measured image coordinates are knowns. Unknowns include EOPs, that is, coefficients of the polynomials of Equation 5.4. After linearization of Equation 5.2, we have the linearized equations

$$\begin{aligned} v_{GCP}^{xR} = {} & \beta_{11}da_0 + \beta_{12}db_0 + \beta_{13}dc_0 + \beta_{14}dd_0 + \beta_{15}de_0 + \beta_{16}df_0 + \\ & \beta_{17}da_1 + \beta_{18}db_1 + \beta_{19}dc_1 + \beta_{110}dd_1 + \beta_{111}de_1 + \beta_{112}df_1 + \\ & \beta_{113}da_2 + \beta_{114}db_2 + \beta_{115}dc_2 + \beta_{116}dd_2 + \beta_{117}de_2 + \beta_{118}df_2 + \\ & \beta_{119}da_3 + \beta_{120}db_3 + \beta_{121}dc_3 + \beta_{122}dd_3 + \beta_{123}de_3 + \beta_{124}df_3 + l_{GCP}^{x} \\ v_{GCP}^{yR} = {} & \beta_{21}da_0 + \beta_{22}db_0 + \beta_{23}dc_0 + \beta_{24}dd_0 + \beta_{25}de_0 + \beta_{26}df_0 + \\ & \beta_{27}da_1 + \beta_{28}db_1 + \beta_{29}dc_1 + \beta_{210}dd_1 + \beta_{211}de_1 + \beta_{212}df_1 + \\ & \beta_{213}da_2 + \beta_{214}db_2 + \beta_{215}dc_2 + \beta_{216}dd_2 + \beta_{217}de_2 + \beta_{218}df_2 + \\ & \beta_{219}da_3 + \beta_{220}db_3 + \beta_{221}dc_3 + \beta_{222}dd_3 + \beta_{223}de_3 + \beta_{224}df_3 + l_{GCP}^{y} \end{aligned} \tag{5.5}$$

where

$$l_{GCP}^{x} = -x_R + z_R \frac{r_{11}\left(X_G - X_S^0(t)\right) + r_{12}\left(Y_G - Y_S^0(t)\right) + r_{13}\left(Z_G - Z_S^0(t)\right)}{r_{31}\left(X_G - X_S^0(t)\right) + r_{32}\left(Y_G - Y_S^0(t)\right) + r_{33}\left(Z_G - Z_S^0(t)\right)}$$

$$l_{GCP}^{y} = -y_R + z_R \frac{r_{21}\left(X_G - X_S^0(t)\right) + r_{22}\left(Y_G - Y_S^0(t)\right) + r_{23}\left(Z_G - Z_S^0(t)\right)}{r_{31}\left(X_G - X_S^0(t)\right) + r_{32}\left(Y_G - Y_S^0(t)\right) + r_{33}\left(Z_G - Z_S^0(t)\right)}$$

The coefficients β_{11} to β_{224} are partial derivatives of Equation 5.2 with respect to the polynomial parameters and the details of coefficients β_{11} to β_{224} can be found out in G. Zhou and L.I. Ron (2000).

- ***Unknown ground points and checkpoints.*** A few image coordinates are measured and a few ground coordinates of unknown points and checkpoints are calculated. (In fact, the 3D coordinate of checkpoint is known.) Compared to the GCPs, the additional unknown parameters are the ground coordinates.

$$\begin{aligned}
v_M^x = {} & \gamma_{11}da_0 + \gamma_{12}db_0 + \gamma_{13}dc_0 + \gamma_{14}dd_0 + \gamma_{15}de_0 + \gamma_{16}df_0 + \\
& \gamma_{17}da_1 + \gamma_{18}db_1 + \gamma_{19}dc_1 + \gamma_{110}dd_1 + \gamma_{111}de_1 + \gamma_{112}df_1 + \\
& \gamma_{113}da_2 + \gamma_{114}db_2 + \gamma_{115}dc_2 + \gamma_{116}dd_2 + \gamma_{117}de_2 + \gamma_{118}df_2 + \\
& \gamma_{119}da_3 + \gamma_{120}db_3 + \gamma_{121}dc_3 + \gamma_{122}dd_3 + \gamma_{123}de_3 + \gamma_{124}df_3 + \\
& \gamma_{125}dX + \gamma_{126}dY + \gamma_{127}dZ + l_M^x \\
v_M^y = {} & \gamma_{21}da_0 + \gamma_{22}db_0 + \gamma_{23}dc_0 + \gamma_{24}dd_0 + \gamma_{25}de_0 + \gamma_{26}df_0 + \\
& \gamma_{27}da_1 + \gamma_{28}db_1 + \gamma_{29}dc_1 + \gamma_{210}dd_1 + \gamma_{211}de_1 + \gamma_{212}df_1 + \\
& \gamma_{213}da_2 + \gamma_{214}db_2 + \gamma_{215}dc_2 + \gamma_{216}dd_2 + \gamma_{217}de_2 + \gamma_{218}df_2 + \\
& \gamma_{219}da_3 + \gamma_{220}db_3 + \gamma_{221}dc_3 + \gamma_{222}dd_3 + \gamma_{223}de_3 + \gamma_{224}df_3 + \\
& \gamma_{225}dX + \gamma_{226}dY + \gamma_{227}dZ + l_M^y
\end{aligned} \tag{5.6}$$

where

$$l_M^x = -x_M + z_R \frac{r_{11}\left(X - X_S(t)\right) + r_{12}\left(Y - Y_S(t)\right) + r_{13}\left(Z - Z_S(t)\right)}{r_{31}\left(X - X_S(t)\right) + r_{32}\left(Y - Y_S(t)\right) + r_{33}\left(Z - Z_S(t)\right)}$$

$$l_M^y = -y_M + z_R \frac{r_{21}\left(X - X_S(t)\right) + r_{22}\left(Y - Y_S(t)\right) + r_{23}\left(Z - Z_S(t)\right)}{r_{31}\left(X - X_S(t)\right) + r_{32}\left(Y - Y_S(t)\right) + r_{33}\left(Z - Z_S(t)\right)}$$

Similarly, the coefficients γ_{11} to γ_{227} are explained in Zhou and Li (2000), and Li and Zhou (2002).

- ***Navigation data as observations.*** The exposure center coordinates at OLs from GPS can be built with the observations below:

$$\begin{pmatrix} v_{GPS}^{X_1} \\ v_{GPS}^{X_2} \\ \vdots \\ v_{GPS}^{X_{N3}} \end{pmatrix} = \begin{pmatrix} 1 & t_1 & t_1^2 & t_1^3 \\ 1 & t_2 & t_2^2 & t_2^3 \\ & & \vdots & \\ 1 & t_{N3} & t_{N3}^2 & t_{N3}^3 \end{pmatrix} \begin{pmatrix} da_0 \\ da_1 \\ da_2 \\ da_3 \end{pmatrix} + \begin{pmatrix} X_S^0(t_1) - X_{GPS}^1 \\ X_S^0(t_2) - X_{GPS}^2 \\ \vdots \\ X_S^0(t_{N3}) - X_{GPS}^{N3} \end{pmatrix} \tag{5.7a}$$

$$\begin{pmatrix} v_{GPS}^{Y_1} \\ v_{GPS}^{Y_2} \\ \vdots \\ v_{GPS}^{Y_{N3}} \end{pmatrix} = \begin{pmatrix} 1 & t_1 & t_1^2 & t_1^3 \\ 1 & t_2 & t_2^2 & t_2^3 \\ & & \vdots & \\ 1 & t_{N3} & t_{N3}^2 & t_{N3}^3 \end{pmatrix} \begin{pmatrix} db_0 \\ db_1 \\ db_2 \\ db_3 \end{pmatrix} + \begin{pmatrix} Y_S^0(t_1) - Y_{GPS}^1 \\ Y_S^0(t_2) - Y_{GPS}^2 \\ \vdots \\ Y_S^0(t_{N3}) - Y_{GPS}^{N3} \end{pmatrix} \quad (5.7b)$$

$$\begin{pmatrix} v_{GPS}^{Z_1} \\ v_{GPS}^{Z_2} \\ \vdots \\ v_{GPS}^{Z_{N3}} \end{pmatrix} = \begin{pmatrix} 1 & t_1 & t_1^2 & t_1^3 \\ 1 & t_2 & t_2^2 & t_2^3 \\ & & \vdots & \\ 1 & t_{N3} & t_{N3}^2 & t_{N3}^3 \end{pmatrix} \begin{pmatrix} dc_0 \\ dc_1 \\ dc_2 \\ dc_3 \end{pmatrix} + \begin{pmatrix} Z_S^0(t_1) - Z_{GPS}^1 \\ Z_S^0(t_2) - Z_{GPS}^2 \\ \vdots \\ Z_S^0(t_{N3}) - Z_{GPS}^{N3} \end{pmatrix} \quad (5.7c)$$

Also, observations for the three rotation angles from INS are

$$\begin{pmatrix} v_{INS}^{\phi_1} \\ v_{INS}^{\phi_2} \\ \vdots \\ v_{INS}^{\phi_{N3}} \end{pmatrix} = \begin{pmatrix} 1 & t_1 & t_1^2 & t_1^3 \\ 1 & t_2 & t_2^2 & t_2^3 \\ & & \vdots & \\ 1 & t_{N3} & t_{N3}^2 & t_{N3}^3 \end{pmatrix} \begin{pmatrix} dd_0 \\ dd_1 \\ dd_2 \\ dd_3 \end{pmatrix} + \begin{pmatrix} \phi_S^0(t_1) - \phi_{INS}^1 \\ \phi_S^0(t_2) - \phi_{INS}^2 \\ \vdots \\ \phi_S^0(t_{N3}) - \phi_{INS}^{N3} \end{pmatrix} \quad (5.8a)$$

$$\begin{pmatrix} v_{INS}^{\omega_1} \\ v_{INS}^{\omega_2} \\ \vdots \\ v_{INS}^{\omega_{N3}} \end{pmatrix} = \begin{pmatrix} 1 & t_1 & t_1^2 & t_1^3 \\ 1 & t_2 & t_2^2 & t_2^3 \\ & & \vdots & \\ 1 & t_{N3} & t_{N3}^2 & t_{N3}^3 \end{pmatrix} \begin{pmatrix} de_0 \\ de_1 \\ de_2 \\ de_3 \end{pmatrix} + \begin{pmatrix} \omega_S^0(t_1) - \omega_{INS}^1 \\ \omega_S^0(t_2) - \omega_{INS}^2 \\ \vdots \\ \omega_S^0(t_{N3}) - \omega_{INS}^{N3} \end{pmatrix} \quad (5.8b)$$

$$\begin{pmatrix} v_{INS}^{k_1} \\ v_{INS}^{k_2} \\ \vdots \\ v_{INS}^{k_{N3}} \end{pmatrix} = \begin{pmatrix} 1 & t_1 & t_1^2 & t_1^3 \\ 1 & t_2 & t_2^2 & t_2^3 \\ & & \vdots & \\ 1 & t_{N3} & t_{N3}^2 & t_{N3}^3 \end{pmatrix} \begin{pmatrix} df_0 \\ df_1 \\ df_2 \\ df_3 \end{pmatrix} + \begin{pmatrix} k_S^0(t_1) - k_{INS}^1 \\ k_S^0(t_2) - k_{INS}^2 \\ \vdots \\ k_S^0(t_{N3}) - k_{INS}^{N3} \end{pmatrix} \quad (5.8c)$$

All of these heterogeneous data of position and attitude should combine into the Equations 5.8a and 5.9 after being transformed into a common coordinate system.

5.3.7 Adjustment Computation

- ***Vector of observation equations:*** Suppose that we have N_1 *OLs*, N_2 *GCPs*, and N_3 unknown points/checkpoints. The vector of observation equations is

  GCPs:

$$\underset{4N_2\times 1}{V_{GCP}} = \underset{4N_2\times(24+3N_3)}{A_3} \cdot \underset{(24+3N_3)\times 1}{\delta X} + \underset{4N_2\times 1}{L_3} \quad (5.9)$$

  Unknown points/checkpoints:

$$\underset{4N_3\times 1}{V_M} = \underset{4N_3\times(24+3N_3)}{A_4} \cdot \underset{(24+3N_3)\times 1}{\delta X} + \underset{4N_3\times 1}{L_4} \quad (5.10)$$

Navigation data:

$$\underset{3N_1\times 1}{V_{GPS}} = \underset{3N_1\times(24+3N_3)}{A_1} \cdot \underset{(24+3N_3)\times 1}{\delta X} + \underset{3N_1\times 1}{L_1} \tag{5.11}$$

$$\underset{3N_1\times 1}{V_{\omega,\varphi,\kappa}} = \underset{3N_1\times(24+3N_3)}{A_2} \cdot \underset{(24+3N_3)\times 1}{\delta X} + \underset{3N_1\times 1}{L_2} \tag{5.12}$$

The details of the structures of the observation equations are described in Li and Zhou (2002).

- ***Blunders.*** The data measured unavoidably contain errors and/or even blunders. Blunders must be located and eliminated by robust estimation methods. The data snooping method is applied to remove these blunders.

The solutions are carried out through an iterative process. In each iteration, the increments for all unknown parameters are obtained and used to update the approximations for the next iteration.

- ***Accuracy evaluation.*** The standard deviations of unknown points/checkpoints are typically used to evaluate the quality of the adjustment and the accuracy of adjusted unknowns. It is expressed as

$$\sigma_{X_i} = \sigma_0 {N^{-1}}_{X_{ii}} \tag{5.13}$$

where ${N^{-1}}_{Xii}$ denotes the partial matrix of the inverse of the normal matrix, σ_0 is the standard deviation of the unit weight, which is computed by

$$\sigma_0 = \sqrt{\frac{V^T PV}{r}} \tag{5.14}$$

V is the residual vector and r is redundancy of the observation equation. In addition, the root-mean square (RMS) error of checkpoints is usually computed independently as a measure of external accuracy. It is calculated by

$$RMS_{x,y,z} = \sqrt{\frac{\Delta^T \Delta}{n-1}} \tag{5.15}$$

where Δ is the difference of known and estimated coordinates of checkpoints and n is number of checkpoints used.

- ***Implementation.*** The first version of the bundle adjustment software has been implemented in C programming language on a Silicon Graphics O2/UNIX system. The interface is programmed by Motif and Xwindow, and some of the functions in the algorithms of image processing need support of GL/OpenGL.

5.4 TEST FIELD ESTABLISHMENT

5.4.1 The High-Altitude Test Range – the First Test Field

The High-Altitude Test Range for evaluating the potential of accuracy of new generation high-resolution satellite images was established by cooperation with the Ohio Department of Transportation (ODOT), Space Imaging Inc. and the Ohio State University in Madison County,

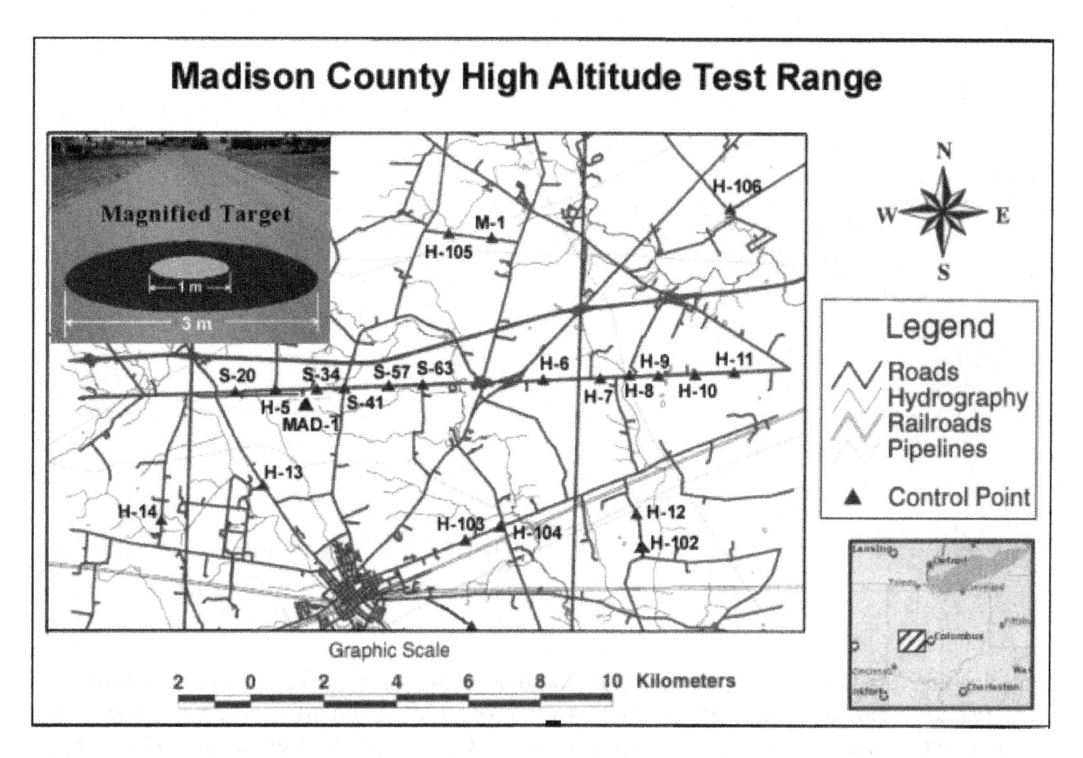

FIGURE 5.6 The distribution of GCP and target point pattern at the High-Altitude Test Range – the First Test Field, Madison County.

Central Ohio, USA. The control network, which consisted of 21 ground target points plus the three higher-order control points, was specifically distributed in a flat area approximately 16 × 11 km, centered at latitude 39° 56′24″ north, and longitude 83° 24′42″ west along east–west direction. The target points were spaced at least 1 km apart. All target points were painted with concentric circles, a 1 m flat white circle and a 3 m flat black circle as background, centered on a monument (see Figure 5.6). Additionally, 24 checkpoints were also surveyed for the purpose of checking the attainable accuracy from the linear satellite data.

Twenty-four control points were observed by *Trimble* for at least two hours with a static mode in order to ensure that at least four GPS were locked. In fast static mode, the observation time of each station of 23 checkpoints and 18 feature points would be different, with various numbers of GPS locked. In this case, MAD-1 was used as the reference. Feature points were mainly selected at the intersection of road, bridge, and distinct permanent marks. Finally, geographic coordinates of all observed points including GCPs, checkpoints, and feature points in the WGS-84 and in UTM system were obtained. The standard errors of these control points could reach higher than 0.02, 0.02 and 0.10 m in latitude, longitude, and elevation respectively. This level of accuracy is comparable with geodetic accuracy standard for Order C (1.0 cm plus ppm).

5.4.2 The Test Field 1185 – The Second Test Field

The Second Test Field chosen was located at Madison County and covered 2 km by 2 km and lay within aerial image AIMS 1185 (see Figure 5.7), so it was called *test field 1185*. In this test field, 14 target control points (in the same pattern as the above test field) were measured by the ODOT. All of the GCPs could clearly be recognized in aerial image AIMS 1185.

In fast static mode, 21 feature points, which specifically appeared at the intersection of roads, or permanent marks, were measured using two Trimble 4000SSI receivers. The higher-level

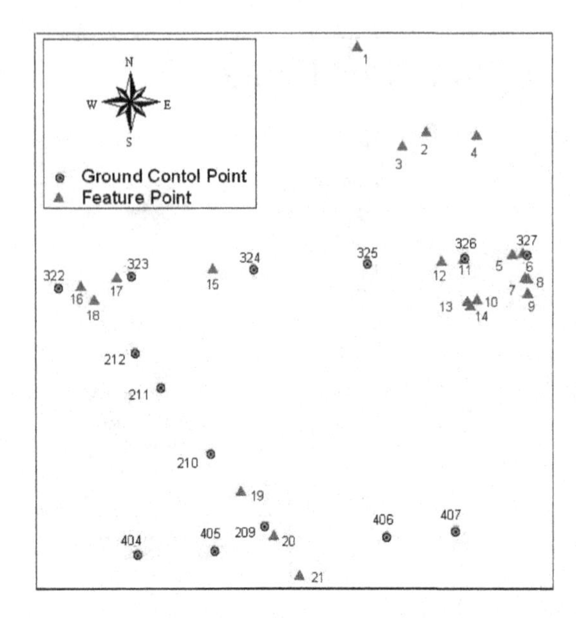

FIGURE 5.7 The distribution of GCPs and feature points at the Second Test Field.

control point, MAD-1 (PID: AB6042 from National Geological Survey-NGS), was linked to these feature points and was considered as a reference station. Finally, the coordinates of 21 feature points in various coordinate systems, UTM, SPC, and WGS84 were obtained by computation with software GPSurvey V. 2.30a. The accuracy of all baselines achieved millimeter level. This level of accuracy sufficiently met the accuracy requirement as the checking points were at centimeter level.

5.5 IMAGE SIMULATION OF SATELLITE IKONOS

Satellite IKONOS-1 is a typical representative of linear stereo imaging mode. Unfortunately, it fell into the Pacific Ocean on April 27, 1999 in spite of deliberate design for several years. For the purpose of further experiments, the satellite images looking at the nadir, aft, and fore have to be simulated first. The generation of synthetic satellite images is based on the principle of "projection" and "back-projection" (see Figure 5.8). The projection indicates here data processing from aerial images to DEM, and back-projection from DEM onto satellite images at fore-, nadir- and aft-looking. The brightness in the simulated satellite images is from the aerial images. Both of them share the identical DEM. The simulation of IKONOS satellite is conducted as follow.

5.5.1 Simulation of IKONOS Imaging

- **DEM data**
  A data set of DEM in Madison County was computed from hypsographic features (contour lines) in DLG files from 7.5′ USGS quad sheets, downloaded from the USGS website. ARC/INFO software provided the functions to create raster DEM data with 1 m by 1 m spacing.

- **Aerial photograph**
  The Center for Mapping of the Ohio State University provided the aerial images acquired by a digital camera with focal length of 50 mm and image size of 4096 by 4096 pixel (60 by 60 mm imaging area resulting in a 15 micron pixel size) in 1997 for the Airborne Integrated Mapping System (AIMS) project. The flying height was about 1366.0 m and GSD was around 0.409 m. The image AIMS 1185 (see Figure 5.12) was chosen for our experiments.

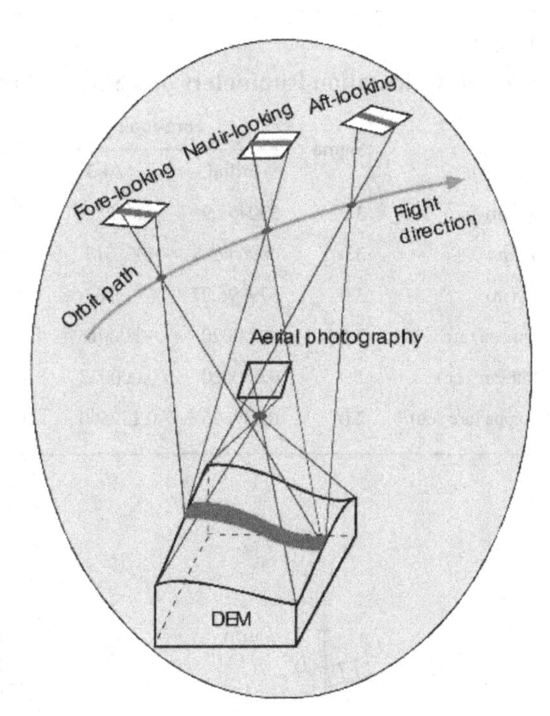

FIGURE 5.8 Principle of synthetic satellite images, which is based on aerial image, DEM, and flight parameters of IKONOS.

- **Specification of IKONOS**
  The details of the system characteristics of IKONOS and imaging geometry of IKONOS have been described in Zhou and Li (2000). Table 5.1 lists part of specifications of IKONOS used for this simulation.

- **Scene design**
  The aim was to select an area within the established test field in which the aerial photography and corresponding DEM were available. Also, a reference point was selected as the starting position of the satellite trajectory. From the reported specification of IKONOS, the orbit inclination was 98.1°, the angle *kappa*, 98.1°, which would be the azimuth, and would be used during the image generation. Considering that the satellite would be traveling from north to south at speeds of 7000 m per second, the velocity vector must be decomposed into the components along the x-axis and y-axis, neglecting any component in the *z* direction. Table 5.2 shows the simulated exterior parameters at an initial time and after five seconds at fore-, nadir- and aft-looking. With altitude of 680 km and focal length of 10 m, the GSD was around 0.816 m. Considering the coverage of an aerial photograph of 1676.6 m by 1676.6 m, the size of satellite image is designed to 2000 pixel by 2000 pixel. Figure 5.9 shows the geometric relationships between the orientation of the satellite orbit, ground coordinate system, and the image coordinate system.

TABLE 5.1
Technical Specifications of IKONOS Used for this Experiment

Parameter	Values	Parameter	Values	Parameter	Values
Altitude (km)	680	Convergence angle (deg)	45	Ground resolution (m)	0.82
Inclination (deg)	98.1	Pixel size (μm)	12	Swath width (km)	11

TABLE 5.2
Exterior Orientation Parameters of a Segment of a Simplified Orbit

Parameter	Sigma	Fore-look		Nadir-look		Aft-look	
		Initial	At 5″	Initial	At 5″	Initial	At 5″
X_S (m)	3.0	330459.9	365125.3	284315.7	318981.1	238171.5	272830.6
Y_S (m)	3.0	4096476.8	4091613.1	4424815.0	4419951.1	4753153.2	4748289.4
Z_S (m)	3.0	679996.87	679996.6	679996.87	679999.6	679996.8	679998.9
Omega (arc sec)	2.0	0.000020	−0.000157	0.000020	−0.006215	0.000010	−0.000633
Phi (arc sec)	2.0	0.000020	0.000022	0.000020	−0.000931	0.000010	0.00001
Kappa (arc sec)	2.0	0.139636	0.139560	0.139647	0.139749	0.139647	0.144059

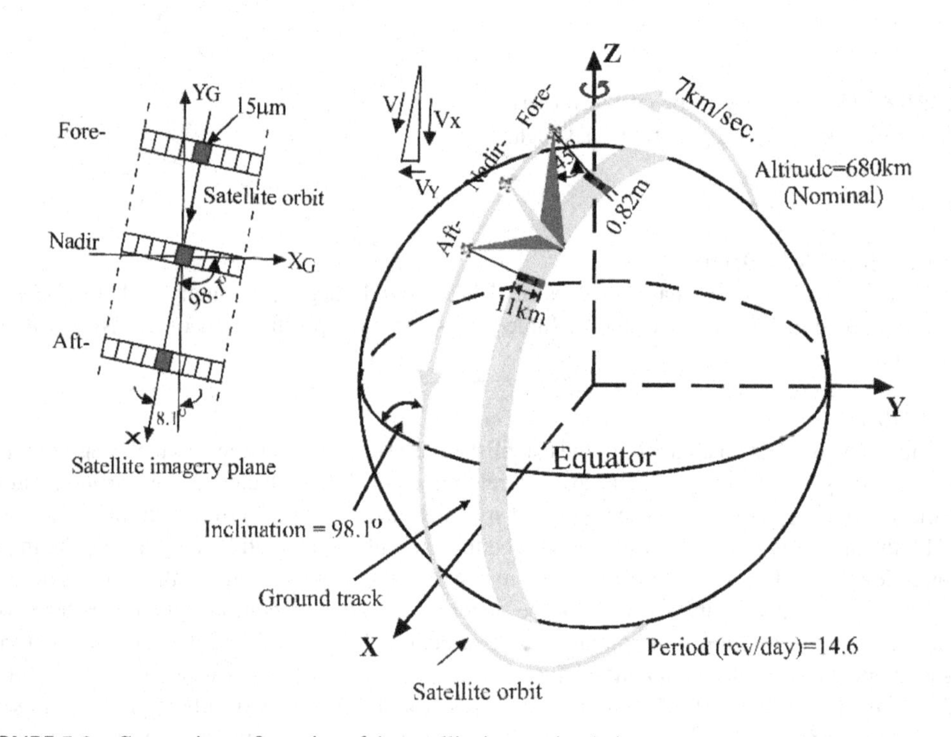

FIGURE 5.9 Geometric configuration of the satellite image simulation.

5.5.2 The Brightness on Satellite Imagery

Assuming that the coordinates of any point in a one-dimensional image plane at nadir-looking is (x_1, y_1), its corresponding ground coordinates are determined by

$$v_{Z_k} = \hat{Z}_k - Z_k\left(\hat{X}_k, \hat{Y}_k\right) \quad (5.16)$$

v_{Z_k} is residual of Z_k, $Z_k(\hat{X}_k, \hat{Y}_k)$ denotes coordinates from DEM, $\hat{X}_k$, $\hat{Y}_k$, $\hat{Z}_k$ unknown coordinates. The observation Z_k is given by the planimentric position of the point P_k and the parameters describing DEM. In case of a raster DEM, these are the coordinates of raster points and

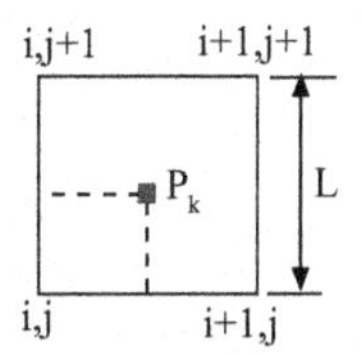

FIGURE 5.10 Bilinear brightness interpolation from the surrounding four raster points.

the algorithm for the interpolation of the height of an arbitrary point from the raster heights. If a bilinear interpolation is used, this height is calculated from the surrounding four raster points (see Figure 5.10), that is,

$$Z_k\left(\hat{X}_k, \hat{Y}_k\right) = \left(1 - C_1\right)\left(1 - C_2\right)Z_{i,j} + C_1\left(1 - C_2\right)Z_{i+1,j} + \left(1 - C_1\right)C_2 Z_{i,j+1} + C_1 C_2 Z_{i+1,j+1} \tag{5.17}$$

where $C_1 = \left(\hat{X}_k - X_i / L\right)$, $C_2 = \left(\hat{Y}_k - Y_i / L\right)$, $L = \left(X_{i+1} - X_i\right) = \left(Y_{j+1} - Y_j\right)$ is a constant of raster width. The linearization of Equation 5.18 using the initial values, X_k^0 and Y_k^0, yields

$$v_{Z_k} = a_1 \Delta X_k + a_2 \Delta Y_k + a_3 \Delta Z_k - l_k \tag{5.18}$$

where:

$$a_1 = -\frac{\left(1 - C_2^0\right)}{L} Z_{i,j} + \frac{\left(1 - C_2^0\right)}{L} Z_{i+1,j} - \frac{C_2^0}{L} Z_{i,j+1} + \frac{C_2^0}{L} Z_{i+1,j+1}$$

$$a_2 = -\frac{\left(1 - C_1^0\right)}{L} Z_{i,j} - \frac{C_1^0}{L} Z_{i+1,j} + \frac{\left(1 - C_1^0\right)}{L} Z_{i,j+1} + \frac{C_1^0}{L} Z_{i+1,j+1}$$

$$a_3 = 1.0,\ l_k = Z_k^0 - Z_k\left(X_k^0,\ Y_k^0\right),\ C_1^0 = -\frac{\left(X_k^0 - X_i\right)}{L},\ C_2^0 = -\frac{\left(Y_k^0 - Y_i\right)}{L}$$

The value $Z_k\left(X_k^0, Y_k^0\right)$ is derived from approximate values X_k^0 and Y_k^0, which change in each iteration. Therefore, the observation also varies from iteration to iteration.

When knowing the ground coordinate $\hat{X}_k, \hat{Y}_k, \hat{Z}_k$, the corresponding image coordinates (x'_1, y'_1) in the aerial image can be determined by a back-projection (from DEM to the aerial image), whose interior parameters, EOPs, and calibration parameters are available. The brightness $G(x'_1, y'_1)$ at (x'_1, y'_1) of the aerial image is assigned to that at (x_1, y_1) of the satellite image plane. In this way, the brightness of point-by-point on the one-dimensional linear array image plane can be obtained by repeating the same process above.

Based on the pushbroom imaging principle and traveling at a speed of 7000 m per second over 680 km flying height, a successive one-dimensional image is simulated along the direction of flight. In this way, a simulated scene at nadir-looking is simulated and the result is depicted in Figure 5.13. Similarly the images at fore-, and aft-looking are simulated as well and listed in Figures 5.14 and 5.15.

5.5.3 Error Analysis

The simulation of the satellite imaging process is here only for the purpose of photogrammetric positioning. In practice, Earth self-rotation, moon and planet perturbation, atmospheric refraction, and nonlinear mathematical modeling of orbital dynamics truly impact the positional accuracy of each pixel in the image plane. Besides, the accuracy of raster DEM from DLG data format also

FIGURE 5.11 Original aerial image (GSD: 0.4 m).

FIGURE 5.12 Simulated nadir-looking image (GSD: 0.82 m).

FIGURE 5.13 Simulated fore-looking image (GSD: 0.82 m).

FIGURE 5.14 Simulated aft-looking image (GSD: 0.82 m).

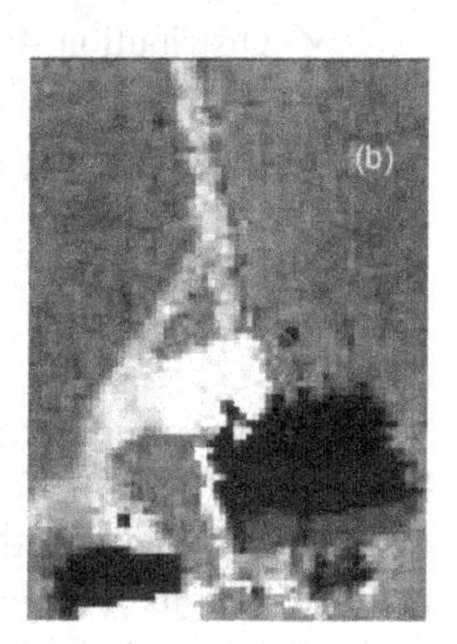

FIGURE 5.15 Feature point error and image distortion caused by DEM (a) original aerial photography with GSD of 0.41 m, and (b) simulated satellite image with GSD of 0.82 m at nadir-looking.

influences the positional accuracy of each pixel. This type of error results in the edges of some objects in satellite images not being smooth instead of rough (see Figure 5.15). On the other hand, for the downloaded DLG data from USGS 1:24,000 scale, 7.5-minute topographic quadrangle maps were derived in 1979, while the aerial photography was taken in 1997. The 18-year interval between DEM and aerial photography will probably result in a big positional error of a few features on the satellite image plane because of new building and road renewal.

5.6 POTENTIAL ACCURACY ATTAINABLE FOR GROUND POINTS OF IKONOS

5.6.1 Accuracy Assessment Based on the First Test Field

The 24 GCPs (21 high-altitude ground target points plus three higher-order control points) and 14 checkpoints at the established first test field are used for experiments. Their 2D image coordinates on satellite image planes at nadir-, aft-, and fore-looking are computed by the rigorous mathematical model of bundle adjustment, and then random errors (1 sigma) are added to the following simulated parameters:

- Position of OLs: 3 m
- Attitude of OLs: 2 arc second
- Measured image coordinates of GCPs: 0.5 pixel

The accuracy assessment of the ground points is conducted based on the assumption above.

- ***Accuracy versus various number of GCPs.*** The ground coordinates of 14 checkpoints are computed by a bundle adjustment program. Differences between the computed and known ground coordinates of 14 checkpoints are depicted in Figure 5.16 with a variation in the numbers of GCPs used. The average planimetric and vertical absolute error can reach 11 to 12 m without GCPs, and around 2.8 m with 24 GCPs. When four GCPs are used, the accuracy is largely improved. However, more than four GCPs do not contribute significantly. Thus, it can be concluded that it is not encouraged to improve the geometric accuracy by increasing the number of GCPs.
- ***Accuracy versus distributions of GCPs.*** Figure 5.17 shows the GPS network. The six groups of GCPs are chosen to form six distributions in order to examine the impact of GCP distribution on accuracy. The six distributions are:

  ✓ Distribution 1: a triangle consisting of points MAD-1, BLT-0 and H-34;
  ✓ Distribution 2: a trapezoid formed by points H-34, BLT-0, H-105 and H-106;
  ✓ Distribution 3: approximately a straight line (cross-track) formed by points H-105, M-1 and H-106;
  ✓ Distribution 4: Points S-20, H-5, S-34, MAD-1, S-41, S-57, S-63, H-6, H-7 and H-11 spreading in the area;
  ✓ Distribution 5: Points H-34, H-14, H-13, H-12 and BLT-0 approximately on a straight line (cross-track);
  ✓ Distribution 6: Points S-20, S-34, S41 and S-57 approximately on a straight line (cross-track).

The accuracy of ground coordinates of 14 checkpoints versus 6 distributions of GCPs is depicted in Figure 5.18. It can be found that the strength of GCP distributions affects the planimetric and vertical accuracy significantly. For example, Distributions 1 and 3 have the same number of GCPs, but not distributions. The geometric accuracy for Distribution 1 achieves 1.84 m and 3.43 m in X and Y, 3.21 m in Z, while Distribution 3 has 53.4 m and 8.11 m in X and Y, and 5.88 m in Z. Distributions 2 and 6 present a similar situation. It should be noted that GCPs distributed on a straight line across

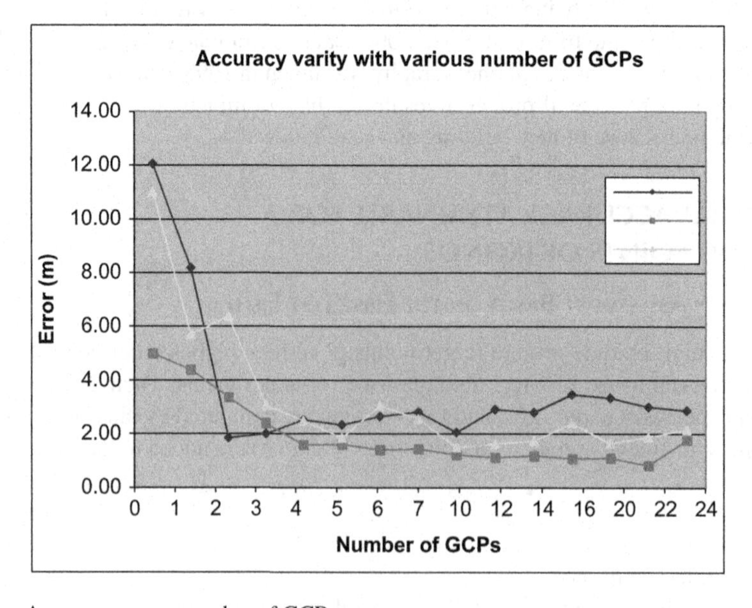

FIGURE 5.16 Accuracy versus number of GCPs.

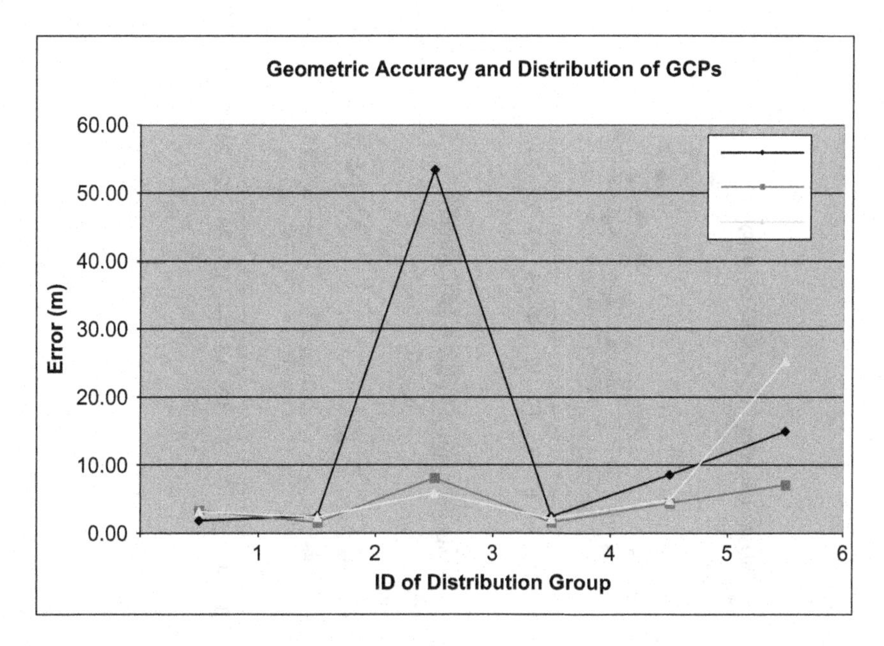

FIGURE 5.17 Accuracy versus GCP distribution.

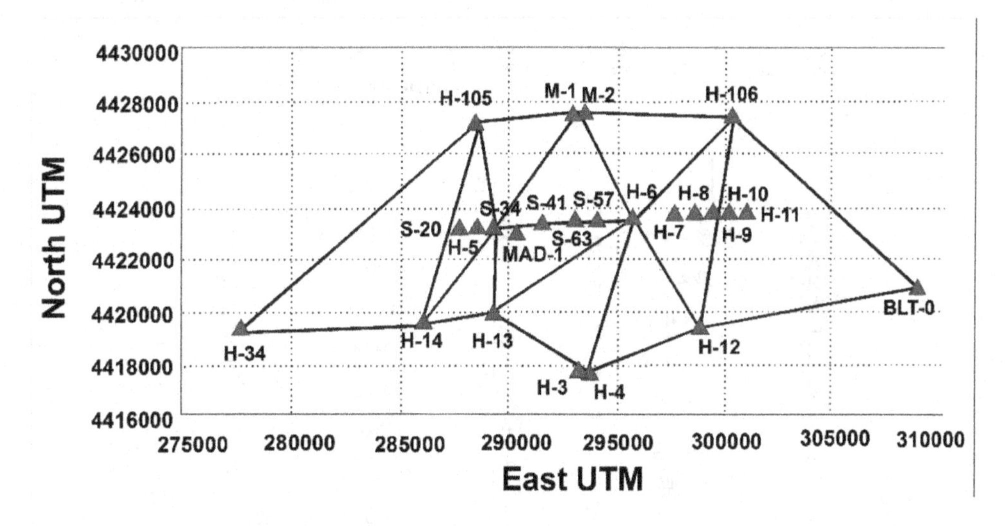

FIGURE 5.18 The distribution of GCPs at Madison Test Range.

the track constitute a weak geometric configuration (Distributions 3, 5, and 6). Thus, GCP distribution is a critical key to achieve high accuracy.

- ***Accuracy versus image measuring errors of checkpoints.*** When measuring image coordinates, errors are unavoidable. The impact of these errors on ground coordinates should be assessed. Suppose that image coordinate measurement errors range from 6 μm (0.5 pixel) to 24 μm (2 pixel), and all other parameters are noise-free. The ground coordinates accuracy of 14 checkpoint points versus errors in image coordinates is depicted in Figure 5.19. The result demonstrates that large measurement errors in image coordinates affect accuracy of ground coordinates. Indeed, the planimetric coordinate errors increase 0.5 m for each error of 0.5 pixels in image coordinates.

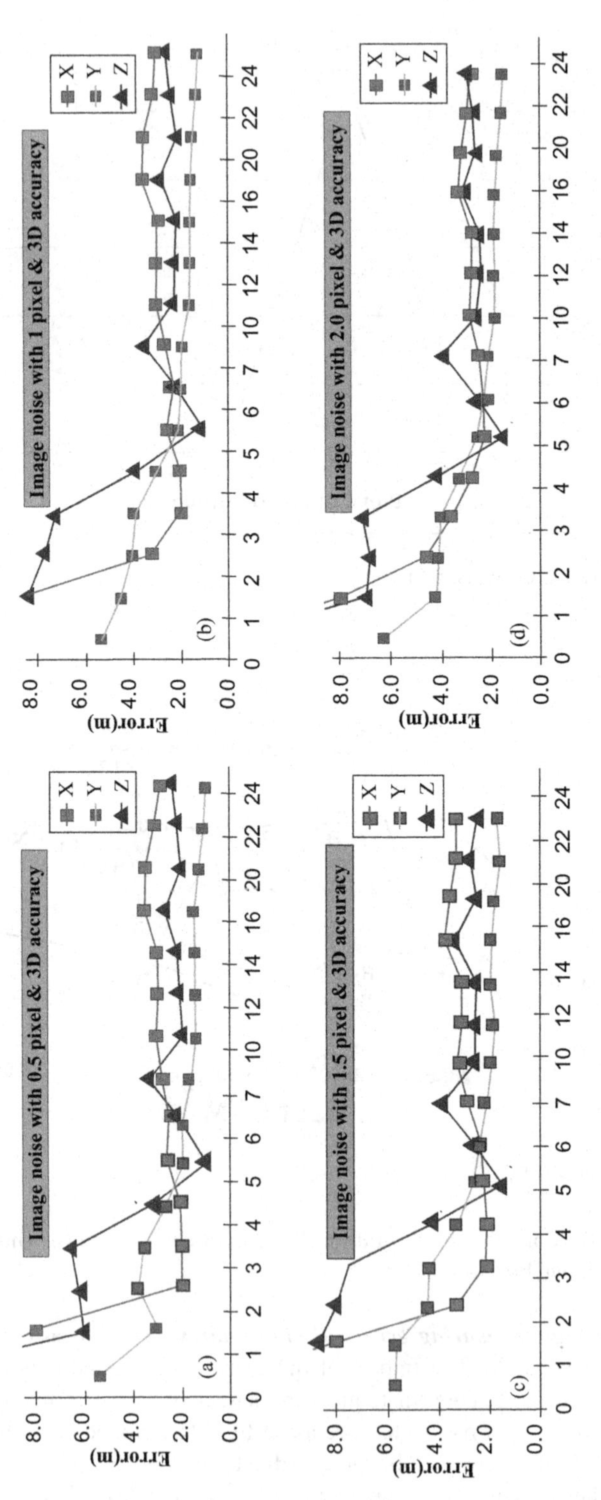

FIGURE 5.19 Accuracy versus errors in image coordinates.

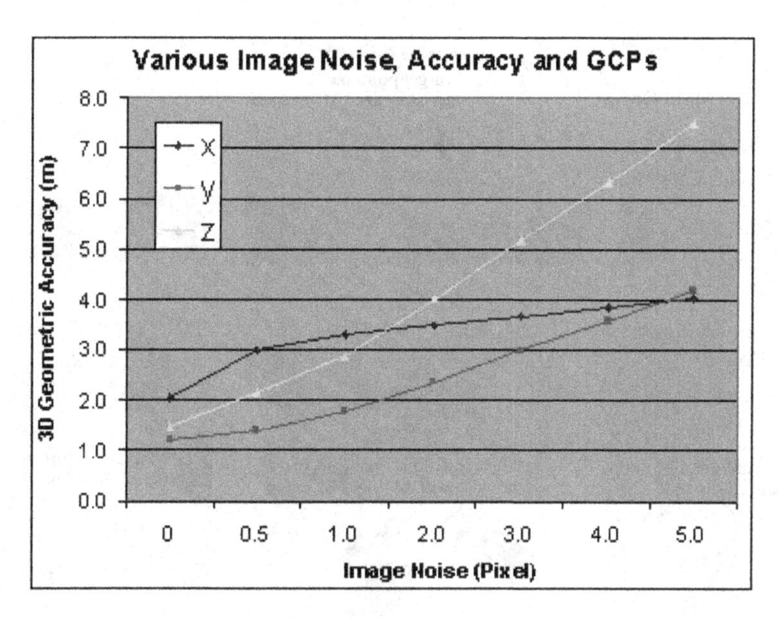

FIGURE 5.20 Accuracy versus errors in image coordinates.

- ***Accuracy versus errors of image coordinates of GCPs.*** Under the same condition as above, the impact of image coordinate errors of GCPs on ground coordinate accuracy is examined as well. The errors of ground coordinates for 14 checkpoints are shown in Figure 5.20. When image measuring errors are 5 pixels, z coordinate error reaches 7.7 m and x and y errors are around 4 m, even though all 24 GCPs are taken as control information. Obviously, accurately locating and measuring GCPs is helpful in improving accuracy.

5.6.2 Accuracy Assessment Based on the Second Test Field

In the Second Test Field, all target points and feature points on aerial image AIMS 1185 can clearly be recognized. It means that all image coordinates can be located. In order to undertake further processing, each target point and feature point on the aerial image is marked by a cross of 5 by 5 pixels (see Figure 5.21) in order that all target points and feature points can still be recognized in the simulated satellite images. Unfortunately, the target point 212 and feature points 3, 7, and 20 cannot be located correctly because of errors in the out-of-date DEM (created in 1979), which would not correspond to the aerial image taken in 1997. The following experiments conducted are based on the above condition. No random noise is added to any parameters.

- ***Accuracy versus various numbers of GCPs.*** The ground coordinates of 18 feature points are computed by the bundle adjustment method. The differences between the computed and ground coordinates measured by GPS are listed in Table 5.3. The averages of the RMSs can reach 11 m in x, 38 m in y, and 13 m in z without GCPs, and around 2.7 m in x, 1.6 m in y, and 1.99 m in z with 13 GCPs. When three spare GCPs, which cover the whole of image 1185, are used, the accuracy is largely improved. However, more than seven GCPs do not contribute to the accuracy significantly. Thus, it is not encouraged to improve the geometric accuracy by increasing the number of GCPs. This result is similar to the above experimental results.
- ***Accuracy versus distributions of GCPs.*** Six distributions constructed by six different point IDs are formed in order to examine the impact of GCP distribution on accuracy. The accuracy of 18 feature points corresponding to distribution are listed in Table 5.4.

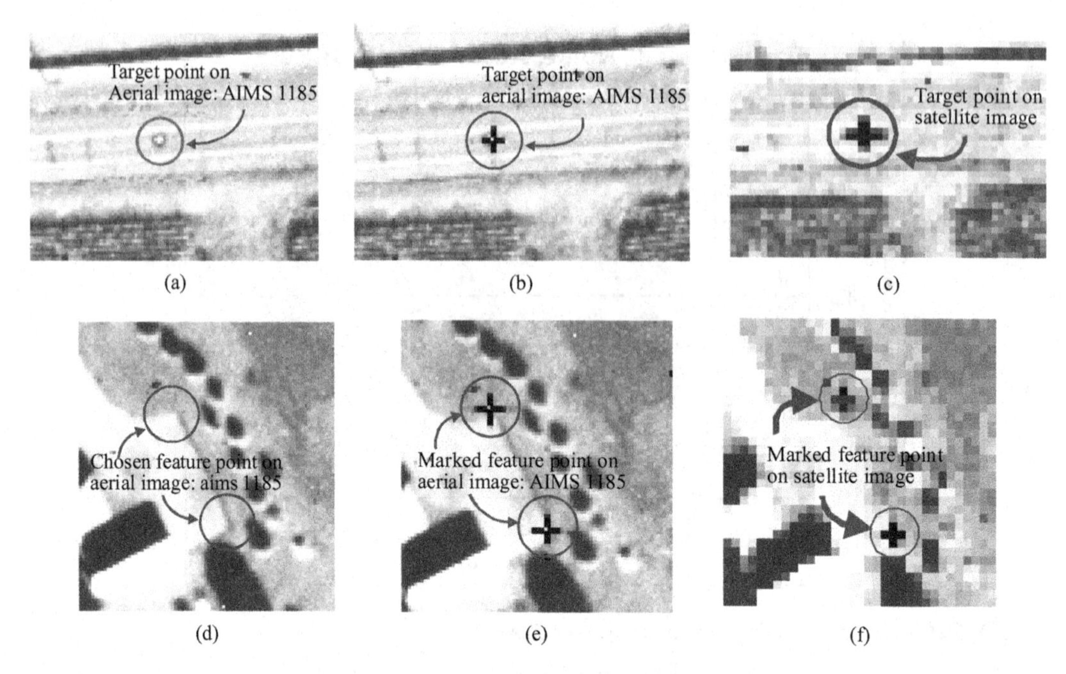

FIGURE 5.21 Target point and feature point as well as their marks on aerial image AIMS 1185 and simulated satellite image.

TABLE 5.3
Experimental Result of Accuracy Versus Number of GCPs in the Second Test Field

Number	Point ID (see Figure 5.7)	RMS_x	RMS_y	RMS_z
0	–	10.67	38.55	13.41
1	(324)	4.97	30.63	2.35
2	(324, 209)	5.07	30.63	2.11
3	(324, 209, 327)	4.89	1.85	3.98
7	(322, 209, 327, 1, 323, 407, 404)	3.18	1.58	2.06
13	(above 7 points + 405, 406, 210, 212, 2, 325)	2.74	1.57	1.99

The accuracy versus various distributions of GCPs is depicted in Figure 5.11. it can be found that the strength of GCP distributions affects the planimetric and vertical accuracy significantly. For example, Distributions 1 and 3 have the same number of GCPs, but not distributions. The accuracy for Distribution 1 achieves 1.84 m and 3.43 m in X and Y, 3.21 in Z, while Distribution 3 has 53.4 m and 8.11 m in X and Y, and 5.88 m in Z. Distributions 2 and 6 present a similar situation. It should be noted that GCPs distributed on a straight line across the track constitutes a weak geometric configuration (Distributions 3, 5 and 6). Thus, GCP distribution is critical to achieving high accuracy.

TABLE 5.4
Experimental Result of Accuracy Versus Distribution of GCPs in the Second Test Field

Point ID for Distribution of GCP (see Figure 5.7)	RMS_x	RMS_y	RMS_z
327, 404	6.77	39.06	4.58
209, 327, 322	3.82	1.79	2.19
407, 209, 404	25.46	15.96	34.47
407, 326, 1	5.74	1.74	4.44
327, 322, 407, 404	3.42	1.59	2.07
322, 323, 324, 325, 326, 327	75.92	1.91	36.07

5.7 AIRBORNE HRSC AND SPACE SHUTTLE MOMS-2P THREE-LINE SENSOR

5.7.1 Data Processing and Accuracy Evaluation for HRSC Imaging Data

- ***The HRSC data set.*** In the mission, DLR provided the data set including five strips (nadir channel, stereo aft-, and fore-panchromatic (S1, S2), photometry aft- and fore-panchromatic), 109 GCPs in WGS-84, and EOPs for each linear array images. The airborne sensor flew from southeast to northwest with a flying height of 2500 m. The images are recorded in the VICAR format developed by JPL/NASA. In this test field, 69 GCPs are covered. Unfortunately, only 49 GCPs are available since the image coordinates of other GCPs cannot be identical in the image plane.
- ***Image coordinates.*** The geometric relationship between image coordinate and reference coordinate system is drawn in Figure 5.2. DLR provided the following data: (1) screen coordinate and image coordinate, (2) principal point offsets for S1 and S2, and (3) distortion rectification coefficients, including CCD line curvature and lens distortion, image coordinates of GCPs and checkpoints.
- ***Bundle adjustment and accuracy assessment.*** The geometric conditions of HRSC are quite similar to those of other pushbroom scanners such as IKONOS and MOMS-2P. Therefore, the bundle adjustment model described above is appropriate. The camera, reference, and ground coordinate systems are defined in Figure 5.2. Table 5.5 indicates the attainable accuracy of 49 checkpoints versus the number of GCPs ranging from 0 to 49.

The experimental results demonstrate that increasing the number of GCPs does not improve the geometric accuracy even slightly. The interior accuracy and exterior RMS appear to be the same. This is because the EOPs of each linear image and the image coordinates of GCPs and checkpoints provided by DLR are bundle adjustment results already. A further bundle adjustment does not improve the result, but this fact demonstrates that our bundle adjustment program is correct.

Another experiment concerns the accuracy improvement versus order of polynomial and number of orientation images. Various combinations were tried and the results are shown in Table 5.6. The results show that more than 10 and fewer than 10 OLs will not be helpful to the interior accuracy of ground points and will have little effect on the exterior accuracy of ground points. The order of polynomial depends on the real flight path. The experimental results show that various orders of polynomial have little effect on the accuracy significantly. This may be because the flight path of this mission is an approximate straight line (Li and Zhou 2002).

TABLE 5.5
Accuracy of Ground Points Versus Number of GCPs

GCP	Check Points	σ_0 (μm)	σ_x (cm)	σ_y (cm)	σ_z (cm)	RMS_x (cm)	RMS_x (cm)	RMS_x (cm)
0	49	6.990	15.140	7.324	18.010	15.144	8.276	18.758
1	49	6.952	16.025	8.230	18.654	15.140	7.324	18.010
2	49	6.914	15.939	8.186	18.553	15.140	7.324	18.010
3	49	6.877	15.853	8.142	18.454	15.140	7.324	18.010
4	49	6.840	15.769	8.099	18.355	15.140	7.324	18.010
5	49	6.805	15.686	8.059	18.259	15.140	7.324	18.010
49	49	5.628	12.973	6.663	15.101	15.140	7.324	18.009

TABLE 5.6
Accuracy of Ground Points Versus the Number of OLs and Order of Polynomial

Order of Polyn.	# of OLs	σ_0 (μm)	σ_x (cm)	σ_y (cm)	σ_z (cm)	RMS_x (cm)	RMS_x (cm)	RMS_x (cm)
1	5	11.756	27.109	13.924	31.561	15.143	7.3253	18.0122
	9	8.817	20.326	10.439	23.661	15.140	7.3243	18.0093
	10	**8.632**	**19.899**	**10.220**	**23.165**	**15.140**	**7.3241**	**18.0089**
	13	9.768	22.517	11.564	26.211	15.140	7.3242	18.0091
	21	10.794	24.880	12.778	28.961	15.140	7.3242	18.0091
	29	11.481	26.463	13.590	30.802	15.140	7.3242	18.0091
	73	13.384	30.846	15.841	35.904	15.142	7.3242	18.0095
2	5	17.037	39.440	20.316	137.653	15.021	7.481	32.714
	9	9.322	21.535	11.076	25.338	15.026	7.542	18.054
	10	**8.671**	**20.031**	**10.303**	**23.569**	**15.026**	**7.542**	**18.054**
	13	9.407	21.726	11.175	25.565	15.026	7.542	18.054
	21	9.845	22.735	11.694	26.754	15.026	7.542	18.054
	29	10.497	24.240	12.468	28.526	15.026	7.542	18.054
	73	11.565	26.704	13.734	31.424	15.026	7.542	18.054
3	5	5.118	11.811	6.069	13.756	15.141	7.324	18.00
	9	4.660	10.748	5.521	12.513	15.143	7.324	18.00
	10	**4.491**	**10.357**	**5.320**	**12.058**	**15.141**	**7.324**	**18.00**
	13	5.011	11.555	5.945	13.451	15.141	7.324	18.00
	21	5.397	12.441	6.389	14.482	15.141	7.324	18.00
	29	5.623	12.973	6.663	15.101	15.141	7.324	18.00
	73	6.400	14.751	7.576	17.170	15.141	7.324	18.00

5.7.2 Data Processing and Accuracy Evaluation for MOMS-2P

- ***Test field.*** A typical scene collected from a flying height of 39 km covers an area of about 178 km long and 50 km wide, comprised of scenes 27 to 30 (see Figure 5.22), from southeast Germany to about 160 km beyond the Austrian border. The ground pixel size of the imagery is 5.9 m and 17.7 m. A base-to-height-ratio of 0.8 is expected. The GSD for the stereo lenses 6 and 7 is about 18 m, and the image strip has an extent of 2976 pixels.
- ***Calibration data.*** The geometric calibration of the MOMS imaging system was performed at the laboratories of the German aerospace company DASA, where MOMS was developed and is manufactured. A rigorous model of the geometry of each linear sensor is composed of five parameters (see Table 5.7). The model and the derivation of the model parameters from the calibration measurements are comprehensively described in Kornus (1996, 1997).
- ***Navigation data.*** MOMS-2P has its own navigation system, MOMS-NAV, consisting of a Motorola Viceroy GPS receiver with dual antennas and a LITEF gyro system. The eight OLs cover the whole strip. The distance between the OLs was 3330 image lines, corresponding to 8.2 seconds' flight time.
- ***Ground control points and checkpoints.*** The 10 GCPs and the 24 checkpoints within the test field and along the strip are chosen. The ground coordinates of GCPs and checkpoints were

TABLE 5.7
The Calibrated Camera Parameters in-Lab and In-Flight for MOMS-2P (Kornus 1996, 1997)

	Lab-calibrated Parameters				In-flight-calibrated Parameters by Photogrammetry			
	HR5A	HR5B	ST6	ST7	HR5A	HR5B	ST6	ST7
f (mm)	660.256	660.224	237.241	237.246	–	–	237.1936	237.2509
x_0 (pixel)	0.1	0.2	−7.2	−0.5	–	–	46.7152	−10.5293
y_0 (pixel)	−0.4	0.1	8.0	19.2	–	–	−110.4370	−166.339
K_c (pixel)	−0.3	−0.4	−1.1	1.7	–	–	0.01111	−0.01153
k (mdeg)	−2.9	5.4	−1.5	−1.4	–	–	–	–

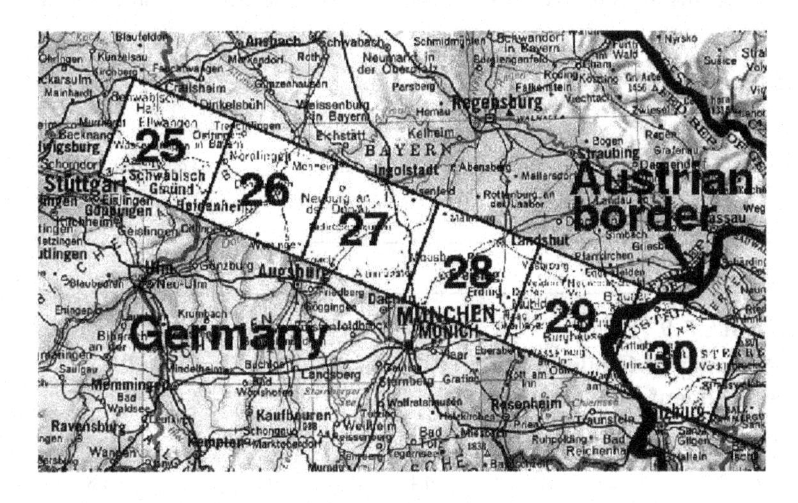

FIGURE 5.22 The test area (Kornus and Lehner 1998).

obtained by topographic maps of scale 1:50,000 with an accuracy of 1.5 m in x, y, and z, which are provided by the University of Stuttgart. They are in the fourth zone of Gauss-Krueger coordinate system.

- ***Interior orientation.*** For each linear sensor, there are a focal length f, principal point offset (x_o, y_o), and a curvature parameter K_c. Thus, the interior orientation includes transforming the screening coordinate to image coordinate and correcting all distortions. In this experiment, the interior parameters are treated as constants and no self-calibration is performed in the bundle adjustment.
- ***Access of orientation parameters at OLs from fore- and aft-looking strips.*** The navigation data are available only at nadir-looking strips. The fore- and aft-looking strips access the navigation data through the imaging geometry. In this data set, the first line belongs to the first line of scene 25 of the nadir-looking strip (5A and 5B). To get the correct orientation of aft- and fore-looking strips, it is necessary to add/subtract 8600 lines respectively (see Zhou and Li 2000; Li and Zhou 2002).
- ***Transformation from Gauss-Krueger to WGS-84.*** The provided navigation data of OLs are in the geocentric WGS-84 coordinate system. However, the coordinates of GCPs and checkpoints are in the Gauss-Krueger coordinate system with (x, y, h). We performed a coordinate transformation from Gauss-Krueger coordinate system to WGS-84. Usually, geoid undulations have to be considered in transforming the height h above the geoid to ellipsoidal heights H with respect to WGS-84.
- ***Input data information.*** Table 5.8 gives an overview of all observations and their *a priori* standard deviations and weight assignment. These were fed into the bundle adjustment. For the entire segment of the orbit, eight OLs were employed (the distance between the OLs was 3330 image lines). This proved to be sufficient to model the temporal course of the EOPs.
- ***Adjustment computation and accuracy.*** In the bundle adjustment, the EOPs are estimated only at OLs, while those between OLs are modeled by third-order polynomials. The parameters and navigation data are treated as uncorrelated. Table 5.9 shows that accuracy of 10.4 m for X, 9.8 m for Y, and 13.8 m in Z was achieved as verified by 24 independent checkpoints when using 10 GCPs measured with 0.5 pixel precision (equal to 8.5 m in object space). This result is close to the result reported by Fritsch et al. (1998) at the University of Stuttgart with 11 m in X and Y, and 14 m in Z.

TABLE 5.8
Observations and Weights Introduced into the Bundle Adjustment

Observations	Type	Prior σ	Weight Assigned
24 checkpoints	Image coord.	0.5 pixel	1.0
10 GCPs	Image coord.	0.5 pixel	1.0
3 × 8 position	Ext. orient.	6.0 m	2.6e–6 (mm²/m²)
3 × 8 attitude	Ext. orient.	0.11e–2°	100.0 (mm²/deg²)

TABLE 5.9
Accuracy of Ground Coordinates Calculated by Our Bundle Adjustment from Checkpoints

GCP	Checkpoints	RMS_x (m)	RMS_y (m)	RMS_z (m)
10	24	10.4	9.8	13.8

5.8 CONCLUSIONS

The mathematical model of 3D coordinate determination from linear array stereo imaging system has been derived. Two real data sets acquired by linear array stereo imaging systems (MOMS-2P and HRSC) and a simulated satellite, IKONOS data, have been used for checking the model. The results demonstrate that the first version of the developed bundle adjustment software system for creation of a digital surface model (DSM) from a linear array stereo imaging data is appropriately designed and implemented. The model and system were first used in NOAA due to a contracted project. On the other hand, the simulation study provides a tool for examining overall accuracy potential of the 1 m resolution satellite imagery from IKONOS. The suggestions for improving accuracy of the ground points are:

- One to four GCPs are recommended for computing 3D coordinate with reduced costs and without losing the accuracy.
- Distribution of GCPs along a straight line does not increase accuracy of 3D coordinates of ground points.
- Estimated ground point accuracy through the simulation is 3 m (X and Y) and 2 m (Z) with 24 GCPs, and 12 m (X and Y) and 12 m (Z) without GCPs.
- Higher-precision navigation data are largely helpful to the accuracy of ground points in the case of weak distributions of GCPs and/or few or no GCPs. A well-spread distribution of even a few GCPs is still more beneficial to accuracy improvement than a dense but poorly spread distribution.

REFERENCES

Albertz, J., Ebner, H., and Neukum, G., The HRSC/WAOSS Camera Experiment on the MARS'96 Mission – A Photogrammetric and Cartographic View of the Project, *Int. Arch. of Photogrammetry and Remote Sensing*, Vol. 31, Part B4, pp. 58–63, 1996.

Corbley, K., One Meter Satellites: Practical Applications by Spatial Data Users, *GeoInfo System*, Vol. 6, No. 7, pp. 28–33, 1996,

Ebner, H., Kornus, W., and Ohlhof, T., A Simulation Study on Point Determination for the MOMS-02/D2 Space Project Using an Extended Functional Model, *Int. Arch. of Photogrammetry and Remote Sensing*, Vol. 29, Part B4, pp. 458–464, 1992.

Ebner, H., Kornus, W., and Strunz, G., A Simulation Study on Point Determination Using MOMS-02/D20 Imagery, *Int. Journal of PE & RS*, Vol. 57, No. 10, pp. 1315–1320, 1991.

Ebner, H., and Strunz, G., Combined Point Determination Using Digital Terrain Models as Control Information, *Int. Arch. of Photogrammetry and Remote Sensing*, Vol. 27, Part B11, pp. III/578–587, 1988.

Fraser, C., and Shao, J., Exterior Orientation Determination of MOMS-02 Three-Line Imagery: Experiences with the Australian Testified Area, *Int. Archives of P&RS*, Vol. XXXI, Part B3, pp. 207–214, 1996.

Fritsch, D., et al. Improvement of the Automatic MOMS02-P DTM Reconstruction, *Int. Arch. of Photogrammetry and Remote Sensing*, Vol. 32, Part 4, GIS – Between Visions and Applications, Stuttgart, Germany, pp. 170–175, 1998.

Fritz, L.W., Commercial Earth Observation Satellite, *Int. Arch. of Photogrammetry and Remote Sensing*, Vol. XXXI, Part B4, pp. 273–282, 1996.

Gupta, R., and Hartley, R.I., Linear Pushbroom Cameras, *IEEE PAMI*, Vol. 19, No. 9, September, pp. 963–974, 1997.

Hofmann, O., Nave, P., and Ebner, H., DPS – A Digital Photogrammetric System for Producing Digital Elevation Models and Orthophotos by Means of Linear Array Scanner Imagery. *Int. Arch. of Photogrammetry and Remote Sensing*, Vol. 24, Part B3, pp. 216–227, 1982.

Kiefer, R., *Remote Sensing and Image Interpretation*, 3rd edition, John Wiley & Sons, Inc., New York, 1997, pp. 486.

Kornus, W., *MOMS-2P Geometric Calibration Report (Version 1.1) – Results of Laboratory Calibration*, Institute of Optoelectronics, DLR, Wessling, Germany, 1996.

Kornus, W., *MOMS-2P Geometric Calibration Report (Version 1.3) – Results of Channel 5A/5B Registration*, Institute of Optoelectronics, DLR, Wessling, Germany, 1997.

Kornus, W., and Lehner, M., Photogrammetric Point Determination and DEM Generation Using MOMS-2P/PRIRODA Three-Line Imagery, *Int. Arch. of Photogrammetry and Remote Sensing*, Vol. 32, Part 4, Stuttgart, Germany, pp. 321–328, 1998.

Li, R., and Zhou, G., Photogrammetric Processing of High-Resolution Airborne and Satellite Linear Array Stereo Images for Mapping Applications, *Int. J. of Remote Sensing*, Vol. 23, No. 20, pp. 4451–4473, 2002.

McGlone, C., Block Adjustment of Linear Pushbroom Imagery with Geometric Constraints, *Int. Arch. of Photogrammetry and Remote Sensing*, Vol. XXXII, Part B2, Cambridge, London, UK. pp. 198–205, 1998.

Wewel, F., Determination of Conjugate Points of Stereoscopic Three Line Scanner Data of MARS'96 Mission. *Int. Arch. of Photogrammetry and Remote Sensing*. Vol. XXXI, Part B3. Vienna, pp. 936–939, 1996.

Wu, J., *Geometrische Analyse fuer Bilddaten stereoskopischer Dreifach-Linearzeilenabtaster*, Wiss. Arbeiten der Fachrichtung Vermessungswesen der Univ. Hannover, Heft Nr. 146, 1986.

Zhou, G., and Li, R., Accuracy Evaluation of Ground Points from High-Resolution Satellite Imagery IKONOS, *Photogrammetry Engineering & Remote Sensing*, Vol. 66, No. 9, pp. 1103–1112, 2000.

6 Urban 3D Building Extraction Through LiDAR and Aerial Imagery

6.1 INTRODUCTION

To meet the increasing requirement for accurate three-dimensional models of buildings, continuous updates for urban planning, cartographic mapping, and civilian and military emergency responses, the efforts for building extraction using a variety of data sources have been made in the past in remote sensing, photogrammetry, image processing, and computer vision communities. The previous methods can be grouped into three categories in terms of the employed data sources. The first category uses optical imagery, such as aerial imagery and high-resolution satellite imagery (e.g., IKONOS). The second category utilizes active imagery data, such as LiDAR data and radar imagery. The third category integrates both high-resolution optical and active imagery data.

For the first category, the earliest effort can be traced back to the 1980s. For example, McKeown (1984, 1989) presented a knowledge-based interpretation method to recognize buildings from aerial imagery. Afterwards, Matsuyama (1987) developed a so-called Expert System (ES) for building extraction. Mohan and Nevatia (1989) proposed a matching multiple-image technique to extract 3D urban buildings. Lee and Lei (1990) extracted house roofs through assumed gray matter on a house's surface, and matching house corners and edges. Kim and Muller (1996), Paparoditis et al. (1998), Fischer et al. (1998), Baillard and Maître (1999), and Mayer (1999) also presented their methods. In the twenty-first century, methods for the building extraction from optical imagery are still continuously explored. For example, Cord et al. (2001), Jaynes et al. (2003), and Fradkin et al. (2001) presented the methods for urban buildings using multiple aerial images in dense urban areas. Jung (2004), Khoshelham et al. (2010), and Suveg and Vosselman (2004) proposed detecting complex buildings from multitemporal aerial stereopairs. Peng et al. (2005) used snake models for building detection from aerial images. Zebedin (2006), Besnerais et al. (2008), Sirmacek and Unsalan (2009), and Ahmadi et al. (2010) presented their methods to extract buildings. These early methods attempted to extract the boundary of a building to identify the buildings, but separating building boundaries from non-building objects, such as parking lots, is difficult. It is especially difficult to extract the complete boundaries of a building from a single optical image, due to unavoidable occlusions, poor contrasts, shadows, and disadvantageous image perspectives (Ekhtari et al. 2009).

The second category employs active imagery data, such as microwave imagery (Simonetto et al. 2005) and LiDAR data (Gamba and Houshmand 2002). Particularly, the emergence of the LiDAR technique in the middle of 1990s resulted in considerable efforts for 3D building extraction in the past decades. The methods can be grouped into two categories (Zhou et al. 2004): the classification approach and the adjustment approach. The classification approach detects the ground points using certain operators designed based on mathematical morphology (Morgan and Tempfli 2000), terrain slope (Axelsson 1999), or local elevation difference. The refined classification approach uses a Triangulated Irregular Network (TIN) data structure (Axelsson 1999) and iterative calculation (Sithole and Vosselman 2004) to consider the discontinuity of the LiDAR data or terrain surface.

The purpose of these methods is to first form a grid LiDAR elevation into a depth imagery, and then use image segmentation techniques to detect building footprints. The adjustment approach essentially uses a mathematical function to approximate the ground surface, which is determined in an iterative least squares process, while the outliers of non-ground points are detected and eliminated. For example, Pu and Vosselman (2009) proposed a knowledge-based reconstruction of building models from laser scanning data. Rutzinger et al. (2009) made an experimental comparison and analysis for evaluating several techniques of building extraction. Zhang et al. (2006) and Sampath and Shan (2007) separated the ground and the non-ground LiDAR measurements using a progressive morphological filter. Building points were labeled via a region-growing algorithm, and the building boundaries were derived by connecting the boundary points. The major shortcoming of applying single active data is that texture and boundary information cannot be effectively reflected.

The third category uses both optical and active data, since the building extraction from only monocular imagery (e.g., optical imagery, or LiDAR data) cannot reach a satisfactory result, due to the strengths and weaknesses of each data source. Therefore, efficient fusion of different types of data sources can utilize the natural complementary properties of each data source, and compensate for the weaknesses found from one data source to another. For instance, Zhou et al. (2004), Zhou (2010), Hermosilla et al. (2011), Kabolizade et al. (2010), and Yu et al. (2009) presented the method for extraction of urban houses and road networks through the integration of LiDAR and aerial images. Schenk and Csatho (2002) and Habib et al. (2005) proposed feature-based fusion of LiDAR data and aerial imagery to obtain a better building surface description. Sohn and Dowman (2007) focused on the exploitation of synergy of IKONOS multispectral imagery information combined with a hierarchical segmentation of LiDAR DEM. Rottensteiner et al. (2007) proposed a method consisting of building detection, roof plane detection, and the determination of roof boundaries. Fujii and Arikawa (2002) and Hu et al. (2007) proposed integrating LiDAR, aerial images, and ground images for complete urban building modeling. O'Donohue et al. (2008) combined thermal-LiDAR imagery for the extraction of urban man-made objects. Zabuawala (2009), Wang and Neumann (2009), Mastin et al. (2009), and Dong et al. (2008) suggested automatic registration of LiDAR and optical imagery of urban scenes for the extraction of houses and roads.

Recently, the methods of the third category have been widely used since the efficient fusion of the two types of data sources can compensate for the weaknesses of one data type or the other. However, the most current efforts have still been focusing on the study of the "gray-level" fusion, with few efforts on utilization of the geometric or chromatic features of buildings to support the deep data fusion for building extraction. To confront these issues, this chapter describes an aspect graph-driven method to explore how the features of houses (e.g., geometry, structures, and shapes) are utilized for the data fusion between aerial imagery data and LiDAR data.

6.2 PRINCIPLE OF ASPECT CODE AND CREATION OF ASPECT CODE DATABASE

The concept of aspects and aspect graph techniques was first presented by Biederman (1985), and then applied by Dickinson et al. (1992) for 3D shape recovery of industrial objects. This chapter extends the early idea through developing aspects and aspect graphs of urban houses.

6.2.1 Selection of the 3D Primitives (Houses)

Since the structures of houses in urban areas are intricate, the initial paradigm only considered nine types of houses with flat roofs (Figure 6.1) and two types of houses with convex roofs (Figure 6.2) as 3D house primitives.

6.2.2 Creation of 2D Aspects

Dickinson et al. (1992) employed an aspect to represent the projection of an entire 3D industrial object, but this representation is not appropriate for 3D houses due to occlusion caused by the

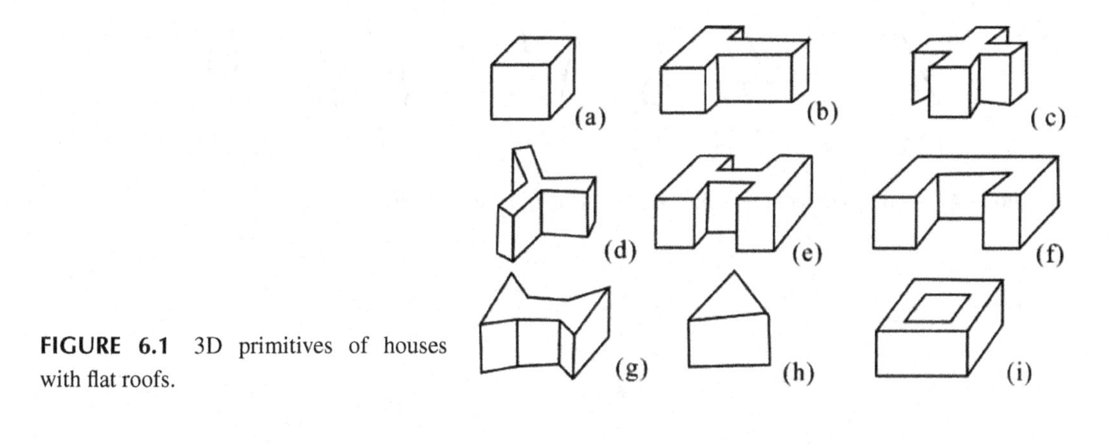

FIGURE 6.1 3D primitives of houses with flat roofs.

FIGURE 6.2 3D primitives of houses with convex roofs.

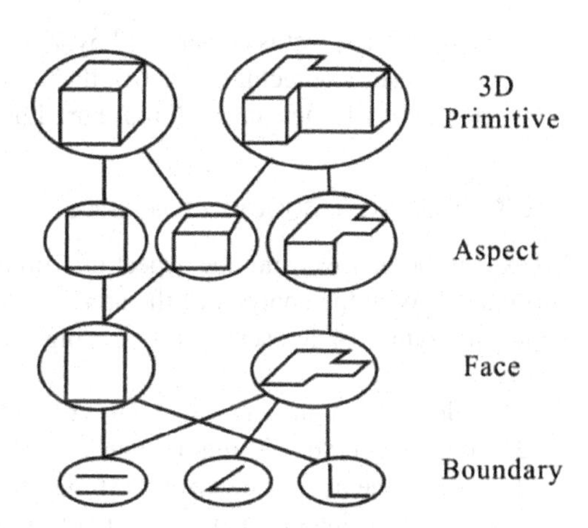

FIGURE 6.3 Hierarchical aspect representation.

perspective projection in aerial imagery. This chapter expands this method through the use of an aspect to represent the set of topological distinct views of a 3D house primitive. The basic idea is that an aspect represents either an entire projection or part of a projection of a 3D house primitive. Each aspect consists of a set of 2D faces, which correspond to the surface of a primitive. An object can be modeled by a set of aspects. The relationship between 2D aspects and 3D primitives is depicted through an aspect hierarchical graph, which consists of three layers on the basis of the faces' appearance in the aspect set (Figure 6.3).

- ***Boundary set.*** A boundary set represents all subsets of lines and curves composing the faces. A complete set of nine boundaries is listed in Figure 6.4 on the basis of the possible appearances of the selected 3D house primitives, in which the non-linear edges of houses are considered. The nine boundary set represents non-accidental contour relations such as connecting manner, parallelism, and symmetry occurring within a face.
- ***Faces.*** Faces represent the set of the projected surfaces of houses in the aerial imagery such as contour boundary and the relationship between the contours. Each face differs in the number of contours, the types of contours, or non-accidental relationship between the contours. Houses are segmented into many varied faces. A complete set of ten faces is identified to represent the chosen 3D house primitives (Figure 6.5).

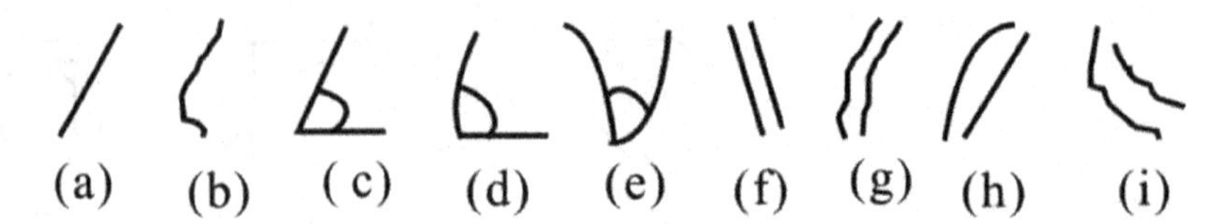

FIGURE 6.4 A complete set of nine boundaries and their aspect coding regulation.

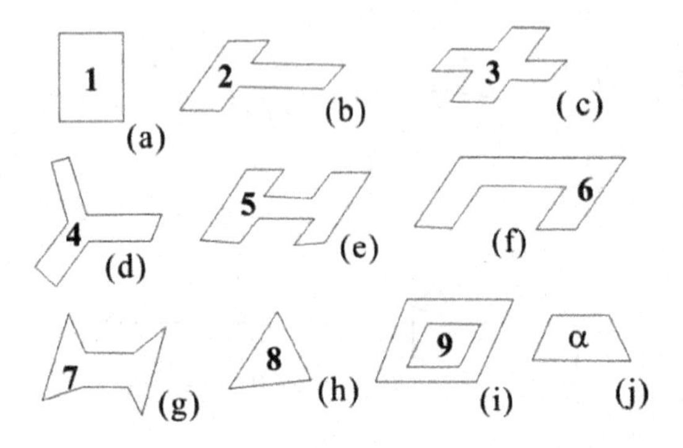

FIGURE 6.5 A complete set of ten faces for the houses selected in Figures 6.1 and 6.2, of which the algebraic number is the assigned code of the aspect.

- ***Aspect.*** An aspect is constructed by a group of faces. The aspects of houses can be formed by any surface connecting with a wall. An aspect also describes the relationship of face adjacencies, and indicates the contours shared by adjacent faces.

6.2.3 Coding Regulation for Aspect

To extract the houses using the aspect interpretation method, an aspect coding regulation has to be formulated. With the analysis of the apparent aspects in Section 2.2, nine types of contour sets, as shown in Figure 6.4, are selected for coding. The coding regulation is formulated as follows:

1. Code of a straight line is **a** (Figure 6.4a).
2. Code of a curve is **b** (Figure 6.4b).
3. Code of the intersection of two straight lines is **aa** (Figure 6.4c).
4. Code of the intersection of a straight line and a curve is **ab** (Figure 6.4d).
5. Code of the intersection between two curves is **bb** (Figure 6.4e).
6. Code of two parallel straight lines is **a-a** (Figure 6.4f).
7. Code of two parallel curves is **b-b** (Figure 6.4g).
8. Code of a straight line opposite to another curve is **a-b** (Figure 6.4h).
9. Code of a curve opposite to another curve is **b-b** (Figure 6.4i).

All the basic aspects in Figure 6.4 are labeled through the codes formulated above. Once these basic aspects are coded, their codes are archived in a so-called *codebase* for use of the aspect interpretation, which will be described in Section 2.7. Using these basic aspects and their codes above, a complex house can be coded. For example, for the house in Figure 6.1b, its probable projections are depicted in Figure 6.6, due to varying the positions of viewpoint. As observed, only the house's roof needs to be considered for aspect coding, since all the walls of a house are perpendicular to the ground in the selected 3D house primitives. Therefore, the following additional rules for coding a house are formulated:

1. The total edge number of a face is defined as the first figure of a code. For example, if a face consists of eight edges, the first figure of a code is **8**. Others are based on this analogy.
2. If all edges of a house consist of straight lines, the second figure is assigned **a**, or otherwise assigned as **b**.

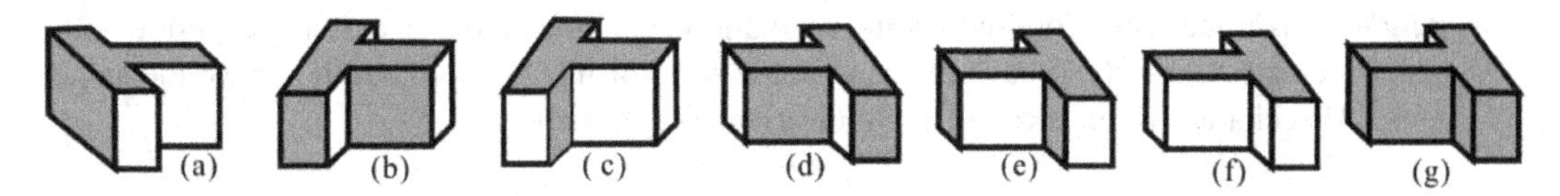

FIGURE 6.6 Code merging regulation for houses with flat roof and wall.

3. The edges of a face are numbered clockwise, starting from any edge, using natural numbers, 1, 2, ..., *n*. If the ordered edges with even numbers are parallel to each other, a letter ***1*** is assigned to the end of the code with a dash line. Otherwise, it is assigned ***0***.

With the coding rules formulated above, a complete set of the code for 3D house primitives selected in Figure 6.2 is as follows:

- **4a-1**; 2. **8a-1**; 3. **12a-1**; 4. **12a-0**; 5. **12a-1**; 6. **7a-1**; 7. **8a-0**; 8. **3a-0**; 9. **2-4a-1**.

6.2.4 Coding Regulation for Aspect Merging

The above coding regulations are for the complete projection of a house in an aerial image. However, a house is rarely in practice projected into a complete 2D aspect due to unavoidable occlusion, poor contrast, shadow, and perspective projection of aerial imagery. Moreover, the houses with a convex roof consist of several surfaces. Thus, the coding regulation for merging multiple aspects is formulated. The basic idea is: for a house with flat roof, its aspect is to merge a roof aspect with multiple wall aspects, whose aspects are **4a** in the selected houses. The coding regulation for merging two aspects is to merge two ordinal aspect codes into a new code with a dash line connection. For example, the emerged code of the house shown in Figure 6.6a is **2-1**, where **2** stands for the aspect **2** in Figure 6.5b, whose code of visible face is **8a-1**, and **1** stands for aspect **1** in Figure 6.5a, whose code of visible face is **4a**. If an entire 2D aspect consists of several visible faces, the coding regulation is the same as that described above. For example, the house shown in Figure 6.6g consists of a flat roof and four visible walls, so its code is **2-1-1-1-1**.

According to the coding regulation of aspect merging formulated above, for a house with a convex roof, as depicted in Figure 6.7a, the merged code is **8-α-8-α**, where **8** stands for the aspect **8** in Figure 6.5h, and **α** stands for the aspect **α** in Figure 6.5j. If these aspects are merged with walls, the coding regulations are the same as those formulated for flat houses.

6.2.5 Discussion of the Proposed Coding Regulation

The proposed coding regulation above has two apparent properties: size invariance and angle invariance.

- ***Size invariance.*** The proposed coding regulation ignores the size of an aspect. For example, Figures 6.8a and 6.8b have different sizes of the aspect, but they have the same code. This characteristic is called size invariance of the aspect.

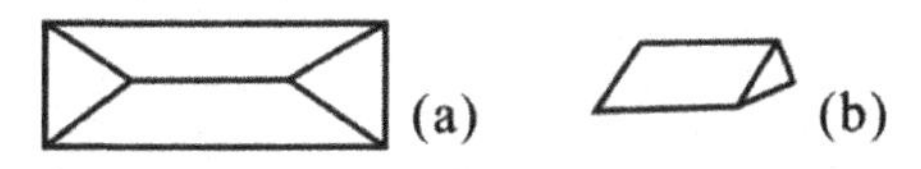

FIGURE 6.7 Code merging regulation for houses with a convex roof.

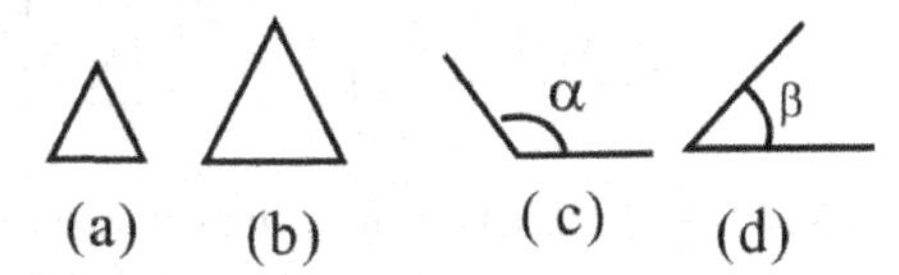

FIGURE 6.8 Size invariance and angle invariance of the aspects.

- ***Angle invariance.*** The proposed coding regulation also ignores the size of angles. For example, Figures 6.8c and 6.8d have different angle sizes of intersection, but they have the same code. This characteristic is called angle invariance of the aspect.

In addition to the above two basic characteristics, it has been observed that the proposed coding regulations have rotation invariance and scale invariance. Therefore, the characteristics of the aspect coding regulation are summarized as follows:

1. Aspect is based on a view centered (see Figures 6.9a through 6.9f);
2. Aspect is a rotate invariance (see Figures 6.9a and 6.9f);
3. Aspect is an angle invariance (see Figures 6.9c, 6.9d, and 6.9e); and
4. Aspect is a scale invariance (see Figures 6.9a and 6.9f).

With the above characteristics, it can be noted that different houses probably have the same codes such as Figures 6.1c and 6.1e. These characteristics do not actually affect the house extraction, since our goal is to extract the houses, rather than recognize the type of house.

6.2.6 Creation of Aspect Graphs

The work above only created the 2D aspects of the houses, but it is not sufficient to correctly interpret the complex houses or partially projected houses caused by the occlusions if only using the above 2D aspects codes. For this reason, creation of a hierarchical aspect graph using nodes and arcs is proposed in this chapter. The hierarchical aspect graph consists of three layers, that is, face-aspect graph, primitive-aspect graph, and aspect hypergraph. In the hierarchical aspect graph, a node represents the 2D aspect of an entire 3D object and an arc represents the relation between two 2D aspects. A face is chosen as a node (element) to construct a face-aspect graph; a primitive-aspect graph is used to describe an entire projection or partial projection of a 3D primitive; and an aspect hypergraph is constructed through the description of the connected relations between primitive graphs.

During the construction of the hierarchical aspect hypergraph, a critical step is the creation of the face-aspect graph. To this end, the following regulations are formulated.

1. A basic pattern, as shown in Figure 6.10a, is taken as a node.
2. A connected relationship between two basic patterns is taken as a connected arc (see Figure 6.10b).

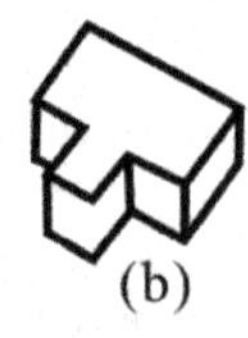

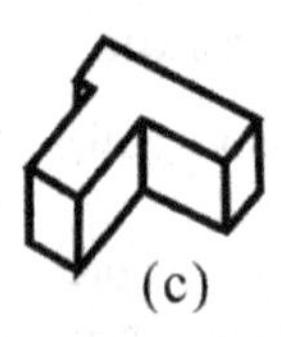

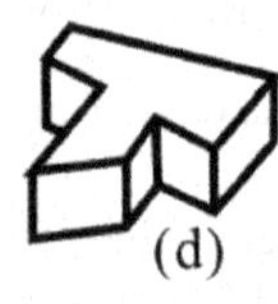

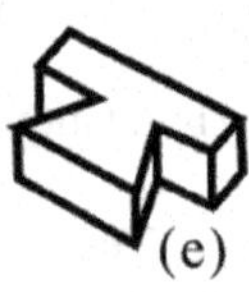

FIGURE 6.9 The properties of aspect invariance in the coding regulation.

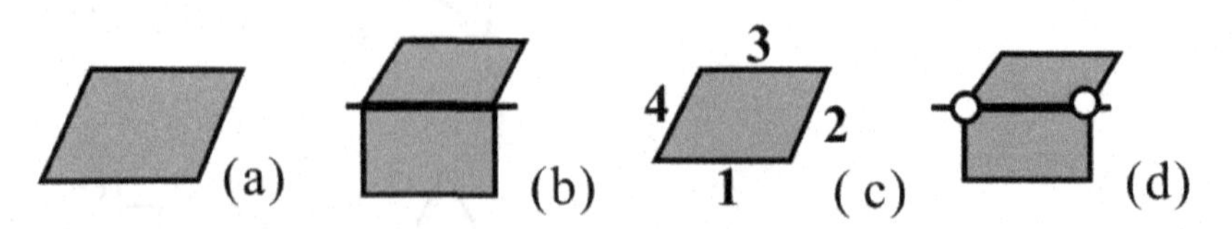

FIGURE 6.10 The basic regulation of aspect graph construction.

3. The information, such as type, edge number, and other factors in the basic pattern, is taken as the attributes of the node (see Figure 6.10c).
4. The connected information is taken as the connection attributes of the arc (see Figure 6.10d).

The two nodes are connected using an arc. The arc is assigned with attributes. The attribute values are suggested as follows.

1. The regulations for attribute value assignment of the nodes are:
   - The area of a face-aspect is taken as an attribute value.
   - The perimeter of a face-aspect is taken as an attribute value.
   - The center coordination (X_c, Y_c) of a face-aspect is taken as an attribute value.
   - The orientation angle θ of a face-aspect is taken as an attribute value, where:

$$\theta = tan^{-1}\left[2 \cdot M_{11} / \left(M_{20} - M_{02}\right)\right] / 2 \tag{6.1a}$$

   Where $M_{i,j}$ is i-th, j-th – order moment, which can be expressed by

$$M_{i,j} = \sum_{(X,Y) \in R} \left(X_c - X\right)^i \left(Y_c - Y\right)^j \tag{6.1b}$$

   The recognized primitive through the above aspect interpretation method is taken as an attribute value.
   - The codes describing an aspect are taken as an attribute value.
2. The regulations for attribute value assignment of the arcs are:
   - The number of the shared edge(s) is taken as an attribute value.
   - The type of the connected edge(s), straight or curved, is taken as an attribute value.
   - The slope of a straight line is taken an attribute value.
   - The curvature of a curve is taken an attribute value.
   - The information of two end points of the connected line is taken an attribute value.

With the regulations formulated above, the face-aspect graph and the primitive-aspect graph correspondingly describing the face-aspect and the primitive-aspect can be constructed. Furthermore, with the description of the connected relationship with the primitive-aspect graph, an aspect hypergraph can be constructed. The implementation of the above method and algorithms is called aspect describer.

6.2.7 Creation of Aspect Codes Database

The codes describing the aspects and the hierarchical aspect graphs (face-aspect graph, primitive-aspect graph, and aspect hypergraph) are taken as knowledge and archived in a database, called the codebase (see Figure 6.11). The data structures for archiving the knowledge, associated with the aspect interpretation, are depicted in Figure 6.11.

As seen from Figure 6.11, with the established aspect codebase, house extraction using the aspect interpretation is carried out through a correlated operation between the aspect codes created from aerial imagers (called image aspect codes) and the aspect codes archived in the codebase.

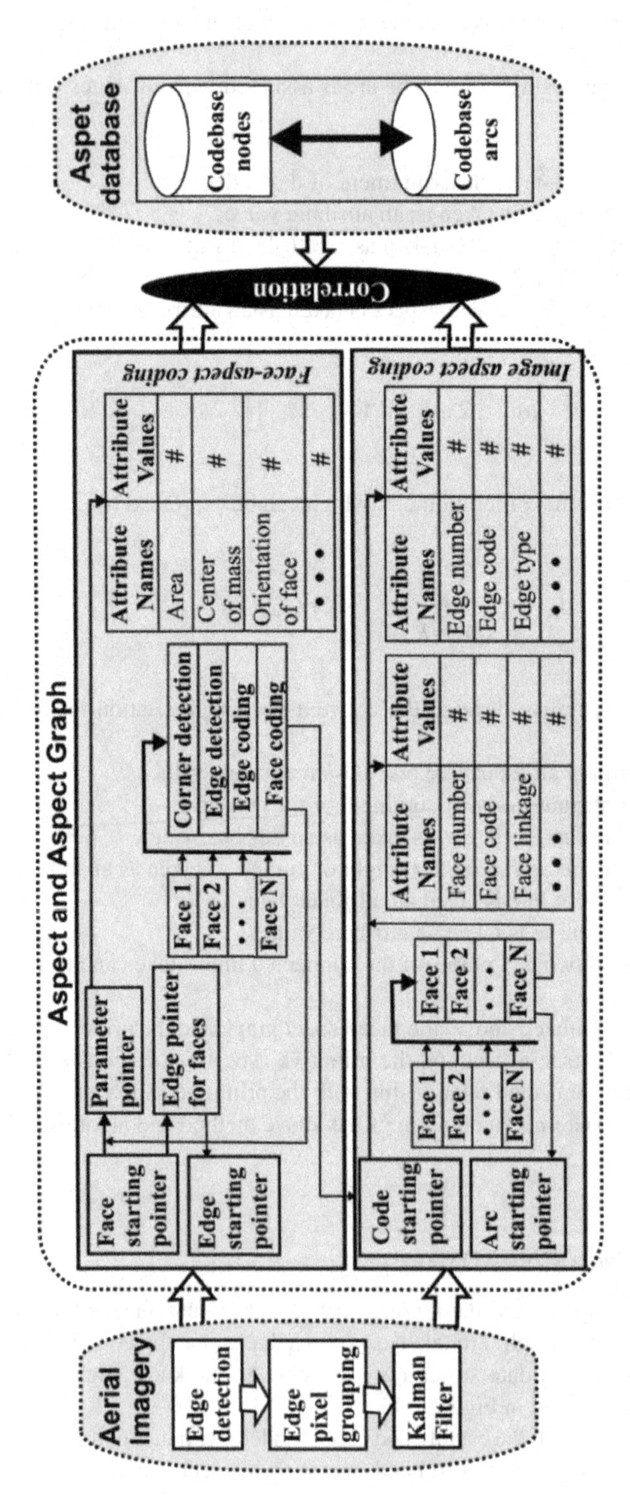

FIGURE 6.11 Data structure of aspect database (codebase) creation.

The basic steps of implementing the operation are as follows:

1. The segmented aerial imagery inputs into the aspect describer.
2. The aspect describer creates the codes for houses and their connection using the aspect and the attributes in accordance with the coding regulations formulated in Sections 6.2 and 6.3.
3. The aspect interpretation is conducted. First, the face-aspect is created and the faces describing houses are interpreted using the face-aspect correlation operation. With the implementation of face-aspect interpretation, a primitive-aspect graph, which describes an entire projection or partial projection of a 3D primitive house, is created. The 3D primitive houses are interpreted through the correlation operation between the image aspect graph and the codebase aspect graph in combination with the LiDAR data processing. Finally, an aspect hypergraph is created and the 3D houses are further interpreted through the correlation operation between the image hypergraph and the codebase hypergraph with assistance of the LiDAR data processing. The corresponding data structures for implementation of the aspect interpretation proposed above are depicted in Figure 6.11.

6.3 EXPERIMENTS AND ANALYSES

6.3.1 Data Set

The Virginia Department of Transportation (VDOT), contracting to Woolpert LLC at Richmond, Virginia, has established a high-accuracy test field in Wytheville, Virginia for the accuracy evaluation of the LiDAR system. The field extends from the west side of Wytheville approximately 11.4 miles east, with a northern and southern extent of approximately 4.5 miles (Figure 6.12). All of the elevation data, including LiDAR and 19 ground control points (GCPs), were referenced to the NAVD88 datum, and horizontal data were referenced to NAD83/93 Virginia State Plane Coordinate system. The accuracy of the 19 GCPs averages 0.02 m, 0.02 m, and 0.01 m in X, Y, and Z, respectively. The density of spare LiDAR data is an average of 7.3 feet. The specification for the airborne LiDAR sensor is listed in Table 6.1 (Zhou et al. 2004).

In the experimental field, aerial images using a Woolpert 5099 camera at a focal length of 153.087 mm were acquired on September 19, 2000. The endlap of the images is about 65%, and the sidelap is approximately 30%. The specification for aerial image collection is listed in Table 6.2 (Zhou et al. 2004).

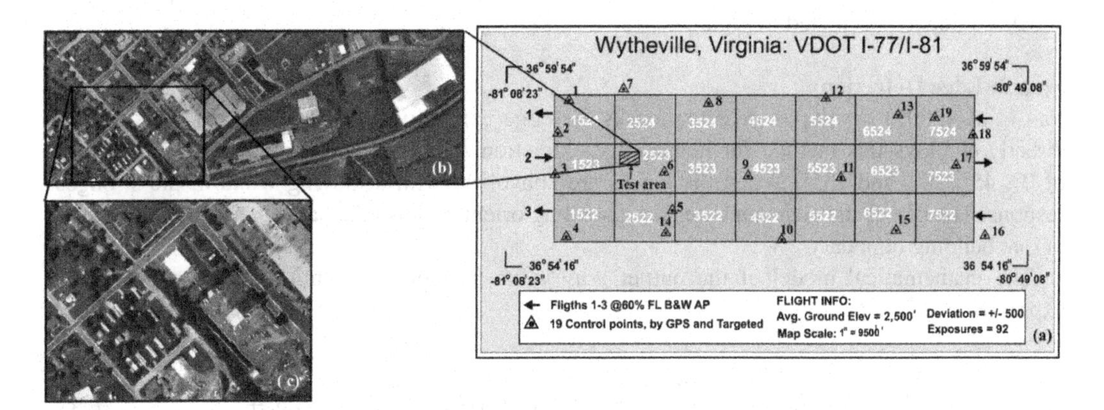

FIGURE 6.12 Geodetic control test field in Wytheville, VA.

TABLE 6.1
Specification for LiDAR Data Collection

- **System name:** Optech 1210 LiDAR system
- **Date:** September 19, 2000
- **Accuracy:** 2.0 ft
- **Density:** an average spacing of 7.3 ft
- **Aircraft speed:** 202 ft/s
- **Flying height:** 4500 ft above ground level
- **Scanner field of view (half angle):** ±16 degrees
- **Scan frequency:** 14 Hz
- **Swath width:** 2581 ft (1806 ft with a 30% sidelap)
- **Pulse repetition rate:** 10 KHz

TABLE 6.2
Specification for Aerial Image Data Collection

- **Camera:** Woolpert camera 5099
- **Film:** Kodak 2405
- **Focal length:** 153.087 mm
- **Total exposures:** 96
- **Scale:** 1:1000
- **Image type**: black-and-white aerial imagery
- **Pixel resolution:** 2.0 pixels
- **Date:** September 19, 2000

In order to evaluate the proposed method, a patch is cut from the original aerial image (Imagery ID: 2523 in Figure 6.12a) for this experiment. The flowchart of the proposed aspect interpretation for house extraction is depicted in Figure 6.13.

6.3.2 Preprocessing of Aerial Imagery

Step 1: Image Enhancement and Noise Removal

Although the aerial image was captured by a high-quality Woolpert aerial camera 5099, it is unavoidable for the image to involve noise. Considering the extraction of house edges in the next step, a traditional 3 × 3 filtering algorithm is used to accentuate edges. The typical high-pass filter mask is:

$$Mask\,E = \begin{Bmatrix} -1 & -1 & -1 \\ -1 & 9 & -1 \\ -1 & -1 & -1 \end{Bmatrix} \tag{6.2}$$

With the proposed mask, the convolution with the original image (see Figure 6.14a) is operated. The enhanced image is depicted in Figure 6.14b.

Step 2: Edge Detection

Yan et al. (2007) developed detector masks for edge extraction (see Figure 6.15). This chapter directly employs this method for house edge extraction. The mask orientations vary at an angle of 0°, 45°, 90°, and 135°. When the masks are convolved with a moving window, the strongest response is taken as the edge, and the corresponding orientation is taken as the direction of an edge of one building boundary.

The mathematical model of the output gray value through convolution computation can be expressed by

$$g(m, n) = max\left\{ \sum_{i=0}^{2}\sum_{j=0}^{2} f(m+i, n+j) \times w_k(i, j) \right\} k = 1, 2, 3, 4 \tag{6.3}$$

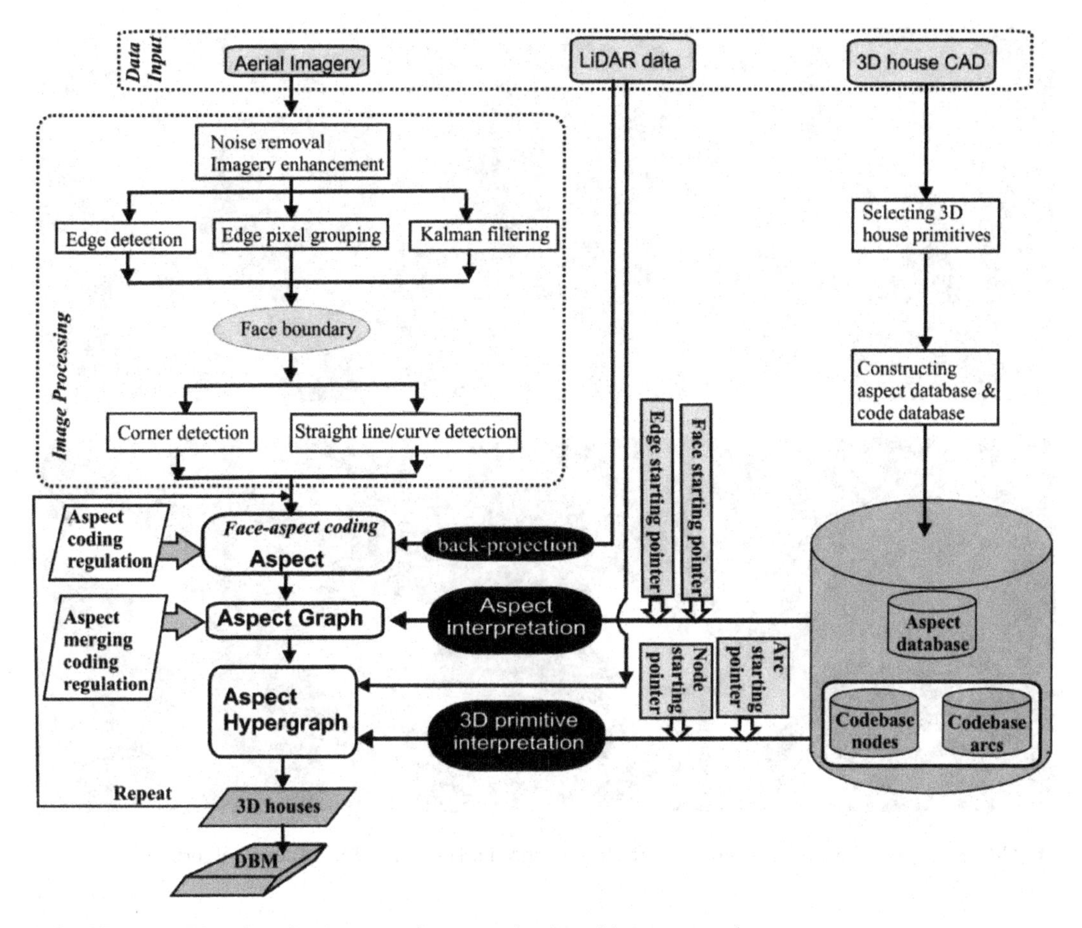

FIGURE 6.13 Flowchart for the aspect interpretation-based house extraction.

where $f(m,n)$ is the gray value of the original image, and $w_k(i,j)$ is the mask whose values are depicted in Figure 6.15. With the convolution computation, the gray values of building edges are enhanced, and the background noise is refrained efficiently. Thus, the candidate pixels of the building's edges can be extracted. After thinning of edges, the results of the extracted edges are shown in Figure 6.16.

Step 3: Edge Pixels Grouping

When detecting the building edge pixels above, a building boundary is segmented into several edge fragments. For this reason, the following task is to trace the edge fragment pixels and then group them. The method to group two neighbor edge fragments into one new edge "fragment" is proposed as follows (see Figure 6.17):

1. The slopes of two neighbor edge fragments meet the condition of $|\alpha - \beta| \le slope\_threshold$, where α, β represent the slopes of two edge fragments, respectively.
2. The distance D between the ending point of an edge fragment and the starting point of another edge fragment should meet the condition of $D \le dist\_threshold$.

FIGURE 6.14 (a) The original image, and (b) the enhanced image using 3 × 3 filtering algorithm.

$$
\begin{matrix}
-1 & -1 & -1 \\
3 & 3 & 3 \\
-1 & -1 & -1 \\
 & w_1 &
\end{matrix}
\qquad
\begin{matrix}
-1 & -1 & 3 \\
-1 & 3 & -1 \\
3 & -1 & -1 \\
 & w_2 &
\end{matrix}
\qquad
\begin{matrix}
-1 & 3 & -1 \\
-1 & 3 & -1 \\
-1 & 3 & -1 \\
 & w_3 &
\end{matrix}
\qquad
\begin{matrix}
3 & -1 & -1 \\
-1 & 3 & -1 \\
-1 & -1 & 3 \\
 & w_4 &
\end{matrix}
$$

FIGURE 6.15 Edge detection masks.

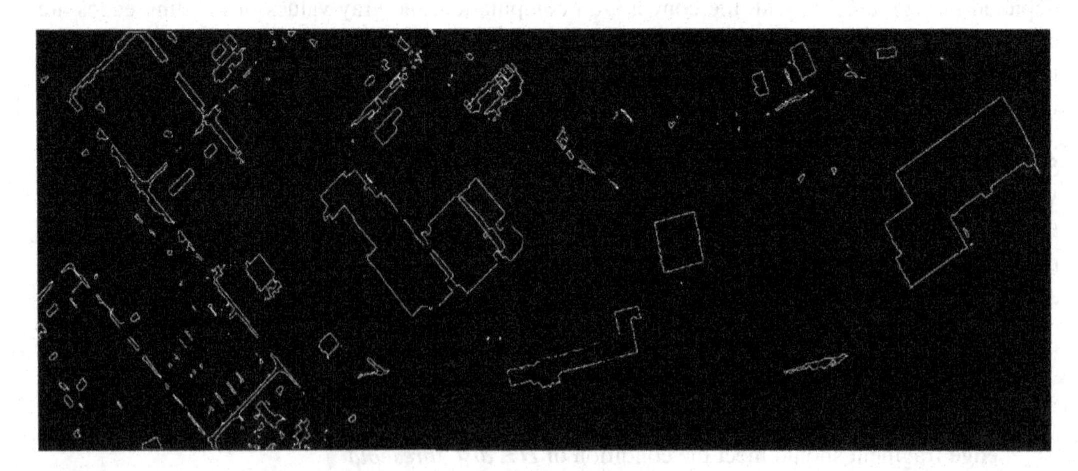

FIGURE 6.16 The result of the extracted edges of the houses.

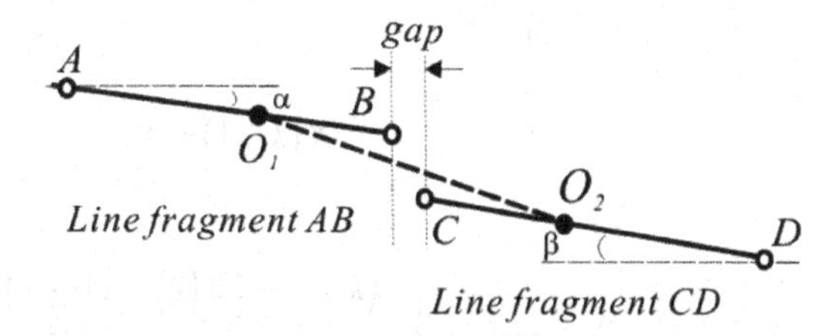

FIGURE 6.17 Merging neighboring edge fragments.

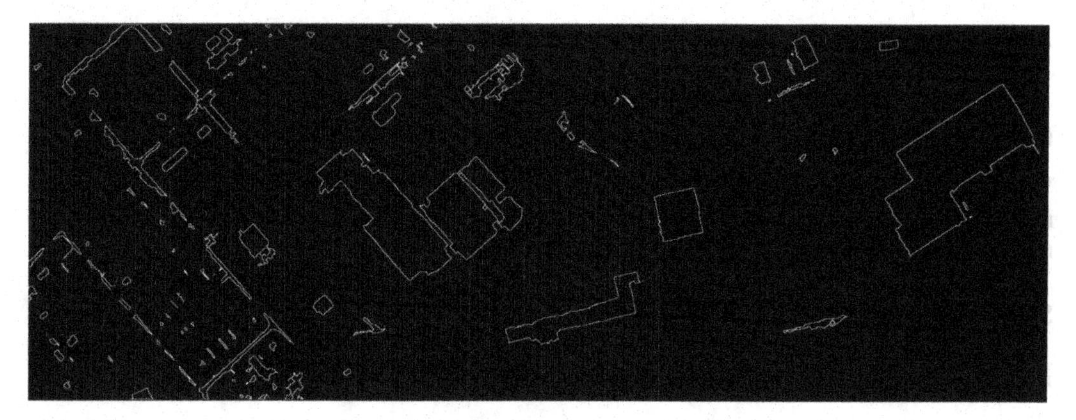

FIGURE 6.18 The result of an extracted house boundary after implementing Step 3.

When the two conditions above are met, two neighbor edge fragments are merged by

$$AD = AO_1 + O_1O_2 + O_2D \tag{6.4}$$

The same operations above are repeated for each edge fragment, until all edge fragments are finished. After this step, most of the edge fragments are smoothly connected, as shown in Figure 6.18.

Step 4: Kalman Filter to Track Line

The above operations unavoidably cause the gaps along edges of a house. The Kalman filter algorithm is proposed to track the broken-off edges to fill the gaps. The basic idea of this method is: each edge is considered as a track of a moving point on an edge. When the broken-off edge fragments are not detected, the Kalman filter keeps track of the next fragment. If there exist two edge fragments, which meet the two conditions in Step 2, they will be merged into one edge fragment. When a broken-off edge is detected, a number of the pixels are prefilled. The filled number is required to assure that the method can effectively track the next edge fragment. This computational process is mathematically described as follows: If an edge is tracked, let $X(k)$ be the state vector of movement, $x(k)$ be moving location, $x'(k)$ be velocity, $y(k)$ be observation location which is the location of the detected edge point, T be step size, k be the tracing time, the state equation of constant velocity of point D is:

$$X\left((k+1)/k\right) = AX(k) \tag{6.5}$$

where $X = [x \;\; x']^T$, $A = \begin{bmatrix} 1 & T \\ 0 & 1 \end{bmatrix}$. The measurement model is simplified as:

$$Y(k) = CX^T(k) \tag{6.6}$$

where $C = \begin{bmatrix} 1 & 0 \end{bmatrix}^T$, and $Y(k) = \begin{bmatrix} y & y' \end{bmatrix}^T$. The predicated state equation is:

$$\hat{X}\left((k+1)/k\right) = A\hat{X}(k) \tag{6.7}$$

The output of the filter is:

$$\hat{X}(k+1) = \hat{X}\left(\frac{k+1}{k}\right) + K(k+1)$$

$$\left(Y(k+1) - C\hat{X}\left((k+1)/k\right)\right) = \hat{X}\left((k+1)/k\right) \tag{6.8}$$

After the Kalman filter in combination with Step 2 is carried out, an entire building boundary is extracted. The result for the study area is depicted in Figure 6.18.

6.3.3 Creation of Aspect Graphs and Extraction of Houses

With the detected edges above, the next task is to create aspect and aspect graphs using the proposed method in Section 6.2, and then create a digital building model (DBM) of the houses. To this end, the following steps are proposed.

Step 1: Face-Coding

The first step of aspect creation is face-coding to the boundary detected in Section 6.2. The following steps are conducted.

- *Face filling.* Using a four-neighbor seed point growing algorithm, each face is filled with gray value, 255. An example filled by the shadow is shown in Figure 6.19. With the face filling, the area of each face is obtained by counting the number of the filled pixels. After being filled, a binary imagery is created.

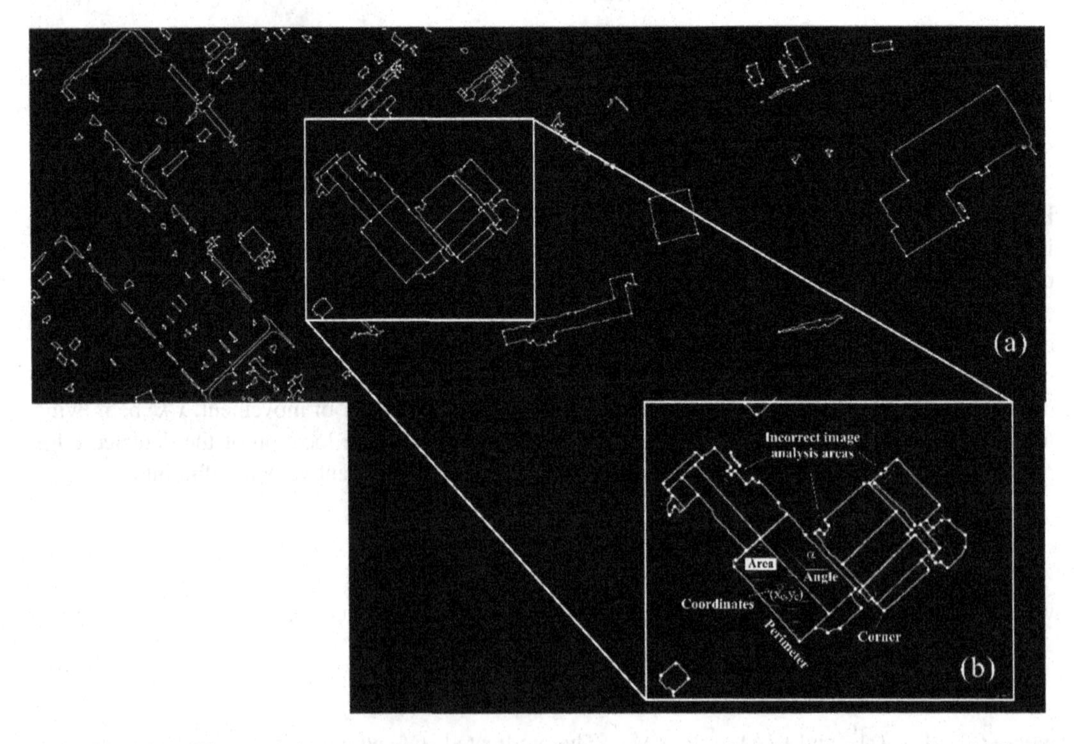

FIGURE 6.19 The procedures of coding a house using the formulated coding regulation, in which the small circles indicate an endpoint or the corner of a line or a curve. The orientation and length of a straight-line segment are calculated. The area, perimeter, and the central coordinates of a face are calculated as well.

- *Boundary detection.* With the binary imagery created above, a gray threshold of 100 is set up to detect the boundary. After that, thinning edges and deleting edge burrs are implemented consequently. Finally, the boundaries with a pixel width are obtained and displayed in Figure 6.18.
- *Boundary vectorization.* With the result obtained above, the boundary of each face is vectorized using an eight-neighbor tracking operator. With the operation, the perimeter of each boundary can be obtained and is used as the attribute value of the face-aspect.
- *Corner detection.* With the vectorized boundary of each face above, the corner detection is conducted using a method called the maximum curvature difference between forward K steps and backward K steps. The basic steps of this method are shown below (see Figure 6.19):

Selecting one pixel along the boundary, which we called the central pixel, and calculating the curvatures of the line located in the selected pixel (noted k_0), the forward 3 pixels (noted k_{+3}), and the backward 3 pixels (noted k_{-3}), respectively. The curvature differences between the selected and the forward pixels (noted k_{0+3}), and between the selected and the backward pixels (noted k_{0-3}) are calculated, respectively. The corner is detected by using the second curvature difference, that is,

$$Curvature_0 = \left|k_{0-3} - k_{0+3}\right| \begin{cases} \textit{if Curvature} > \textit{threshold} \text{ noted as corner} \\ \textit{if Curvature} \leq \textit{threshold} \text{ noted as no corner} \end{cases} \tag{6.9}$$

If the $Curvature_0$ is greater than the given threshold, the pixel is determined as a corner; otherwise, it is not a corner pixel. Sometimes, it is necessary to suppress the local non-maximum since the multiple $Curvature_0$ surrounding the central pixel simultaneously meets the condition of Equation 6.9. With the proposed method here, the result of the detected corners is displayed in Figure 6.19.

- *Determination of straight line or curve.* When using the aspect or aspect graph for house interpretation, the property of a line segmentation, either a straight line or a curve, has to be determined. To this end, the curvature of the line segmentation between two corners is calculated. If the curvature is close to zero, the line segment is considered as a straight line, otherwise, as a curve.
- *Face-coding.* With the operations above, the face-coding is conducted in terms of the coding regulation formulated in Sections 2.3 and 2.4. When the above operations are finished, the face codes are completed. An example is depicted in Figure 6.19.

Step 2: Creation of Aspect Graphs

With the faces coded above, the aspect graph is constructed. The basic steps include the following. First, each face is given an attribute name using natural numbers 1, 2, 3.... Meanwhile, the attribute values are calculated using the parameters listed in Table 6.3. Second, the aspect graph is constructed using nodes and arcs. Each face is taken as a node of the aspect graph, and each of the shared edges is determined by the same imagery coordinates and taken as an arc of the aspect graph (see Figure 6.20b). Third, the primitive-aspect graph is constructed using the aspect merging regulation described in Section 2.4. For example, the face-aspects 5 and 6, face-aspects 9 and 10, and face-aspects 2 and 4 are merged in Figure 6.20b. In addition, the face-aspect 1 is an isolated aspect, which represents an independent house. Fourth, the aspect graphs between those created in this section and archived in the database are correlated. With this operation, the aspect graph using the proposed face-coding regulation can effectively remove the mis-segmented area (see Figures 6.20a and 6.20b).

TABLE 6.3
The Attribute Values of Nodes and Arcs

Attribute Values for Node	Attribute Values for Arc
• Area	• The coordinates of a shared edge
• Perimeter	• The types connected edges
• Central coordinates	• The slope of straight line
• Orientation angle θ of a face-aspect	• The curvature of a curve
• Codes describing face-aspect	• The information of two endpoints

Step 3: Co-Registration Between Aspect and LiDAR Data

The created aspects above unavoidably contain objects which do not belong to houses, such as vehicles or parking lots. For this reason, the application of elevation information driven by LiDAR data will help separate ground objects and non-ground objects. A common method is to co-register the LiDAR data and the created aspects, with which the houses' boundaries and 3D coordinates can be obtained and are used to further create a digital building model (DBM). This chapter develops a method called aspect-based co-registration. The basic idea is: (1) the aspect is selected as a co-registration element, since the aspect consists of the house boundary, which is an important invariant geometric element in the house; (2) LiDAR data is back-projected onto the aspect imagery plan to determine which LiDAR footprints are located within an aspect; and (3) the algorithm for co-registration between the LiDAR data and the aspects is developed to create the DBM. The detailed steps are presented below.

- ***Determination of reference system.*** The LiDAR data with X, Y, and Z coordinates have been referenced to NAVD88 datum at height and the NAD83/93 Virginia State Plane Coordinate system in planimetric coordinates. Thus, the datum of the LiDAR data set is used as the common reference framework.
- ***Back-projection of LiDAR onto aspect imagery plane.*** The LiDAR data are back-projected onto the created aspect plane, also called "aspect imagery," for removal of the ground objects and addition of the objects missed by the operation of aspect creation in Step 2. The basic steps consist of:
  1. Back-projection of LiDAR data onto the aspect plane (called aspect imagery) using the collinearity equations accompanying the known interior and exterior orientation parameters of the aerial images, which are provided by a vendor.
  2. Correlation of the created aspects and LiDAR data, and identification of the following four types of correlation cases (Figure 6.20):
     - Case 1: LiDAR data is correctly matching with the created house aspects.
     - Case 2: LiDAR data is partially matching with the created house aspects.
     - Case 3: LiDAR data does not match with the created house aspects.

For Case 2 and Case 3 above, there are three possibilities: (1) edges of the houses in the aerial imagery possibly have not completely been detected due to traditional optical imagery problems, such as low gray contrast; (2) partial or complete occlusion due to property of the perspective projection; and (3) the created aspect is part of an entire house. For Case 3 above, there are two possibilities: first, size of the house is less than 14 × 14 feet in length and width, that is, fewer than two LiDAR footprints were shot on the house roof. Second, the created aspects are for ground objects such as parking lots, which have no elevation information.

This chapter only considers cases where over three LiDAR footprints are shot on the aspects, that is, Case 1 and Case 2 (Figure 6.21).

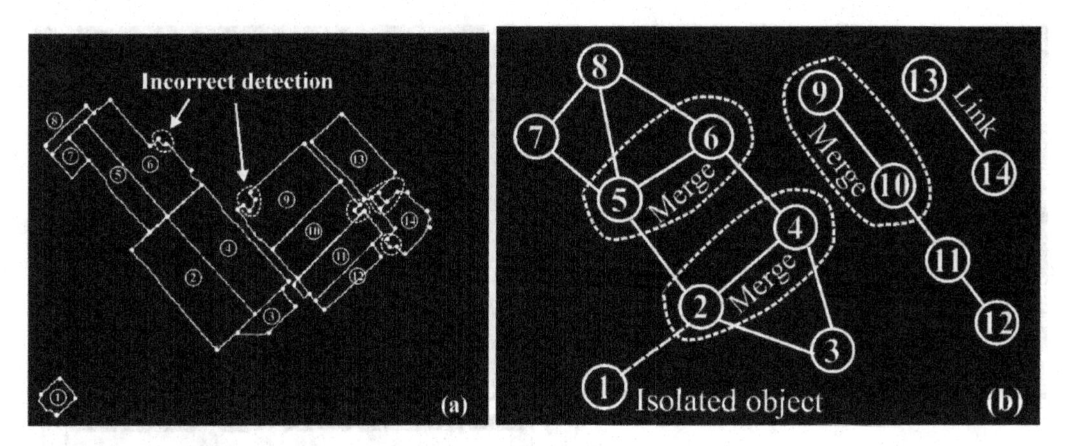

FIGURE 6.20 Illustration of aspect and aspect graph construction, (a) the segmented houses, and (b) their aspect graphs.

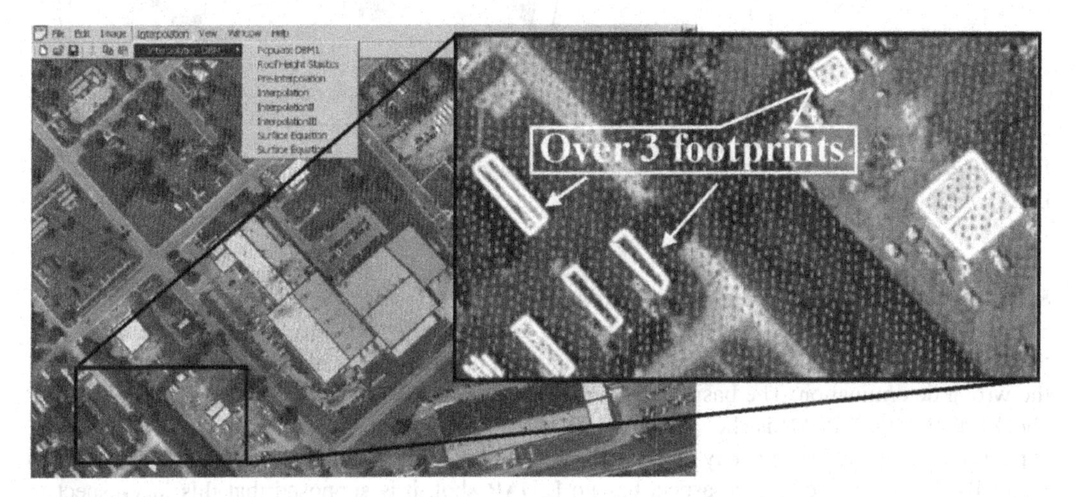

FIGURE 6.21 Co-registration of aerial imagery and LiDAR footprints. The white polygon is the extracted boundary of the houses, and the white node is the point cloud of the LiDAR in the black and white image.

With the above co-registration between aspects and LiDAR data, the roof equation can be established using a local plane fitting equation, that is,

$$aX + bY + cZ = 1 \tag{6.10}$$

Where a, b, and c are unknown coefficients, and X, Y, and Z are coordinates of LiDAR footprints shot within the house's roof. As seen from Equation 6.10, at least three LiDAR footprints are required to determine the equation coefficients, a, b, and c by least squares estimates. With this algorithm, a coarse 3D model of houses can be generated. As observed in Figure 6.22, the boundaries of the 3D model of each house are very coarse, since it is difficult to extract the boundaries of each house from a single piece of LiDAR data. In addition, some objects are wrongly detected as aspect (i.e., non-ground objects) in the operation of the edge detection and the aspect creation using the method described above (see Figure 6.22a). However, they are in fact ground objects, since they have no elevation when co-registered with the LiDAR data (see Figure 6.22b).

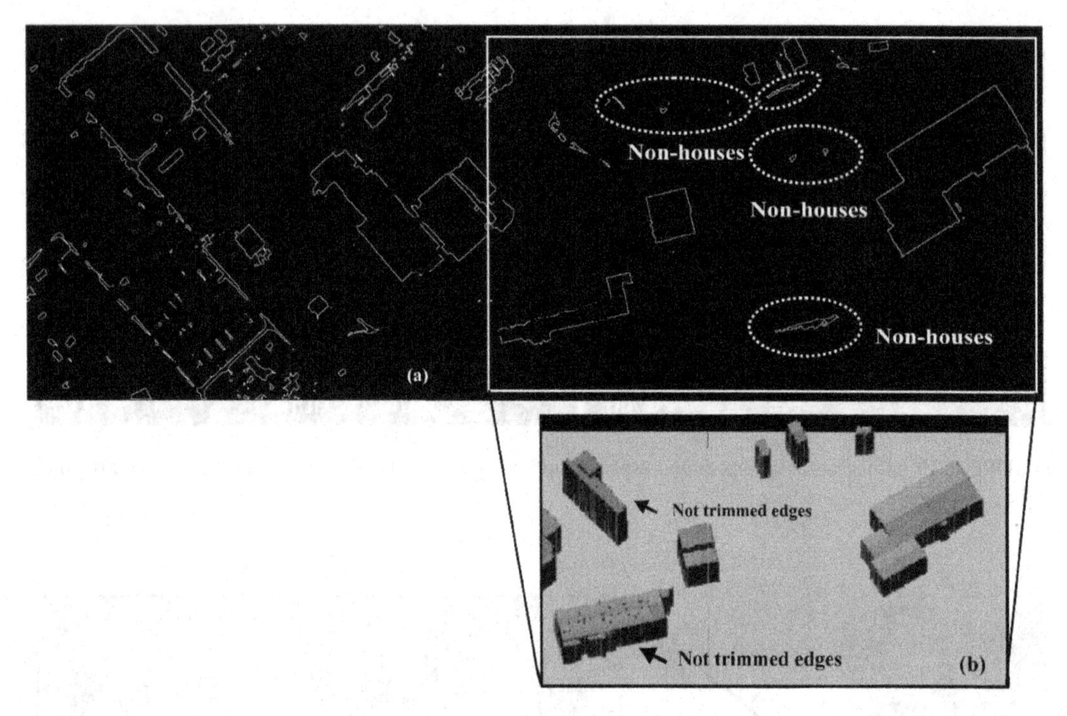

FIGURE 6.22 Coarse 3D models of the houses through co-registration between the aspect and the LiDAR data.

Step 4: Construction of Aspect Hypergraphs

Step 3 unavoidably missed the detection of a few houses, and/or wrongly detected the houses. For this reason, this chapter uses the aspect hypergraph in combination with LiDAR data to overcome the wrong determination. The basic procedures are: for a given face-aspect (see Figure 6.23a), we check whether the LiDAR is shot within the face-aspect. If only one LiDAR footprint is shot, it is supposed that the face-aspect may represent a house. We further examine the connected face-aspect manually. If the connected face-aspect has no LiDAR shot, it is supposed that this face-aspect is probably a wall (Figure 6.23b). Consequently, this wall face-aspect will be merged to the roof face-aspect to construct a 3D volume primitive graph (see Figure 6.23c). If two neighboring face-aspects have no LiDAR shot, it is supposed that the face-aspects represent other objects, such as parking lots. The procedures above are repeated, with manual checking of whether all houses are detected one-by-one. If not, the house has to be manually added.

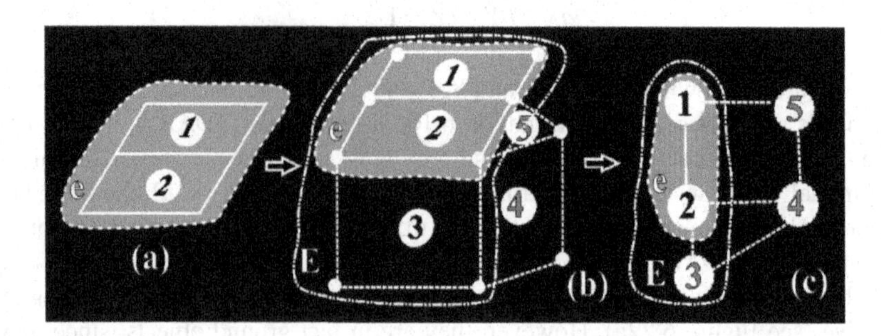

FIGURE 6.23 Illustration of 3D primitive graph creation procedure.

Step 5: House Extraction Using Aspect Interpretation

The primary design for the extraction of houses using aspect interpretation is a correlation operation between the aspect codes created from aerial imagery (called image code) and the aspect codes archived in the codebase. The correlation operation is carried out by the attribute values assigned in both the nodes and the arcs of the aspect graphs. Each node and each arc are composed of many attribute values. A linear combination of these attributes is constructed by

$$node_{attribute} = Area + Perimeter + CentralCoord + Orientation + Code_{aspect} \tag{6.11}$$

$$arc_{attribute} = Coord_{Two-ends} + Type_{edge} + Slope_{line} + Curvature_{curve} \tag{6.12}$$

The correlation coefficient maximum of the node/arc attributes between those obtained from aerial imagery and those archived in the database is taken as a criterion. With the new combined variables in Equations 6.11 and 6.12, the two criteria are employed to determine their correlation, respectively. The first criterion is the correlation coefficient maximum of the node, which is employed for determining the candidate house under a given threshold of correlation coefficient. The mathematical model is expressed by:

$$r_i = \begin{cases} r_i \ if \ r_i \le R_\theta \\ 100 \, if \ r_i > R_\theta \end{cases} \quad r_i \subseteq R \tag{6.13}$$

where R_θ is threshold, and R is the set of r_i. The R_θ is set at 0.95 in this research. The successful operation of this step will be able to interpret and find a significant amount of geometric information about a house, such as aspect property, geometry, gray area, and relationship, and separate the ground objects (e.g., parking lots) and non-ground objects (e.g., houses). The second criterion is a correlation coefficient maximum of the arcs between those obtained from the aerial imagery and those archived in the database. The same operation as the node correlation is carried out. The successful operation of this step will be able to interpret and find a primitive-aspect graph, which describes an entire projection or partial projection of a 3D primitive (house). If an aspect is identified as a house, we will further determine how many LiDAR footprints hit its roof and then calculate the house's model, which will be described in Step 7.

Step 7: Repeat the Operations Above

Repetition of the operations in Steps 4 and 5 is required in order to avoid missing the extraction of houses.

Step 8: Creation of Accurate 3D Model of Houses

An aspect with its correct interpretation is capable of accurately depicting the boundary of a house. With the accurate boundary of house, a precise 3D model can be created using the method proposed in Zhou et al. (2004). The major difference between this step and Step 3 is that boundary information of the aspect interpreted is used to fit the planar equation. Thus, the boundary of the 3D model appears to be trimmed, that is, an accurate DBM describing houses can be established (see Figure 6.24). As compared with Figure 6.22, it is found that the aspect-based 3D model is much more accurate than the other method.

6.3.4 Occlusion Analysis

High trees unavoidably occlude the houses, resulting in missing the house detection. In order to evaluate the impact of the proposed method on the tree occlusion, this study extends the experiment to another aerial image (ID 3523 in Figure 6.12). The experimental results are depicted in

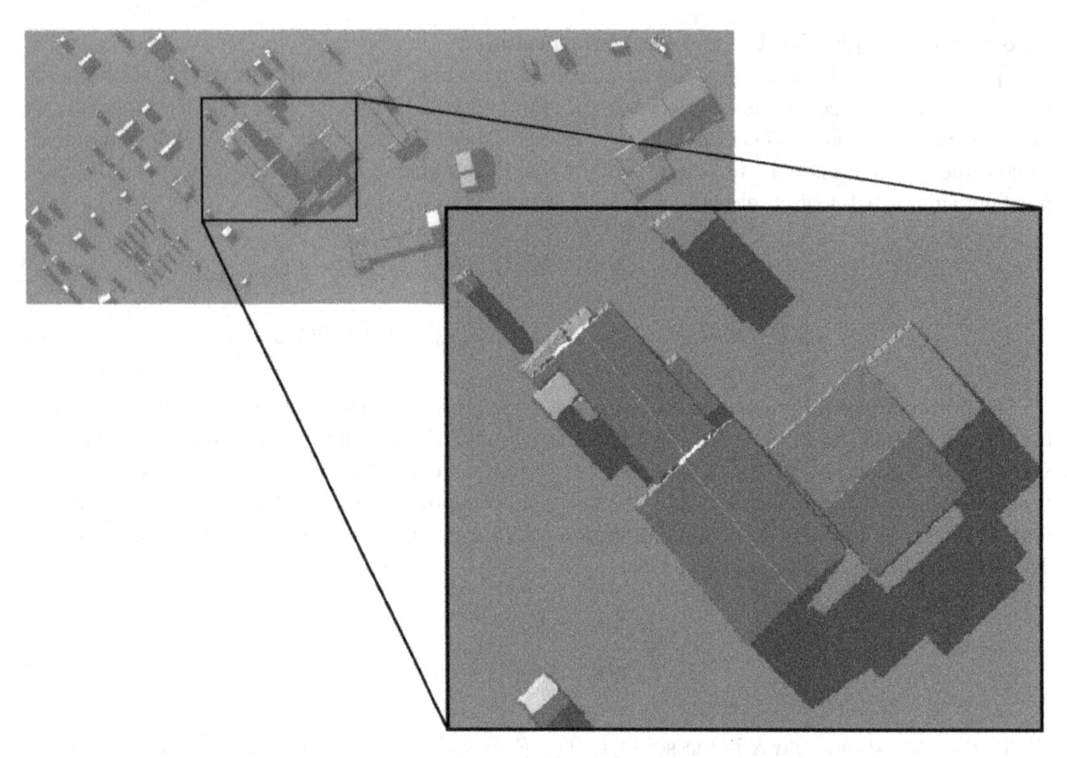

FIGURE 6.24 The houses extracted by the proposed method.

Figure 6.25. When checking the extracted houses (Figure 6.25b) and the corresponding original image (Figure 6.25a) manually, it is found that a few houses cannot be detected when using the proposed method. The undetected houses are mainly from either tree occlusion or the gray values of the houses' roofs are close to those of the trees' (see Figure 6.25b). The rates of successfully extracted houses are 93%, that is, 113 out of 122 houses are detected using the proposed method. The major reasons for missing the detection of a house using our method are probably caused when: (a) a high tree occludes most of a house, so the house cannot be detected, since the house's aspect cannot be formed after the aerial image is processed using the method proposed in this chapter; and (b) fewer than three LiDAR footprints shot a house, so the house cannot be detected, since the height information cannot be obtained. We manually checked the detected and undetected houses and found that eight undetected houses are due to the tree occlusions or the brightness of the house's roof being close to that of the trees, and only one house is due to there being fewer than three LiDAR footprints onto the house. This fact demonstrated that tree occlusion of houses and the brightness of houses' roofs being close to that of the surrendering objects significantly impact the house detection when using the proposed method (Figure 6.26).

6.3.5 Comparison Analysis

The comparison between our method and the hierarchical morphological filtering (called Method 1 hereafter), which was proposed by Rottensteiner (2003), is conducted. The basic idea of Method 1 is to first remove ground footprints using a ground filtering algorithm, and then further remove the remaining non-building pixels using size, shape, height, and the height difference between the first and last return. This section gives a brief description below for Method 1, and its details can be referenced in Rottensteiner (2003).

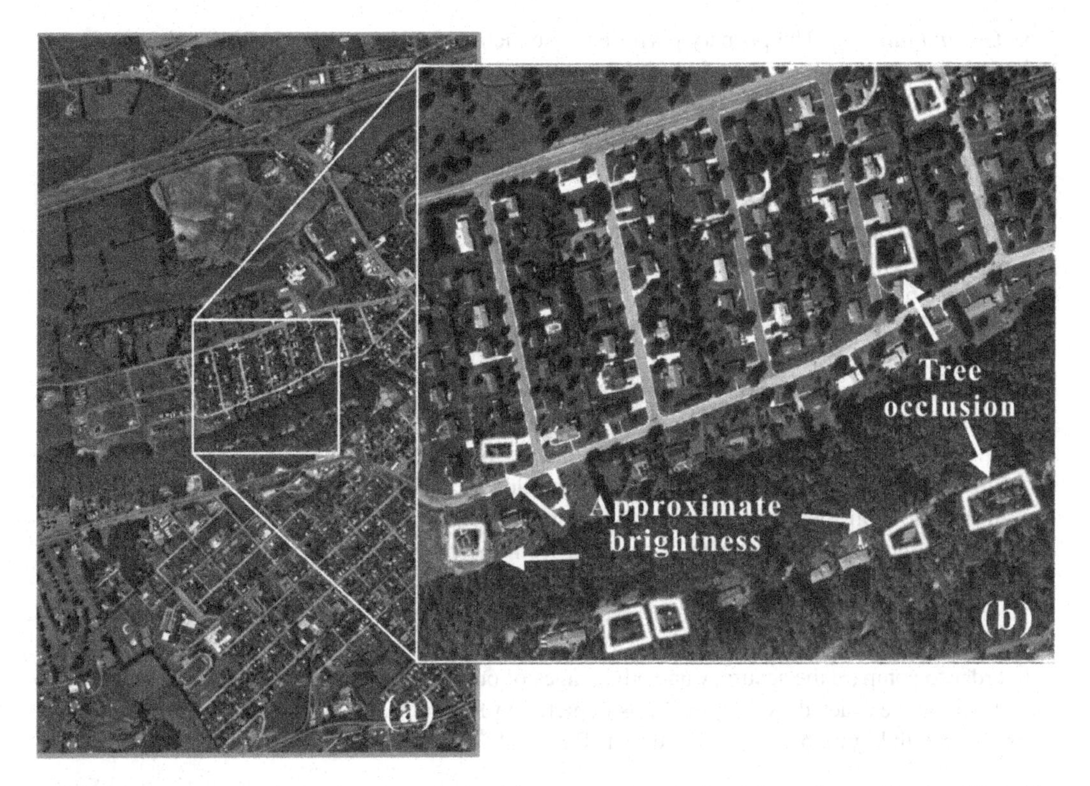

FIGURE 6.25 Tree occlusion impact analysis of the proposed method and the undetected houses. The white polygons represent the extracted boundary of the houses in the black and white image.

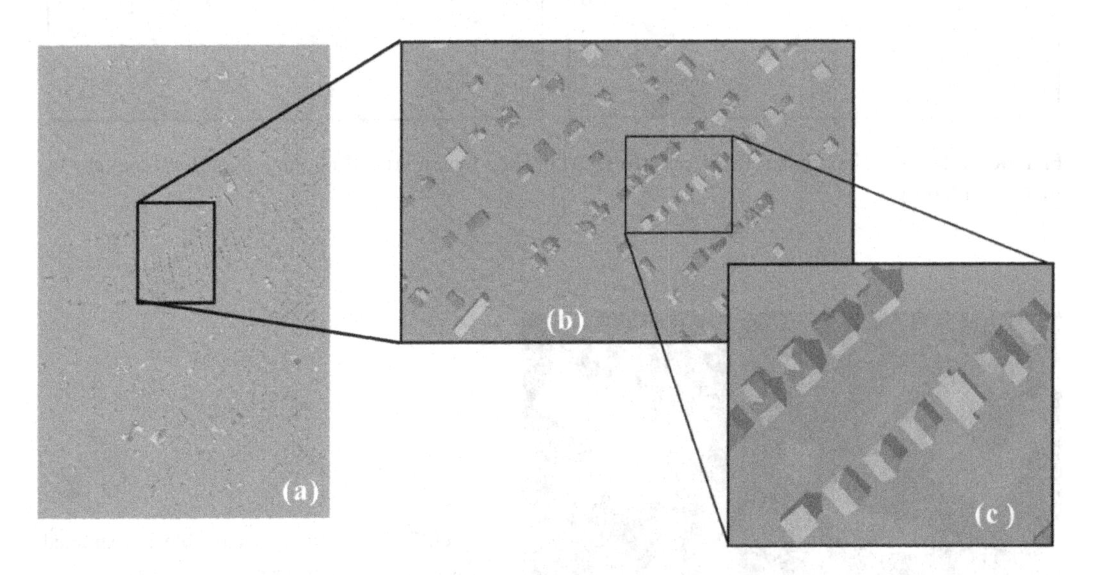

FIGURE 6.26 The houses extracted in the aerial image (ID 3523) by the proposed method.

- ***Ground filtering.*** The primary purpose of the method is to apply the multidirectional ground filtering algorithm to identify LiDAR ground footprints and generates a Digital Elevation Model (DEM). A two-dimensional mesh is first overlaid on the LiDAR point clouds, and the elevation of the nearest point is assigned to the cell. The ground filter algorithm is used to separate ground and non-ground footprints using the given thresholds of slopes, the elevation difference between a point and its local minimum elevation, and the elevation difference between a point and the nearest ground point. With successful implementation of this operation, the DEM, only composed of non-ground objects, is created at the same resolution (see Figures 6.27a and 6.27b).
- ***Removing non-building pixels.*** The coarse DEM created above is mainly composed of vegetation, buildings, and other objects, such as vehicles. Accordingly, three pixel-based operations are used to break non-building blocks into smaller fragments on the basis of the fact that houses differ from other objects in height, size, and shape, and laser lights possess a relatively strong ability to penetrate vegetation and hardly pass through solid house roofs. This operation can remove objects that are shorter and lower than the minimum house in length and in height from the house candidates within the coarse DEM, such as shrubs and vehicles. Meanwhile, vegetation is also filtered out from the coarse DEM. Finally, an accurate DEM only composed of houses is created, and the boundary of each house is vectorized into vectors (see Figure 6.28).

In order to compare the accuracy and advantages of our method compared to Method 1, we visualize the houses extracted by Method 1, as depicted in Figure 6.29. As observed when comparing Figure 6.29 with Figure 6.24, it is found that (Ren et al. 2009):

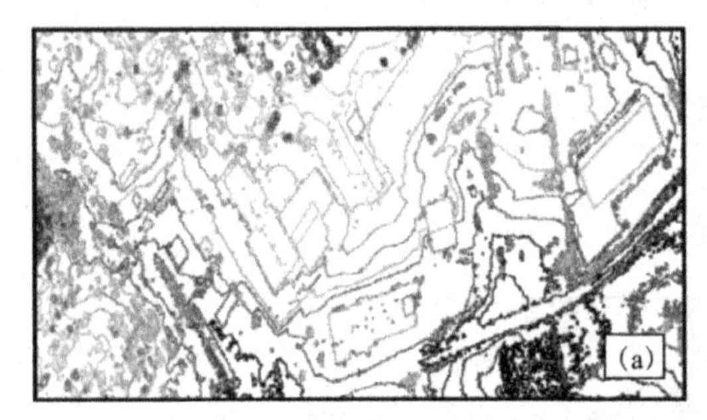

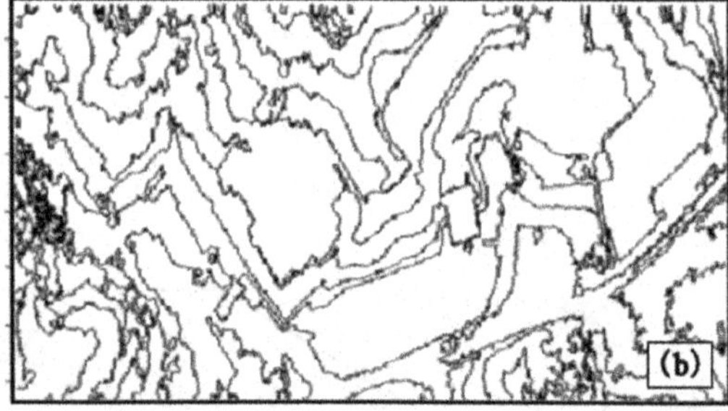

FIGURE 6.27 (a) The original DSM, and (b) DEM created by the ground filtering algorithm (courtesy by Ren et al. (2009)).

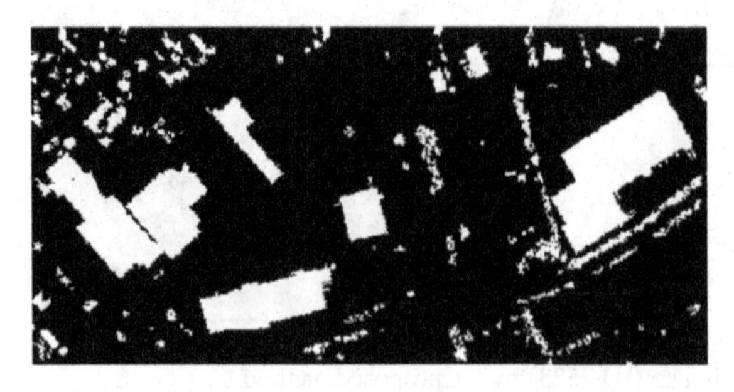

FIGURE 6.28 Non-ground objects extracted using single LiDAR data.

FIGURE 6.29 Visualization of the houses extracted by using single LiDAR data.

1. The rates of extracting houses are 82% (i.e., 100 out of 122 houses) using Method 1. The major reasons for the undetected houses using Method 1 are probably caused by the fact that Method 1 only used height information, that is, LiDAR data, and did not use the aerial image information. When trees surround a house, Method 1 will be blind to either trees or a house.
2. Method 1 only creates a coarse boundary (see Figure 6.29), whereas our method is able to create a higher accuracy of DBM than Method 1 (see Figures 6.24 and 6.26), since our method proposed an aspect interpretation operation, which accurately fused the LiDAR data and the imaged houses.

6.4 CONCLUSIONS

The main contribution of this chapter lies in the development of a seamless fusion between LiDAR data and aerial imagery. This fusion differs from the traditional methods, which simply merged aerial imagery with LiDAR data. The proposed method is (1) to create the aspects and the aspect graphs from aerial imagery, and then (2) merge the created aspects with LiDAR data. The major characteristics of the proposed method lie in: (1) the aspect and the aspect graph contain a significant amount knowledge, such as a building's geometric features, shapes, and structures, therefore the proposed method is a type of geometric information-driven method for the extraction of houses, and (2) the data fusion is carried out at each process of the house detection and the house description, so it is thus called a seamless fusion in this chapter.

The main work of the proposed method includes: (1) aerial image preprocessing, including edge detection using a detector mask and edge pixel grouping, and gap filling is developed; (2) eleven types of 3D house primitives are selected and their projections are represented by the aspects. The coding regulations are formulated for creation of the (hierarchical) aspect graphs on the basis of the results of aerial image processing and the results of LiDAR data processing. In the aspect graphs, the nodes are represented by the face-aspect, and the arcs are described by the attributes using the formulated coding regulations; and (3) the co-registration between the aspects and the LiDAR data is implemented using the correlation operation. As a consequence, the aspects and/or the aspect graphs are interpreted for the extraction of houses, and the extracted houses are fitted using a planar equation for the creation of a DBM.

The experimental field, which was established by the Virginia Department of Transportation, contracting to Woolpert LLC at Wytheville, VA, is used to evaluate the proposed method. The experimental results demonstrated that the proposed method is capable of effectively extracting houses at a rate of 93%, compared to another method, which is 82% effective with LiDAR spacing of approximately 7.3 by 7.3 feet square in the experiment area. The accuracy of the 3D DBM is higher than the method using only single LiDAR data.

REFERENCES

Ahmadi, S., et al. Automatic urban building boundary extraction from high resolution aerial images using an innovative model of active contours, *International Journal of Applied Earth Observation and Geoinformation*, vol. 12, no. 3, June, pp. 150–157, 2010.

Axelsson, P., Processing of laser scanner data – Algorithms and applications, *ISPRS Journal of Photogrammetry & Remote Sensing*, vol. 54, pp. 138–147, 1999.

Baillard, C., and Maître, H., 3-D reconstruction of urban scenes from aerial stereo imagery: A focusing strategy, *Computer Vision and Image Understanding*, vol. 76, no. 3, December, pp. 244–258, 1999.

Besnerais, G.L., Sanfourche, M., and Champagnat, F., Dense height map estimation from oblique aerial image sequences, *Computer Vision and Image Understanding*, vol. 109, no. 2, February, pp. 204–225, 2008.

Biederman, I., Human image understanding: recent research and a theory, *Computer Vision, Graph and Image*, vol. 32, pp. 29–73, 1985.

Cord, M., Jordan, M., and Cocquerez, J.P., Accurate building structure recovery from high resolution aerial imagery, *Computer Vision and Image Understanding*, vol. 82, no. 2, May, pp. 138–173, 2001.

Dickinson, S.J., Pentland, A.P., and Rosenfeld, A., 3-D shape recovery using distributed aspect matching, *IEEE Transactions on Pattern Analysis and Machine Intelligence (PAMI)*, vol. 14, no. 2, pp. 174–197, 1992.

Dong, H. L., Kyoung, M. L., and Sang, U. L., "Fusion of lidar and imagery for reliable building extraction," *Photogrammtric Engineering and Remote Sensing*, vol. 74, no. 2, February, pp. 215–225, 2008.

Ekhtari, N., et al. Automatic building extraction from LiDAR digital elevation models and WorldView imagery, *Journal of Applied Remote Sensing*, vol. 3, no. 1, 2009.

Fischer, A., et al. Extracting buildings from aerial images using hierarchical aggregation in 2D and 3D, *Computer Vision and Image Understanding*, vol. 72, no. 2, November, pp. 185–203. 1998.

Fradkin, M., Maître, H., and Roux, M., Building detection from multiple aerial images in dense urban areas, *Computer Vision and Image Understanding*, vol. 82, no. 3, June, pp. 181–207, 2001.

Fujii, K., and Arikawa, T., Urban object reconstruction using airborne laser elevation image and aerial image. *IEEE Transactions on Geoscience and Remote Sensing*, vol. 40, no. 10, pp. 2234–2240, 2002.

Gamba, P., and Houshmand, B., Joint analysis of SAR, LiDAR and aerial imagery for simultaneous extraction of land cover, DTM and 3D shape of buildings, *International Journal of Remote Sensing*, vol. 23, no. 20, pp. 4439–4450, 2002

Habib, A., et al. Photogrammetric and lidar data registration using linear features, *Photogrammetric Engineering and Remote Sensing*, vol. 71, no. 6, pp. 699–707, 2005.

Hermosilla, T., et al. Evaluation of automatic building detection approaches combining high resolution images and LiDAR data, *Remote Sensing*, vol. 3, no. 6, pp. 1188–1210, 2011.

Hu, J., You, S., and Neumann, U., Integrating LiDAR, aerial image and ground images for complete urban building modeling. In *Third International Symposium on 3D Data Processing, Visualization, and Transmission, 3DPVT 2006*, pp. 184–191, 2007.

Jaynes, C., Riseman, E., and Hanson, A., Recognition and reconstruction of buildings from multiple aerial images, *Computer Vision and Image Understanding*, vol. 90, no. 1, April, pp. 68–98, 2003.

Jung, F., Detecting building changes from multitemporal aerial stereopairs, *ISPRS Journal of Photogrammetry and Remote Sensing*, vol. 58, no. 3–4, January, pp. 187–201, 2004.

Kabolizade, M., Ebadi, H., and Ahmadi, S., An improved snake model for automatic extraction of buildings from urban aerial images and LiDAR data. *Computers, Environment and Urban Systems*, vol. 34, no. 5, pp. 435–441, 2010.

Khoshelham, K., et al. Performance evaluation of automated approaches to building detection in multi-source aerial data, *ISPRS Journal of Photogrammetry and Remote Sensing*, vol. 65, no. 1, pp. 123–133, 2010.

Kim, T., and Muller, J.P., Automated urban area building extraction from high resolution stereo imagery, *Image and Vision Computing*, vol. 14, no. 2, pp. 115–130, 1996.

Lee, Y.C., and Lei, K.S., Machine understanding: Extraction and unification of manufacturing features, *IEEE Computer Graphics Application*, vol. 18, pp. 20–32, 1990.

Mastin, A., Kepner, J., and Fisher, J., Automatic registration of LiDAR and optical images of urban scenes. In *IEEE Computer Society Conference on Computer Vision and Pattern Recognition Workshops, CVPR Workshops 2009*, pp. 2639–2646, 2009.

Matsuyama, T., Knowledge-based aerial image understanding system and expert systems for image processing, *IEEE Transactions on Pattern Analysis and Machine Intelligence (PAMI)*, vol. 25, no. 3, pp. 305–316, 1987.

Mayer, H., Automatic object extraction from aerial imagery—A survey focusing on buildings, *Computer Vision and Image Understanding*, vol. 74, no. 2, pp. 138–149, 1999.

McKeown, D.M., Knowledge-based aerial photo interpretation, *Photogrammetria*, vol. 39, no. 3, pp. 91–123, 1984.

McKeown, D.M., Automating knowledge acquisition for aerial image interpretation, *Computer Vision, Graph, and Image*, vol. 46, pp. 37–81, 1989.

Mohan, R., and Nevatia, R., Using perceptual organization to extract 3-D structures, *IEEE Transactions on Pattern Analysis and Machine Intelligence (PAMI)*, vol. 11, no. 11, pp. 1121–1139, 1989.

Morgan, M., and Tempfli, K., Automatic building extraction from airborne laser scanning data, *International Archives of Photogrammetry and Remote Sensing*, vol. 33, Part B3, Amsterdam, pp. 616–623, 2000.

O'Donohue, D., et al. Combined thermal-LiDAR imagery for urban mapping. In *23rd International Conference Image and Vision Computing New Zealand, IVCNZ*, 2008.

Paparoditis, N., et al. Building detection and reconstruction from mid- and high-resolution aerial imagery, *Computer Vision and Image Understanding*, vol. 72, no. 2, pp. 122–142, 1998.

Peng, J., Zhang, D., and Liu, Y., An improved snake model for building detection from urban aerial images, *Pattern Recognition Letters*, vol. 26, no. 5, pp. 587–595, 2005.

Pu, S., and Vosselman, G., Knowledge based reconstruction of building models from terrestrial laser scanning data. *ISPRS Journal of Photogrammetry and Remote Sensing*, vol. 64, no. 6, pp. 575–584, 2009.

Ren, Z., Cen, M., and Zhou, G., Filtering method of LiDAR data based on contours, *Journal of Remote Sensing*, vol. 13, no. 1, pp. 55–62, 2009. (Chinese)

Rottensteiner, F. Automatic generation of high-quality building models from LiDAR data, *IEEE Computer Graphics and Applications*, vol. 23, no. 6, pp. 42–50, 2003.

Rottensteiner, F., et al. Building detection by fusion of airborne laser scanner data and multi-spectral images: Performance evaluation and sensitivity analysis, *ISPRS Journal of Photogrammetry and Remote Sensing*, vol. 62, no. 2, pp. 135–149, 2007.

Rutzinger, M., Rottensteiner, F., and Pfeifer, N., A comparison of evaluation techniques for building extraction from airborne laser scanning, *IEEE Journal of Selected Topics in Applied Earth Observations and Remote Sensing*, vol. 2, no. 1, pp. 11–20, 2009.

Sampath, A., and Shan, J., Building boundary tracing and regularization from airborne lidar point clouds, *Photogrammetric Engineering and Remote Sensing*, vol. 73, no. 7, pp. 805–812, 2007.

Schenk, T., and Csatho, B., Fusion of lidar data and aerial imagery for a more complete surface description, *International Archives of the Photogrammetry, Remote Sensing and Spatial Information Sciences Science*, vol. XXXIV, pp. 301–317, 2002.

Simonetto, E., Oriot, H., and Garello, R., Rectangular building extraction from stereoscopic airborne radar images, *IEEE Transactions on Geoscience and Remote Sensing*, vol. 43, no. 10, pp. 2386–2395, 2005.

Sirmacek, B., and Unsalan, C., Urban-area and building detection using SIFT keypoints and graph theory, *IEEE Transactions on Geoscience and Remote Sensing*, vol. 47, no. 4, pp. 1156–1167, 2009.

Sithole, G., and Vosselman, G., Experimental comparison of filter algorithms for bare-Earth extraction from airborne laser scanning point clouds, *ISPRS Journal of Photogrammetry and Remote Sensing*, vol. 59, no. 1–2, pp. 85–101, 2004.

Sohn, G., and Dowman, I., Data fusion of high-resolution satellite imagery and LiDAR data for automatic building extraction, *ISPRS Journal of Photogrammetry and Remote Sensing*, vol. 62, no. 1, pp. 43–63, 2007.

Suveg, I., and Vosselman, G., Reconstruction of 3D building models from aerial images and maps, *ISPRS Journal of Photogrammetry and Remote Sensing*, vol. 58, no. 3–4, pp. 202–224, 2004.

Wang, L., and Neumann, U., A robust approach for automatic registration of aerial images with untextured aerial LiDAR data. In *IEEE Computer Society Conference on Computer Vision and Pattern Recognition Workshops, CVPR Workshops 2009*, pp. 2623–2630, 2009.

Yan, G., Zhou, G., and Li, C., Automatic extraction of power lines from large-scale aerial images, *IEEE Transactions on Geoscience and Remote Sensing Letter*, vol. 4, no. 6, pp. 387–391, 2007.

Yu, Y., Buckles, B.P., and Liu, X., Residential building reconstruction based on the data fusion of sparse LiDAR data and satellite imagery, *Lecture Notes in Computer Science*, vol. 5876 LNCS, no. 2, pp. 240–251, 2009. *Advances in Visual Computing – 5th International Symposium, ISVC 2009.*

Zabuawala, S., Fusion of LiDAR and aerial imagery for accurate building footprint extraction. In *Proceedings of SPIE – The International Society for Optical Engineering, IS and T Electronic Imaging – Image Processing: Machine Vision Applications*, vol. 7251, 2009.

Zebedin, L., Towards 3D map generation from digital aerial images, *ISPRS Journal of Photogrammetry and Remote Sensing*, vol. 60, no. 6, pp. 413–427, 2006.

Zhang, K., Yan, J., and Chen, S., Automatic construction of building footprints from airborne LiDAR data. *IEEE Transactions on Geoscience and Remote Sensing*, vol. 44, no. 9, pp.2523–2532, 2006.

Zhou, G. Geo-referencing of video flow from small low-cost civilian UAV, *IEEE Transactions on Automation Engineering and Science*, vol. 7, no. 1, pp. 156–166, 2010.

Zhou, G., et al. Urban 3D GIS from LiDAR and digital aerial images, *Computers and Geosciences*, vol. 30, pp. 345–353, 2004.

Zhou, G., and Zhou, X., Seamless fusion of LiDAR and aerial imagery for building extraction. *IEEE Transactions on Geoscience and Remote Sensing*, vol. 52, no. 11, pp.7393–7407, 2014.

7 Urban 3D Building Extraction from LiDAR and Orthoimages

7.1 INTRODUCTION

During the past decade, digital photogrammetric methods for automatic digital surface model (DSM) or digital terrain model (DTM) generation have become widely used due to the efficiency and cost-effectiveness of the production process. The performance of these systems is very good for smooth terrain at small to medium scale when using small and medium resolution imagery. However, it decreases rapidly for complex scenes in dense urban areas using high-resolution imagery. The degradation in the performance of photogrammetric processes is mainly due to the failures of image matching, which are primarily caused by, for example, occlusions, depth discontinuities, shadows, poor or repeated textures, poor image quality, foreshortening and motion artifacts, and the lack of modeling of man-made objects (Zhou et al. 2004). To offset the effect of these problems, the extraction of buildings and DTM generation in urban areas are currently still done by human-guided interactive operations, such as stereo compilation from a screen. The whole process is both costly and time-consuming. Over the past years, a lot of researchers in the fields of photogrammetry and computer vision have been striving to develop a comprehensive and reliable system with a high success rate with either full automation or semi-automation to ease human–computer interactive operations. However, automatically extracting building information is still an essentially unsolved problem. A lot of efforts for overcoming the problems mentioned above are still needed.

LiDAR (**Li**ght **D**etection **A**nd **R**anging) has been widely applied in urban 3D data analysis since the 1990s. A variety of different methods have been proposed for this purpose, some of which can be found in Tao and Hu (2001). Baltsavias et al. (1995) discuss three different approaches for this purpose, namely using an edge operator, mathematical morphology, and height bins for detection of objects higher than the surrounding topographic surface. These main approaches are also used by other authors like Haala (1995), and Eckstein and Muenkelt (1995). They analyzed the compactness of height bins, or used mathematical morphology (Eckstein and Muenkelt 1995; Hug 1997). Hug (1997) applies mathematical morphology in order to obtain an initial segmentation, and the reflectance data are used to discern man-made objects from natural ones via a binary classification. Other building extraction methods include extraction of planar patches, some of which use height, slope and/or aspect images for segmentation (e.g., Haala et al. 1998; Morgan and Tempfli 2000; Morgan and Habib 2002). In general, these methods can be grouped into two categories (Yoon and Shan 2002): a classification approach and an adjustment approach. The classification approach detects the ground points using certain operators designed based on mathematical morphology (Lindenberger 1993; Vosselman 2000) or terrain slope (Axelsson 1999), or local elevation difference (Wang et al. 2001). A refined classification approach uses a Triangulated Irregular Network (TIN) data structure (Axelsson 2000; Vosselman and Mass 2001) and iterative calculation (Axelsson 2000; Sithole 2001) to consider the discontinuity in the LiDAR data or terrain surface. The adjustment approach essentially uses a mathematical function to approximate the ground surface, which is determined in an iterative least adjustment process, while outliers of non-ground points are detected and eliminated (Kraus and Pfeifer 1998, 2001; Schickler and Thorpe 2001). Despite plenty of efforts that have been made in urban 3D data analysis, difficulties still remain. The DTM generation from LiDAR data is not yet mature (Vosselman and Mass 2001; Yoon and Shan 2002). It has been realized, also by many other photogrammetrists, that methods based on a single terrain characteristic or criterion can hardly obtain satisfactory results in all terrain types.

In this chapter, a combination of LiDAR data and orthoimage data for urban 3D DBM, DSM, and DTM generation is presented below.

7.2 BUILDING DETECTION AND EXTRACTION

7.2.1 Edge Detection from Orthoimage

As described above, the building extraction based on images and LiDAR data cannot reach a satisfactory result. One of main causes is because of breaklines for urban building. It is thus very important to extract the breaklines of a building before applying any interpolation technique, because breaklines can be used to identify the sudden change in slope or elevation. Therefore, detecting breaklines will serve both interpolation and building extraction. In urban areas, most of the breaklines represent parts of artificial objects such as buildings. In digital image, a breakline (edge) is a sharp discontinuity in the gray-level profile. Thus, the simplest edge detection method is to inspect the change in the digital number of each pixel in a neighboring region with the first derivative or the second derivative of the brightness. A lot of edge detection methods have been developed in the past decades in the image processing community. However, the situation is complicated by the presence of noise, image resolution, object complexity, occlusion, shadow, and so on. Our implementation of building edge detection is that the zero-cross edge detection operator is first employed, and then some post-processing, such as merging a line segment into a line, or deleting an isolated point and line segment, are carried out. Finally, a human–computer interactive operation is employed for extraction of complete edges of objects. These extracted edges of objects, associated with the horizontal coordinates, are coded and saved in files in vector format for the interpretation and interpolation of objects (see the description in Section 2.2).

7.2.2 Image Interpretation and Building Extraction

After the complete edges of buildings have been detected, the algorithms for extraction of the building, geometrical parameters for interpretation of objects will be performed. The LiDAR data interpretation is based on two facts: (1) the buildings are higher than the surrounding topographic surface; and (2) the ability of the laser to penetrate vegetation, thus giving echoes from several heights, makes it possible to distinguish between the two classes: man-made objects and vegetation. The extraction procedures are based on an implementation of the minimum description length (MDL) criterion for robust estimation (Rissanen 1983; Axelsson 1992). Thus, the main steps are: (1) linking the 2D complete image edges of building with 3D LiDAR data using horizontal coordinates, (2) determining the three-dimensional building breaklines from image edges and exactly estimating the building boundary via integrating image edges and LiDAR, and (3) interpreting the LiDAR data for buildings or vegetation using two facts and MDL. Internal breaklines can be determined by intersecting the adjacent planar facades within the building. It is known that the laser points are not selective, and they do not match the building boundary. Therefore, one cannot determine the building boundary with only height data, unless the density of LiDAR point cloud is like image gray representation. Figure 7.1 shows a portion of a building near its boundary. Some laser data points are located on the building, while others are located on the ground. The segments of LiDAR data therefore are from the image segments, which describe various buildings. Therefore, The georeferenced images, whose 2D geodetic coordinates are known, are chosen. The horizontal coordinates of the boundary edges are directly used to obtain each 3D building model. The building boundary, in addition to the internal façade parameters and the internal 3D breaklines, will be the result of the building extraction process. The topological relationships of building facades are described in Section 7.3.

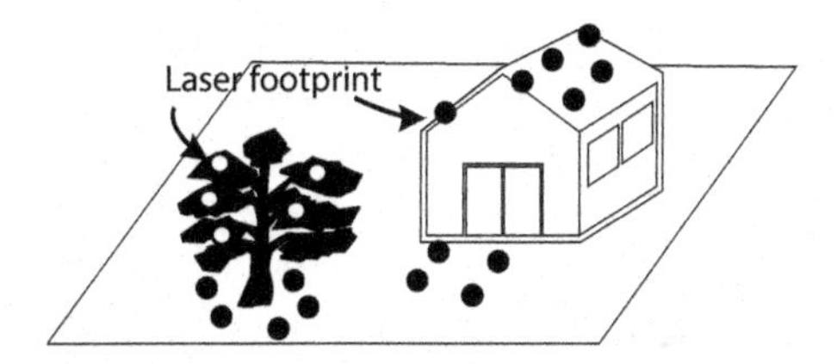

FIGURE 7.1 LiDAR footprints on buildings and vegetation.

7.3 DIGITALLY MODELING BUILDINGS

In our research, an object-oriented data structure has been developed for the description of a digital building model (DBM). During the development of this model, how to make best use of the LiDAR point cloud data sets for creating DBM is seriously analyzed, for instance, the roof, which is obtained from the georeferenced image, provides the boundary and height information for the roof of the building. Therefore, each building is an object of the building class, that is, an entity of the class. One building object consists of the attributes of the building ID, roof type ID, and the series of the roof surfaces. Each surface in the surface series of a building object is also considered an object. The surface's class is comprised of the surface's boundary, the LiDAR footprints within the surface, and planar equation parameters describing the surface by fitting LiDAR footprints. The boundary is composed of a set of points. One of the advantages of this model is its flexibility for future expansion, for example, by adding other building attributes, such as wall surfaces, texture, and so on (see Figure 7.2). This data structure is proposed as follows:

```
typedef struct{
    double dx;
    double dy;
    double dElevation;
} LiDARPoint;
class CBuilding: public CObject
{
protected:
    unsigned m_nBID; //Building ID
    unsigned m_nRoofType; //Roof Type ID
public:
           CTypedPtrList<CObList, CSurface*> m_surfaceList; //Surfaces
           series in one building
........
........
........
};
class CSurface: public Cobject
{
public:
      //Planar equation parameters
      double m_dP1;
      double m_dP2;
      double m_dP3;
public:
       CArray<CPoint, CPoint> m_ptEdgeArray;
       //Point array on behalf of the surface boundary
          CArray<LiDARPoint, LiDARPoint> m_ptLiDARArrayIn; //Series of
          LiDAR points within the footprint of the surface
    ............
    ...........
    ............
    };
```

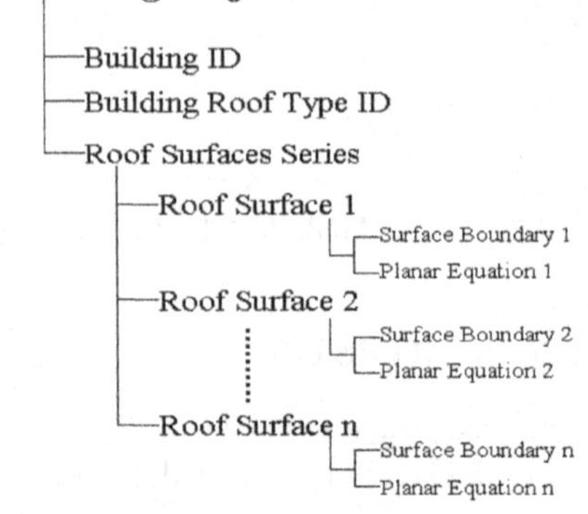

FIGURE 7.2 Object-oriented DBM.

7.4 CREATION OF DIGITAL SURFACE MODEL (DSM)

7.4.1 Establish the Relationship Between Images and LiDAR Point Cloud Data

The orthoimages are stored as raster data, while the LiDAR point cloud is collected along-track. The linkage of the two data sets is implemented by the horizontal coordinates. Thus, the first task is to determine which LiDAR footprints are inside the boundary of the building. To this end, a filling algorithm is used, whose steps are as follows (note that a rectangle is selected as a sample in Figure 7.3):

- *Creation of the polygon of roof's surface.* The edge of a roof surface extracted from the orthoimage is a set of point coordinates like (X_1, Y_1; X_2, Y_2; ……; X_n, Y_n). The surface polygon by connecting the edge points in order can be obtained in this step (see Figure 7.3).
- *Obtain the boundary of the polygon of the roof surface.* For a given roof surface, for example, the coordinates of four corner points of its rectangular boundary can be obtained by (see Figure 7.3):

  Corner 1: (X_1, Y_1)
  Corner 2: (X_2, Y_2)
  Corner 3: (X_3, Y_3)
  Corner 4: (X_4, Y_4)
  ……
  Corner n: (X_n, Y_n)

- *Obtain the reduced LiDAR points within the rectangular boundary.* To speed up the calculation, the LiDAR points are reduced by examining whether these points are in the roof surface or not. By simple comparison of the LiDAR point coordinates and the rectangular corners, the reduced LiDAR points are obtained.

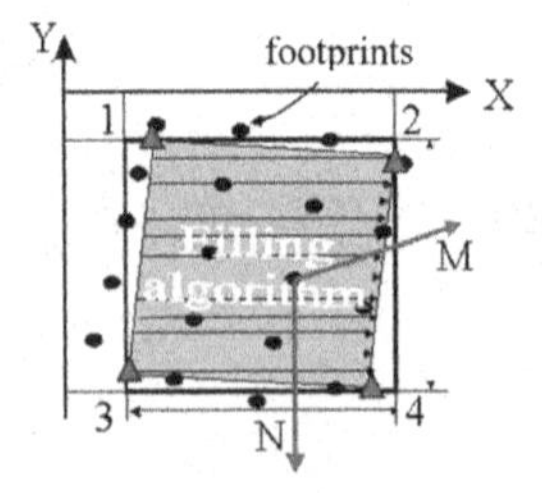

FIGURE 7.3 The determination of inside footprints in a building using a filling algorithm.

- *The determination of the LiDAR footprints in the reduced points.* The determination of whether the LiDAR footprints in the reduced points are inside or outside is carried out by a filling algorithm. This algorithm was realized by Microsoft MFC function, that is, CRgn::PtInRegion in MS VC++.

The above procedure is then repeated for each roof plane of each building until all buildings are implemented.

7.4.2 Interpolation Algorithm via Planar Equation

After a complete extraction of the roof surfaces is obtained, the LiDAR point cloud data within the footprint of the roof surfaces are obtained and stored in an array in a surface object of a building, whose procedure was described in Section 7.1. Now, each building object has its own LiDAR point cloud data, associated with boundary information, with which the DSM of urban areas can be created. There are many interpolation methods available for this purpose. However, these methods cannot meet the accuracy request, such as IDW (Inverse Distance Weight), which calculate the unknown elevation by using the nearby known neighbors, and give them different weight on the basis of the distance between them and the unknown points. An innovative method for LiDAR data interpolation is suggested below. The basic principle is to fit the roof surfaces of the building using a planar equation, which is solved by the LiDAR footprints within the roof boundary that have already been obtained in Section 7.1. The planar equation is:

$$AX + BY + CZ = 1 \tag{7.1}$$

where A, B, and C are unknown parameters, and X, Y, and Z are coordinates of LiDAR data. At least three LiDAR footprints are requested to determine the planar equation (surface of building). However, usually more than three footprints are measured on each surface. The least square method is thus employed to calculate the parameters of the planar equation. The equation is:

$$\begin{bmatrix} X_1, Y_1, Z_1 \\ X_2, Y_2, Z_2 \\ \cdots\cdots\cdots \\ X_m, Y_m, Z_m \end{bmatrix} \times \begin{Bmatrix} A \\ B \\ C \end{Bmatrix} = \begin{Bmatrix} 1 \\ 1 \\ \cdots \\ 1 \end{Bmatrix} \tag{7.2}$$

where m is the number of LiDAR points on a surface. This interpolation method for DSM generation via planar equation and the surface boundary can reach higher accuracy and greater efficiency than the method via LiDAR raw data array interpolation within the boundary.

7.5 CREATION OF DIGITAL TERRAIN MODEL (DTM)

LiDAR data presents two aspects: ground and buildings. Thus, the data could be segmented into two types of regions corresponding, on the one hand, to a surface linked to the ground, and on the other hand, to a surface linked to surface objects. Therefore, the DTM can be generated by separation of the surface objects from the DSM. The DBM has been generated above, and the DTM can be generated by removing the surface objects. The steps are:

1. Based on the extracted boundary of the building in image processing, obtaining the horizontal coordinates of these boundary points.
2. Seeking corresponding LiDAR footprints according to horizontal coordinates.

3. Removing those LiDAR footprints whose horizontal coordinates are the same as one of the building boundaries.
4. Interpolating the DTM via the IDW method.

7.6 EXPERIMENTS

7.6.1 Data Set

The Virginia Department of Transportation (VDOT), contracting to Woolpert LLC at Richmond, Virginia, has established a high-accuracy test field in Wytheville, Virginia. It consists of approximately 21 targeted ground points placed specifically for airborne LiDAR data collection and accuracy test purposes. The field extends from the west side of Wytheville east approximately 114 miles, with a north–south extent of approximately 4.5 miles centered on Wytheville, that is, from latitude, 36° 54′16″ to 36° 59′54″ North, and from longitude, 81° 08′ 23″ to 81° 49′08″ West (see Figure 7.4). The target points are spaced at least several kilometers apart and distributed in a generally east–west direction (see Figure 7.4). The point accuracy can attain standard deviations better than 0.02 m, 0.02 m, and 0.10 m in X, Y, and Z respectively. This level of accuracy is comparable with the geodetic accuracy standard for Order C (1.0 cm plus 10 ppm). This control field will be used for our test of orientation parameter determination.

7.6.1.1 LiDAR Data

The LiDAR data were obtained by using an Optech 1210 LiDAR system in September 2000. The LiDAR data have accuracy (on hard surfaces) of 2.0 feet at least and point sampling density is sufficient to provide an average post spacing of 7.3 feet in the raw DSM (see Figure 7.5). The data were provided in raw text format. The LiDAR parameters used for this project are as follows:

- Aircraft Speed: 202 ft/s
- Flying Height: 4500 ft above ground level
- Scanner Field of View (half angle): ±16 degrees
- Scan Frequency: 14 Hz
- Swath Width: 2581 ft (1806 ft with a 30% sidelap)

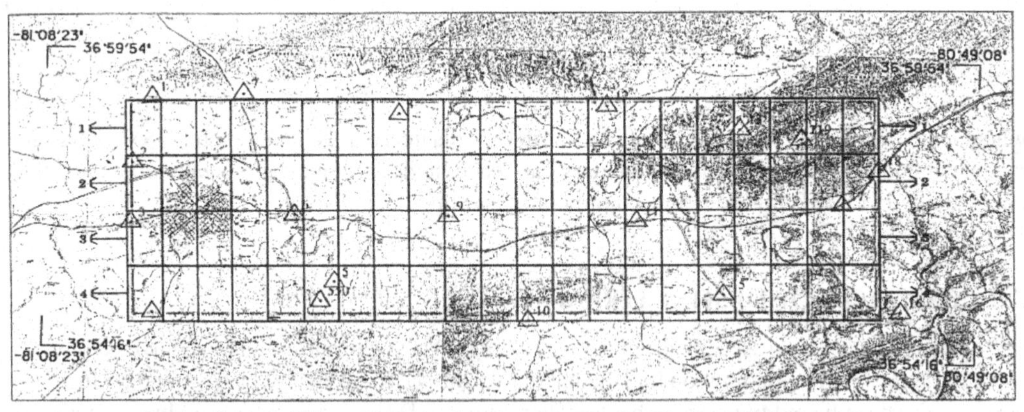

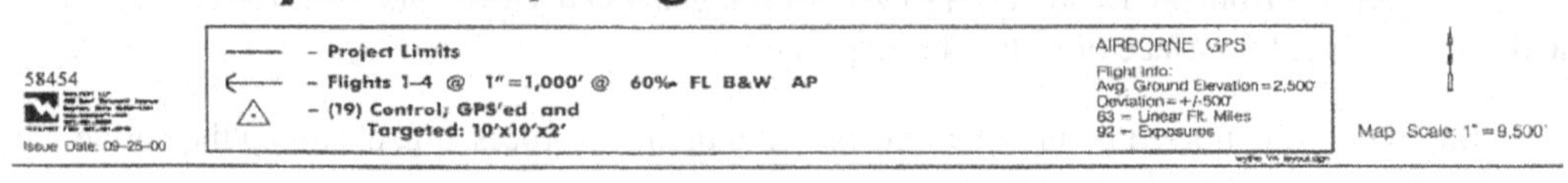

FIGURE 7.4 Geodetic control test field in Wytheville, VA.

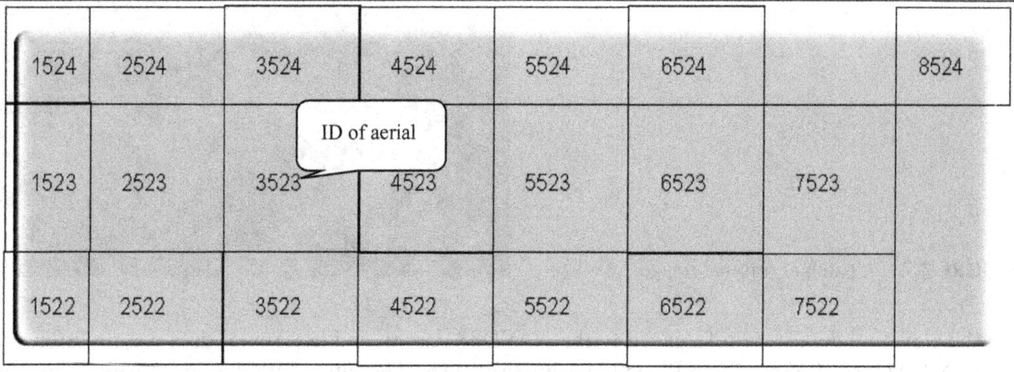

1524	2524	3524	4524	5524	6524		8524
1523	2523	3523	4523	5523	6523	7523	
1522	2522	3522	4522	5522	6522	7522	

ID of aerial

FIGURE 7.5 The semi-automatic urban 3D model generation. (The green points are the LiDAR point cloud, the gray image is aerial images, and the red lines are the detected edges of the building.)

- Pulse Repetition Rate: 10 kHz
- Sampling Density: average 7.3 ft

7.6.1.2 Aerial Image Data

To aid planimetric compilation and quality control of the LiDAR data, the analog black-and-white aerial photography was acquired along east–west flight lines over the project area on September 19, 2000 at a scale of 1:1000. Woolpert camera number 5099 was used. Kodak 2405 file was used with a 525 nm filter and 153.087 mm focal length. A total of 96 exposures were acquired over four equal-length flight lines (see Figure 7.6). The aerial photo has a pixel resolution of 2.0 feet, and the orthoimage was produced using fully differential rectification techniques and the LiDAR DTM.

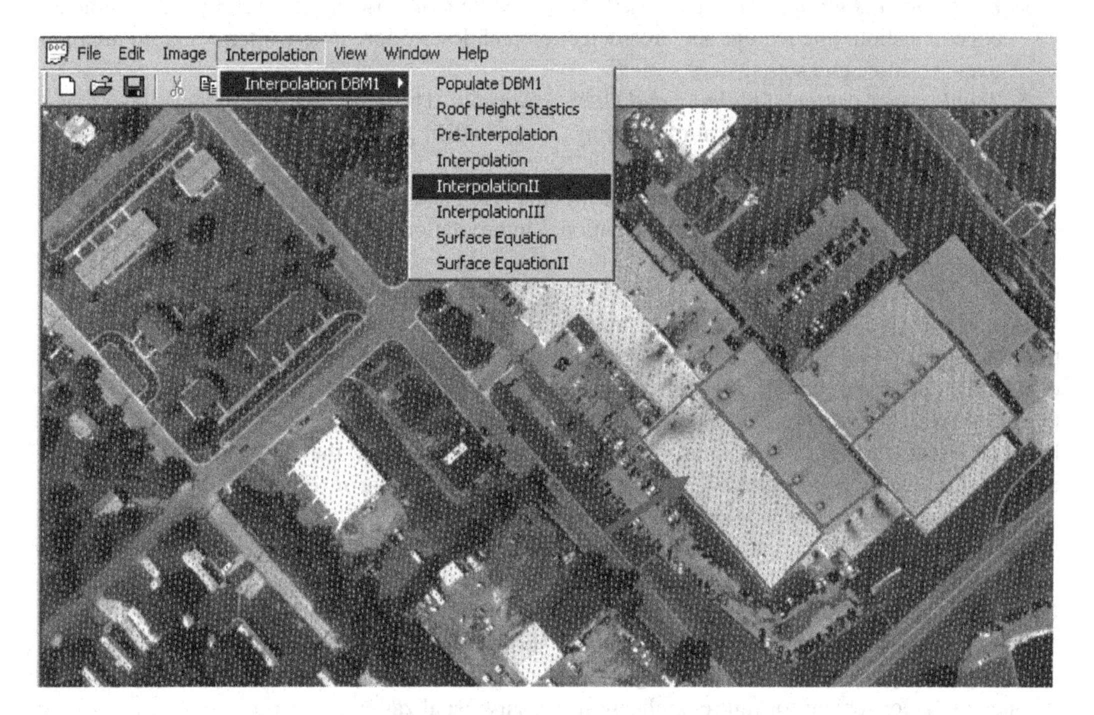

FIGURE 7.6 The configuration for aerial image and LiDAR data collection. (The shadow area indicates the coverage of LiDAR data, and the number indicates the ID of aerial images.)

FIGURE 7.7 A patch of original image.

All the elevation data were referenced to the NAVD88 datum, and horizontal data were referenced to the NAD83/93 Virginia State Plane Coordinate system. The city of Wytheville, Virginia, lies in the west part of the data coverage. According to the availability of data and their precision, the data from the southern part of the city are selected for this chapter's study (see Figure 7.7).

7.6.2 System Development

A software system of semi-automated urban 3D model generation from LiDAR data and image data was developed on the Microsoft Visual C++ platform. The system consists of the following modules.

1. *New/open a project.* This module opens an existing or new project.
2. *LiDAR data check.* This module is to check the systematic errors of LiDAR via various methods, such as overlaying LiDAR data onto georeferenced image, ground control points checks, and so on.
3. *Data input (image and LiDAR).* This module contains LiDAR data input, image display, data format conversion (e.g., for raw image to bmp image, tiff image format, and so on).
4. *Image processing and interactive edit.* This module contains image filtering, enhancement, edge detection, line feature and area detection and description, image interpretation, interactive operation, and so on.
5. *Topology generation of building and DBM generation.* This module is to implement the functions of topologic description of buildings and of DBM generation using an object-oriented data structure.
6. *Urban DSM and DTM generation.* This module is to generate a high-accuracy DSM by applying the surface equation; some conventional interpolation methods, such as IDW, are available. The DTM is generated by removing surface objects.

With this software, a group of experimental results were obtained and are listed in Figure 7.8 through Figure 7.13. Figure 7.8 is the result of automatic detection of building edges, and Figure 7.9 depicts the detected buildings after the human–computer interactive operation. Figure 7.10 depicts the DSM, which is generated by our algorithm. In order to compare the interpolation accuracy between our method and other interpolation methods, for example, the IDW method and Spline method, the results from the IDW and Spline methods are shown in Figures 7.11 and 7.12. As observed from the experimental results, the two interpretation methods, IDW and Spline interpolation, cannot reach high accuracy. The building edges are not very clear. It appears that there are dim slopes to the ground. Also, the roof surfaces are rough, but most of the real roof surfaces are planar. Obviously, the interpolation method proposed in this chapter has much higher accuracy than the IDW and Spline methods. The edges and the roof surfaces are clearer. Figure 7.13 is the DBM. The most important point is that each building is an object in the method proposed in this chapter, which is much more convenient for future application, such as visualization.

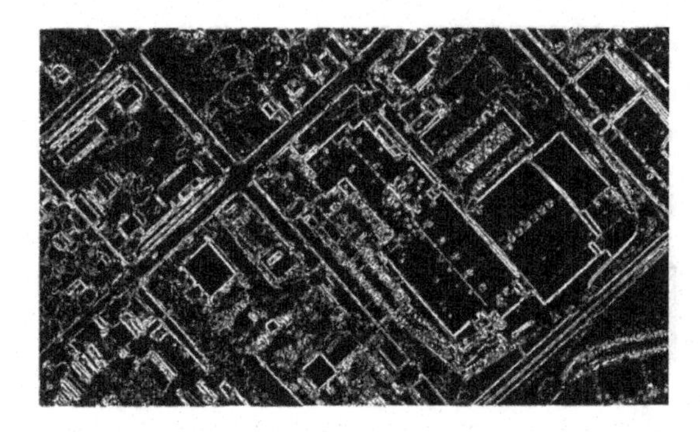

FIGURE 7.8 Automatically detected building edges.

FIGURE 7.9 Detected building edges by human–computer interaction operation (red line in color image).

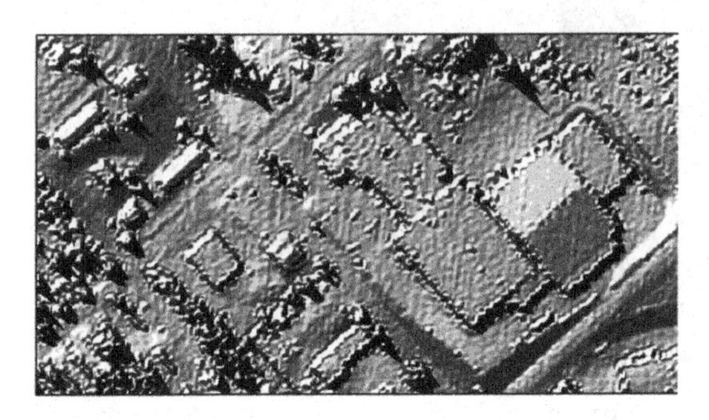

FIGURE 7.10 The result of raw LiDAR data interpolated by our software.

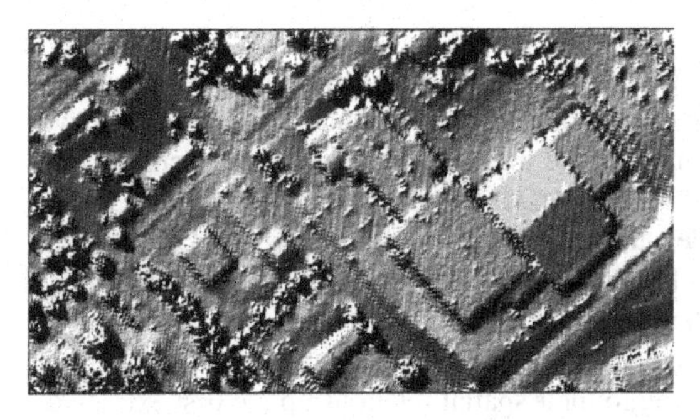

FIGURE 7.11 The result of raw LiDAR data interpolated by IDW.

FIGURE 7.12 The result of raw LiDAR data interpolated by Spline. (The Spline parameters are: Weight = 0.1, number of points = 12, type is regularized.)

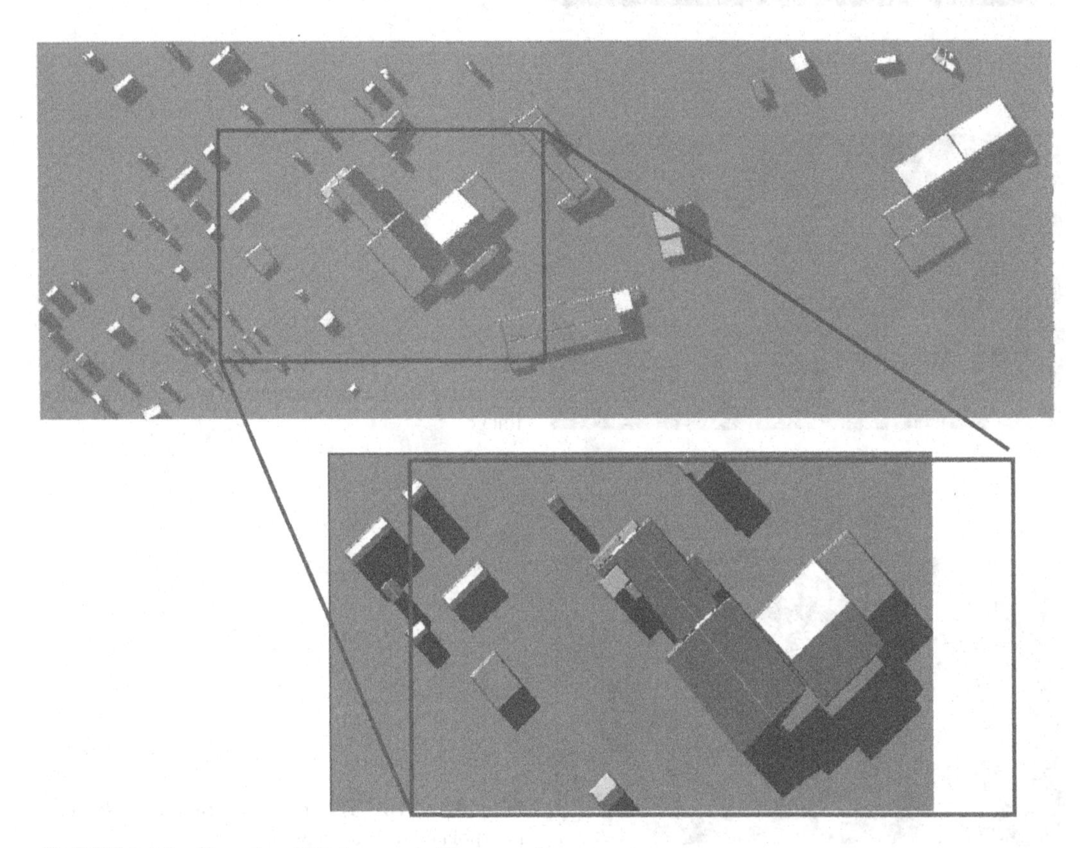

FIGURE 7.13 The urban DBM generated by our software system.

7.7 CONCLUSIONS

In this chapter, the generation of an urban 3D model, including 3D DSM, DBM, and DTM via integrating image knowledge and LiDAR point cloud data, is presented. A human–computer interactive operation system has been developed for this purpose. The main contributions of this chapter are to develop a high-accuracy interpolation method for DBM/DTM/DSM generation and to develop an object-oriented building model. In this model, the roof types, roof surfaces, planar equation parameters, and so on are defined. When considering the roof surfaces in particular, the model consisted of the roof surfaces' boundaries and their planar equations, which are obtained from the combined processing of the LiDAR and orthoimage data. For the planar equation of each roof surface, the LiDAR point data lying within it according to their spatial relationship are first extracted, and the planar equation's parameters with these LiDAR points are calculated using the least-squares method. The planar equation is applied to calculate the grid value within the roofs. The experimental results demonstrated that the high accuracy of the DSM, DBM, and DTM in urban areas has been reached via the software system developed in this chapter.

REFERENCES

Axelsson, P., Minimum description length as an estimator with robust properties. In: Foerstner, W., Ruwiedel, S. (Eds), *Robust Computer Vision*, Wichmann-Verlag, Karlsruhe, pp. 137–150, 1992.

Axelsson, P., Processing of laser scanner data – Algorithms and applications, *ISPRS Journal of Photogrammetry & Remote Sensing*, vol. 54, pp. 138–147, 1999.

Axelsson, P., DEM generation from laser scanner data using adaptive TIN models, *International Archive of Photogrammetry and Remote Sensing*, vol. XXXIII, part B4, pp. 110–117, 2000.

Baltsavias, E., et al., Use of DTMs/DSMs and orthoimages to support building extraction. In: Gruen, Al, Kubler, Ol, Agouris, P. (Eds), *Automatic Extraction of Man-Mad Objects from Aerial and Space Images*, Birkhaeuser, Basel, pp. 199–210, 1995.

Eckstein, W., and Muenkelt, O., Extracting objects from digital terrain models. In: Schenk, T. (Eds), *Remote Sensing and Reconstruction for Three-Dimensional Objects and Scenes. Proceedings of SPIE Symposium on Optical Science, Engineering, and Instrumentation*, vol. 2572, San Diego, 1995.

Haala, N., 3D building reconstruction using linear edge segments. In: Fitsch, D., Hobbie, D. (Eds), *Photogrammetric Week*. Wichmann, Karlsruhe, pp. 19–28, 1995.

Haala, N., Brenner, C., and Anders, K., 3D urban GIS from laser altimeter and 2D map data. *International Archives of Photogrammetry and Remote Sensing*, vol. 32, no. 3/1, pp. 339–346, 1998.

Hug, C., Extracting artificial objects from airborne laser scanner data. In: Gruen, A., Baltsavias, E., Henricsson, O. (Eds), *Automatic Extraction of Man-Made Objects from Aerial and Space Images (II)*. Birkhaeuser, Basel, pp. 203–212, 1997.

Kraus, K., and Pfeifer, N., Advanced DEM generation from LiDAR data. In: Hofton, M.A. (Ed.), *Proceedings of the ISPRS Workshop on Land Surface Mapping and Characterization Using Laser Altimetry, Annapolis, Maryland, International Archives of the Photogrammetry, Remote Sensing and Spatial Information Sciences*, vol. XXXIV, part 3/W4 Commission III, 2001.

Kraus, K., and Pfeifer, N., Determination of Terrain models in wooded areas with airborne laser scanner data, *ISPRS Journal of Photogrammetry and Remote Sensing*, vol. 53, pp. 193–203, 1998.

Lindenberger, J., *Laser-Profilmessungen zur topographischen Gelaedeaufnahme*, Deutsche Geodaetische Kommission, Series C, No. 400, Munich, 1993.

Morgan, M., and Habib, A., Interpolation of LiDAR data and automatic building extraction, *ASPRS Annual Conference*, CD-ROM, Washington, DC, April 19–25, 2002.

Morgan, M., and Tempfli, K., Automatic building extraction from airborne laser scanning data, *International Archives of Photogrammetry and Remote Sensing*, vol. 33, no. B3, Amsterdam, pp. 616–623, 2000.

Rissanen, J., A universal prior for integers and estimation by minimum description length. *The Annals of Statistics*, vol. 11, no. 2, pp. 416–431, 1983.

Schickler, W., and Thorpe, A., Surface estimation based on LiDAR, *Proceedings of ASPRS Annual Conference*, CD-ROM, Washington, DC, 2001.

Sithole, G., Filtering of laser altimetry data using a slope adaptive filter In: Hofton, M.A. (Ed.), *Proceedings of the ISPRS Workshop on Land Surface Mapping and Characterization Using Laser Altimetry, Annapolis, Maryland, The International Archives of the Photogrammetry, Remote Sensing and Spatial Information Sciences*, Vol. XXXIV part 3/W4 Commission III, pp. 203–210, 2001.

Tao, C., and Hu, Y., A review of post-processing algorithms for airborne LiDAR data, *Proceedings of ASPRS Annual Conference*, CD-ROM, Washington, DC, 2001.

Vosselman, G., Slope based filtering of laser altimetry data, *International Archives of Photogrammetry and Remote Sensing*, vol. XXXIII, part B3, Amsterdam 2000, pp. 935–942, 2000.

Vosselman, G., and Mass, H., Adjustment and filtering of raw laser altimetry data, *Proceedings OEEPE Workshop on Airborne Laser Scanning and Interferometric SAR for Detailed Digital Elevation Models, Stockholm*, March 1–3, 2001.

Wang, Y., et al. Automatic generation of bald earth digital elevation models from digital surface models created using airborne IFSAR, *Proceedings of ASPRS Conference*, CD-ROM, Washington, DC, 2001.

Yoon, J., and Shan, J., Urban DEM generation from raw airborne LiDAR data, *ASPRS Annual Conference*, CD-ROM, Washington, DC, April 19–25, 2002.

Zhou, G., et al. Urban 3D GIS from LiDAR and digital aerial images. *Computers & Geosciences*, vol. 30, no. 4, pp. 345–353, 2004.

8 Vehicle Extraction from High-Resolution Aerial Images

8.1 INTRODUCTION

With rapidly increasing traffic volume and road congestion, traffic monitoring, especially in urban areas, is highly required. A lot of ground sensors, such as induction loop detectors (Zheng and Li 2007; Zheng et al. 2012; Hinz et al. 2005), bridge sensors, and stationary cameras (Liu et al. 2011; Yamazaki and Liu 2008; Salehi et al. 2012; Mishra 2012) are mounted. However, these sensors can only partially acquire the traffic information from major arterials in urban areas. The traffic conditions of the majority of highways are seldom collected (Liu and Li 2001; Dubuisson and Jain 1995). Hence, area-wide images of the entire highway networks are required to complement the traffic monitoring, which have more advantages over ground-based monitoring. Moreover, vehicles, even cars, can be clearly identified on high-resolution aerial images with a ground resolution of a ground sampling distance (GSD) of 0.15 m × 0.15 m. Thus, using the high-resolution aerial imagery to detect vehicles for traffic monitoring and mitigation of an area with a large spatial extent is very attractive.

Of the existing vehicle detection methods using aerial imagery (Guigues et al. 1996; Schlosser et al. 2003; Sharma et al. 2006; Papageorgiou and Poggio 2000; Burlina et al. 1995; Moon et al. 2002), two types of vehicle models have been basically used: (1) an appearance-based implicit model, and (2) an explicit model. The implicit model typically consists of image intensity or texture features computed using a small window or kernel that surrounds a given pixel or a small cluster of pixels. Detection is conducted by examining feature vectors of the immediate surrounding pixels of the image. For example, Guigues et al. (1996) developed a two-step analysis method that is composed of vehicle detection on the aerial imagery with 0.3–0.4 m GSD, and then validated the method through line clustering. In the detection, the intensity values of surrounding pixels were analyzed using a multilayer perceptron (MLP). In the validation, the perceptual grouping theory was used to group the vehicles into lines. In addition, Papageorgiou and Poggio (2000) presented a trainable system for the vehicle detection from aerial imagery taken from a stationary camera. A Haar wavelet transformation was employed to describe the object classes in terms of local, directional, and multi-scale differences of intensity around adjacent regions. The detection model was derived through the training of a support vector machine classifier using numerous positive and negative examples of vehicles. In the explicit models, a vehicle is usually described by a box or wire-frame representation. Detection is carried out by matching the model to the image with a "top-down" strategy, or grouping low-level features to construct structures similar to the model using a "bottom-up" approach. For instance, a vehicle was represented as a 3D box with dimensions in width, length, and height (Burlina et al. 1995; Moon et al. 2002). Site models were used to constrain vehicles in parking lots or on roads.

These early approaches are time-consuming, and although a few methods demonstrate a promising result, such as Liu et al. (2011), Yamazaki and Liu (2008), Salehi et al. (2012), and Mishra (2012), it is still questionable whether or not they are operational, according to Kluckner et al. (2007), Nguyen et al. (2007), and Grabner et al. (2008). In addition, the overall accuracy in vehicle detection can be low or unsatisfactory (Zheng and Li 2007; Hinz et al. 2005; Salehi et al. 2012). For this reason, this chapter proposes a method to detect vehicles using gray-scale opening transformation and gray-scale top-hat transformation.

8.2 VEHICLE DETECTION FROM AERIAL IMAGERY

8.2.1 Structure Element Identification

A vehicle's length varies from a couple of meters to tens of meters for a car with a long trailer or a semi-truck. However, the width of a vehicle is around 2 m. On an image with a spatial resolution of 0.15 m × 0.15 m (see Figure 8.1), the width dimension is 13–14 pixels. On the other hand, because the vehicle in width dimension (e.g., a car) is basically considered as symmetrical, a disc with a radius of about 5 pixels is considered as a structure element. Thereby, its diameter is a little smaller than the width of the vehicle.

8.2.2 Gray-Scale Morphological Method

The detailed discussions of the gray-scale morphological method can be found in Jin and Davis (2007, 2004), Serra (1988), Chi et al. (2009), Otsu (1979), Wiedemann et al. (1998), Niu (2006), and Zheng et al. (2013). This section directly applies the theory, that is, the gray-scale opening transformation and gray-scale closing transformation are directly presented here as:

$$\text{Opening}: f \circ b = \left(f \Theta b\right) \oplus b \tag{8.1}$$

$$\text{Closing}: f \bullet b = \left(f \oplus b\right) \Theta b \tag{8.2}$$

where f presents the original image, b is the structure element, ∘ presents the gray-scale opening transformation, and • presents the gray-scale closing transformation, while Θ and ⊕ represent the erosion operator and dilation operator, respectively. The gray-scale top-hat transformation and gray-scale bot-hat transformation are defined as follows.

$$\textit{Top-hat}: T = f - f \circ b \tag{8.3}$$

$$\textit{Bot-hat}: B = f \bullet b - f \tag{8.4}$$

where T presents the image after the gray-scale top-hat transformation, and B is the image after the gray-scale bot-hat transformation.

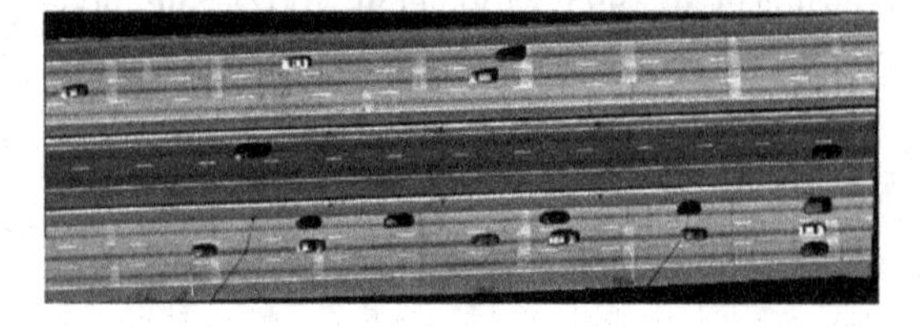

FIGURE 8.1 An image collected on a highway with the GSD of 0.15 m × 0.15 m.

8.2.3 Background Estimation

On the one hand, after the gray-scale opening transformation, bright pixels representing dashed white lines for lane division and concrete road dividers with a bright surface background, as well as a little target of bright color can be filtered out as noise. Thus, the gray-scale opening transformed image can be considered as an estimation of the background (see Figure 8.2). Then, the top-hat transformed image (see Figure 8.3) is derived with the original image (see Figure 8.1) minus the gray-scale opening transformed image.

On the other hand, after the gray-scale closing transformation, pixels with low intensity or a small digital number (DN) can be filtered out as noise. The pixels could be new pavement with tar or a target with a dark or black color. Thus, the image after the gray-scale closing transformation can also be considered as a background estimation as well (see Figure 8.4). The transformed gray-scale bot-hat image (see Figure 8.5) is derived through the transformed gray-scale closing image (see Figure 8.4) minus the original image (see Figure 8.1).

8.2.4 Vehicle Detection

The transformed gray-scale top-hat image (see Figure 8.3) can be converted into a black and white or binary image (see Figure 8.6) with the Otsu threshold method (Otsu 1979). Since the dimensions of most of family cars are 5 m in length × 2 m in width or less than this size, the number of pixels

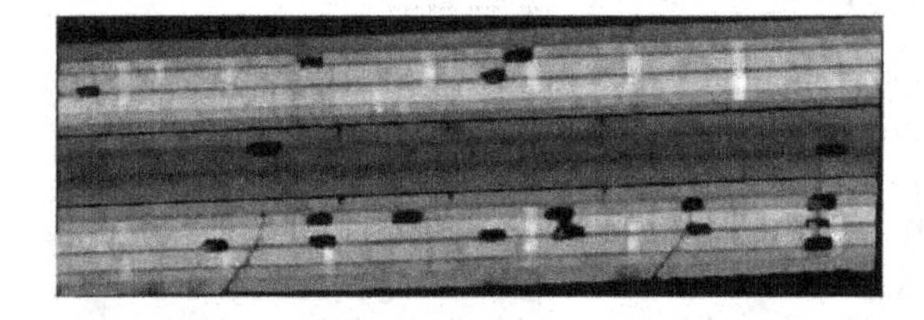

FIGURE 8.2 The gray-scale opening transformation image.

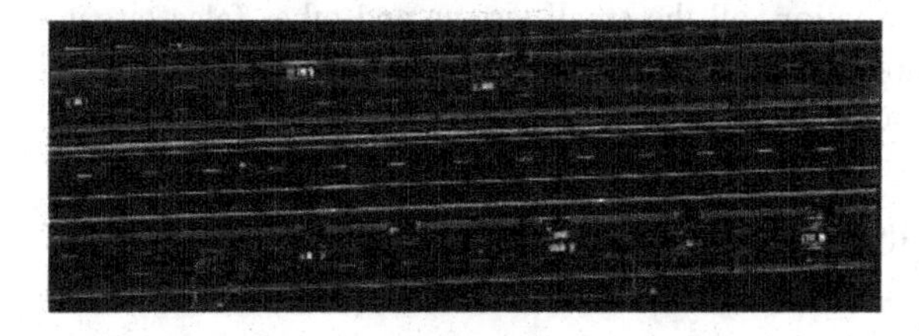

FIGURE 8.3 The gray-scale top-hat transformation image.

FIGURE 8.4 The gray-scale closing transformation image.

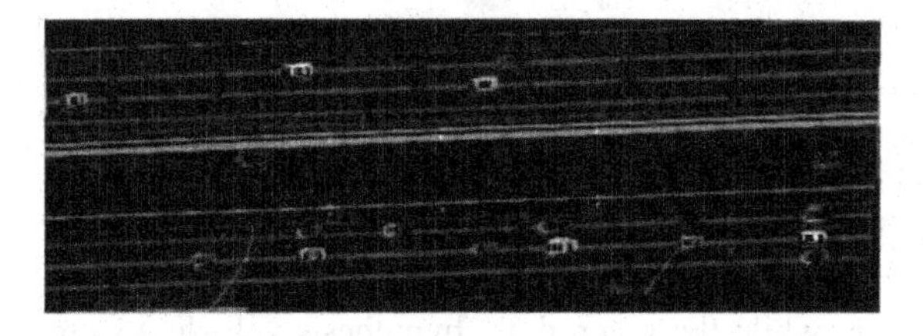

FIGURE 8.5 The gray-scale bot-hat transformation image.

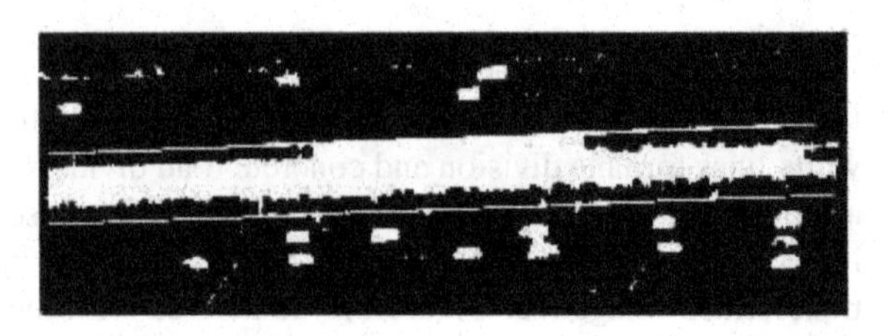

FIGURE 8.6 The gray-scale top-hat transformation image after the partitioning using the Otsu's method.

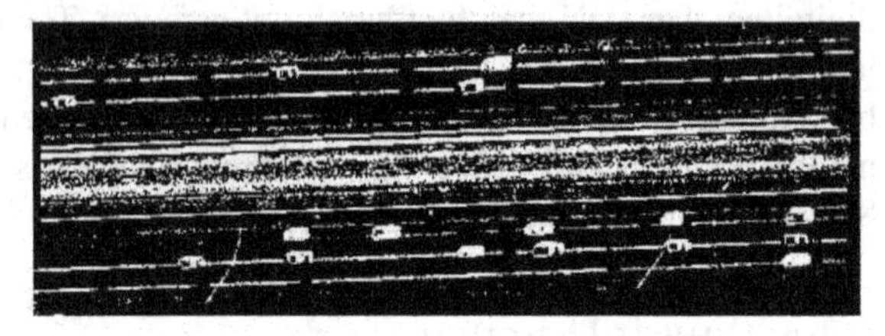

FIGURE 8.7 The gray-scale bot-hat transformation image after the partitioning using the Otsu's method.

representing a family car is around 34 × 14 pixels in the imagery with a GSD of 0.15 m × 0.15 m. When considering a car with a long trailer or a semi-truck, an area threshold with 2000 pixels is set. For example, a long vehicle may be 15 m in length, which corresponds to approximately 100 pixels, and 3 m in width, which corresponds to approximately 20 pixels. Then, a target with an area larger than 2000 pixels is sieved in the detection. In addition, a small car, whose width is less than 10 pixels, can be filtered out using the morphological opening transformation with the initial structure element of a 5-pixel radius. Thus, the vehicles on a light background are detected.

The gray-scale bot-hat transformation image (see Figure 8.6) can also be converted into a black and white or binary image (see Figure 8.7). Because the vehicle targets adhere to the background together (see Figure 8.7), the false big area targets are not sieved when using the area threshold. Otherwise, the vehicles on a dark background can be sieved together with the background. However, when using a morphological opening transformation, all the small targets and other false targets whose width is less than 10 pixels can be smoothed out, even though their area could be very big. Thus, the vehicles on a dark background and some vehicles on a light background are detected.

8.3 EXPERIMENTS AND DISCUSSION

8.3.1 Data Set

The study area was the city of Norfolk, Virginia, USA. The GIS road vector map covering streets and highways of the city was downloaded from the Hampton Roads Transportation Planning Organization (http://hrtpo.org/). The nadir-view aerial images were acquired at the same time of day in spring of 2007. The GSD of the imagery is 0.15 m × 0.15 m. Seventeen highway scenes were selected. With a few exceptions, such as a car with a long trailer or a semi-truck, the size of most vehicles is approximately 5 m × 2 m, which corresponds to about 34 pixels in length and 14 pixels in width (see Figure 8.1).

8.3.2 Vehicle Extraction Results

With the method described above, vehicles were detected from Scene 3 (see Figure 8.1), where the background is light and dark, respectively. The results are presented in Figures 8.8, 8.9, 8.10, and 8.11, respectively. Afterward, the hypothesis vehicles detected from the two cases were overlaid (see Figure 8.12). The red dots (hypothesis vehicles) were detected using gray-scale opening transformation and gray-scale top-hat transformation, while the green dots (hypothesis vehicles) were

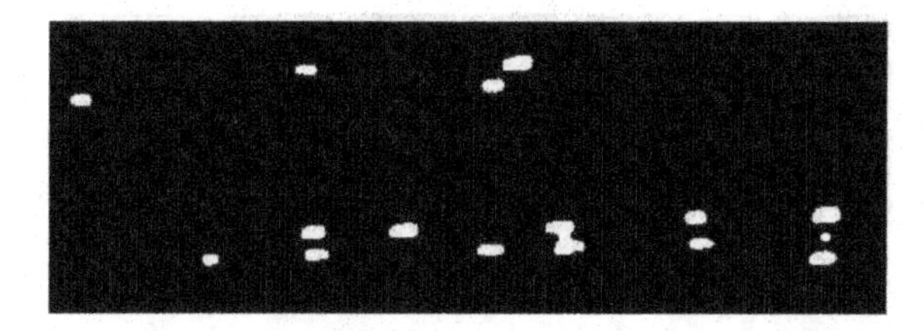

FIGURE 8.8 The results of vehicle detection using gray-scale top-hat transformation.

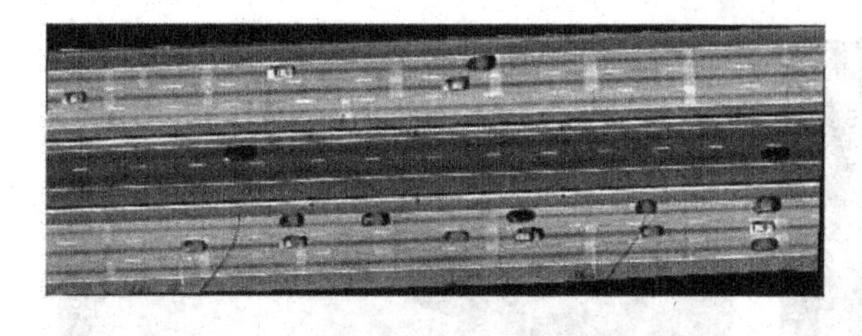

FIGURE 8.9 The results of vehicle detection on a light background.

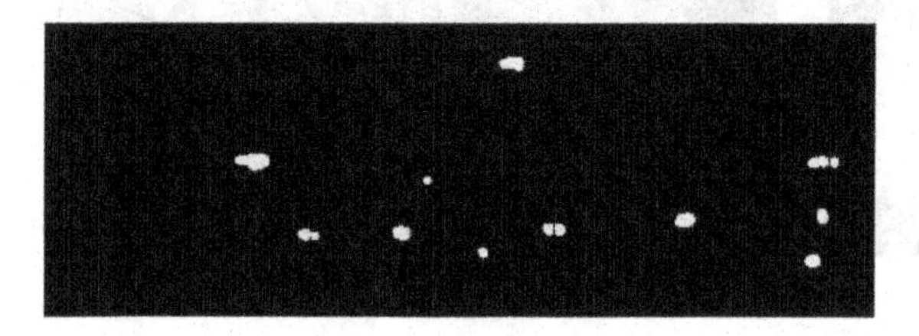

FIGURE 8.10 The results of vehicle detection using gray-scale bot-hat transformation.

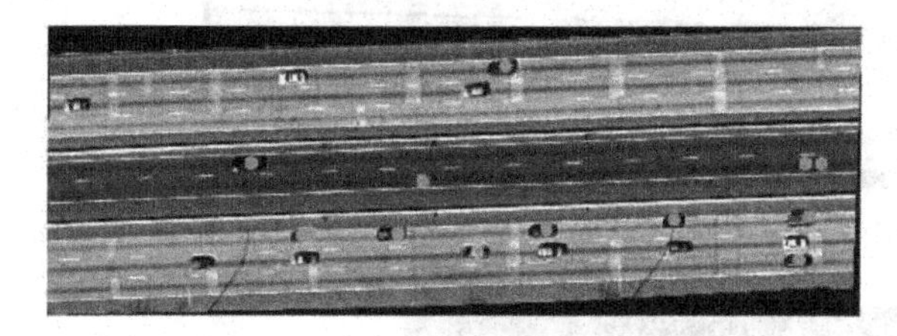

FIGURE 8.11 The results of vehicle detection on a dark background.

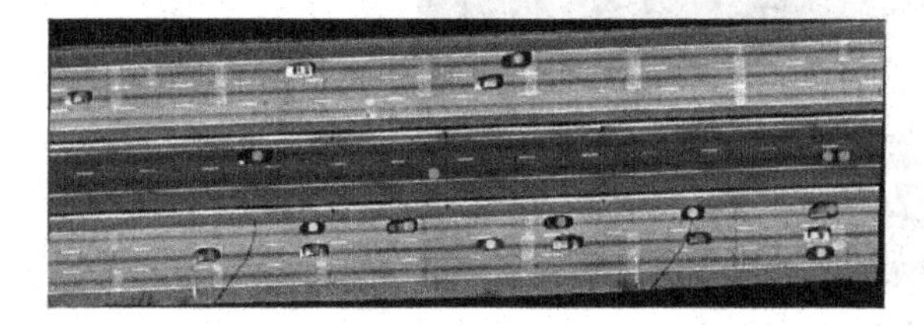

FIGURE 8.12 Overlay of vehicle detection results from two cases.

detected using gray-scale closing transformation and gray-scale bot-hat transformation. The yellow dots (hypothesis vehicles) were detected twice by the two cases, and the two identical hypothesis vehicles were subsequently amalgamated to a single one with closing transformation during the accuracy assessment. Figure 8.13 showed the final result of the vehicle detection from Scene 3. Similarly, the hypothesis vehicles on the 16 other highway scenes were detected using the same method, and the results for vehicle detection from the five typical scenes are shown in Figure 8.14.

In order to evaluate the results of vehicle detection using the method proposed above, a numerical accuracy assessment was conducted by comparing the number of vehicles identified manually and

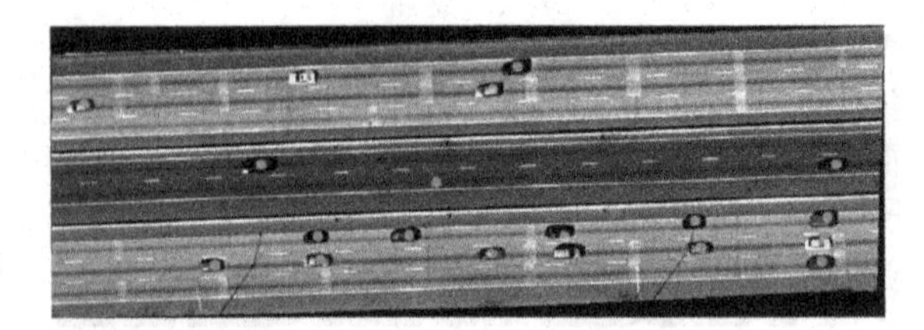

FIGURE 8.13 The results of vehicle detection from two cases.

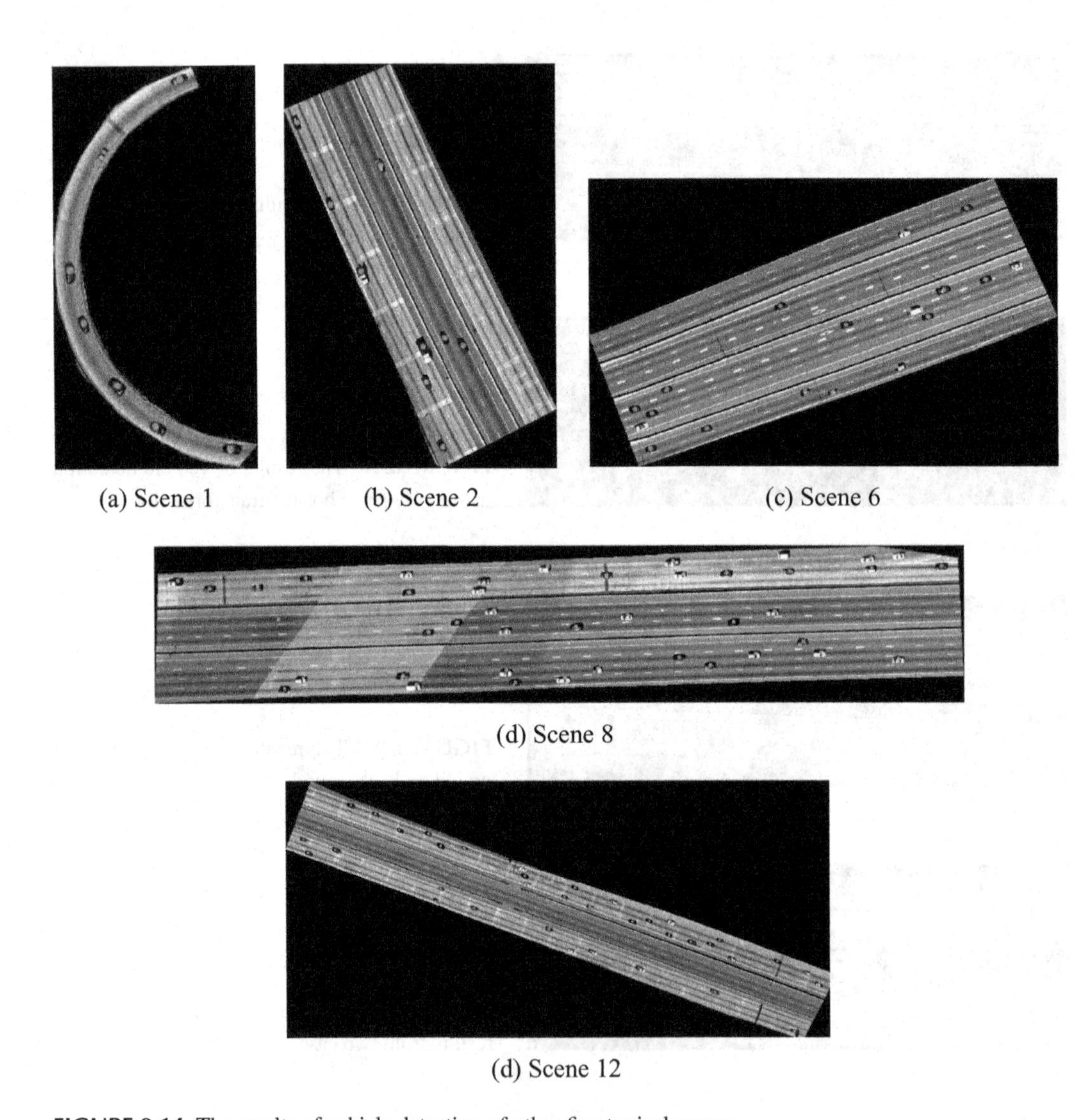

(a) Scene 1 (b) Scene 2 (c) Scene 6

(d) Scene 8

(d) Scene 12

FIGURE 8.14 The results of vehicle detection of other five typical scenes.

automatically by the proposed method. According to Wiedemann et al. (1998), three categories of extraction results are defined as follows:

- True Positive (TP): the number of correctly extracted true vehicles.
- False Positive (FP): the number of incorrectly extracted false vehicles.
- False Negative (FN): the number of omitted vehicles.

With the three definitions above, three statistical measures are computed by

$$Correctness = \frac{TP}{TP + FP} \times 100\% \tag{8.5}$$

$$Completness = \frac{TP}{TP + FN} \times 100\% \tag{8.6}$$

$$Quality = \frac{TP}{TP + FP + FN} \times 100\% \tag{8.7}$$

The correctness is a measure ranging between 0 and 1 that indicates the detection accuracy rate relative to ground truth. The correctness and completeness are the converse of commission and omission errors, respectively. The two measures are complementary and need to be interpreted simultaneously (Niu 2006). The quality shows the overall accuracy of the extraction method, and the quality value can never be higher than either the completeness or correctness value. Table 8.1 reports the three calculated measures for vehicle detection results from the 17 highway scenes.

TABLE 8.1
Quality Measures for Vehicle Detection Results from the 17 Highway Scenes

Highway	TP	FP	FN	Correctness (%)	Completeness (%)	Quality (%)
Scene 1	7	0	0	100.00	100.00	100.00
Scene 2	9	0	0	100.00	100.00	100.00
Scene 3	17	1	1	94.44	94.44	89.47
Scene 4	7	1	0	87.50	100.00	87.50
Scene 5	9	0	2	100.00	81.82	81.82
Scene 6	17	0	1	100.00	94.44	94.44
Scene 7	11	0	2	100.00	84.62	84.62
Scene 8	39	1	2	97.50	95.12	92.86
Scene 9	26	0	3	100.00	89.66	89.66
Scene 10	24	0	2	100.00	92.31	92.31
Scene 11	18	0	2	100.00	90.00	90.00
Scene 12	37	0	0	100.00	100.00	100.00
Scene 13	18	0	3	100.00	85.71	85.71
Scene 14	16	1	0	94.12	100.00	94.12
Scene 15	14	0	1	100.00	93.33	93.33
Scene 16	17	0	1	100.00	94.44	94.44
Scene 17	29	1	4	96.67	87.88	85.29
Total	315	5	24	98.44	92.92	91.57

8.3.3 Discussion

As observed from Figure 8.14, 17 vehicles were successfully detected from Scene 3. Only one vehicle was omitted, and one noisy area was incorrectly extracted as a vehicle. The omitted vehicle is near to one vehicle, with a white color like the background. In fact, the omitted vehicle was detected, and it can be seen from Figures 8.6 and 8.8, where the two vehicles adhere to each other. Thus, one vehicle with a light background was omitted using the gray-scale closing transformation and gray-scale top-hat transformation methods (see Figure 8.9). As observed from Figure 8.9, no false vehicles were detected from the imagery with a light background. As observed from Figure 8.11, one false vehicle was detected from the imagery with a dark background. Since the vehicle targets on the imagery with a dark background adhered to the background together (see Figure 8.8), the false hypothesis vehicles with big area could not be sieved using an area threshold. Otherwise, the vehicles on the imagery with a dark background could be sieved. When using a morphological opening transformation, all the small targets and other targets whose widths are less than 10 pixels are smoothed out, but only one false hypothesis vehicle happened. As observed from Scene 3, the correctness and completeness achieved up to 94.44%, and the quality of assessment achieved up to 89.47% (see Table 8.1).

As observed from Table 8.1, the correctness, completeness, and quality for Scene 1, Scene 2, and Scene 12 can achieve up to 100%, and a car with a trailer was detected in Scene 2 (see Figure 8.14). As observed from Scene 6, the correctness, completeness, and quality can achieve up to 100%, 94.44%, and 94.44%, respectively (see Table 8.1). A vehicle was omitted because of its low contrast to imagery with the light background, and a truck was detected (see Figure 8.14). As observed from Scene 8, the correctness, completeness, and quality can achieve 97.5%, 95.12%, and 92.86%, respectively (see Table 8.1). One noisy area was incorrectly extracted as a vehicle. One vehicle near to another one was omitted, and one vehicle was omitted because of its low contrast on the imagery with the light background (see Figure 8.14).

In summary, of the 17 images covering highway scenes, the correctness, completeness, and quality of the proposed method for vehicle extraction can achieve an average of 98%, 93%, and 92%, respectively.

8.4 CONCLUSIONS

The gray-scale morphological algorithm was developed in this chapter to detect vehicles from high-resolution aerial photos covering highway scenes. The major components of the algorithm included the gray-scale opening transformation, gray-scale top-hat transformation, gray-scale closing transformation, and gray-scale bot-hat transformation. The GIS road vector layer was used to constrain the detection algorithm to the highway networks. Of the 17 highway scenes validated, the correctness, completeness, and quality of the proposed method can achieve up to 98%, 93%, and 92%, respectively. Therefore, the proposed vehicle detection method is robust and efficient when detecting vehicles from high-resolution aerial images.

The further work should include: (1) optimizing the length of the disc's radius, and (2) detecting the vehicles partially occluded by trees or bridges. It should be admitted that the proposed method cannot be applicable in a complex environment, since the correctness, completeness, and quality can largely decrease with noise.

REFERENCES

Burlina, P., et al. "Context-based exploitation of aerial imagery," in *IEEE Workshop on Context-Based Vis.*, pp. 38–49, 1995.

Chi, H., et al. "Vehicle detection from satellite images: Intensity, chromaticity, and lane-based method," *J. Transp. Research Board, National Research Council*, pp. 109–117, 2009.

Dubuisson, M., and Jain, A.K., "Contour extraction of moving objects in complex outdoor scenes," *Int. J. Comput. Vis.*, vol. 14, no. 1, pp. 83–105, 1995.

Grabner, H., et al. "On-line boosting-based car detection from aerial images," *ISPRS J. Photogramm. Remote Sens.*, vol. 63, no. 3, pp. 382–396, 2008.

Guigues, R.L., Airault, S., and Jamet, O., "Vehicle detection on aerial images: A structural approach," in *Proc. Int. Pattern Recognition*, pp. 900–904, 1996.

Hinz, S., Leitloff, J., and Stilla, U., "Context-supported vehicle detection in optical satellite images of urban areas," in *Proc. IEEE IGARSS 2005*, vol. 4, pp. 2937–2941, 2005.

Jin, X., and Davis, C.H., "Vector-guided vehicle detection from high-resolution satellite imager," in *Proc. IEEE IGARSS 2004*, vol. 2, pp. 1095–1098, 2004.

Jin, X., and Davis, C.H., "Vehicle detection from high-resolution satellite imagery using morphological shared-weight neural networks," *Image Vis. Comput.*, vol. 25, no. 9, pp. 1422–1431, 2007.

Kluckner, S., et al. "A 3-D teacher for car detection in aerial images," in *Proc. IEEE Int. Conf. Comput. Vis.*, pp. 1–8, 2007.

Liu, G., and Li, J., "Moving target detection via airborne HRR phased array radar," *IEEE Trans. Aerosp. Electron. Syst.*, vol. 37, no. 3, pp. 914–924, 2001.

Liu, W., Yamazaki, F., and Vu, T.T., "Automated vehicle extraction and speed determination from Quickbird satellite images," *IEEE J. Sel. Topics Appl. Earth Observ. Remote Sens.*, vol. 4, no. 1, pp. 75–82, 2011.

Mishra, R.K., "Automatic moving vehicle's information extraction from one-pass WorldView-2 satellite imagery," *Int. Arch. Photogramm. Remote Sens. Spatial Inform. Sci.*, vol. XXXIX, part B7, 2012 XXII ISPRS Congress, pp. 323–328, 2012.

Moon, H., Chellappa, R., and Rosenfeld, A., "Performance analysis of a simple vehicle detection algorithm," *Image Vis. Comput.*, vol. 20, no. 1, pp. 1–13, 2002.

Nguyen, T., et al. "On-line boosting for car detection from aerial images," in *Proc. IEEE Int.Conf. Res., Innov. Vis. Future (RIVF07)*, pp. 87–95, 2007.

Niu, X., "A semi-automatic framework for highway extraction and vehicle detection based on a geometric deformable model," *ISPRS J. Photogramm. Remote Sens.*, vol. 61, no. 3–4, pp. 170–186, 2006.

Otsu, N., "A threshold selection method from gray-level histograms," *IEEE Trans. Syst., Man and Cybernetics*,vol. 9, no. 1, pp. 62–66, 1979.

Papageorgiou, C., and Poggio, T., "A trainable system for object detection," *Int. J. Comput. Vis.*, vol. 38, no. 1, pp. 15–33, 2000.

Salehi, B., Zhang, Y., and Zhong, M., "Automatic moving vehicles information extraction from single-pass WorldView-2 imagery," *IEEE J. Sel. Topics Appl. Earth Observ. Remote Sens.*, vol. 5, no. 1, pp. 135–145, 2012.

Schlosser, C., Reitberger, J., and Hinz, S., "Automatic car detection in high resolution urban scenes based on an adaptive 3-D-model," in *Proc. 2nd GRSS/ISPRS Joint Workshop on Data Fusion and Remote Sens. Over Urban Area*, pp. 167–170, 2003.

Serra, J., *Image Analysis and Mathematical Morphology*, Vol. 2, Academic Press, New York, 1988.

Sharma, G., et al. "Vehicle detection in 1-m resolution satellite and airborne imagery," *Int. J. Remote Sens.*, vol. 27, no. 4, pp. 779–797, 2006.

Wiedemann, C., et al. "Empirical evaluation of automatically extracted road axes," in Bowyer, K., Phillips, P. (Eds.), *Empirical Evaluation Methods Comput. Vis.. IEEE Comput. Soc. Press*, pp. 172–187, 1998.

Yamazaki, F., and Liu, W., "Vehicle extraction and speed detection from digital aerial images," in *Proc. IEEE IGARSS 2008*, vol. 3, pp. 1334–1337, 2008.

Zheng, H., and Li, L., "An artificial immune approach for vehicle detection from high resolution space imagery," *Int. J. Comput. Sci. Netw. Secur.*, vol. 7, no. 2, pp. 67–72, 2007.

Zheng, Z., et al. "Vehicle detection based on morphology from highway aerial images," in *Proc. IEEE IGARSS 2012*, pp. 5997–6000, 2012.

Zheng, Z., et al. "A novel vehicle detection method with high resolution highway aerial image." *IEEE J. Sel. Top. Appl. Earth Observ. Remote Sens.*, vol. 6, no. 6, pp. 2338–2343, 2013.

9 Single Tree Canopy Extraction from LiDAR Point Cloud Data

9.1 INTRODUCTION

With increasing needs in annual forest resource inventory for accurate and effective investigation and management of forest data, many methods have been proposed in the past decades. For example, traditional passive remote sensing, such as multispectral imagery, are affected by weather and surface coverage (Gougeon and Leckie 2001; Sajdak et al. 2014; Unger et al. 2014). Airborne LiDAR technology has unique advantages over passive remote sensing for the accurate extraction of single tree canopy (Zhou et al. 2004, 2013; Zhou and Zhou 2018; Mu et al. 2020). Over the past decades, many algorithms have been developed for extraction of single tree canopy, which can be categized as follows (Jakubowski et al. 2013; Lu et al. 2014):

1. *Triangulated Irregular Network (TIN) method.* This type of method first involves a few preprocessing stages, including the point cloud denoising, filtering, classification, and so on to separate vegetation point cloud data and ground point cloud data. With the segmentation above, the TIN interpolation method is applied in the vegetation point cloud data to generate a digital surface model (DSM), and in the ground point cloud data to generate a digital elevation model (DEM). Through calculating the difference between the DSM and DEM, a canopy height model (CHM) is obtained. The CHM is further processed using a computer graphics algorithm into single tree segmentation.
2. *Spatial structure relationship between LiDAR point cloud data and the characteristics of trees.* This type of method sets up classification rules to directly segment LiDAR point cloud data. For example, Lee et al. (2010) proposed a novel method, which first obtains the feature points with local maximum values using the circle feature detection algorithm, and then takes the center of the circle as a base to judge the distance away from other point clouds for clustering the other points from the LiDAR point cloud data. With the clustered points, the position, height, and DBH of a single tree are extracted. In combination with the aerial imagery of the same position, the single tree position is extracted. Coops et al. (2007) applied the iterative watershed method to identify the low canopy in the forests dominated by coniferous trees. It has been demonstrated by the experiments that this method encountered challenges for small and dense canopy with the same height.
3. *Combining the CHM gray imagery and spatial structure of the point cloud of the tree.* This type of method segments the point cloud data into single trees. For example, Li (2015) presented the label-controlled area growth method to extract single tree canopy from the combined LiDAR data, orthophotos, and in-field survey data. These experimental results demonstrated that this method can effectively extract the single tree from the heavily coniferous forests with big leaves and high canopy density. By adjusting the threshold, the algorithm is more flexible and the processing speed is faster than other methods, such as the TIN method. Guo et al. (2016) used the label-control watershed segmentation method to extract the canopy from the sparse forest and dense forest area, respectively. Khosravipour et al. (2014) proposed a new algorithm for extraction of single trees. This method first uses a subset of LiDAR point cloud data to fit enclosed concave pits for extraction of the CHM grid without concave pits, and then uses the window local maximum method for extraction of single tree canopy.

TABLE 9.1
Algorithm Comparison for Single Tree Extraction and Segmentation Methods

Types of Data	Segmentation Methods	References	Forest Type	Point Density (pts/m²)	Accuracy (%)
Segmentation based on CHM	Regional growth	Hyyppa et al. (2001)	Coniferous forest	8–10	7
	Variable window local maximum method	Popescu et al. (2003)	Deciduous, broad-leaved, mixed forest	1.35	82.87
	"Water injection" algorithm	Koch et al. (2006)	Coniferous forest, broad-leaved forest		62
	Watershed	Jing et al. (2012)	Coniferous forest	45	69
	Mark control watershed	Chen (2006)	Coniferous forest	95	64
	Local maximum method	Smits et al. (2012)	Coniferous forest, broad-leaved forest	9	87.50
	Pittree + variable window maximum	Anahita et al. (2015)	Coniferous forest	160	74
	Normalized cut	Reitberger et al. (2009)	Coniferous forest, broad-leaved forest	25/10	66
Based on point cloud	Regional growth combined with threshold judgment	Li et al. (2012)	Coniferous forest	>6	90
	Bottom-up area growth	Lu et al. (2017)	Mixed broad-leaved forest	10.28	84
	Area growth	Hamraz et al. (2017)	Broadleaf forest	25/1.5	72
	Layerstacking	Elias et al. (2017)	Coniferous forest, broad-leaved forest	25/1.5	72
	Iterative watershed	Duncanson et al. (2014)	Broadleaf forest	18	70
CHM + point cloud	Watershed + K-means	Tochon et al. (2015)	Coniferous forest, broad-leaved forest		69.86
	Hierarchical approach	Claudia et al. (2016)	Coniferous forest, multilayer forest	50/8	92–97

A brief overview of the previous algorithms for single tree extraction is listed in Table 9.1.

With the methods listed in Table 9.1, the watershed transformation algorithm, which combines a morphological algorithm and regional growth, has advantages over other methods. In the image segmentation method, it is usual to first divide a whole region into several sub-regions or an entire point cloud data set into several subsets. The watershed transformation algorithm has been widely applied in imagery segmentation and processing with advantages due to its simple calculation. However, this method is sensitive to noise. For this reason, this chapter proposes a K-means clustering watershed algorithm to segment the single tree.

9.2 K-MEANS CLUSTERING WATERSHED ALGORITHM

Watershed is a region-based widespread technique for image segmentation. This method originated from mathematical morphology (Serra 1986), and a breakthrough in applicability was achieved by Vincent and Soille (1991), since it can effectively extract the area of interest (AOI) of the imagery. Moreover, the watershed algorithm has the characteristics of being less time-consuming and highly accurate at locating at imagery edges, having simple formulas and operation, and being able to carry out the parallel processing using a computer relative to morphological segmentation algorithms.

The basic idea of the watershed algorithm is that an imaged LiDAR data set is regarded as a topographic landscape with ridges and valleys (Preim and Botha 2013). The elevation values of the LiDAR point cloud data are quantified by the gray values of the respective pixels. Based on such a gray (3D) representation, the watershed transform decomposes the imagery into catchment basins. For each of the local minimum values, a catchment basin comprises all points whose path of steepest descent terminates at this minimum value. The watershed transform algorithm separates the basins from each other, and decomposes an image completely and thus assigns each pixel either to a region or a watershed. Due to noisy image data, the watershed algorithm could cause segmentation of a large number of small regions. This phenomenon is known as the "over-segmentation" problem. However, this method has the following two obvious shortcomings:

1. *Sensitive to noise.* This is because the watershed algorithm generally uses the gradient image as the reference image for segmentation, and the gradient image is extremely susceptible to noise, which will not only cause the image to be over-segmented, but also cause the local segmentation line offset to cause the wrong segmentation.
2. *Over-segmentation.* For general natural color distribution images, the watershed algorithm will produce over-segmentation. This is because such images often have more detailed texture features, resulting in too many local minimums.

Since the watershed algorithm is sensitive to noise, and the phenomenon of over-segmentation occurs as a result, the CHM imagery extracted from LiDAR point cloud data suffers from effects such as blurring and unevenness. In order to overcome the shortcomings above, many scholars have proposed many algorithms to improve the watershed algorithm (see Figure 9.1). The improvement methods can be roughly divided into: image preprocessing before watershed transformation and region merging after watershed transformation. The preprocessing before segmentation generally focuses on the noise reduction of the image. Its purpose is to filter out the irrelevant small targets in the image, ignore the small details inside the target, and then perform a watershed transformation to obtain a better segmentation. Most of the current improved watershed algorithms are the method of preprocessing before segmentation. This chapter presents the K-means clustering algorithm to improve the over-segmentation caused by the watershed algorithm.

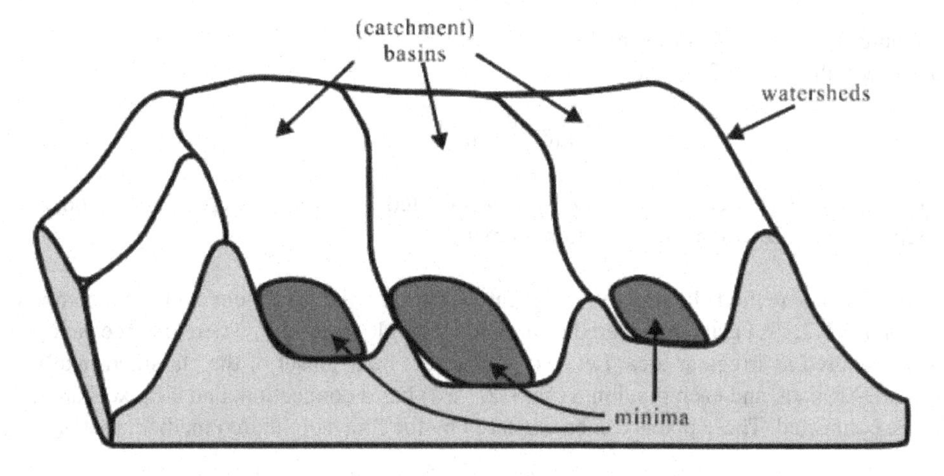

FIGURE 9.1 Watershed algorithm (courtesy of Preim and Botha 2013).

9.2.1 K-Means Clustering

The K-means clustering algorithm performs initial clustering with similar gray levels in the image to remove noise (Chen et al. 2007; Ren et al. 2013). The highest point of the local pixel maximum detection combined with the crown maximum model is taken as the highest point of the tree, and then each local highest point is used as the seed point of K-means clustering for clustering (Morsdorf et al. 2004). The K-means clustering algorithm can be described as follows:

Suppose that (x, y) represents the pixel coordinates of the digital image, and $f(x, y)$ is its gray function. $Q_l^{(i)}$ represents the l-th region after the i-th clustering, and $m_j^{(i+1)}$ represents the average value of the l-th category (cluster center) after the $i + 1$th clustering. The steps of K-means clustering algorithm are:

Step 1. Suppose that the initial clustering centers are $m_1, m_2, \ldots, m_k$, which represent the positions of the variable window at the treetop;

Step 2. For each point (x, y), find the closest clustering center according to a given criterion, and assign it into the clustering by:

$$\left|f(x, y) - m_l^{(i)}\right| < \left|f(x, y) - m_j^{(i)}\right| \tag{9.1}$$

Where $m_l^{(i)}(i = 1, 2, \ldots,)$ represents the average value of the l-th class after the i-th clustering, $f(x, y) \in Q_l^{(i)}$. After finishing this step, update the clustering center position at $i + 1$ to get a new clustering center position, $m_l^{(i+1)}$;

Step 3. The new clustering center position of $Q_l^{(i)}$ in Step 2 is $m_l^{(i+1)}$, which can be determined by Let

$$\xi^2 = \sum_{j=1}^{K} \sum_{(m,n) \in Q_j^{(i)}} \left|f(x, y) - m_l^{(i+1)}\right|^2 = min \tag{9.2}$$

Where $l = 1, 2, \ldots, k$, then

$$m_l^{(i+1)} = \frac{1}{N_l} \sum_{(x,y) \in Q_l^{(i)}} f(x, y) \tag{9.3}$$

Where N_l is the number of samples in $Q_l^{(i)}$;

Step 4. For all of $l = 1, 2, \ldots, k$,

$$\text{if}\, m_l^{(i+1)} = m_l^{(i)} \tag{9.4}$$

the iteration of the computation above is terminated. Otherwise, Step 1–Step 3 above are repeated unit Equation 9.4 (condition) is met.

After the K-means method classifies an image into K clusters, the maximum connected components, noted as $S_i(i = 1, 2, \ldots, k)$, are considered as individual single tree areas, where the tree canopy contour is considered as irregular area. Let $S_i(i = 1, 2, \ldots, k)$ be a subset of the classified pixels in the imaged LiDAR data, and each pixel in $S_i(i = 1, 2, \ldots, k)$ has a connection, and all pixels are considered to be connected. These pixels can be clustered by the four-neighbored method, that is,

$$S_i = argmax\left(S_1, \ldots, S_i, S_j \ldots, S_k\right) \tag{9.5}$$

9.2.2 Watershed Segmentation Combining with K-means Clustering

With the K-means clustering, the watershed algorithm is conducted and the specific steps are as follows:

Step 1. Let the K-means method segment the image into N clusters, named regions, each of the regions be represented by $S_i(x_i, y_i) \in S(x, y)$, the gray value be $I(x_i, y_i)$, and the size of each region be n_i, $0 < i < N$, calculate the gray variance of each region, noted as $M_i^2(i = 1, 2, \ldots, k)$.

Step 2. Let S_i and S_j be the two adjacent regions, and define the gray variance of the two adjacent regions as

$$M_{ij}^2 = \left(M_i - M_j\right)^2 \left(i \neq j; i, j = 1, 2, \ldots, k\right) \tag{9.6}$$

Step 3. Let S_i and S_j be adjacent segment regions, $(x_i, y_i) \in S_i$, $(x_j, y_j) \in S_j$, $(x_i, y_i)(i \neq j; i, j = 1, 2, \ldots, k)$ are adjacent points on the edge of the segment region, (x_i, y_i) and (x_j, y_j) satisfy the four-neighborhood relationship. Support that N_{ij} is the number of pixels on the edge of the divided region that satisfy the four-neighborhood relationship, and the mean value of the standard deviation of the gray value of the four-neighbored pixels on the edge of the individual region can be calculated by

$$P_{ij} = \frac{1}{N_{ij}} \sum_{(x_i, y_i),(x_j, y_j)} \sqrt{\left(I\left(x_i, y_i\right) - I\left(x_j, y_j\right)\right)^2} \left(i \neq j;\ i,\ j = 1,\ 2,\ \ldots,\ k\right) \tag{9.7}$$

Step 4. Define the similarity between two adjacent regions by

$$C_{ij} = \frac{\sqrt{\left(M_{ij} - P_{ij}\right)^2}}{2} \left(i \neq j;\ i,\ j = 1,\ 2,\ \ldots,\ k\right) \tag{9.8}$$

Suppose that a threshold is initially given to maximize the variance between C_1 and C_2, C_1 is the pixel with gray level $\{0, 1, \ldots, k\}$, C_2 is a group of pixels with gray level $[k + 1, \ldots, L - 1]$, the ratio of the variance between the clusters to the overall gray variance is a measure that can classify the gray of an image into two categories by

$$\eta(k) = \frac{\sigma_B^k(k)}{\sigma_G^2} \tag{9.9}$$

The value range is $[0, 1]$, and the threshold is T, that is,

$$\begin{cases} if\ C_{ij} > T\ S_i \text{ and } S_j \text{ are not merged} \\ if\ C_{ij} < T\ S_i \text{ merged with } S_j \end{cases} \tag{9.10}$$

9.3 VALIDATION AND ANALYSIS

9.3.1 Study Area

The study area is located in the Saihanba forest area, which is at the northernmost tip of Hebei Province, and is divided into two geomorphic units: the upper plateau of the dam and the mountainous area of northern Hebei. Saihanba extends 58.6 km long from north to south and 65.6 km wide from east to west. The geographical coordinates are 116° 51′~117° 39′ East, and 42° 2′~42° 36′ N. The forest area of Saihanba is 68708.05 m², and the forest coverage rate is approximately 72.3%.

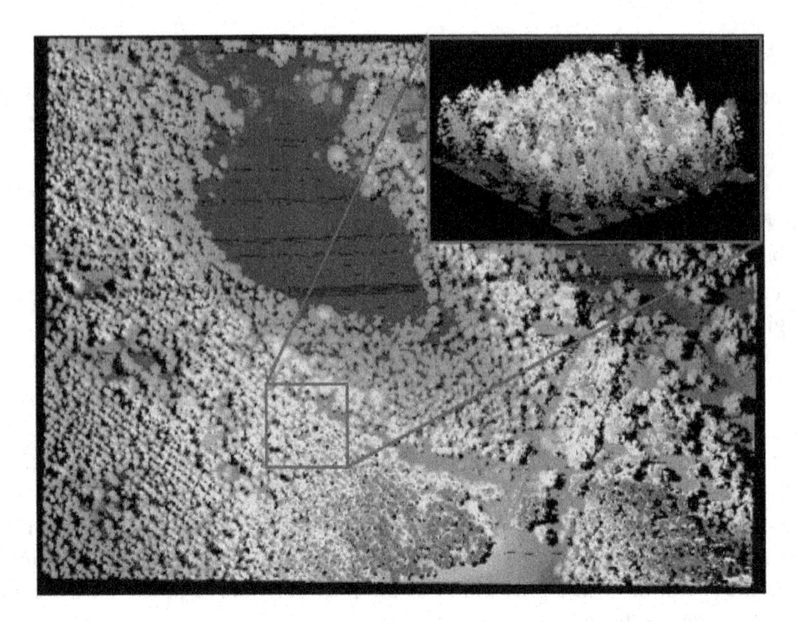

FIGURE 9.2 Point cloud data in the study area.

The UAV-based LiDAR collected the data in 2016 with an average flight speed of 61.7 m/s, the flight altitude was 150–200 m, and the overlap rate was 67%. The LiDAR data includes four echoes, and the point cloud density is 50 (pts/m^2). The tree species are mainly pines with an average height of 20–30 m (see Figure 9.2).

9.3.2 Data Preprocessing

9.3.2.1 Point Cloud Filtering

Due to various errors in the LiDAR point cloud data, the radius filtering method is applied to denoise the errors. A circle with radius of 15 m is selected around a point to account for the number of points that fall inside the circle. When the number is greater than a given value, delete the points; when the number is less than a given value, keep the points. The results after filtering are shown in Figure 9.3.

9.3.2.2 Generation of CHM

CHM is indeed obtained by subtracting the DEM from the DSM of the forest area. Therefore, it is easy to produce many pits, that is, pixels in the local range of the CHM gray image are smaller than the pixels in the neighborhood during the creation of the CHM. Thereby, a semi-automatic interaction is developed to set the empirical threshold for pit removal, and to visually judge the optimal effect after pit removal. The steps include:

1. Apply the Laplace algorithm to remove the pits from the original CHM image;
2. Generate the cumulative histogram of the CHM on the basis of the results produced in Step 1;
3. Generate a binary image (gray CHM image) with 1 for the pit pixels and 0 for non-pit pixels;
4. Extract the gray values of the pit pixels from the gray CHM image after median filtering;
5. Multiple experiments have found that the optimal median filter window is 5 × 5, filling the original CHM image;
6. Replace the pit pixels of the gray CHM image using the pixel (gray values) generated by 5 × 5 median filtering.

The results are depicted in Figure 9.4.

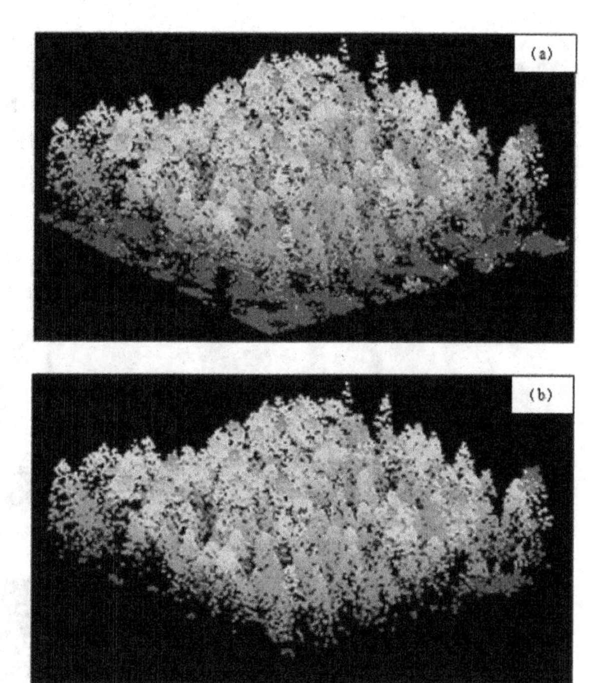

FIGURE 9.3 Point cloud filtering: (a) before filtering, and (b) after filtering.

9.3.3 Treetop Detection

Popescu et al. (2003) proposed a variable window method for detection of a tree height and crown size from LiDAR point cloud data. This section presents how the tree height is manually measured, and the crown size is measured through their average crown diameter along two vertical profiles.

The window size can be calculated by

$$Window\ Size = 29.77 - 1.46 \times TC + 0.03 \times TH^2 \tag{9.11}$$

where *TC* is the crown diameter of each individual tree, *TH* is the average tree height of individual single tree. The analysis for window size and tree height is depicted in Figure 9.5.

The window size for tree crown detection is predicted by Equation 9.11 and is used to accurately detect the position of the treetop. In order to obtain a canopy surface model, a named "Erosion-Expansion" operation is carried out on the CHM image generated by the above operation. That is, the image is first eroded and then expanded, to eliminate small objects, smooth the shape boundary, but not change their area. Moreover, this operation can efficiently remove small particle noise and separate the adhesion between neighboring objects, in which each grid unit value is replaced by the height of the point cloud data in the neighborhood. The maximum value is the canopy surface model, which is called the CMM model, and is expressed by

$$H_{x,y}^{CMM} = max_{(x-1)(y-1) \le i \times j \le (x+1)(y+1)} Z_{i,j}^{CHM} \tag{9.12}$$

The experimental result for CMM model generation using Equation 9.12 is depicted in Figure 9.6a. A comparison analysis with the CHM model is conducted and the result is depicted in Figure 9.6b. The experimental results demonstrate that the false treetops can be significantly eliminated, and the phenomenon of the missing treetops is largely reduced.

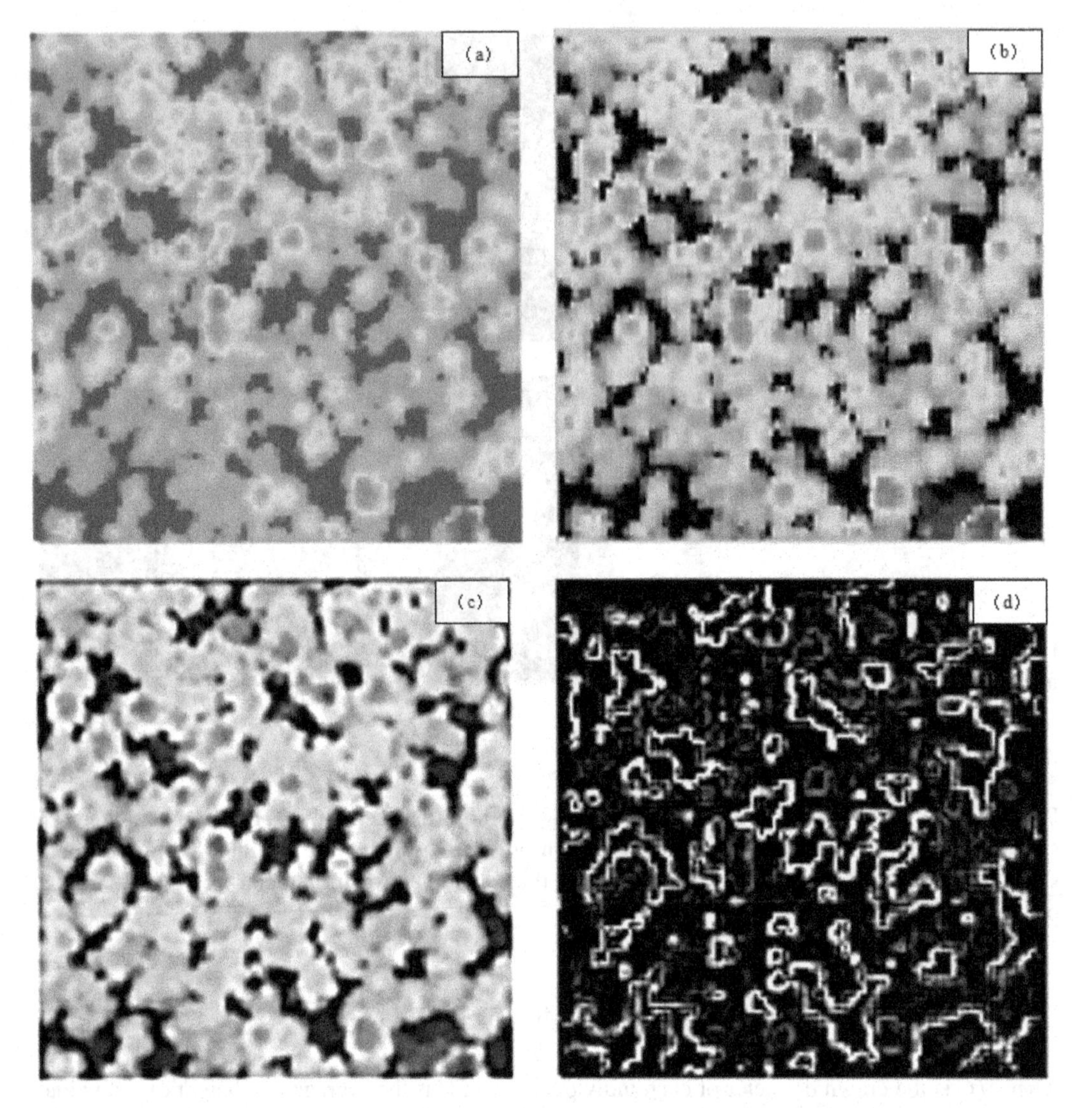

FIGURE 9.4 (a) Original CHM image, (b) gray CHM image, (c) the gray CHM image after median filter, and (d) the image with the pit boundary extracted.

9.3.4 Accuracy Assessment and Comparison Analysis

In order to quantitatively evaluate the accuracy of the proposed K-means watershed algorithm for single tree extraction from LiDAR point cloud data, the tree recognition rate (noted as *R*), recognition accuracy rate (noted as *P*), and *F* value (*F*-score) were used as indexes, and are expressed by

$$F = 2 \times \frac{R \times P}{R + P} \tag{9.13}$$

$$R = \frac{TP}{TP + FN} \tag{9.14}$$

$$P = \frac{TP}{TP + FP} \tag{9.15}$$

where the extents of the *F*, *R*, and *P* values are between 0 and 1.

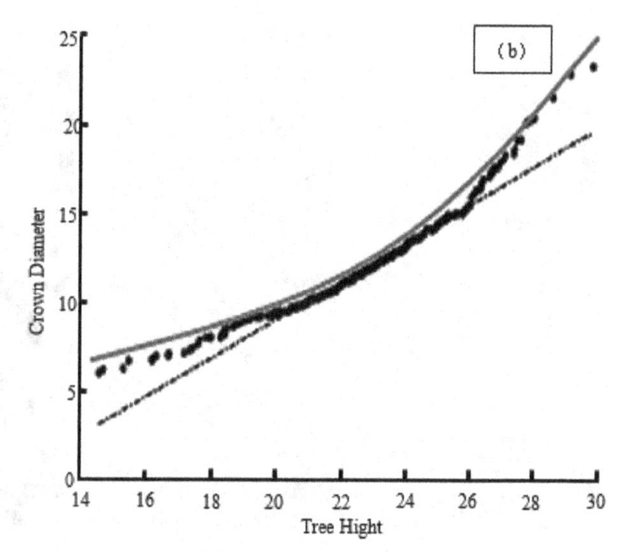

FIGURE 9.5 Relationship between canopy size and tree height, where the solid line is the regression curve, and the dashed line is the limit of the prediction interval.

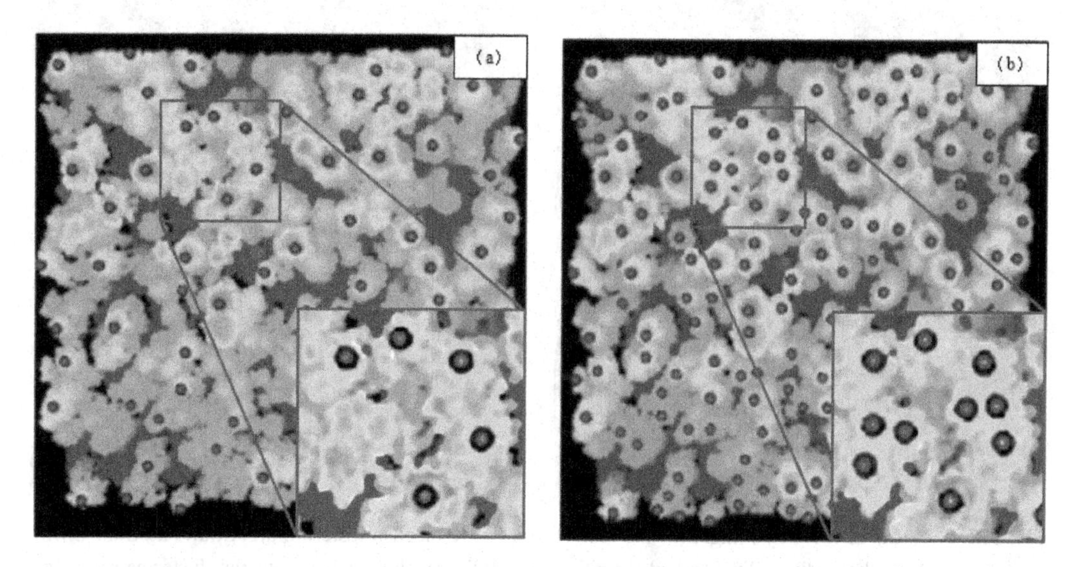

FIGURE 9.6 Comparison analysis for the 175 treetops extraction by two models: (a) CHM model, and (b) CMM model.

Figure 9.7a is the CMM gray image. The K-means clustering algorithm is used to segment the initial coarse CMM gray image. The result is shown in Figure 9.7b. The traditional watershed algorithm is used for classification of the CMM gray image. The result is depicted in Figure 9.7c. Figure 9.7d is the result from the K-means clustering watershed algorithm proposed in this chapter. As a comparison of the four methods, the proposed K-means clustering watershed algorithm has the highest accuracy out of them.

It can be seen from Table 9.2 that the F value with the K-means watershed algorithm can reach 0.9 in the experimental area, which is better than in the traditional watershed algorithm, which is only 0.82. This means that the K-means watershed method has a higher single treetop recognition rate and segmentation rate than the traditional watershed method. In addition, with the F-score (F),

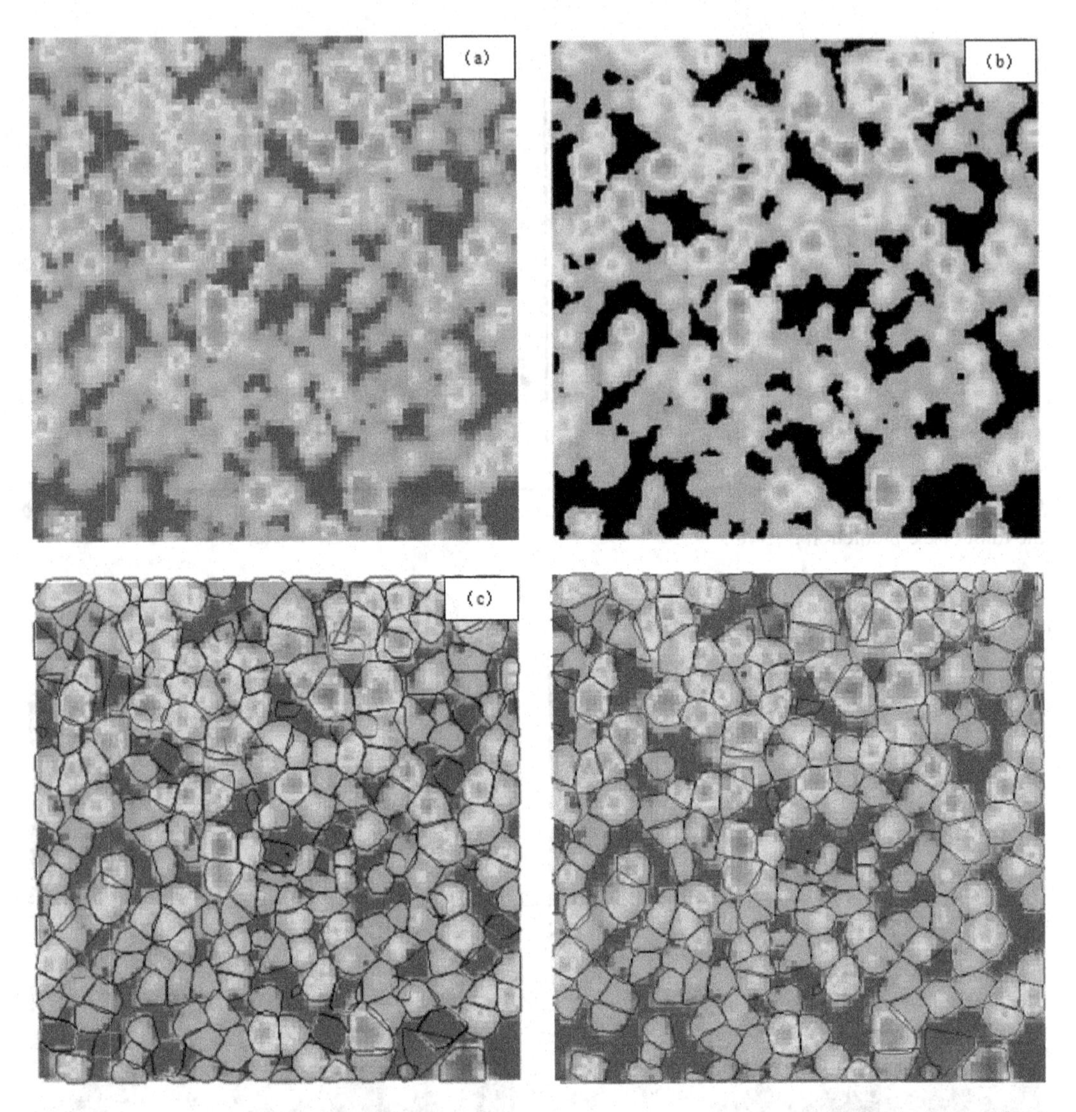

FIGURE 9.7 (a) CMM gray image, (b) the result from the K-means clustering algorithm based on CMM, (c) the result from the traditional watershed algorithm and (d) the result from the K-means clustering watershed algorithm proposed in this chapter.

TABLE 9.2
***F*-score Comparison Analysis**

Index for Accuracy Assessments	Traditional Watershed Algorithm	K-means Watershed Algorithm
R	0.85	0.91
P	0.8	0.89
F	0.82	0.9

Recall (*R*), and Precision segmentation accuracy judgment parameters (*P*), it can be concluded that the K-means watershed method can effectively segment crowns, and avoid the over-segmentation of the traditional segmentation.

9.4 CONCLUSIONS

In order to solve the problems of over-segmentation and noise that the traditional watershed algorithm have encountered, this chapter presents the K – means watershed algorithm for extraction of single treetop from airborne LiDAR point cloud data. In this method, K-means initializes clustering, then threshold segmentation is further conducted, before finally the watershed transformation is applied to segment the single treetop regions.

The airborne-based LiDAR data in the test field located in the Saihanba forest area, which is 58.6 km long from north to south and 65.6 km wide from east to west, are used to validate the method proposed. The experimental results demonstrate that the *P* and *R* values calculated by the traditional watershed algorithm are lower than the K-means cluster watershed algorithm. *F* value is an obvious index for measuring the comprehensive segmentation accuracy. In addition, the traditional watershed algorithm exposes the shortcomings of frequent over-segmentation and miss-segmentation, while the K-means clustering watershed algorithm is able to overcome this problem to a great extent. On the other hand, the significant increasing of the *P* value, *R* value, and *F* value demonstrate that the method proposed in this chapter is able to improve the over-segmentation and miss-segmentation. Furthermore, the CHM image cannot better represent the covered trees, resulting in missed and misclassified phenomena of the trees, while the proposed method has overcome this shortcoming, since the method proposed combines K-means clustering and the watershed algorithm.

REFERENCES

Anahita, K., Skidmore, A. K., Wang, T., et al. (2015). Effect of slope on treetop detection using a LiDAR canopy height model. *ISPRS Journal of Photogrammetry & Remote Sensing*, 104 (June): 44–52.

Chen, Q. (2006). Isolating individual trees in a savanna woodland using small footprint LiDAR data. *Photogrammetric Engineering & Remote Sensing*, 72(8): 923–932.

Chen, T. W., Hsu, S. C., Chien, S. Y. (2007). *Robust video object segmentation based on k-means background clustering and watershed in ill-conditioned surveillance systems. IEEE International Conference on Multimedia & Expo.*

Claudia, P., Valduga, D., Bruzzone, L. (2016). A hierarchical approach to three-dimensional segmentation of LiDAR data at single-tree level in a multilayered forest. *IEEE Transactions on Geoscience and Remote Sensing*, 54(7): 1–14.

Coops, C., Hilker, T., Wulder, M.A., St. Onge, B., Newnham, G., Siggins, A. (2007). Estimating canopy structure of Douglas-fir forest stands from discrete-return LiDAR. *Trees*, 21(3): 295–310.

Duncanson, L. I., Cook, B. D., Hurtt, G. C., et al. (2014). An efficient, multi-layered crown delineation algorithm for mapping individual tree structure across multiple ecosystems". *Remote Sensing of Environment*, 154: 378–3861.

Elias, A, Fraver, S., Kershaw Jr, J. A., et al. (2017). Layer stacking: A novel algorithm for individual forest tree segmentation from LiDAR point clouds. *Canadian Journal of Remote Sensing*, 43(1): 16–27.

Gougeon, F.A., Leckie, D.G. (2001). *Individual tree crown image analysis – A step towards precision forestry. First International Precision Forestry Symposium (CD-ROM)*, June 17–20, 2001, Seattle, Washington, USA.

Guo, Y., Liu, Q., Liu, G. (2016). Single tree canopy extraction of high-resolution remote sensing image based on marker control watershed segmentation method. *Journal of Earth Information Science*, 18(9): 1259–1266. (Chinese)

Hamraz, H., Contreras, M. A., Zhang, J. (2017). Vertical stratification of forest canopy for segmentation of understory trees within small-footprint airborne LiDAR point clouds. *ISPRS Journal of Photogrammetry and Remote Sensing*, 130: 385–392.

Hyyppa, J., Kelle, O., Lehikoinen, M., et al. (2001). A segmentation-based method to retrieve stemvolume estimates from 3-D tree height models produced by laser scanners. *IEEE Trans. on Geoscience & Remote Sensing*, 39(5): 969–975.

Jakubowski, M., Jakubowski, K., Li, W., Guo, Q. (2013). Delineating individual trees from LiDAR data: A comparison of vector- and raster-based segmentation approaches. *Remote Sensing*, 5(9) (September): 4163–4186.

Jing, L., Hu, B., Li, J., et al. (2012). Automated delineation of individual tree crowns from LiDAR data by multi-scale analysis and segmentation. *Photogrammetric Engineering & Remote Sensing*, 78(12): 1275–1284.

Khosravipour, A., Skidmore, A.K., Isenburg, M. (2014). Generating pit-free canopy height models from airborne LiDAR. *Photogrammetric Engineering & Remote Sensing*, 80(9): 863–872.

Koch, B., Heyder, U., Weinacker, H. (2006). Detection of individual tree crowns in airborne LiDAR data. *Photogrammetric Engineering & Remote Sensing*, 72(4): 357–363.

Lee, H., Slatton, K. C., Roth, B. E., et al. (2010). Adaptive clustering of airborne LiDAR data to segment individual tree crowns in managed pine forests. *International Journal of Remote Sensing*, 31(1–2): 117–139.

Li, C.. (2015). *Single Tree Canopy Extraction Based on LiDAR Data and Marker-Controlled Area Growth Method*. Ph.D. Dissertation, Northeast Forestry University, Harbin City, China.

Li, W., Guo, Q., Jakubowski, M. K., et al. (2012). A new method for segmenting individual trees from the LiDAR point cloud. *Photogrammetric Engineering & Remote Sensing*, 78(1): 75–84.

Lu, X., Guo, Q., Li, W., Flanagan, J. (2014). A bottom-up approach to segment individual deciduous trees using leaf-off lidar point cloud data. *ISPRS Journal of Photogrammetry & Remote Sensing*, 94(8): 1–12.

Lu, X., Ji, H., Ye, Q. (2017). Research on single tree extraction method based on LiDAR point cloud. *Computer Measurement and Control*, 25(6): 142–147. (Chinese)

Morsdorf, F., Meier, E., Kötz, B., Itten, K., Dobbertin, M., Allgöwer, B. (2004). LiDAR-based geometric reconstruction of boreal type forest stands at single tree level for forest and wildland fire management. *Remote Sensing of Environment*, 92(3): 353–362.

Mu, Y., Zhou, G., Zhou, X., Gao, J., Peng, X. (2020). HPR and OPK angle element conversion method based on airborne LiDAR alignment axis error calibration. *The International Archives of the Photogrammetry, Remote Sensing and Spatial Information Sciences*, XLII-3/W10: 71–75.

Popescu, S.C., Wynne, R.H., Nelson, R.F. (2003). Measuring individual tree crown diameter with LiDAR and assessing its influence on estimating forest volume and biomass. *Canadian Journal of Remote Sensing*, 29(5): 564–577.

Preim, B., Botha, C. (2013). *Visual Computing for Medicine: Theory, Algorithms, and Applications*, 2nd ed., ISBN: 9780124158733, Morgan Kaufmann, Burlington, MA, November 25, 2013, 836 pp.

Reitberger, J., Schnorr, C., Krzystek, P., et al. (2009). 3D segmentation of single trees exploiting full waveform LiDAR data. *ISPRS Journal of Photogrammetry & Remote Sensing*, 64(6): 561–574.

Ren, L., Lai, H., Chen, Q. (2013). Cotton segmentation algorithm based on improved K-means clustering and HSV model. *Computer Engineering and Design*, 34(5): 1772–1776.

Sajdak, M., Velázquez-Martí, B., López-Cortés, I., Fernández-Sarría, A., Estornell, J. (2014). Prediction models for estimating pruned biomass obtained from *Platanus hispanica* Münchh. used for material surveys in urban forests. *Renewable Energy*, 66 (June): 178–184.

Serra, J. (1986). Introduction to mathematical morphology. *Computer Vision, Graphics, and Image Processing*, 35: 283–305.

Smits, I., Prieditis, G., Dagis, S., et al. (2012). Individual tree identification using different LiDAR and optical imagery data processing methods. *Biosystems Inform Tech*, 2 1: 19–24.

Tochon, G., Féret, J.B, Valero, S., et al. (2015). On the use of binary partition trees for the tree crown segmentation of tropical rainforest hyperspectral images. *Remote Sensing of Environment*, 159: 318–331.

Unger, D.R., Hung, I.K., Brooks, R., Williams, H. (2014). Estimating number of trees, tree height and crown width using LiDAR data. *GIScience & Remote Sensing*, 51(3): 227–238.

Vincent, L., Soille, P. (1991). Watersheds in digital spaces: An efficient algorithm based on immersion simulations. *IEEE Transactions on Pattern Analysis and Machine Intelligence*, 13(6) (June): 583–598. http://dx.doi.org/10.1109/34.87344

Zhou, G., Song, C., Simmers, J., Cheng, P. (2004). Urban 3d GIS from LiDAR and digital aerial images. *Computers & geosciences*, 30(4): 345–353.

Zhou, G., Yang, B., Zhang, W., Tao, X., Yang, C. (2013). *Simulation study of new generation of airborne scannerless LiDAR system. IEEE International Geoscience and Remote Sensing Symposium, July 21–26, 2013*, Melbourne, Australia.

Zhou, G., Zhou, X. (2018). *Technology and Applications for Array LiDAR Imager*. Wuhan University Press, ISBN: 978-7-307-19683-4. (Chinese)

10 Power Lines Extraction from Aerial Images

10.1 INTRODUCTION

From a traditional point of view, aerial photogrammetric technique is not suited to surveying power lines because the power lines in the aerial images are too small to be detected (Baltsavias 1999). Although the recent airborne light detection and ranging (LiDAR) system is being applied in power line corridor monitoring, not all small private companies can afford the costs of a laser scanning system. With the increasing resolution of the aerial digital camera, it is possible for photogrammetrists to use the existing digital photogrammetric technique to monitor and periodically inspect the status of the power lines. To this end, a flight mission must be designed with consideration of variables such as flight height, the imagery resolution, and so on to ensure that the resolution of the ground sample distance (GSD) is higher than 0.05 m and the imaged power line can be higher than at least one pixel wide.

This chapter presents the method and algorithm for the power line extraction from the aerial images acquired from a helicopter. The related research work from active sensors has been investigated by a few authors. For example, Melzer and Briese (2004) extracted the power lines from the scattered 3D laser point clouds on the basis of the principle of linear object extraction. They projected a group of parallel power lines (corridors) described by 3D laser point cloud onto the x–y-plane using a 2D Hough transform, and then locally fit each 3D power line within its corresponding corridor. For the reason of the resolution of 3D laser point clouds, the accuracy of estimating power lines' location is low, and the extraction of the power line cannot reach full automation. Sarabandi and Park (2000) used a number of low-grazing incidence polarimetric SAR images at 35 GHz to detect power lines. The coherence between the co- and cross-polarized backscatter components is used as the detection parameter in the statistical polarimetric power line detection algorithm, and the detection criteria are based on clutter backscattering coefficients, power line size, and aspect angle with respect to the radar, as well as the number of independent samples that are formulated. Yamamoto and Yamaa (1997) applied an infrared (IR) camera and a color video camera to detect and identify power lines. Several imaging filters are applied to the IR images to enhance the target-to-background contrast and to suppress the image noise. Fusion of IR and color images is conducted to generate a virtually enhanced image for detecting and identifying the power lines. Blazquez (1994) analyzed the problems of detecting high-voltage power lines when the IR imagery is used because the IR imagery is subject to error caused by background thermal patterns such as clouds and tall trees. He thus suggested the application of the Probe Eye Scanner/Normal Color Video System (PESNVS) onboard a Piper Twin Comanche to detect the power lines in east Florida.

It has been demonstrated that automatic extraction of the power lines from aerial imagery with a clutter background is a rather challenging task. There are few investigators to develop the algorithm for this purpose, since the early aerial images had a very low resolution such that the power line cannot be recognized. This chapter presents a method for power line extraction from high-resolution aerial images.

10.2 POWER LINE EXTRACTION METHOD

10.2.1 Imaged Properties of Power Lines

When considering the complexity of the background of the power line image, two methods for the background removal are developed on the basis of the line detector mask (Chan and Yip 1996) and ratio operator (Tupin et al. 1998). Their performances are also compared and evaluated.

10.2.1.1 Line Detector Mask

The line detector masks presented by Chan and Yip (1996) for linear feature extraction are depicted in Figure 10.1a. The mask orientations vary at angles of 0°, 45°, 90°, and 135°. When the masks are operated and revolved with a moving window along the image from left to right and from up to down, the strongest response is taken as the power lines, and the corresponding orientation is taken as the direction of a power line. According to the photogrammetric mission planning, the power lines are usually imaged along the direction of flight, that is, approximately parallel to the margin of image plane, and the imaged power lines are usually straight. Thus, the mask w_1 is adopted. Further, we extend the mask w_1 into mask w in order to constrain the noise (Figure 10.1b).

The above mask can be mathematically expressed by

$$W = \left(w\left(-4,\ -1\right),\ \left(-3,-1\right),\ \ldots,\ w\left(4,\ 1\right)\right)$$

After the mask is revolved with a moving window of the aerial image, the output is

$$g\left(m,\ n\right) = \sum_{i=-1}^{1}\sum_{j=-4}^{4} f\left(m+i,\ n+j\right) \times w\left(i,j\right) \tag{10.1}$$

where $f(m,n)$ is the gray value of the image. After the image is revolved, the gray value of the power line along the image margin direction is enhanced. Because the mean of the mask is 0, the background noise is refrained efficiently. The candidate pixels of power lines can be extracted through an automatic threshold segmentation method, which was proposed by Otsu (1979).

10.2.1.2 Ratio Line Detector

Touzi et al. (1988) first presented an operator called the ratio edge detector. Tupin et al. (1998) applied this algorithm for linear object extraction. This chapter slightly modifies this algorithm as follows.

Let the amplitude of pixel s be denoted A_s, the radiometric empirical mean μ_i of a given region i having n_i pixels is $\mu_i = 1/n\ {}_i\sum_{s \in i} A_s$. The response of the edge detector between the regions i and j is (see Figure 10.2):

$$r_{ij} = 1 - \min\left(\mu_i / \mu_j,\ \mu_j / \mu_i\right) \tag{10.2}$$

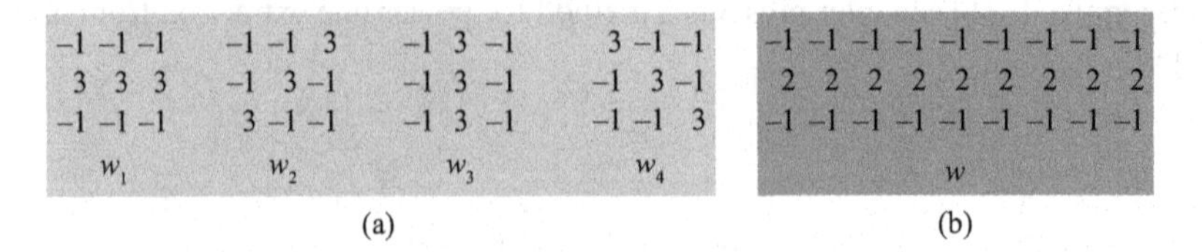

FIGURE 10.1 (a) Horizontal line detector mask, and (b) line detector masks.

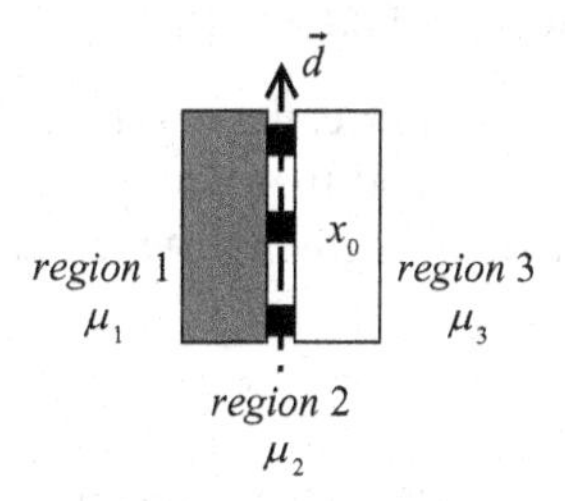

FIGURE 10.2 Vertical line model by ratio detection.

Because an imaged power line is one-dimensional, the response of the line detector is

$$r = \min\left(r_{21}, r_{23}\right) \tag{10.3}$$

where the subscripts 1, 2, and 3 denote the central region, and two lateral regions (see Figure 10.2). If a pixel's response r is large enough, that is, higher than a prior given threshold r_{min}, it would be considered as a line object. For a 550 kv power line, for example, the material usually has a lower reflectance than its surrounding background. So the constrain condition below is added

$$\mu_2 < \mu_1 \text{ and } \mu_2 < \mu_3 \tag{10.4}$$

If Equation 10.4 is satisfied, the pixel is extracted, and taken as one of the power line pixels.

When comparing the two operators, it has been widely accepted that the two operators are both effective for linear object extraction with a strong anti-noise capability. Moreover, both are capable of detecting line objects with one pixel wide from a complex natural background. On the other hand, the line mask detector produces noise if the contrast of edge is large, but meanwhile it enhances the image edges of a linear object, and as a result, the candidate line pixels can be obtained using the automatic threshold segmentation method. The ratio line detector can detect the line objects more efficiently than the line mask detector, but usually a prior experimental threshold must be set up.

10.2.2 Power Line Pixel Detection

With the observation of the aerial image, the power line has the following properties (Vosselman and Knecht 1995):

- Power lines are "linear" objects.
- The size of the power line on the aerial images can achieve a width of more than one pixel.
- Power lines are made of special metal and thus have a uniform brightness on aerial images.
- The power lines on the aerial image are close to a straight line, that is, with a very small curvature.
- The topology of power lines is simple, that is, they are continuous with one-dimensional extension between two power towers.
- Power lines are approximately parallel to each other.
- There is no occlusion over the power lines and the background is natural landscape, which is more complex than the road on the aerial images.

The above characteristics of the power lines on aerial images are very useful and helpful for us to develop the methods of automatically extracting power lines. The main problem of power lines extraction is how to detect the power line pixels accurately and connect the segments of power lines efficiently. As the helicopter flies along power lines, the power lines usually are imaged into straight lines on aerial images, and the backgrounds of the image are natural and man-made buildings, such as trees, bushes, houses, and so on. The algorithm for detecting the power lines is developed as follows.

10.2.3 Line Segments Detecting and Grouping

After the power line pixels are detected as above, the next task is to trace candidate lines and describe power line segments with a vector form, and then group them. To this end, the Radon transform (Radon 1971) is applied to detect the line fragments, which are further grouped into line segments to form an entire power line.

10.2.3.1 Radon Transform

The Radon transform is the projection of the image intensity along a radial line oriented at a specific angle (Trad et al. 2003). Figure 10.3 illustrates the geometry of the Radon transform, in which the projection of a two-dimensional function $f(x, y)$ is considered as line integrals. The Radon function computes the line integrals from multiple sources along parallel paths in a certain direction. The projections can be computed along any angle, θ. In general, the Radon transform of $f(x, y)$ is the line integral of $f(x, y)$ parallel to the y' axis, that is,

$$R_\theta(x') = \int_{+\infty}^{-\infty} f(x'\cos\theta - y'\sin\theta, x'\sin\theta + y'\cos\theta)\,dy' \tag{10.5}$$

where

$$\begin{bmatrix} x' \\ y' \end{bmatrix} = \begin{bmatrix} \cos\theta & \sin\theta \\ -\sin\theta & \cos\theta \end{bmatrix} \begin{bmatrix} x \\ y \end{bmatrix} \tag{10.6}$$

With the Radon transform function, a 2D matrix is acquired with a row vector containing the corresponding coordinates along the x' axis and a column vector containing the Radon transform of image for each angle.

10.2.3.2 Line Segments Grouping

After the power line pixels are detected, an imaged power line is segmented into several sections along the approximate orientation of power lines. The segmented sections ensure that each line segment is a straight line. Thus, the Radon transform is employed to detect the line fragments of each part (see Figure 10.4). The following steps describe how to group line segments of neighboring sections:

The slopes of line segments in the neighboring parts meet the following condition, that is,

$$|\alpha - \beta| \le angle\_threshold \tag{10.7}$$

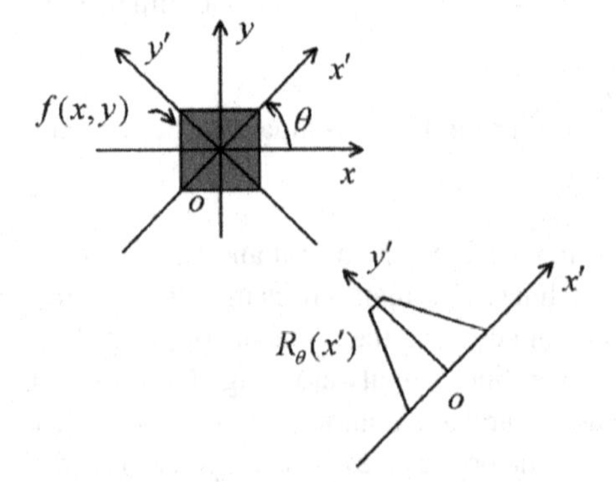

FIGURE 10.3 The geometry of the Radon transform.

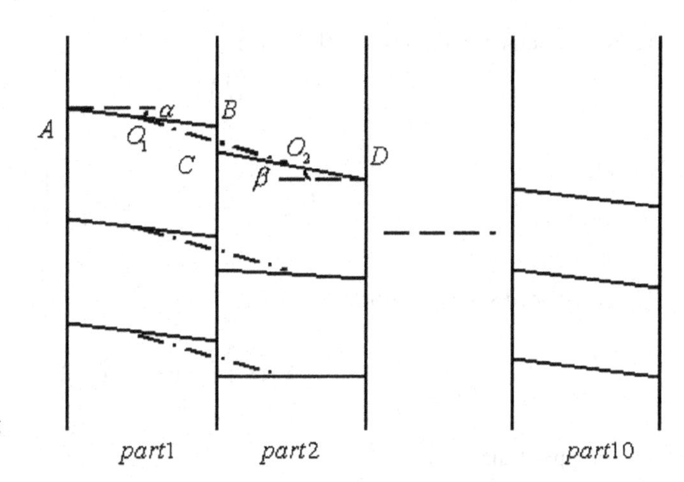

FIGURE 10.4 Line segment grouping procedure.

The distance of the endpoints of a line should meet the following condition, that is,

$$BC \leq distance\_threshold \tag{10.8}$$

The symbols in Equations 10.7 and 10.8 are illustrated in Figure 10.4. When Equations 10.7 and 10.8 are met, a line is fitted using the middle points, that is,

$$AD = AO_1 + O_1O_2 + O_2B \tag{10.9}$$

Repeat the same operation for each section, until all sections are finished. After this step, most of the line fragments are smoothly connected. If there are gaps (break-off) in the lines, a Kalman filter algorithm is employed to track lines to fill them out.

10.2.4 Kalman Filter to Track Line

Traditionally, the Kalman filter has been used to track roads (e.g., Vosselman and Knecht 1995). This section also employs this method to track the broken-off power lines. However, because the computation of the traditional Kalman filter is time-consuming, a slight improvement is made as follows: each line is considered as a track of a moving point on a straight line. When the broken-off line fragments are not detected, the Kalman filter is employed to track the broken-off next section. If there are line segments which meet Equations 10.7 and 10.8, they will be connected. The operation for tracking fragments starts from the middle of a long line segment, and moves up and down. When a broken-off line is encountered, a number of the broken-off pixels along the tracking direction are preassigned. The assigned number is required to assure that the method can effectively track the next section. Thus, the two neighbor pixels are detected, and the detected line segments are connected.

The computational process is described as follows. Tracking a line along the horizontal orientation can be mathematically described. Let $X(k)$ be the state vector of movement, $x(k)$ be moving location, $x'(k)$ be velocity, $y(k)$ be observation location which is the location of the detected line point, T is step size, k is the tracing time, the state equation of constant velocity of point D is as follows:

$$X\left(\left(k+1\right)/k\right) = AX\left(k\right) \tag{10.10}$$

where

$$X = \begin{bmatrix} x \\ x' \end{bmatrix}, \; A = \begin{bmatrix} 1 & T \\ 0 & 1 \end{bmatrix}$$

The measurement model is simplified as:

$$Y(k) = CX^T(k) \tag{10.11}$$

where,

$$C = \begin{bmatrix} 1 \\ 0 \end{bmatrix}, \text{ and } Y(k) = \begin{bmatrix} y \\ y' \end{bmatrix}$$

The predicated state equation is:

$$\hat{X}((k+1)/k) = A\hat{X}(k) \tag{10.12}$$

The output of the filter:

$$\hat{X}(k+1) = \hat{X}((k+1)/k) + K(k+1)\left(Y(k+1) - C\hat{X}((k+1)/k)\right) = \hat{X}((k+1)/k) \tag{10.13}$$

The observation takes the predicated value as the output of the filter. After the Kalman filter is carried out, an entire power line is extracted. The flowchart is depicted in Figure 10.5.

10.2.5 Other Cases

The above developed detector only considers the case where the flight path is approximately parallel to the power lines. If the flight path deviates from the power line at an angle of α, and the flight path is still a straight line (see Figure 10.6a), an image rotation transformation will be conducted so that the margin of the rotated image is "parallel" to the power line (see Figure 10.6b). Thus, the developed algorithm above is still applicable. However, in this case, the size of the rotated image increases, and the computation load is increased because some areas contain no gray information (see Figure 10.6b).

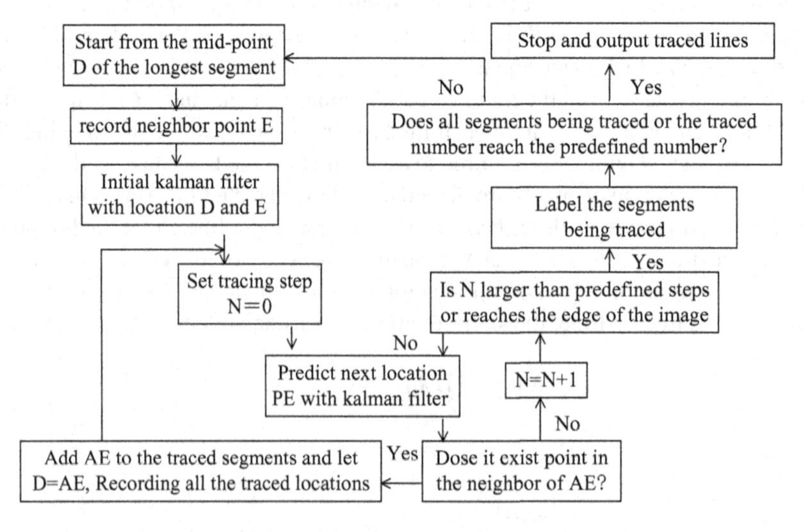

FIGURE 10.5 The workflow of the Kalman filter line tracing.

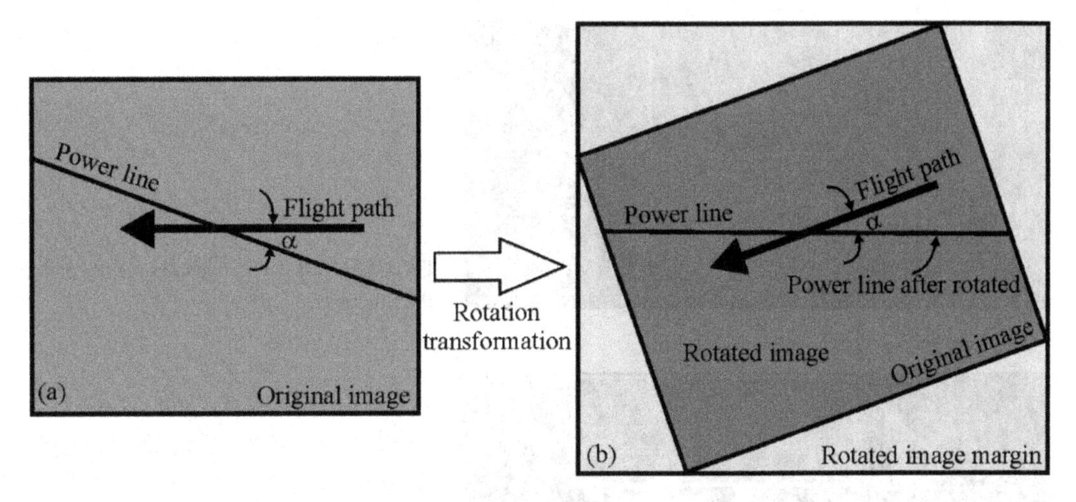

FIGURE 10.6 Flight path deviation from power lines.

10.3 EXPERIMENTAL RESULTS AND ANALYSIS

10.3.1 Experimental Results

The experimental field is located in Xiangfan in Hubei Province, China. The aerial images for high-voltage transmission power lines were acquired in March 2005 using a Kodak Pro 760 camera onboard a helicopter. The transmission power lines are about 35 m high. Figure 10.7 is part of one original aerial image in which the power line with one pixel wide was captured and the background is clutter natural landscape, mainly trees and bushes. In May 2005, a helicopter flight for the inspection of power lines was done in Jixangxia in Hubei Province. Three-view CCDs were used to acquire images. The resolution of each CCD is 1600 × 1200 pixel^2. The flight height was about 150 m and the four 550 kv transmission power lines were about 50 m high over the ground. As a result, each power line with one pixel wide was imaged. The proposed method is applied in the extraction of the power lines, and some results are given as follows (see Figures 10.7, 10.8, and 10.9).

From Figure 10.10, it has been demonstrated that the developed approach can exactly extract the power lines automatically. When the extracted power lines are superimposed onto the original images, as depicted in Figures 10.10b and 10.10c, the minimum position difference is 0.5 pixel, but the maximum one reaches 5–6 pixels. This is probably because an entire power line is considered as a straight line when using the algorithm above, while the true power line has a slight curvature. For this reason, an entire power line is divided into many sections so that each power line section is approximately a straight line. For example, Figure 10.11 was divided into ten sections when

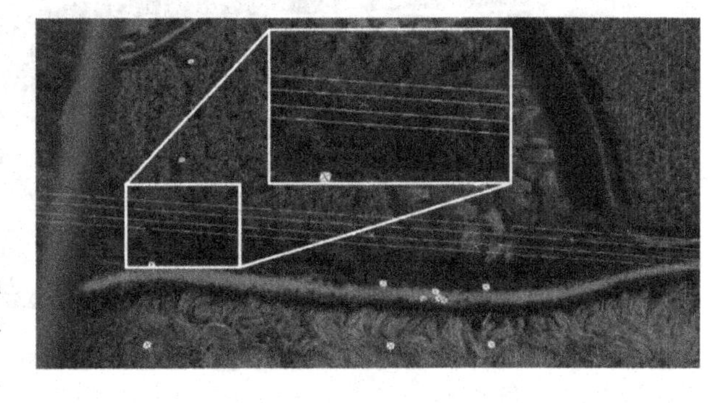

FIGURE 10.7 Original image with four power lines, which is magnified in sub-window.

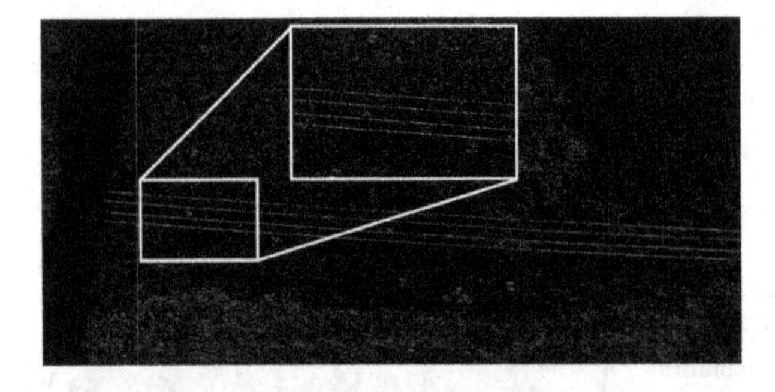

FIGURE 10.8 The detected results using the line mask.

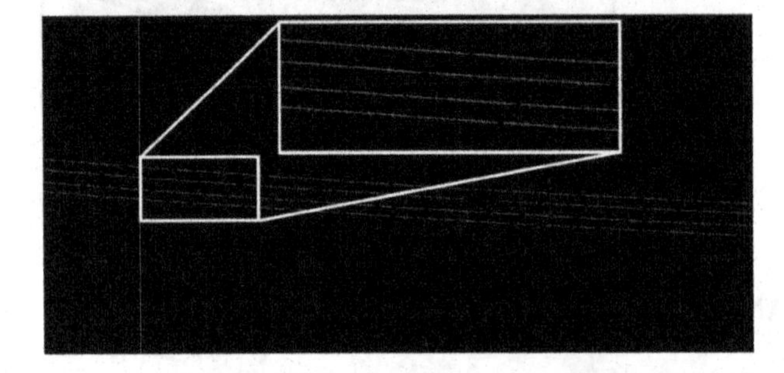

FIGURE 10.9 The detected results using the Radon transform and line grouping.

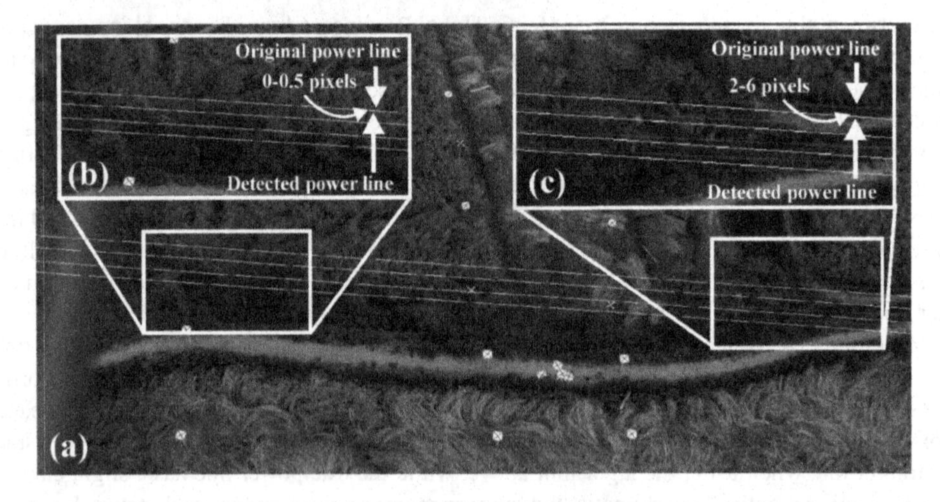

FIGURE 10.10 The final extracted power lines (bright lines), which are superimposed on the original images.

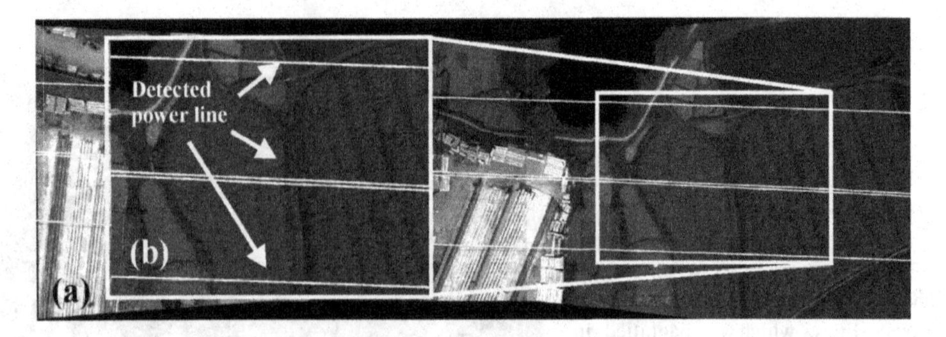

FIGURE 10.11 Experimental results using the developed method in another scene.

conducting the Radon transform. After the Radon transform was carried out, the power lines were extracted. With the Kalman filter, the gap of two neighbor end-to-end lines was filled, and an entire power line was extracted. In order to further validate the effectiveness of our method, another scene was tested. The result demonstrates that the developed method has a significant success in extraction of the power line even though the background is complex (see Figure 10.11).

The proposed method was tested in different background areas: one was in a suburban area, in which the background was very complex; the other one was in a rural area, in which the background brightness was great. The tested results are depicted in Figures 10.12 and 10.13. As seen from Figures 10.12 and 10.13, the proposed method also gave good results in the two areas.

FIGURE 10.12 Evaluation of the proposed method in the suburban area.

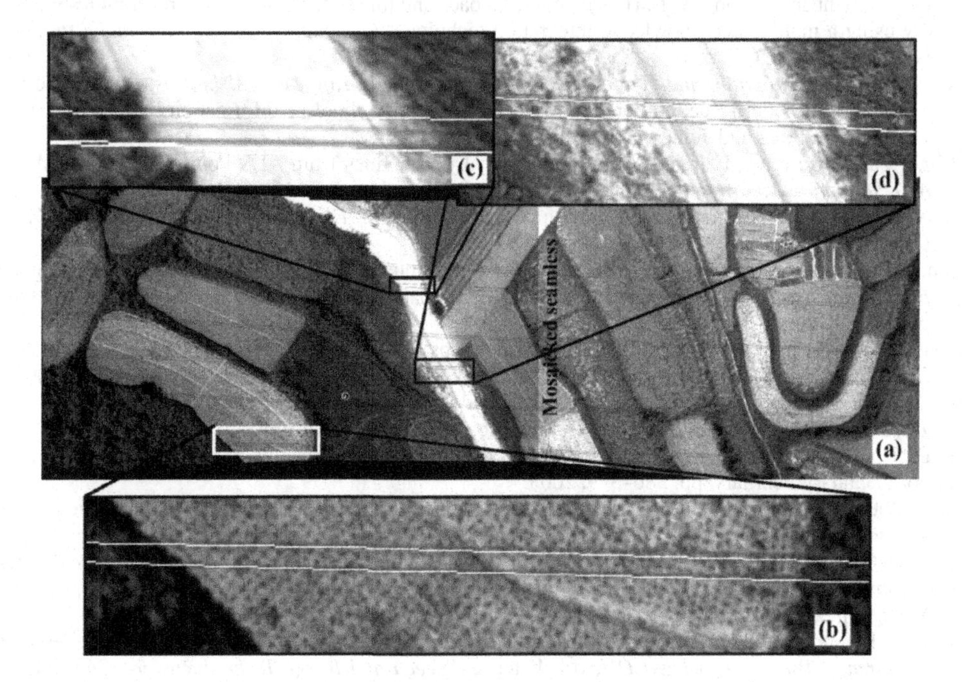

FIGURE 10.13 Evaluation of the proposed method with high brightness background.

10.3.2 Discussion

The application of the traditional Hough transform has been applied for the power line extraction under the different background images, but the results are not satisfying. This may be caused by the fact that the imaged power lines are only one pixel wide, and the background is noisy. The Duda Road Operators (DRO) (Bajesy and Tavakoli 1976; Fishler et al. 1981) have been tested as well, and the results show that this operator largely depends on the gray difference between power lines and the background (Li 2006). Moreover, the computation loads are very great because it needs a convolution calculation on the original image.

10.4 CONCLUSIONS

In this chapter, a method of extracting the power lines from large-scale aerial images is presented. This proposed method considers the properties of power line imaging, such as that they are linear, there may be two parallel power lines, and they are a homogeneous gray. The anti-noise linear detection operator is designed to detect the pixels of a power line. The section Radon transform is presented to extract and group the line segments to form a straight line. The Kalman filter method is applied to fill the gap between two lines. The experimental results demonstrated that the developed methods are capable of successfully identifying the power lines.

REFERENCES

Bajesy, R., and Tavakoli, M., Computer recognition of roads from satellite pictures, *IEEE Transactions on Systems, Man, and Cybernetics*, vol. 6, no. 9, pp. 623–637, 1976.

Baltsavias, E.P., A comparison between photogrammetry and laser scanning, *ISPRS Journal of Photogrammetry and Remote Sensing*, vol. 54, no. 2–3, pp. 83–94, 1999.

Blazquez, C.H., Detection of problems in high-power voltage transmission and distribution lines with an infrared scanner/video system, *Proceedings of SPIE – The International Society for Optical Engineering*, vol. 2245, pp. 27–32, ISSN: 0277-786X, CODEN: PSISDG, ISBN: 0-8194-1549-9, 1994.

Chan, T.S., and Yip, R., Line detection algorithm, In: *Proceedings of the International Conference on Pattern Recognition, August 25–29, 1996, Vienna, Austria*, ISBN: 0-8186-7282-X, pp. 126–130, 1996.

Fishler, M., Tenenbaum, J., and Wolf, H., Detection of roads and linear structures in low-resolution aerial imagery using a multisource knowledge integration technique. *Computer Graphics and Image Processing*, vol. 5, pp. 201–223, 1981.

Li, C., *Power Line Extraction and Height Computation from Multi-Angle Aerial Images*, Ph.D. dissertation, Laboratory of Pattern Recognition and Intelligent System, School of Information Engineering, Beijing University of Posts and Telecommunications, 108pp., 2006.

Melzer, T., and Briese, C., Extraction and modelling of power lines from ALS Point Clouds, *28th Austrian Association for Pattern Recognition (OAGM/AAPR) Workshop*, June 17–18, Hagenberg, Austria, 2004.

Otsu, N.A., Threshold selection method from gray level histogram, *IEEE Transactions on Systems, Man, and Cybernetics*, vol. SMC 9, no. 1, pp. 62–69, 1979.

Radon, J., Über die Bestimmung von Funktionen durch ihre Integralwerte längs gewisser Mannigfaltigkeiten, *Berichte Sächsische Akademie der Wissenschaften, Leipzig, Mathematisch-Physikalische Klasse*, vol. 69, pp. 262–277, 1971.

Sarabandi, K., and Park, M., Extraction of power line maps from millimeter-wave polarimetric SAR images, *IEEE Transactions on Antennas and Propagation*, vol. 48, no. 12, December, pp. 1802–1809, 2000,

Touzi, R., Lopes, A., and Bousquet, P., Statistical and geometrical edge detector for SAR images, *IEEE Transactions on Geoscience and Remote Sensing*, vol. 26, no. 6, November, pp. 764–773, 1988.

Trad, D., Ulrych, T., and Sacchi, M., Latest views of the sparse Radon transform source. *Geophysics*, vol. 68, no. 1, January/February, pp. 386–399, 2003.

Tupin, F., et al., Detection of linear features in SAR images: Application to road network extraction, *IEEE Transactions on Geoscience and Remote Sensing*, vol. 36, no. 2, March, pp. 434–453, 1998.

Vosselman, G., and Knecht, J.D., Road tracing by profile matching and Kalman filtering, *Automatic Extraction of Man-Made Objects from Aerial and Space Images (I) Ascona*, April 24–28, Switzerland, Birkhäuser Verlag, pp. 265–275, 1995.

Yamamoto, K., and Yamaa, K., Analysis of the infrared images to detect power lines, *IEEE Annual International Conference, Proceedings/TENCON*, vol. 1, no. 2, *Speech and Image Technologies for Computing and Telecommunications*, December 2–4, Brisbane, Australia, pp. 343–346, 1997.

Yan, G., Li, C., and Zhou, G., Automatic extraction of power lines from aerial images. *Journal of Image & Graphics,* vol. 4, no. 4, pp. 387–391, 2007.

Section III

Urban Orthophotomap Generation

11 The Basic Principle of Urban True Orthophotomap Generation

11.1 INTRODUCTION

The definitions of DTM, DBM, and DSM are first given here to avoid confusion:

1. *Digital Terrain Model (DTM).* This is an elevation model that describes the surface of the terrain without buildings and vegetation.
2. *Digital Building Model (DBM).* This describes the surface of man-made object details. A real three-dimensional representation that leads to a much more complex data structure than is generally used for DTM is necessary.
3. *Digital Surface Model (DSM).* This is here defined as the representation of the entire surface of the observed region. Therefore, the DSM is a combination of the DTM and DBM. A DTM and one or more DBMs may concurrently exist at the same geographical position, for instance at a bridge (DTM plus bridge DBM).

The National Map includes a large-scale digital orthophoto map (DOM) (also called a digital orthoimage map), since the DOMs have been a critical component of the National Spatial Data Infrastructure (NSDI) and the National Map (Federal Geographic Data Committee 1997; Kelmelis et al. 2003; Kelmelis 2003; Maitra 1998). The DOMs contain both the image characteristics of a photograph and the geometric properties of a map and thus usually serve as a base map upon which an organization can add vector data, attach attribute information to describe the details of features, and provide accurate geodetic registry to other themes of data (Federal Geographic Data Committee 1995).

The National Digital Orthophoto Program (NDOP) was first proposed in 1990. The primary goal of this program is to ensure the public domain availability of data for the Nation (USGS 1998, 1996). The NDOP was based on quarter-quadrangle-centered aerial photographs (3.75 minutes of longitude and latitude in geographic extent) obtained at a nominal flying height of 20,000 feet above mean terrain with a 6-inch focal length camera (photo scale = 1:40,000). The US Geological Survey (USGS) began to produce digital orthophoto quadrangles (DOQ) as a national program in 1991. In 2001, the USGS introduced the National Map, a digital version of its topographic mapping program that upgrades and links the digital topographic data that it had been collecting for the previous 20 years. This program is the product of a consortium of Federal, State, and local partners who provide geospatial data to enhance the US's ability to access, integrate, and apply geospatial data at national and local scales. The information layers planned for distribution include structures in urban areas and are listed on the website at http://nationalmap.usgs.gov/.

It has been demonstrated by many researchers that the early procedures and algorithms for DOM generation were based on earlier USGS mapping operations, such as field control, aerotriangulation (using photogrammetric equations derived in the early 1920s) and 2.5D digital elevation models (DEMs). It has been demonstrated that the generation of urban large-scale (*e.g.*, 1:2000) DOMs using the traditional method developed in the 1990s has encountered the following problems (a detailed analysis was given by Zhou et al. 2005):

Ghost images. In an orthoimage map, the roofs of buildings are superimposed onto the bottom of the buildings, and appear twice. This phenomenon is called "ghost image" (Rau et al. 2002). The ghost images are due to the fact that the higher buildings occlude other objects, and the traditional orthorectification model cannot compensate for the occlusion, leading to the result that the occluding objects remain and the occluded objects cannot be detected and filled. Consequently, the occluding pixels are doubly present.

Occlusion and building lean. Features in the DOM are offset from their true locations. The offset makes a tall building appear to lean over a street or other lower features. This results in occluded features like street, manholes, utility poles, and lower buildings. The traditional orthorectification model is not able to identify occlusions and compensate for them.

Shadows of buildings. Shadows of buildings are another important concern in urban areas. Because the traditional orthorectification model cannot remove a building shadow, the shadow of a tall building can hide features within it. Building shadows thus affect the appearance and usefulness of orthoimages. In order to produce high-quality orthoimages, an operation for automatic detection and radiometric balance for shadowed areas must be carried out.

Incomplete orthorectification. The traditional orthorectification method is based on a DSM, which represents the buildings. It has been found in the orthoimage map that: (1) small objects on top of a building are not completely orthorectified due to the lack of the model; (2) the wall of a building is incompletely orthorectified due to inaccuracy of the DSM; and (3) the boundary of a building is inaccurately orthorectified due to the inaccuracy of the DSM. Thus, the accuracy of the DSM will greatly affect the quality of orthorectification.

Incomplete refilling to occlusion. Occlusions are unavoidable in urban large-scale DOMs. Traditional photogrammetric flying missions require an approximately 70% endlap along-strip and 35% sidelap cross-strip. This specification presents a few challenges in urban photogrammetry, since the buildings are the major objects, especially in those areas containing tall buildings located in the margins of images, where the occlusions happen frequently due to the inherent property of perspective projection. Therefore, it is imperative that the design of photogrammetric flight missions must be conducted to acquire high-quality aerial photographs. This includes the choice of an appropriate endlap, sidelap, flying height, and the camera focal length and field of view (FOV).

Different geometric accuracy of adjacent orthoimage maps. It has been validated that the geometric accuracy of the adjacent orthoimage maps, which were generated using the same method, the same DSM, the same ground control points (GCPs), and the same iterative times, is different. It is found that the building in one orthoimage map was incompletely and inaccurately orthorectified and the wall was still visible, whereas the wall in the other orthoimage was invisible, which means the building had been completely orthorectified in its upright position. This phenomenon is probably caused by the incomplete correction of lens distortion, because the magnitude of the lens distortion affects object locations in an image plane.

Radiometric differences in adjacent orthoimages. Brightness differences between the feature in different scenes is, in large part, due to different reflection angles between the sun, the feature, and the imaging system. This can give a patchy appearance when the images are assembled into a mosaic. In addition, the generation of an orthoimage map in an urban area usually requires filling the occlusions with "slave" orthoimage patches. The radiometric differences between and among different orthoimages should be kept to a minimum to improve appearance and ensure that the best images are available for interpretation and use (Zhou et al. 2002).

The existence of the above problems indicates that the conventional orthorectification method (procedures and algorithms) used in the 1990s are not able to orthorectify the objects into their correct and upright positions and remove sufficient radiometric differences for large-scale urban aerial images. As a result, the usefulness of digital orthoimages in industry, government, and elsewhere is significantly reduced. Because the errors in these incompletely rectified large-scale city orthoimage maps can no longer be tolerated when used for updating and planning urban tasks, the generation of so-called true orthoimages (TOMs) has been of increasing interest.

Skarlatos (1999) and Joshua (2001) demonstrated that building occlusions did significantly influence not only image quality but also the accuracy of orthoimages. Amhar et al. (1998) and Schickler and Thorpe (1998) considered the hidden effects introduced by abrupt changes of surface height (e.g., buildings and bridges). Schickler and Thorpe (1998) and Mayr (2002) considered seamless mosaicking around fill-in areas in order to reduce gray-value discontinuities. Rau et al. (2002) treated enhancements of image radiometry, demonstrating a suitable enhancement technique to restore information within building shadow areas. Jauregui et al. (2002) presented a procedure for orthorectifying aerial photographs to produce and update terrain surface maps. Vassilopoulou et al. (2002) used IKONOS images to generate orthoimages for monitoring volcanic hazards on Nisyros Island, Greece, and Siachalou (2004) used IKONOS images to generate the urban orthoimage. Cameron et al. (2000) analyzed orthorectified aerial photographs to measure changes in native pinewood in Scotland, and Passini and Jacobsen (2004) analyzed the accuracy of orthoimages from very high-resolution imagery. Biason et al. (2004) further explored the automatic generation of true orthoimages. Despite these great efforts, a thorough address regarding the aforementioned issues of true orthorectification in urban areas is not available. For the reasons above, this chapter will introduce the principle for creation of an urban large-scale orthophoto map (DOM), including introduction of the algorithms, methodologies, and data processing procedures.

11.2 PRINCIPLE OF URBAN TRUE ORTHOIMAGE MAP GENERATION

11.2.1 Basic Steps of True Orthoimage Map Generation

The purpose of orthoimage map (DOM) creation is indeed to orthorectify the buildings to their correct, upright positions. Although the DSM can represent the building surface (roof), the orthoimage map generated using the DSM cannot achieve satisfactory accuracy, as described in the introduction (also see the detailed analysis in Zhou et al. 2005). To compensate for the occluded areas, occlusion detection must be conducted. One method is to trace the imaging rays from the top of the surface back to the projection center of the photograph. Only those objects whose rays do not intersect any other object before arriving at the projection center are not occluded, and produce a correctly rectified image pixel (Amhar et al. 1998). This method requires an exact DBM, which describes the building structure, 3D coordinates, topologic relationship, and so on. On the other hand, the relief displacement caused by terrain must be corrected in true orthoimages. For this reason, an exact representation of terrain should be given, that is, DTM, which does not contain buildings. Therefore, the basic steps of true orthoimage map (TOM) generation must include both DTM-based orthoimage map generation and DBM-based orthoimage map generation, as well as the merging of DTM- and DBM-based orthoimage maps (see Figure 11.1).

11.2.2 DBM-Based Orthoimage Map Generation and Occlusion Detection

DBM-based orthoimage map generation orthorectifies only the displacement caused by buildings and does not account for displacement caused by terrain. During this process, the occluded buildings need to be detected. At present, an effective and commonly used method for occlusion detection is to apply the DBM to calculate the distance between object surfaces to the projection center. This distance is called the Z distance, and the method is called the Z-buffer algorithm (Amhar et al. 1998). In this algorithm, a matrix is designed to store the Z distance of each pixel, and another matrix is designed to record the location of an object point. Both are defined on an image plane. Based on the DBM, the Z-buffer is generated by projecting each DBM surface polygon onto the first image plane

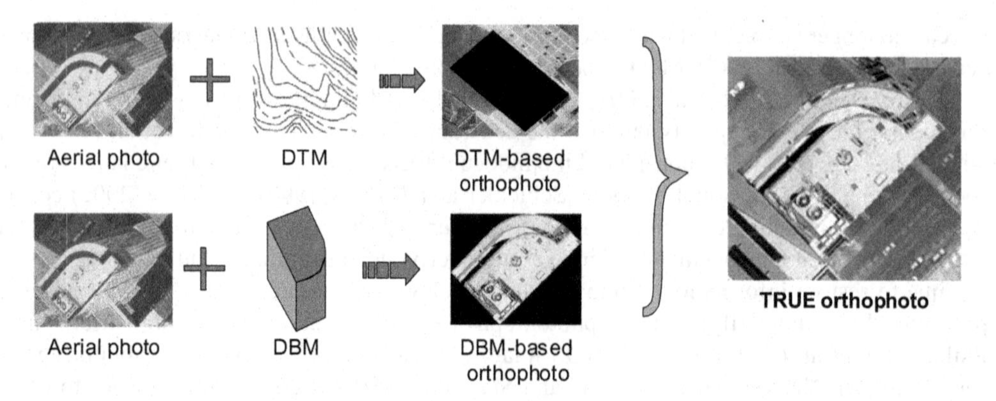

FIGURE 11.1 The procedure for true orthophoto generation.

by using collinearity equations. The projected polygon is rasterized and is filled with the Z distance and the polygon's identification (ID) code. The identification code is used to distinguish between walls and roofs. The process of filling by means of resampling the original image pixel will first check the ID codes and then will rectify only those pixels that are from the "roof," that is, the wall will not be orthorectified. In this step, the displacements caused by visible buildings are orthorectified into their upright planimetric position.

Additionally, because occluded buildings cannot be orthorectified, this step leaves holes in the orthoimage. The algorithm suggests marking these occluded areas by simply back-projecting the individual DBM into the individual image and marking occluded areas as black, that is, setting their brightness value to zero. This means that areas occluded by objects need to be filled with other orthoimages, called a "slave" orthoimage map (SOM) (Rau et al. 2002; Amhar et al. 1998). The above generated orthoimage map contains buildings only, and is thus called a DBM-based orthoimage map creation.

11.2.3 DTM-Based Orthoimage Map Generation

DTM-based orthoimage map generation for correcting relief displacement caused by terrain was traditionally implemented by either top-down or bottom-up methods (Chen and Lee 1993; Chen and Rau 1993). In urban areas consisting of extremely tall buildings, it is inappropriate to use the bottom-up method, which traces a ray from the image plane to the terrain. This is because in doing so, lower objects, which can be major features, are frequently occluded. Therefore, the top-down back-projection method is employed in this section. The basic mathematical model of this algorithm is the collinearity equation, which requires a DTM and the interior and exterior orientation elements of an image. The collinearity equations are used to calculate the image pixels corresponding to a groundel. The gray values of these pixels are assigned from the original images. A detailed description has been given by Zhou et al. (2005), Zhou et al. (2002), Chen and Lee (1993), and Hohle (1996).

11.2.4 Near True Orthoimage Map Generation

Based on the DBM- and DTM-based orthoimage maps obtained above, the TOM is created by merging the two types of orthoimage maps. Amhar et al. (1998) recommended a very simple operation, called logical <OR> operation, to carry out this merging operation when the background values are set to zero. This method has been validated, and as a result, it is demonstrated that this logical <OR> operation is very effective. However, the resulting orthoimage map still contains black patches, which represent those occluded objects. They therefore need to be filled by using additional orthoimages, the SOM. This intermediate orthoimage is thus called Near True Orthoimage Map (NTOM).

11.2.5 Occlusion Compensation

An NTOM contains black patches and needs to be refilled from "slave" orthoimages. The process requires finding conjugate areas in adjacent orthoimages and then filling the patches. For this purpose, in addition to needing a considerable overlap between aerial images, the following factors should be considered:

1. *Optimization of seam lines.* The above occlusion identification can detect the boundary of the occluded areas. The boundary might fall on ground features which are not modeled in the DBM, such as trees, cars or shadows, which probably have different brightness values in the individual NTOM (Schickler and Thorpe 1998) because the occlusion identification is based on the geometry of the imaging system, rather than on image radiometry. Therefore, the least-squares correlation algorithm is employed to minimize seam lines to obtain an apparently seamless mosaic when compensating for occluded areas. The basic idea of this algorithm is to calculate the similarity of the image contents between "master" and "slave" orthoimage maps. The pull range and search space are determined on the basis of the image texture and quality. The optimal seam line is chosen along a path where the two neighbor images are mostly similar.
2. *Optimization of orthoimage patches.* The same ground object has different displacements in different images. Therefore, the qualities of orthorectification of an identical ground object in different NTOIs are different. An algorithm to automatically "pick the best" orthoimage patch from the "slave" NTOMs to compensate for the occluded areas has been developed. The algorithm is based on the fact that the relief displacement at image nadir is zero. Thus, this method first calculates the distances of each patch to its nadir and then chooses that patch with the shortest distance.

In summary, the procedure of TOM generation is depicted in Figure 11.2. As observed from Figure 11.2, the input data include aerial images from multiple strips with over 65% overlap, exterior orientation elements including three position elements and three attitude elements of each image, and interior orientation elements, including focal length, principal point coordinates, length of opposite fiducial marks, DBM, and DTM data.

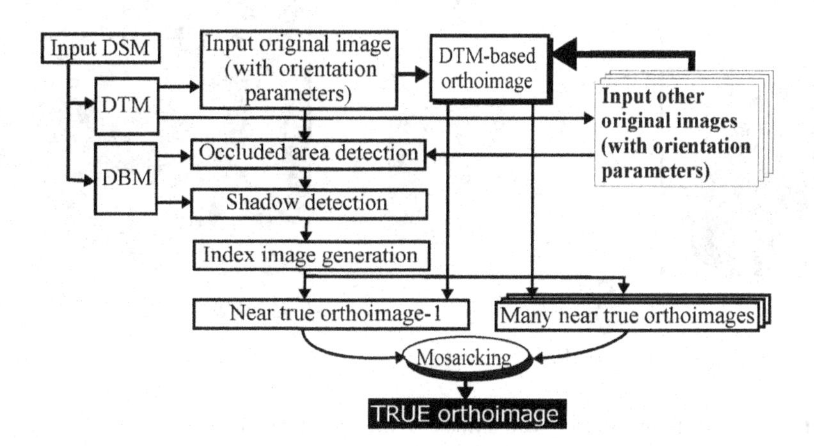

FIGURE 11.2 The procedure for true orthoimage generation.

11.3 EXPERIMENTS AND ANALYSES

11.3.1 Data Set

The geospatial data needed for TOM generation include: (1) digital terrain data, (2) aerial images, and (3) spatial objects (e.g., buildings). The vector Triangular Irregular Network (TIN) data structure is capable of precisely representing the ground surface even though the urban terrain is complex (Schickler and Thorpe 1998). The digital images, including original images, generated orthoimages, and generated DTM or DBM images, are usually represented in grid format with rows and columns. These data structures are described in this section.

(1) *Aerial Images*
The experimental field is located in downtown Denver, Colorado, where the highest building is 125 m, and many others are around 100 m. The six original aerial images from two flight strips were acquired using an RC30 aerial camera at a focal length of 153.022 mm on April 17, 2000. The flying height was 1650 m above the mean ground elevation of the imaged area. The aerial photos were originally recorded on film and later scanned into digital format at a pixel resolution of 25 µm. Figure 11.3 illustrates the configuration of the photogrammetric flight mission and the available DSM. Two of the six aerial images (DV1119 and DV1120) covered the downtown area. The endlap of the images is about 65%, and the sidelap is approximately 30%.

(2) *Digital Surface Model (DSM)*
A DSM, obtained *via* a Z/I photogrammetric workstation with computer-interactive operation in the central part of the downtown area is available (Figure 11.4). The accuracy of planimetric and vertical coordinates in the DSM is approximately 0.1 m and 0.2 m, respectively. The horizontal datum is GRS 1980, and the vertical datum is NAD83. The visualization of the DSM is depicted in Figure 11.4.

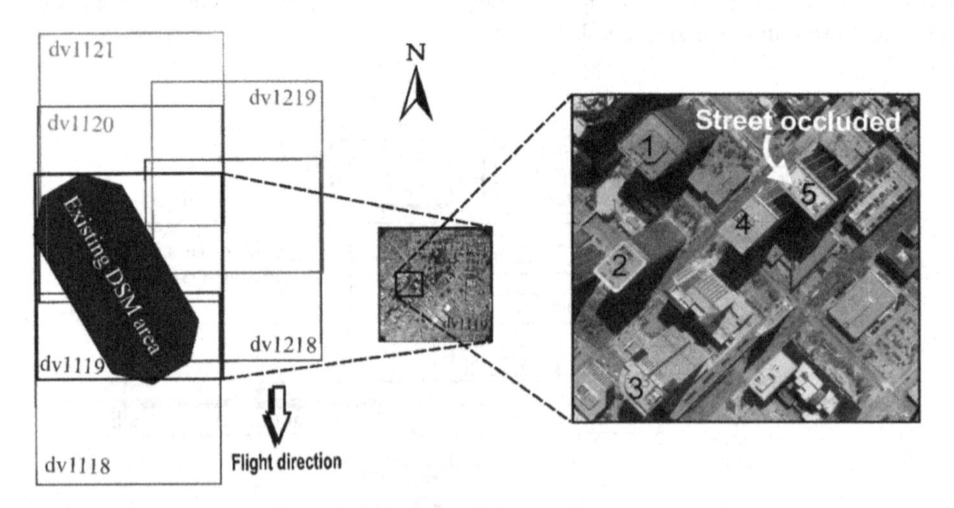

FIGURE 11.3 Six aerial images from two strips covering downtown Denver, Colorado. The heights of the five numbered buildings are 100, 103, 125, 100, and 107 m, respectively.

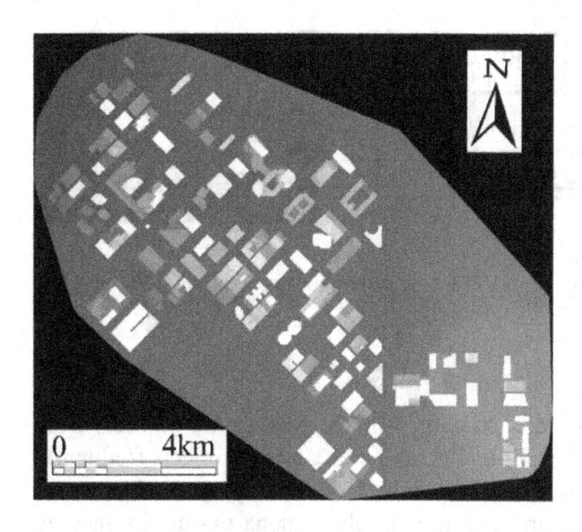

FIGURE 11.4 The digital surface model (DSM) in Downtown Denver, Colorado used for the experiment. The brightness in the DSM represents surface (ground or buildings) height.

(3) *Digital Building Model (DBM)*

- Data Structure

DBM-based orthoimage map generation requires an effective representation of urban buildings. The generation of a 3D urban (building) model is a rather challenging task, from both a practical and a scientific point of view, because different applications (e.g., city planning, communication design, tourism, pollution distribution, military security operation, etc.) require different data types and manipulation functions. Many models have been proposed, for example, Breunig (1996), Graz (1999), Grün and Wang (1998), Zhou et al. (2000), and Zlatanova (2000). For the purpose of TOM generation, the data structure to be developed in this section requires:

- Generating a high-quality DBM-based orthoimage map, and
- Easily creating, storing, designing, analyzing, and querying city buildings for TOM-based urban applications.

Spatial objects are abstractly understood as the following four types of geometric objects (Zhou et al. 2000; Grün and Wang 1998):

- *Point objects.* Described by x and y planimetric coordinates, like benchmarks.
- *Line objects.* One-dimensional objects, length being the only measurable one-dimensional spatial extension.
- *Surface objects.* Two-dimensional or two and half-dimensional objects with area and perimeter as measurable spatial extensions.
- *Body objects.* Three-dimensional objects with volume and surface area as measurable spatial extensions, which are bordered by facets.

Of the above four types of objects, point is the basic geometric element. For example, a point can present a point object, or the starting or ending point of an edge. An edge is a line segment, which is an ordered connection between two points: beginning point and ending point. A facet is completely described by the ordered edges that define the border of the facet. A polyhedron is described by the ordered facets. Image data and thematic data can be attached to a facet. Each facet is related to an image patch through a corresponding link (Figure 11.5). In the proposed data model, each object is identified by a defined attribute data, such as Type Identifier (TI), and thus the four types of objects are referred to as PI (point identifier), LI (line identifier), FI (facet identifier), and PHI (polyhedron identifier), respectively. The other attribute data (e.g., thematic data and geometric data) can be attached to each type of object. The geometric data of 3D objects contain the information of position, shape, size, structure definition, and image identification.

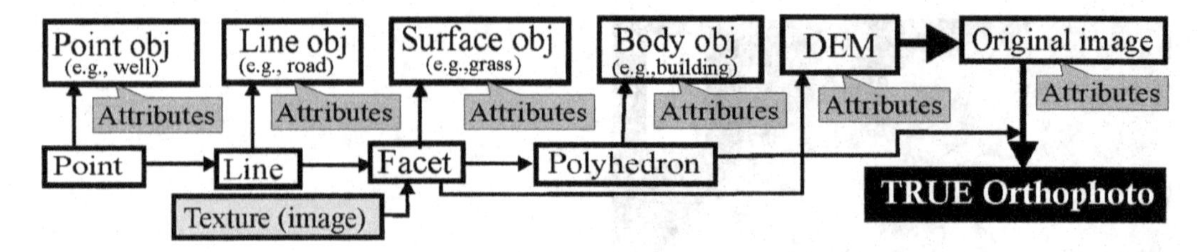

FIGURE 11.5 Data structure of the proposed 3D urban model for true orthophoto generation.

The defined object types are taken as the basic classes of primary geographic entities, from which other geographic entities are derived. For example, line and area types of objects frequently consist of several arcs, each of which has two terminal points. Collectively, these four types of objects can represent most of the tangible natural and human phenomena that a city usually encounters. This data model not only inherits primary operators, but also defines its own special operators. Thus, a spatial object can be extracted into one of the object types according to its attributes (Figure 11.5).

(4) *Topological Relationship*
The topological relationships between geometrical elements are implicitly defined by the data structure. For example, a point object is represented by a distinct point element. The line object is described by ordered edges. Similarly, the body object is described by a polyhedron that is described by the ordered facets. Thus, the topological relationships between point and edge, edge and facet, and facet and polyhedron are registered by the links between the geometrical elements.

The data structure described above is in fact a relational model, which can be implemented by relational database technology. In this relational model, each type of object is defined as a table. For example, a table describing a point object includes three terms: Point Identification (PID), Attribute Linkage (AL), and XYZ coordinates. The PID is for identification of a point object. The AL is a linkage to the attribute table describing point properties, such as point names and features. Different types of objects may have different attribute data. The XYZ coordinates are for geometric description of a point. Similarly, other types of objects, like lines, can be referenced. On the basis of the above relational structure, the query of a geometrical description for a type of object is easily realized. A detailed discussion for a relational database model of 3D city data for true orthoimage generation has been given by Zhou et al. (2003) (see Figure 11.6).

(5) *Urban 3D Model*
Figure 11.7 is the visualization of the DBM model of Denver at a ground resolution of about 25.4 cm for each pixel. Meanwhile, the DTM is generated by removing the digital buildings from the DSM.

11.3.2 Shadow Detection and Restoration

(1) Shadow Detection
Urban large-scale TOM generation requires detecting and removing shadow. Many attempts to detect and restore the shadowed areas from aerial images have been made in the past years for example, Salvador et al. (2001), Peng and Liu (2004), Liow and Theo (1990), Noronha and Nevatia (2001), Stauder et al. (1999), Irvin and McKeown Jr. (1989), Jiang and Ward (1992), Jaynes et al. (2004), Gonzalez et al. (2004), and Tsai (2003). Most of these efforts did not consider the self-shadow, and typically focused on umbra, considering the penumbra as a particular case of umbra. Moreover, they only took advantage of the image gray information and 2D object geometric shapes

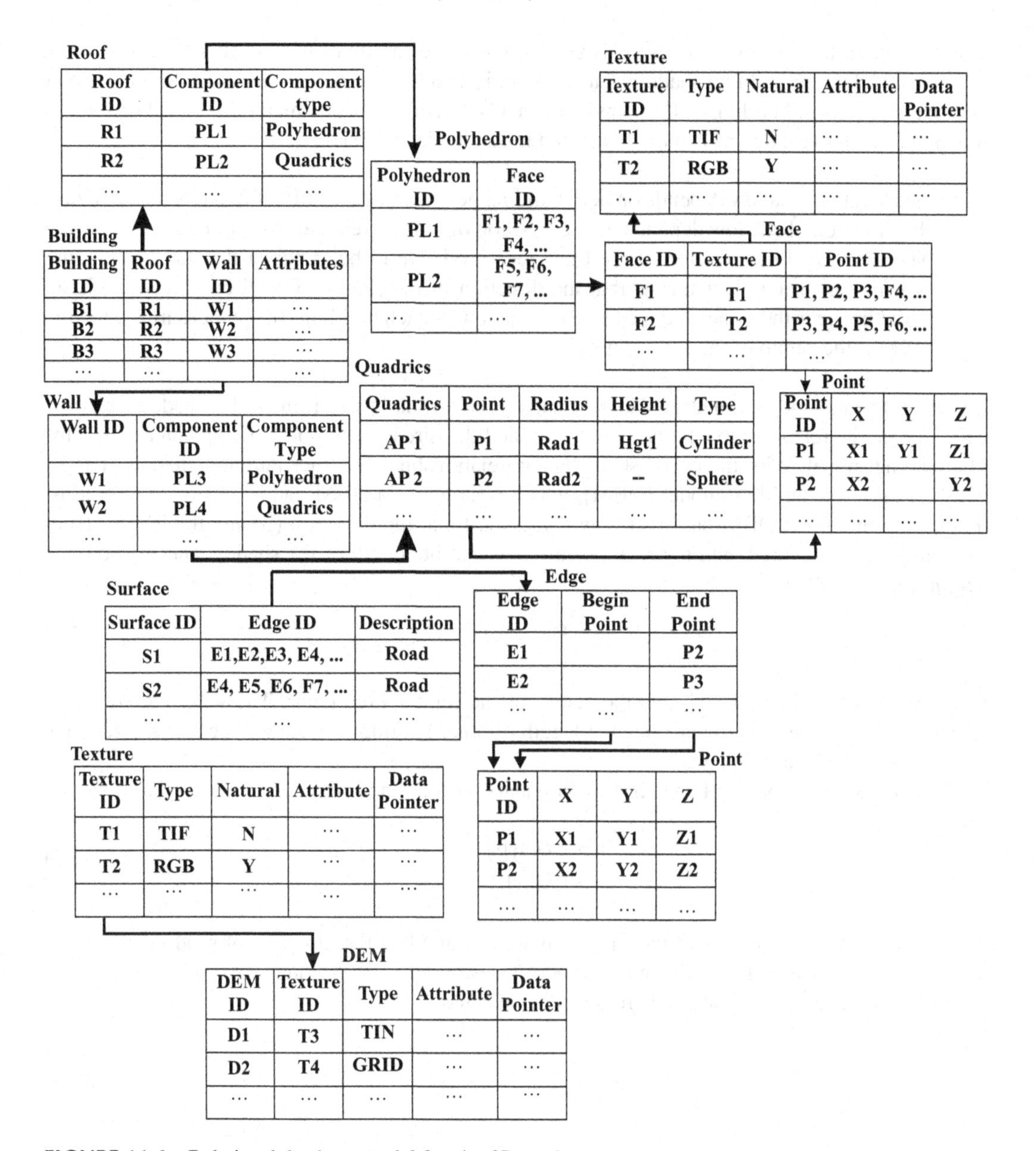

Roof

Roof ID	Component ID	Component type
R1	PL1	Polyhedron
R2	PL2	Quadrics
…	…	…

Polyhedron

Polyhedron ID	Face ID
PL1	F1, F2, F3, F4, ...
PL2	F5, F6, F7, ...
…	…

Texture

Texture ID	Type	Natural	Attribute	Data Pointer
T1	TIF	N	…	…
T2	RGB	Y	…	…
…	…	…	…	…

Face

Face ID	Texture ID	Point ID
F1	T1	P1, P2, P3, P4, ...
F2	T2	P3, P4, P5, F6, ...
…	…	…

Building

Building ID	Roof ID	Wall ID	Attributes
B1	R1	W1	…
B2	R2	W2	…
B3	R3	W3	…
…	…		…

Wall

Wall ID	Component ID	Component Type
W1	PL3	Polyhedron
W2	PL4	Quadrics
…	…	…

Quadrics

Quadrics	Point	Radius	Height	Type
AP 1	P1	Rad1	Hgt1	Cylinder
AP 2	P2	Rad2	--	Sphere
…	…	…	…	…

Point

Point ID	X	Y	Z
P1	X1	Y1	Z1
P2	X2		Y2
…	…	…	…

Surface

Surface ID	Edge ID	Description
S1	E1,E2,E3, E4, ...	Road
S2	E4, E5, E6, F7, ...	Road
…	…	…

Edge

Edge ID	Begin Point	End Point
E1		P2
E2		P3
…	…	…

Texture

Texture ID	Type	Natural	Attribute	Data Pointer
T1	TIF	N	…	…
T2	RGB	Y	…	…
…	…	…	…	…

Point

Point ID	X	Y	Z
P1	X1	Y1	Z1
P2	X2	Y2	Z2
…	…	…	…

DEM

DEM ID	Texture ID	Type	Attribute	Data Pointer
D1	T3	TIN	…	…
D2	T4	GRID	…	…
…	…	…	…	…

FIGURE 11.6 Relational database model for city 3D mode.

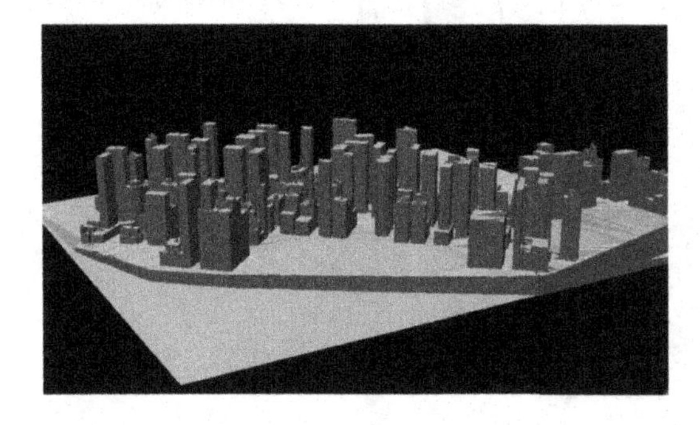

FIGURE 11.7 The city model of Denver created by our system.

when detecting the shadowed areas. This section introduces an innovative method that makes full use of DBM data for building shadow detection but is also little interested in detecting self-shadow because the walls of buildings will be invisible in TOM. This proposed method is based on the following principles (the details can be referenced in Zhou et al. (2005)):

1. The length of shadow depends on both building height and the zenith of the light source (sun); the direction of shadow depends on the position of the sun relative to buildings.
2. The zenith and the position of the light source relative to buildings at the time of imaging remain a constant. This means that the direction of the shadow for all buildings within an aerial image is the same, and the length of shadow of each building can be measured using the DBM if the zenith of the sun is known.

Based on the method outlined above, it usually first manually determines the shadow relationship, that is, the relationship between the length and the direction of one building shadow and the 3D building model. With the established shadow relationship, the other building shadows can be detected using the building model (DBM), which is stored in a DBM database. Figure 11.8 illustrates the principle of DBM-based shadow detection. First, the direction (β) and the length (L) of the shadow are measured, and the relationship between the length of the shadow and the building height is established by:

$$tan\left(\alpha\right) = L/h \tag{11.1}$$

where h is building height and α is the zenith of the light source (sun). The solved zenith from Equation 11.1 can be used to calculate the length of other buildings' shadows because the zenith is constant within an aerial image.

Further, the coordinates of the shadow boundary in a defined local coordinate system can be calculated by:

$$\begin{aligned} x &= L\cos\left(\beta\right) \\ y &= L\sin\left(\beta\right) \end{aligned} \tag{11.2}$$

where x and y are coordinates of the shadow boundary and β is the direction of shadow in a given local coordinate system. Because the local coordinates of the shadow boundary can be transformed to image coordinates, the shadowed areas can be detected.

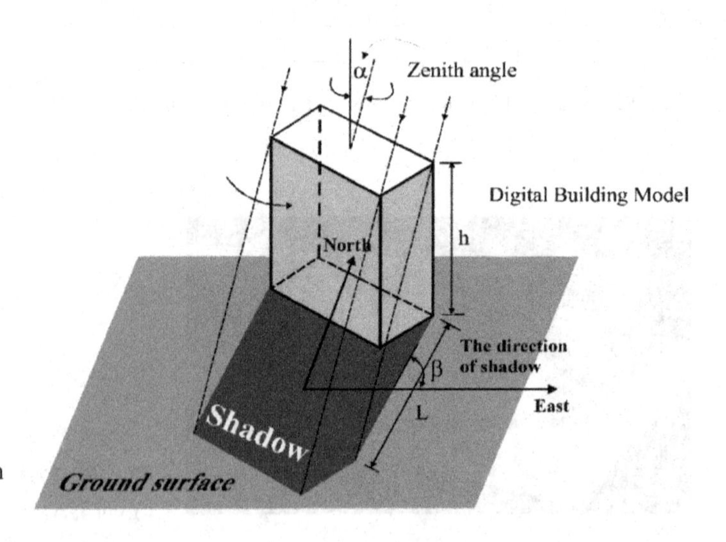

FIGURE 11.8 Shadow detection using the DBM.

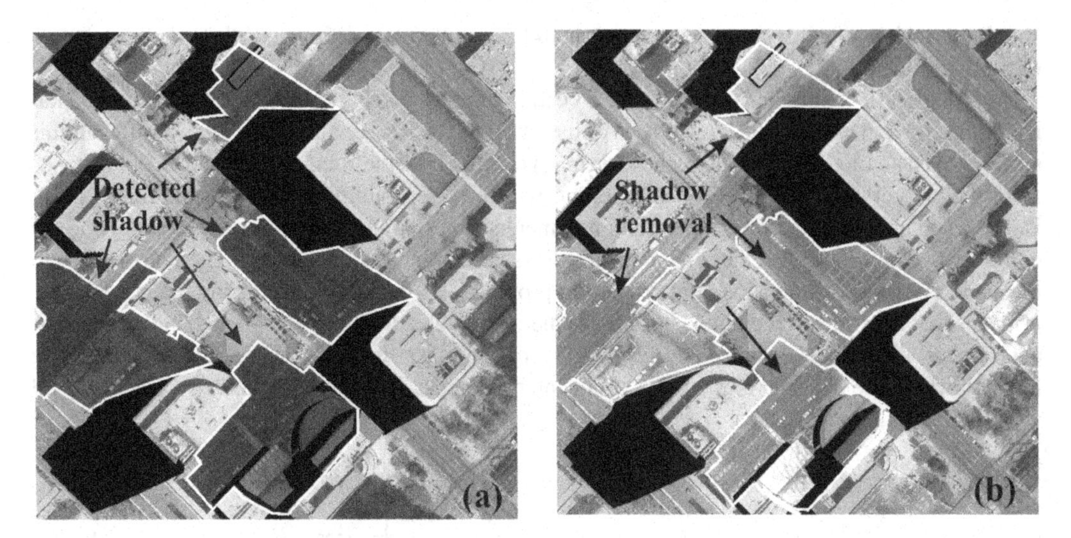

FIGURE 11.9 Comparison between (a) before shadow treatment and (b) after shadow treatment.

The calculated zenith α and the measured direction β remain constant for all buildings within an aerial image. The DBM provides 3D geometric information about the building, which will be used to calculate the length and the boundary of the shadow. Ideally, this proposed method can achieve a satisfactory result, but the real scene is more complex than imagined. It is thereby suggested combining the proposed method with the region-growing method, which makes full use of gray information, to detect the shadowed area. Figure 11.9 is a result of shadow detection from the proposed method.

(2) Shadow Restoration

In order to restore the shadowed areas, Schott et al. (1988) used a histogram adjustment algorithm. This approach is applied to change the mean and the variance of the shadowed areas by matching the mean and the variance of an image with no shadows. The mathematical model can be written as:

$$Y = \sigma_l/\sigma_s * (X - \mu_s) + \mu_l \tag{11.3}$$

where Y is the output of original pixel, X is the shadowed areas after histogram adjustment, and σ and μ are the standard deviation and mean. The subscripts l and s stand for non-shadowed and shadowed areas, respectively.

Figure 11.9 shows the experimental results before and after the shadow treatment. As can be seen, the shadowed objects caused by tall buildings in Figure 11.9a have been substantially restored and the image quality of these shadowed areas is now comparable to the ones around them. The effectiveness of shadow removal can also be demonstrated by their histograms, as shown in Figure 11.10. The mean and dynamic ranges of the data in the shadowed areas become comparable to those in the non-shadowed areas after shadow treatment.

11.3.3 Occlusion Detection and Compensation

In a TOM, objects such as buildings and bridges should be presented in their true upright planimetric positions, that is, what one can see in a true orthoimage is only roofs of the objects. Thus, walls should be invisible and the occluded areas should be refilled by conjugate orthoimage patches, if they are available.

In order to effectively generate a TOM, a visibility analysis applying the Z-buffer technique has been presented in Section 11.2 (also see Amhar et al. 1998; Rau et al. 2002). The Z-buffer in this section consists of two fictitious planes (or matrixes). One in the image plane stores the distance from the projection center to the surface point, whose corresponding position in the image plane is first calculated with the imaging system parameters (interior and exterior orientations) using the collinearity condition. The other one in the ground surface plane is used as a binary index map with the same resolution, dimension, and projection properties as the DBM. When an area is occluded, it means that several surface points have the same coordinates in the image plane. Under these circumstances, the shorter distance is selected to store and mask the index map at the corresponding location where the distance to the projection center is longer. This method allows us to identify all

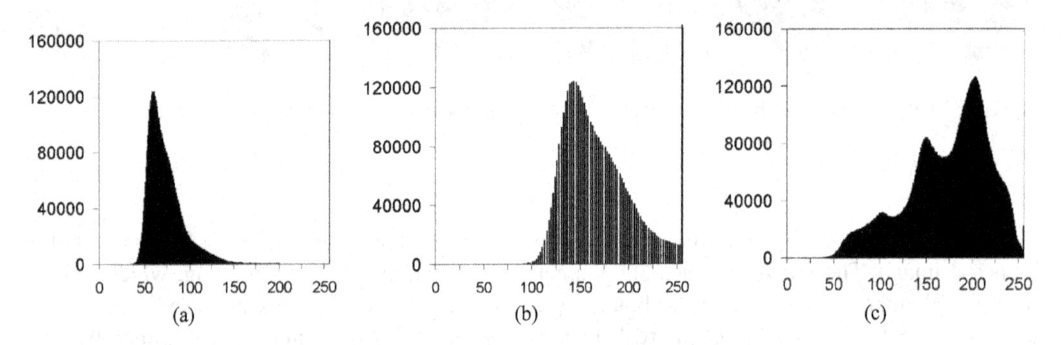

FIGURE 11.10 Comparison of histograms of the shadowed areas: (a) before adjustment; (b) after adjustment; and (c) no shadow in this area.

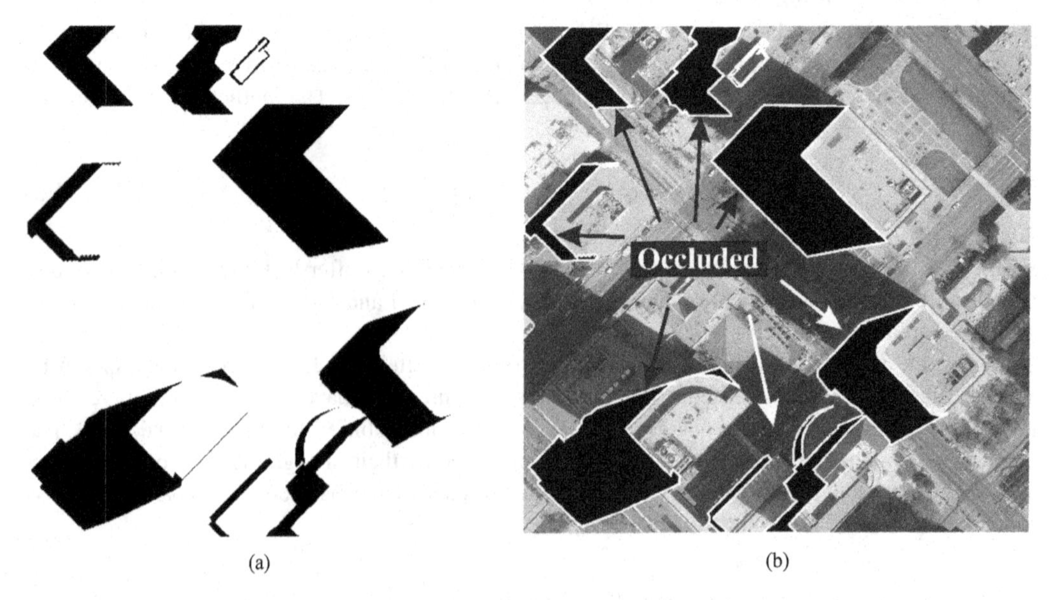

FIGURE 11.11 (a) Index map generated using the Z-buffer algorithm; (b) orthoimage in which the occlusion has been identified by using the index map.

the hidden areas in the index map (Figure 11.11a), and when the actual orthorectification begins, it is necessary to find the brightness values for the surface points that are not masked in the index map. All masked areas can therefore be set as blank or other background values to indicate that they are invisible or occluded in the resulting orthophoto (Figure 11.11b).

Occlusion compensation is implemented by refilling the occluded areas from neighbor "slave" orthoimages. This process involves mosaicking, for which the automatic selection of seam lines from "slave" orthoimages has been described in Section 11.2. The occluded area must be visible in one or even several "slave" orthoimages in order to completely refill an occlusion. To reduce radiometric discrepancy, it is better to use images acquired during the same flight mission because of the similarity in imaging conditions. Figure 11.11 shows that the master image was compensated by conjugate orthoimage patches.

11.3.4 Radiometric Balancing

Significant scene-to-scene radiometric variations are observed in urban aerial images. To overcome this problem, radiometric balancing and blending operations are applied to individual scenes to prevent a patchy or quilted appearance in the final mosaic. An Arc/Info-based tool in combination with a number of available GIS functions, including line and polygon buffering, distance calculation, and grid algebraic operations has been developed to perform the radiometric balancing. This tool allows us to calculate the weights for blending individual scenes along the specified buffer zone using the cubic Hermite function (a detailed description has been given by Zhou et al. (2002)).

11.3.5 True Orthoimage Map (TOM) Generation

After the above procedures are implemented, a TOM can be created. Figure 11.12b depicts a TOM that is generated by using the algorithms and methodologies above. For comparison purposes, Figure 11.12a is a DOM generated by using a conventional differential method. As the "doubling" of building roofs, Figure 11.12a shows that the occlusions caused by buildings still exist in the resulting orthoimage. The building occlusions in Figure 11.12b have been completely removed and refilled. The buildings are correctly orthorectified in their true planimetric positions, and the shadows of buildings have been removed. The results in Figure 11.12b demonstrate that the proposed methodology is capable of generating TOM from large-scale images in urban area. There is, however, room for improvement (e.g., identifying conjugate areas of multiple "slave" images and compensating for the spectral difference between refilled and surrounding areas).

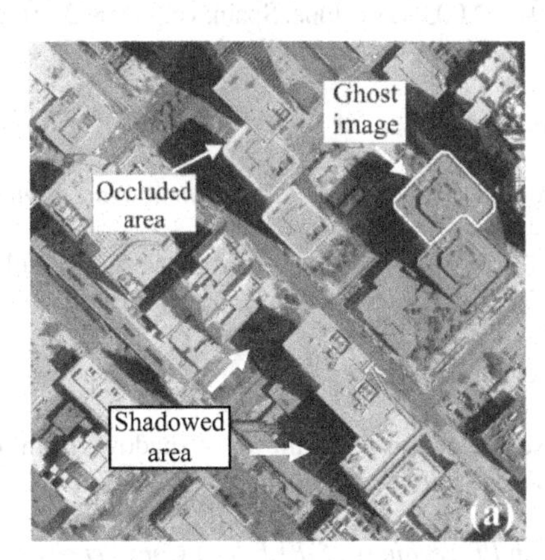

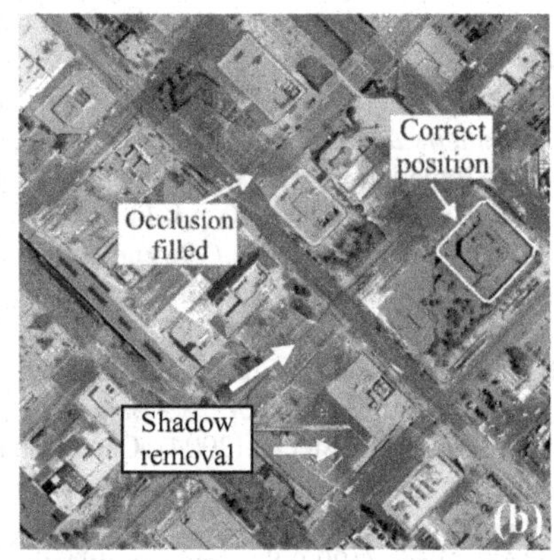

FIGURE 11.12 The comparison of orthorectified large-scale aerial images by (a) traditional differential method and (b) our method.

11.4 CONCLUSIONS

This chapter introduces the principle of urban large-scale orthophoto map generation, including its theories, algorithms, and methods. The major algorithms and methodologies involve: (1) creation of a DBM and a DTM with highly accurate modeling of urban buildings and terrains; (2) generation of DTM-based and DBM-based orthoimage maps and their mosaics; (3) detection and compensation of the shadowed areas and occluded areas; and (4) balance of the radiometric differences between scenes for seamless mosaics.

All of the methodologies are validated using aerial images located in the city of Denver, Colorado, USA. The experimental results demonstrated that the algorithms and methodologies presented in the chapter are correct, since: (1) buildings in the DOM are placed in their proper and accurate upright, planimetric locations; (2) sidewalks and roads are completely visible and building walls are invisible; and (3) the seam lines of mosaicked images are eliminated.

REFERENCES

Amhar, F., J. Josef, and C. Ries (1998). The generation of true orthophotos using a 3D building model in conjunction with a conventi onal DTM, *International Archives of Photogrammetry and Remote Sensing*, vol. 32, Part 4, pp. 16–22.

Biason A., S. Dequal, and A. Lingua (2004). A new procedure for the automatic production of true orthophotos, *The International Archives of the Photogrammetry, Remote Sensing and Spatial Information Sciences*, edited by Orthan Altan, vol. XXXV, July 12–23, DVD.

Breunig, M. (1996). *Integration of Spatial Information for Geo-information Systems*, Springer Corp. Berlin.

Cameron, A. D, D. R. Miller, F. Ramsay, I. Nikolaou, and G. C. Clarke (2000), Temporal measurement of the loss of native pinewood in Scotland through the analysis of orthorectified aerial photographs. *Journal of Environmental Management*, vol. 58, pp.33–43.

Chen, L.C. and L.H. Lee (1993). Rigorous generation of digital orthophotos from SPOT images, *Photogrammetric Engineering & Remote Sensing,* vol. 59, no. 3, pp. 655–661.

Chen, L.C. and J.Y. Rau (1993). A unified solution for digital terrain model and orthoimage generation from SPOT stereopairs, *IEEE Transactions on Geoscience and Remote Sensing*, vol. 31, no. 6, pp. 1243–1252.

Federal Geographic Data Committee (1995). *Development of a National Digital Geospatial Data Framework*, April, 1995, http://www.fgdc.gov/framework/framdev.html.

Federal Geographic Data Committee (1997). *Fact Sheet: National Digital Geospatial Data Framework: A Status Report*, Federal Geographic Data Committee, Washington, D.C., July 1997, 37p. http://www.fgdc.gov/framework/framdev.html

Gonzalez, J. A., M. L. Docampo, and I. C. Guerrero, (2004). Detection of buildings through automatic extraction of shadows in IKONOS imagery, in *Proceedings of The International Society for Optical Engineering: Image and Signal Processing for Remote Sensing IX*, Sep 9–122,003, Barcelona, Spain, vol. 5238, 2004, pp. 36–43.

Graz, M. G. (1999). Managing large 3D urban databases, *Photogrammetric Week*, Stuttgart University, 1999 pp. 341–349.

Grün, A. and X. Wang (1998). CC-Modeler: a topology generator for 3-D city models, *ISPRS Journal of Photogrammetry and Remote Sensing*, vol. 53, pp. 286–295.

Hohle, J. (1996). Experiences with the production of digital orthophotos, *Photogrammetric Engineering and Remote Sensing,* vol. 62, no. 10, pp. 11189–11194.

Irvin, R.B. and D.M. McKeown Jr., (1989). Methods for exploiting the relationship between buildings and their shadows in aerial imagery, *IEEE Transactions on Systems, Man, and Cybernetics*, vol. 19, pp. 1564–1575.

Jauregui, M., J. Vílchez, and L. Chac6n (2002). A procedure for map updating using digital mono-plotting. *Computers & Geosciences*, vol. 28, no. 4, pp. 513–523.

Jaynes, C., S. Webb, and R. Steele, (2004). Camera-based detection and removal of shadows from interactive multiprojector displays, *IEEE Transactions on Visualization and Computer Graphics*, vol. 10, no. 3, May/June, pp. 290–301.

Jiang, C. and M.O. Ward, (1992) Shadow identification, in *Proceedings of IEEE Int'l Conference on Computer Vision and Pattern Recognition*, Champaign, IL, USA, pp. 606–612.

Joshua, G., (2001). Evaluating the accuracy of digital orthophotos quadrangles (DOQ) in the context of parcel-based GIS, *Photogrammetric Engineering & Remote Sensing,* vol. 67, no. 2, pp. 199–205.

Kelmelis, J. A., (2003). To the national map and beyond, *Cartography and Geographic Information Science*, vol. 30, no. 2, pp. 185–198.

Kelmelis, J. A., M. Demulder, C. Ogrosky, N. VanDriel, and B. Ryan, (2003). The National map, from geography to mapping and back again, *Photogrammetric Engineering and Remote Sensing*, vol. 69, no. 10, pp. 1109–1118.

Liow, Y.-T. and P. Theo, (1990). Use of shadows for extracting buildings in aerial images, *Computer Vision, Graphics, and Image Processing*, vol. 49, no. 2, Feb, pp. 242–277.

Maitra, J. B. (1998). *The National Spatial Data Infrastructure in the United States: Standards, Metadata, Clearinghouse, and Data Access*, Federal Geographic Data Committee c/o U.S. Geological Survey, 12,201 Sunrise Valley Drive, 590 National Center Reston, Virginia 20,192.

Mayr, W. (2002). True orthoimages, *GIM International*, vol. 37, April, pp. 37–39.

Noronha, S. and R. Nevatia (2001). Detection and modeling of buildings from multiple aerial images, *IEEE Transactions on Pattern Analysis and Machine Intelligence*, vol. 23, no. 5, May, pp. 501–518.

Passini, R. and K. Jacobsen (2004). Accuracy analysis of digital orthophotos from very height resolution imagery, *The International Archives of the Photogrammetry, Remote Sensing and Spatial Information Sciences*, edited by Orthan Altan, vol. XXXV, July 12–23 (DVD).

Peng, J. and Y. Liu, (2004). *The Role of Context and Model in Urban Aerial Image Interpretation Focusing on Buildings, IEEE International Conference on Networking, Sensing and Control*, Taipei, Taiwan, vol. 1, Mar 21–23 2004, pp. 1–6.

Rau, J.Y., N.Y. Chen, and L.C. Chen (2002). True orthophoto generation of built-up areas using multi-view images, *Photogrammetric Engineering and Remote Sensing,* vol. 68, no. 6, June, pp. 581–588.

Salvador, E., A. Cavallaro, and T. Ebrahimi (2001). *Shadow Identification and Classification Using Invariant Color Models, in Proceedings of IEEE International Conference on Acoustics, Speech and Signal Processing,* May 7–11 2001, Salt Lake, vol. 3, pp. 1545–1548.

Schickler, W. and A. Thorpe (1998). Operational procedure for automatic true orthophoto generation, *International Archives of Photogrammetry and Remote Sensing,* vol. 32, Part 4, pp. 527–532.

Schott, J. R., C. Salvaggio, and W. J. Volchok, (1988). Radiometric scene normalization using pseudoinvariant features. *Remote Sensing of Environment*, vol. 26, pp. 1–16.

Siachalou, S. (2004). Urban orthoimage analysis generated from IKONOS data. *The International Archives of the Photogrammetry, Remote Sensing and Spatial Information Science*, edited by Orthan Altan, Vol. XXXV, July 12–23, (DVD).

Skarlatos, D. (1999). Orthophotograph production in urban areas, *The Photogrammetric Record,* vol. 16, no. 94, pp. 643–650.

Stauder, J., R. Mech, and J. Ostermann (1999). Detection of moving cast shadows for object segmentation, *IEEE Transactions on Multimedia*, vol. 1, no. 1, pp. 65–76, Mar. 1999.

Tsai, J. D. (2003). Automatic shadow detection and radiometric restoration on digital aerial images, *International Geoscience and Remote Sensing Symposium (IGARSS): Learning From Earth's Shapes and Colours*, Toulouse, France, July 21–25, vol. 2, pp. 732–733.

USGS (1996). Digital orthophoto standards, *National Mapping Program-Technical Instructions, Part I, General; Part II, Specifications.* US Department of the Interior, U.S. Geological Survey, National Mapping Division, Virginia, USA. http://rmmcweb.cr.usgs.gov/public/nmpstds/doqstds.html, December 1996.

USGS (1998). *Digital Orthophoto Program*, http://mapping.usgs.goc/www/ndop/index.html, U.S. Department of the Interior, U.S. Geological Survey, Virginia, USA.

Vassilopoulou, S., L. Hurni, V. Dietrich, E. Baltsavias, M. Pateraki, E. Lagios, and I. Parcharidis (2002). Orthophoto generation using IKONOS imagery and high-resolution DEM: a case study on volcanic hazard monitoring of Nisyros Island (Greece). *ISPRS Journal of Photogrammetry and Remote Sensing*, vol. 57, no. 1–2, pp. 24–38.

Zhou, G., W. Chen, and J. Kelmelis, A comprehensive study on urban true orthorectification, *IEEE Transaction on Geosciene and Remote Sensing,* vol. 43, no. 9, pp. 2138–2147, 2005.

Zhou, G., K. Jezek, W. Wright, J. Rand, and J. Granger (2002). Orthorectifying 1960's declassified intelligence satellite photography (DISP) of Greenland. *IEEE Geoscience and Remote Sensing,* vol. 40, no. 6, pp. 1247–1259.

Zhou, G., Z. Qin, S. Benjamin, and W. Schickler (2003). *Technical Problems of Deploying National Urban Large-scale True Orthoimage Generation, The 2nd Digital Government Conference*, Boston, May 18–21, 2003, pp. 383–387.

Zhou G., M. Xie, and J. Gong, (2000). Design and implementation of attribute Database management system for GIS (GeoStar), *International Journal of Geographical Information Science,* vol. 6, no. 2, pp. 170–180.

Zlatanova, S. (2000). *3D GIS for Urban Development*, ISBN: 90–6164–178-0, ITC Dissertation Number 69. Doktor der technischen Wissenschaften an der Technischen Universitate Graz, Enschede, Netherlands.

12 Orthophotomap Creation with Extremely High Buildings

12.1 INTRODUCTION

For urban areas with very high buildings, the traditional method is not capable of orthorectifying the relief displacements caused by high buildings. Therefore, a so-called true orthoimage map (TOM) generation method was presented by a few researchers, such as Amhar et al. (1998), Rau et al. (2002), Nielsen (2004), Biason et al. (2004), Mayr (2002), Palà and Arbiol (2002), Siachalou (2004), Habib et al. (2006), Zhou et al. (2005), Sheng et al. (2003) and Xie and Zhou (2006). Briefly, the TOM generation is composed of digital terrain model (DTM)-based orthoimage map generation, digital building model (DBM)-based orthoimage map generation, and their merging. The associated work for TOM creation includes occlusion detection and refilling, shadow detection and compensation, and radiometric difference balance. However, when a TOM is created for an urban area with very high buildings, such as over 150 m high, it has been found that some of high buildings cannot be completely orthorectified into their upright and correct position, that is, their walls can still be viewed in the orthoimage map (Figure 12.1). In order to further examine which type of errors causes this phenomena, the same buildings in two different aerial images are orthorectified using the same DBM and the same method. It has been found that the buildings are completely orthorectified in one image (Figure 12.1a), but not in the other image (Figure 12.1b). This fact discovered that the DBM is not a bane. In other words, the incomplete orthorectification is probably directly caused by the inexact exterior orientation parameters, because the traditional orthorectification algorithm largely depends on both highly accurate and well-distributed ground control points (GCPs). Moreover, the GCPs are usually laid out in the ground. However, buildings occluding buildings, and/or occluding the ground frequently happens in urban areas, resulting in the fact that the good distribution of the GCPs is impractical. As a result, the orthorectified images are undoubtedly incomplete. Therefore, this chapter introduces a method to enhance the accuracy of TOM creation through use of the relative constraint conditions.

The relevant research efforts include, for instance, demonstration by Skarlatos (1999) and Joshua (2001) that building occlusions did significantly influence not only image quality but also the accuracy of orthoimages. Amhar et al. (1998), Schickler and Thorpe (1998), and Mayr (2002) not only considered the hidden effects introduced by abrupt changes of surface height (e.g., buildings), but also considered seamless mosaicking around fill-in areas in order to reduce gray-value discontinuities. Rau et al. (2002) presented a suitable enhancement technique to restore information within building shadow areas. Jauregui et al. (2002) presented a procedure for orthorectifying aerial photographs to produce and update terrain surface maps. Passini and Jacobsen (2004) analyzed the accuracy of orthoimages from very high-resolution imagery. Biason et al. (2004) further explored the automatic generation of true orthoimages. Zhou et al. (2005) comprehensively discussed the true orthoimage generation.

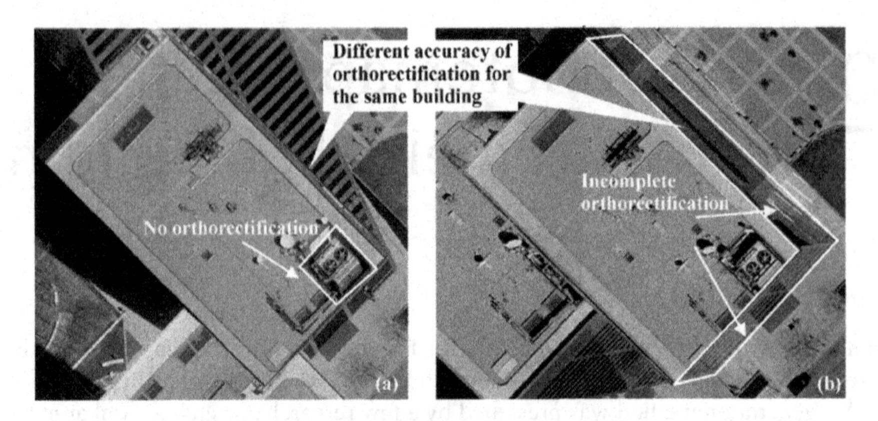

FIGURE 12.1 Incomplete orthorectification and different accuracy for the same building in the two aerial images.

12.2 RELATIVE CONSTRAINT FOR ORTHORECTIFICATION

12.2.1 Traditional Orthorectification Model

The mathematical model of traditional orthorectification is a called the photogrammetry differential model, with which, for any given ground point such as G, it is expressed by

$$x_g - x_0 = -f\frac{a_1(X_G - X_S) + b_1(Y_G - Y_S) + c_1(Z_G - Z_S)}{a_3(X_G - X_S) + b_3(Y_G - Y_S) + c_3(Z_G - Z_S)} \tag{12.1a}$$

$$y_g - y_0 = -f\frac{a_2(X_G - X_S) + b_2(Y_G - Y_S) + c_2(Z_G - Z_S)}{a_3(X_G - X_S) + b_3(Y_G - Y_S) + c_3(Z_G - Z_S)} \tag{12.1b}$$

where f, x_0, y_0 are the camera's interior orientation parameters (IOPs), X_S, Y_S, Z_S are the position of exposure station, a_1, …, c_3 are the elements of rotation matrix, which is a function of three rotation angles ω, ϕ, κ; x_g, y_g are image coordinates of the ground point G, and X_G, Y_G, Z_G are the coordinates of the ground point G in the ground coordinate system.

Equation 12.1 can further be expressed by a linear equation after it is linearized using Taylor's series, that is,

$$\boldsymbol{V} = \boldsymbol{AT} - \boldsymbol{L} \tag{12.2}$$

Where $\boldsymbol{T} = (\phi\ \omega\ \kappa\ X_S\ Y_S\ Z_S)^T$, $\boldsymbol{V} = (v_{gx_1}\ v_{gy_1}\ v_{gx_2}\ v_{gy_2})^T$, $\boldsymbol{L} = (l_{gx_1}\ l_{gy_1}\ l_{gx_2}\ l_{gy_2})^T$, A is coefficient (see Mikhail et al. 2001). With a number of GCPs, the nominal IOPs provided by the vendor can be calibrated, and the EOPs in Equation 12.2 can be solved using least-squares estimation. Once the IOPs and EOPs are determined, any ground point can be orthorectified onto a pre-defined orthogonal plane. The detailed process of orthorectification can be referred to in Zhou et al. (2005).

In order to increase the accuracy of an orthoimage map, one traditionally used to increase the number of GCPs and/or select well-distributed GCPs. This method has its limitation because most of the current GCPs are distributed on the ground, which means the tops of high buildings have no sufficient control information. Thus, this chapter plans using relative constraint conditions, such as perpendicular and collinear conditions, to improve the accuracy.

12.2.2 Perpendicular Control Condition

Suppose that AB and BC are two edges of a flat house roof, and their corresponding edges in the image plane are ab and bc (see Figure 12.2). Suppose that the coordinates of A, B, and C are (X_A, Y_A, Z_A), (X_B, Y_B, Z_B) and (X_C, Y_C, Z_C) in the object coordinate system. For a flat-roof cube house, the heights (i.e., Z coordinates) of A, B, and C are the same, and the line segments AB and BC are perpendicular to each other (see Figure 12.2).

If AB is not perpendicular to BC at an intersection angle θ, that is, B deviates from its correct position at B′, a line from C to O is drawn to make the CO perpendicular to AB′ (see Figure 12.2). l is the distance of B and O, and can be expressed by

$$l = (X_C - X_B)cos\theta + (Y_C - Y_B)sin\theta = (X_C - X_B)\frac{(X_B - X_A)}{s_{AB}} + (Y_C - Y_B)\frac{(Y_B - Y_A)}{s_{AB}} \tag{12.3}$$

where s_{AB} is the distance of the segment AB. Theoretically, the distance l should be zero. Thus, the differential form for Equation 12.3 is:

$$\begin{aligned} l_0 + \Delta l = & \left[(X_C - X_B)\frac{(X_B - X_A)}{s_{AB}} + (Y_C - Y_B)\frac{(Y_B - Y_A)}{s_{AB}} \right] + \\ & \frac{1}{s_{AB}}\Big[(X_B - X_A)\Delta X_C + (Y_B - Y_A)\Delta Y_C + (X_B - X_C)\Delta X_A + (Y_B - Y_C)\Delta Y_A + \\ & (X_C - 2X_B + X_A)\Delta X_B + (Y_C - 2Y_B + Y_A)\Delta Y_B \Big] = 0 \end{aligned} \tag{12.4}$$

Rewrite Equation 12.4 in matrix form:

$$\begin{bmatrix} X_B - X_C \\ Y_B - Y_C \\ X_C - 2X_B + X_A \\ Y_C - 2Y_B + Y_A \\ X_B - X_A \\ Y_B - Y_A \end{bmatrix}^T \begin{bmatrix} \Delta X_A \\ \Delta Y_A \\ \Delta X_B \\ \Delta Y_B \\ \Delta X_C \\ \Delta Y_C \end{bmatrix} + \left[(X_C - X_B)(X_B - X_A) + (Y_C - Y_B)(Y_B - Y_A) \right] = 0 \tag{12.5}$$

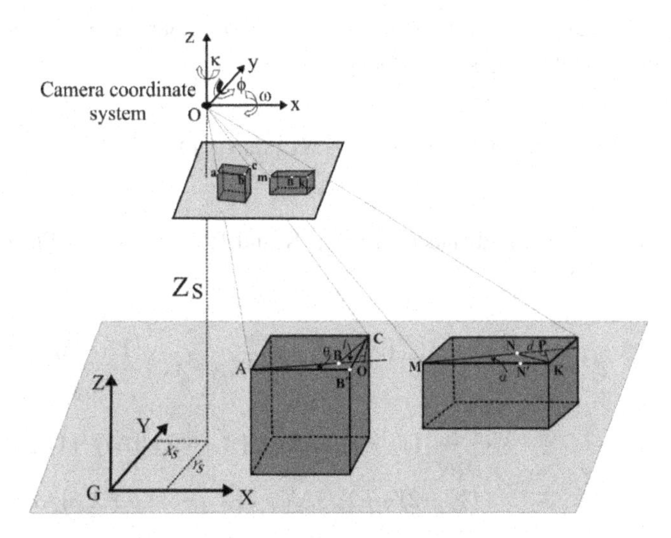

FIGURE 12.2 The geometry for perpendicular constraint and collinear constraint conditions.

Equation 12.5 is in fact a perpendicular constraint condition, which describes line AB perpendicular to BC on the X–Y plane. Similarly, if the points B and C are on the plane X–Z, or Y–Z, the perpendicular relative constraint conditions can be expressed by, respectively,

$$\begin{bmatrix} X_B - X_C \\ Z_B - Z_C \\ X_C - 2X_B + X_A \\ Z_C - 2Z_B + Z_A \\ X_B - X_A \\ Z_B - Z_A \end{bmatrix}^T \begin{bmatrix} \Delta X_A \\ \Delta Z_A \\ \Delta X_B \\ \Delta Z_B \\ \Delta X_C \\ \Delta Z_C \end{bmatrix} + \left[(X_C - X_B)(X_B - X_A) + (Z_C - Z_B)(Z_B - Z_A) \right] = 0 \tag{12.6}$$

$$\begin{bmatrix} Y_B - Y_C \\ Z_B - Z_C \\ Y_C - 2Y_B + Y_A \\ Z_C - 2Z_B + Z_A \\ Y_B - Y_A \\ Z_B - Z_A \end{bmatrix}^T \begin{bmatrix} \Delta Y_A \\ \Delta Z_A \\ \Delta Y_B \\ \Delta Z_B \\ \Delta Y_C \\ \Delta Z_C \end{bmatrix} + \left[(Z_C - Z_B)(Z_B - Z_A) + (Y_C - Y_B)(Y_B - Y_A) \right] = 0 \tag{12.7}$$

The vector form for Equation 12.5 can be rewritten as follows (Equations 12.6 and 12.7 are similar)

$$\mathbf{C}_1 \hat{\mathbf{X}}_1 + \mathbf{W}_1 = \mathbf{0} \tag{12.8}$$

where $\mathbf{C}_1 = \begin{bmatrix} X_B - X_C \\ Y_B - Y_C \\ X_C - 2X_B + X_A \\ Y_C - 2Y_B + Y_A \\ X_B - X_A \\ Y_B - Y_A \end{bmatrix}^T$, $\hat{\mathbf{X}}_1 = \begin{bmatrix} \Delta X_A \\ \Delta Y_A \\ \Delta X_B \\ \Delta Y_B \\ \Delta X_C \\ \Delta Y_C \end{bmatrix}$, $\mathbf{W}_1 = (X_C - X_B)(X_B - X_A) + (Y_C - Y_B)(Y_B - Y_A)$

12.2.3 Collinear Constraint Condition

Also, as depicted in Figure 12.2, supposing that the segments MN and NK are not collinear, at an intersection angle of α, it has

$$d = (Y_K - Y_N)\cos\alpha + (X_K - X_N)\sin\alpha = (Y_K - Y_N)\frac{(X_N - X_M)}{s_{MN}} + (X_K - X_N)\frac{(Y_N - Y_M)}{s_{MN}} \tag{12.9}$$

where d is the distance between K and P. Similarly, its linearized equation and matrix form are, respectively

$$\begin{aligned} d_0 + \Delta d = & \left[(Y_K - Y_N)\frac{(X_N - X_M)}{S_{MN}} + (X_K - X_N)\frac{(Y_N - Y_M)}{S_{MN}} \right] \\ & + \frac{1}{s_{MN}} \left[(Y_N - Y_M)\Delta X_K + (X_N - X_M)\Delta Y_K + (Y_N - Y_K)\Delta X_M + (X_N - X_K)\Delta Y_M \right. \\ & \left. + (Y_K - 2Y_N + Y_M)\Delta X_N + (X_M - 2X_N + X_K)\Delta Y_N \right] \end{aligned} \tag{12.10}$$

$$\begin{bmatrix} Y_N - Y_K \\ X_K - X_N \\ Y_K - 2Y_N + Y_M \\ X_M - 2X_N + X_K \\ Y_M - Y_N \\ X_N - X_M \end{bmatrix}^T \begin{bmatrix} \Delta X_M \\ \Delta Y_M \\ \Delta X_N \\ \Delta Y_N \\ \Delta X_K \\ \Delta Y_K \end{bmatrix} + \left[(Y_K - Y_N)(X_N - X_M) + (X_K - X_N)(Y_N - Y_M) \right] = 0 \tag{12.11}$$

Equation 12.11 is, in fact, a collinear constraint condition, which describes line MN collinear with NK on the X–Y plane. Similarly, if the points M, N, and K are on the planes X–Z, or Y–Z, the corresponding collinear constraint conditions are, respectively,

$$\begin{bmatrix} Z_N - Z_K \\ X_K - X_N \\ Z_K - 2Z_N + Z_M \\ X_M - 2X_N + X_K \\ Z_M - Z_N \\ X_N - X_M \end{bmatrix}^T \begin{bmatrix} \Delta X_M \\ \Delta Z_M \\ \Delta X_N \\ \Delta Z_N \\ \Delta X_K \\ \Delta Z_K \end{bmatrix} + \left[(Z_K - Z_N)(X_N - X_M) + (X_K - X_N)(Z_N - Z_M) \right] = 0 \tag{12.12}$$

$$\begin{bmatrix} Z_N - Z_K \\ Y_K - Y_N \\ Z_K - 2Z_N + Z_M \\ Y_M - 2Y_N + Y_K \\ Z_M - Z_N \\ X_N - X_M \end{bmatrix}^T \begin{bmatrix} \Delta Y_M \\ \Delta Z_M \\ \Delta Y_N \\ \Delta Z_N \\ \Delta Y_K \\ \Delta Z_K \end{bmatrix} + \left[(Z_K - Z_N)(Y_N - Y_M) + (Y_K - Y_N)(Z_N - Z_M) \right] = 0 \tag{12.13}$$

The vector form for Equation 12.11 can be written as follows (Equations 12.12 and 12.13 are similar)

$$\boldsymbol{C_2} \boldsymbol{\hat{X}_2} + \boldsymbol{W_2} = \boldsymbol{0} \tag{12.14}$$

where $\boldsymbol{C_2} = \begin{bmatrix} Z_N - Z_K \\ Y_K - Y_N \\ Z_K - 2Z_N + Z_M \\ Y_M - 2Y_N + Y_K \\ Z_M - Z_N \\ X_N - X_M \end{bmatrix}^T$, $\boldsymbol{\hat{X}_2} = \begin{bmatrix} \Delta Y_M \\ \Delta Z_M \\ \Delta Y_N \\ \Delta Z_N \\ \Delta Y_K \\ \Delta Z_K \end{bmatrix}$, $\boldsymbol{W_2} = (Z_K - Z_N)(Y_N - Y_M) + (Y_K - Y_N)(Z_N - Z_M)$

From Figure 12.2, if points, A, B, and C and their corresponding imaged points, a, b, and c are simultaneously GCPs, this means that one perpendicular or collinear constraint condition will add six additional unknown parameters (i.e., ΔX_A, ΔY_A, ΔZ_A, ΔX_B, ΔY_B, ΔZ_B for perpendicular constraint). Thus, Equation 12.2 can be extended into:

$$\boldsymbol{V} = \boldsymbol{AT} + \boldsymbol{B\hat{X}} - \boldsymbol{L} \tag{12.15}$$

where, $\boldsymbol{T} = (\phi\ \omega\ \kappa\ X_S\ Y_S\ Z_S)^T$,
$\hat{\boldsymbol{X}} = \left(\Delta X_A\ \Delta Y_A\ \Delta Z_A\ \Delta X_B\ \Delta Y_B\ \Delta Z_B\ \Delta X_M\ \Delta Y_M\ \Delta Z_M\ \Delta X_N\ \Delta Y_N\ \Delta Z_N\right)^T$,
A and **B** are coefficient matrix, $\boldsymbol{V}$ and $\boldsymbol{L}$ are similar to Equation 12.2.

Combine Equations 12.8, 12.14, and 12.15, we have

$$\begin{cases} \boldsymbol{V} = \boldsymbol{AT} + \boldsymbol{B}\hat{\boldsymbol{X}} - \boldsymbol{L} \\ \boldsymbol{C}_1\hat{\boldsymbol{X}}_1 + \boldsymbol{W}_1 = \boldsymbol{0} \\ \boldsymbol{C}_2\hat{\boldsymbol{X}}_2 + \boldsymbol{W}_2 = \boldsymbol{0} \end{cases} \tag{12.16}$$

Further, rewrite Equation 12.16 as

$$\begin{cases} \boldsymbol{V} = \boldsymbol{D}\delta\boldsymbol{X} - \boldsymbol{L} \\ \boldsymbol{C}_x\delta\boldsymbol{X} + \boldsymbol{W}_x = \boldsymbol{0} \end{cases} \tag{12.17}$$

where $\boldsymbol{C}_x = \left(\boldsymbol{C}_1 \quad \boldsymbol{C}_2\right)^T$, and $\boldsymbol{W}_x = \left(\boldsymbol{W}_1 \quad \boldsymbol{W}_2\right)^T$, $\boldsymbol{D} = \left(\boldsymbol{A} \mid \quad \boldsymbol{B}\right)$

$\delta\boldsymbol{X} = \left(\Delta\phi\ \Delta\omega\ \Delta\kappa\ \Delta X_S\ \Delta Y_S\ \Delta Z_S \Delta X_A\ \Delta Y_A\ \Delta Z_A \ldots \Delta X_N\ \Delta Y_N\ \Delta Z_N\right)^T$

With the least-squares method (Lawson and Hanson 1995), Equation 12.17 can be written by

$$\Phi = \boldsymbol{V}^T\boldsymbol{V} + 2\boldsymbol{K}_s^T\left(\boldsymbol{C}_x\delta\boldsymbol{X} + \boldsymbol{W}_x\right)$$

The corresponding normal equation is:

$$\begin{cases} \boldsymbol{D}^T\boldsymbol{D}\delta\boldsymbol{X} + \boldsymbol{C}_x{}^T\boldsymbol{K}_S + \boldsymbol{D}^T\boldsymbol{L} = \boldsymbol{0} \\ \boldsymbol{C}_x\delta\boldsymbol{X} + \boldsymbol{0}\boldsymbol{K}_S + \boldsymbol{W}_X = \boldsymbol{0} \end{cases} \tag{12.18}$$

where $\boldsymbol{K}_S$ is an introduced unknown matrix. If number of the total observation equations is m, $\boldsymbol{K}_S$ is a $m \times 1$ matrix, that is, $\boldsymbol{K}_S = (K_{S_1}, K_{S_2}, \ldots, K_{S_m})^T$.

Let $N_{dd} = D^TD$, Equation 12.18 is rewritten by

$$\begin{pmatrix} \boldsymbol{N}_{dd} & \boldsymbol{C}_x{}^T \\ \boldsymbol{C}_x & \boldsymbol{0} \end{pmatrix}\begin{pmatrix} \delta\boldsymbol{X} \\ \boldsymbol{K}_s \end{pmatrix} + \begin{pmatrix} \boldsymbol{D}^T\boldsymbol{L} \\ \boldsymbol{W}_x \end{pmatrix} = \boldsymbol{0} \tag{12.19}$$

The solution of unknown parameters would be

$$\begin{cases} \delta\boldsymbol{X} = -\left(\boldsymbol{Q}_{11}\boldsymbol{D}^T\boldsymbol{L} + \boldsymbol{Q}_{12}\boldsymbol{W}_X\right) \\ \boldsymbol{K}_S = -\left(\boldsymbol{Q}_{21}\boldsymbol{D}^T\boldsymbol{L} + \boldsymbol{Q}_{22}\boldsymbol{W}_X\right) \end{cases} \tag{12.20}$$

where

$$\begin{pmatrix} \boldsymbol{N}_{dd} & \boldsymbol{C}_x{}^T \\ \boldsymbol{C}_x & \boldsymbol{0} \end{pmatrix}^{-1} = \begin{pmatrix} \boldsymbol{Q}_{11} & \boldsymbol{Q}_{12} \\ \boldsymbol{Q}_{21} & \boldsymbol{Q}_{22} \end{pmatrix} \tag{12.21}$$

Equation 12.20 is a mathematical model to be used for orthorectification. As seen, this model combines the building relative controls and the traditional GCP. Thus, higher accuracy of the orthoimage should be achieved.

In traditional photogrammetry, the number and distribution of GCPs significantly impact the accuracy of orthorectification. Similarly, the building size, for instance length and width of a building, probably also impacts the accuracy of this proposed method. As seen in Figure 12.2, the longer the line AB, the smaller the error. On the other hand, the relief displacements caused by high buildings occur along radial lines from the nadir point. Thus, the relief displacements are zero from imaged objects at the nadir point and this increases with increased radial distances from the nadir.

12.3 EXPERIMENTS AND ANALYSES

12.3.1 Experimental Data

A detailed description for the experimental data can be referred to in Zhou et al. (2005). Briefly, the experimental field is located in downtown Denver, Colorado, where the highest building is 125 m, and many others are around 100 m. Six original aerial images are acquired from two flight strips using RC30 aerial camera at a focal length of 153.022 mm on April 1, 2000. The flying height was approximately 1650 m above the mean ground elevation of the imaged area. The aerial photos were originally recorded on film and later scanned into digital format at a pixel resolution of 25µm. Part of the scanned aerial images is shown as Figure 12.3a. Figure 12.3b is the 2D representation of the digital surface model (DSM).

12.3.2 Control Information

Firstly, 232 GCPs (i.e., 3D coordinates) are manually measured from the DBM using Erdas/Imagine. The distribution of these GCPs covers the entire downtown area of Denver (Figure 12.4). The measurement accuracy of the GCPs is approximately 1.48 pixels in x direction and 1.27 pixels in y direction, on average. The corresponding 2D image coordinates are measured from the original aerial images using the following method (Figure 12.5): a few (e.g., 8) GCPs including their 3D and 2D coordinates are first selected, and then the EOPs of the aerial image is calculated using space intersection. With the calculated EOPs, all other GCPs are back-projected onto the aerial image to obtain their approximate position in the aerial image plane. The precise locations (i.e., 2D image coordinates) are obtained by manually adjusting all these points in a magnified sub-window. The measurement accuracy of the 2D image coordinates is sub-pixel level.

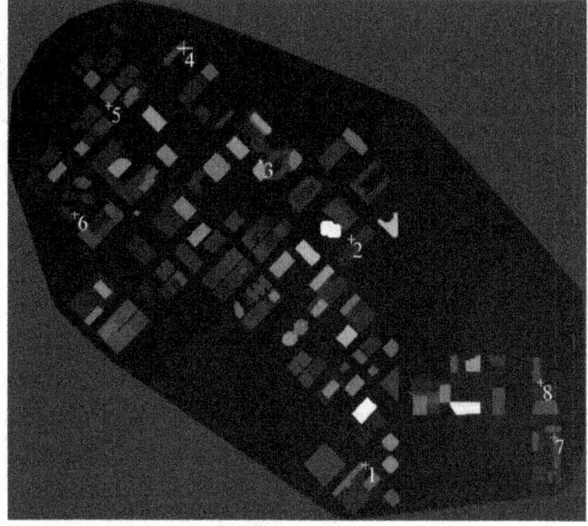

FIGURE 12.3 (a) Part of original aerial images and (b) DSM in downtown Denver, CO, in which the brightness in the DSM represents surface height.

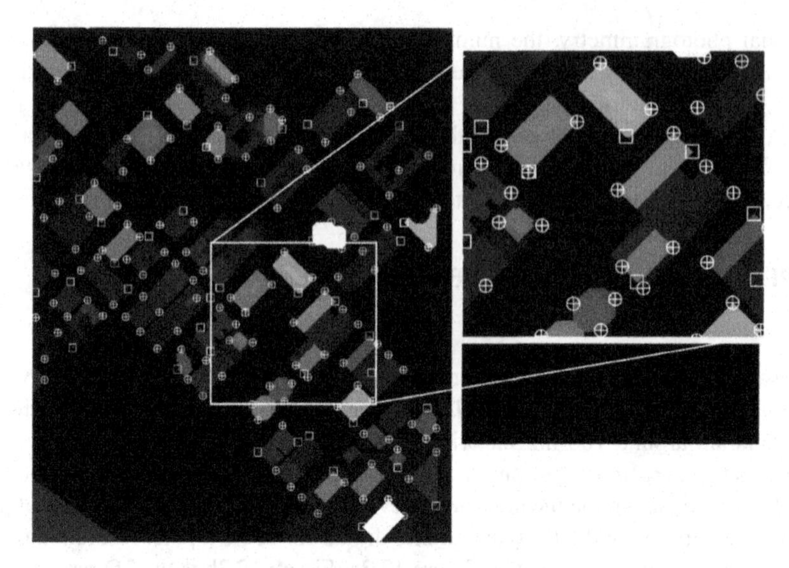

FIGURE 12.4 The measured GCPs and checkpoints as well as their image positions.

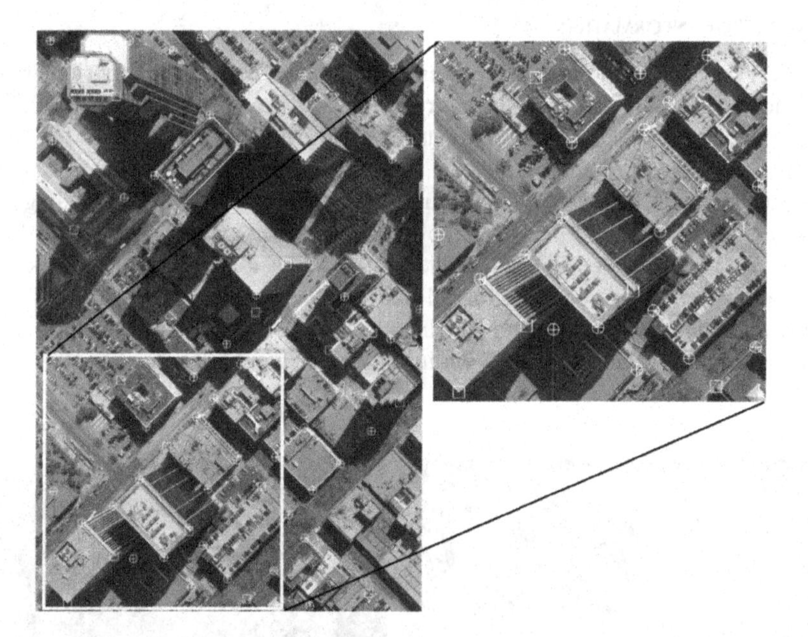

FIGURE 12.5 The 2D image coordinate measurement in the aerial image.

In addition to the 232 GCPs, 89 checkpoints from the DBM are also measured to evaluate the accuracy of the finally orthorectified image, that is, TOM. Also, all the checkpoints are at corners of the buildings. The measurement accuracy is consistent with the GCPs (Figure 12.4).

12.3.3 Constraint Line Extraction

With the measured GCPs, the corresponding lines to be taken as constraint controls are simultaneously extracted. In this experiment, the model of describing each building is realized using the parameterized CSG method (Xie and Zhou 2006). In this model, each element of the CSG primitive has been assigned with its properties, such as wall, roof, bottom, and so on, and each building model has already contained its geometric topology, which describes the relationship of the building edges (Figure 12.6). Thus, the characteristics of topology can automatically be transferred into the relative controls. In other words, the relationship describing the edges of each building, such as perpendicularity and collinearity, can be transformed into Equation 12.8 and/or Equation 12.14. For example, in Figure 12.7, the faces, F_a, F_b, and F_c, the edges, L_1, L_2, and L_3, and their attributes are described

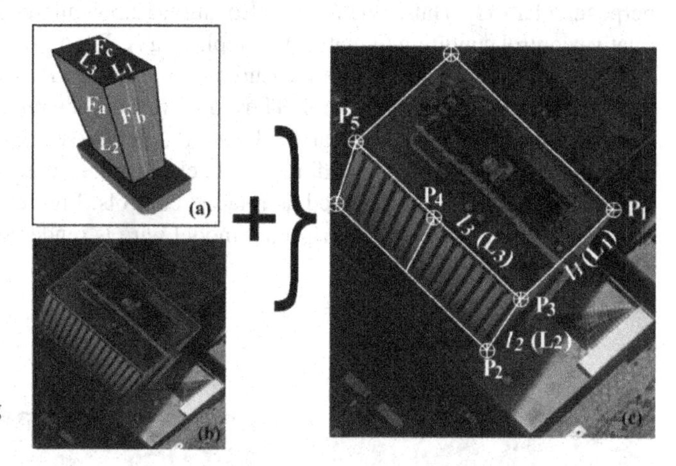

FIGURE 12.6 Data structure of building model.

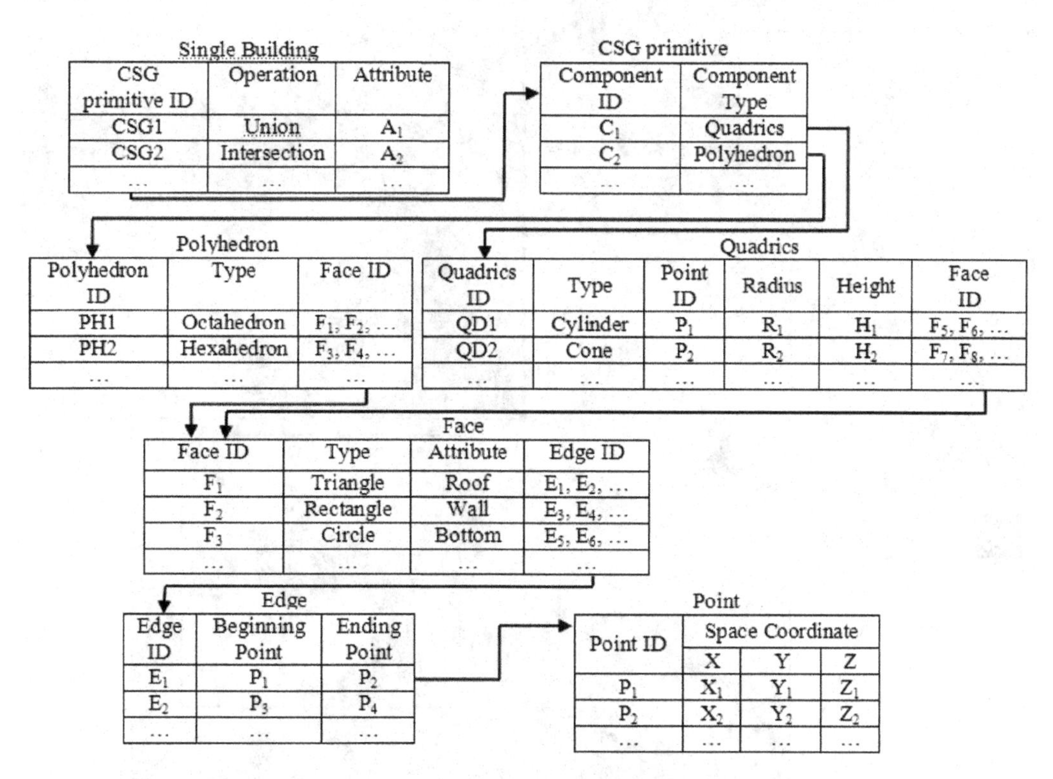

Single Building

CSG primitive ID	Operation	Attribute
CSG1	Union	A_1
CSG2	Intersection	A_2
...	...	...

CSG primitive

Component ID	Component Type
C_1	Quadrics
C_2	Polyhedron
...	...

Polyhedron

Polyhedron ID	Type	Face ID
PH1	Octahedron	$F_1, F_2, \ldots$
PH2	Hexahedron	$F_3, F_4, \ldots$
...	...	...

Quadrics

Quadrics ID	Type	Point ID	Radius	Height	Face ID
QD1	Cylinder	P_1	R_1	H_1	$F_5, F_6, \ldots$
QD2	Cone	P_2	R_2	H_2	$F_7, F_8, \ldots$
...	...	...	...	...	...

Face

Face ID	Type	Attribute	Edge ID
F_1	Triangle	Roof	$E_1, E_2, \ldots$
F_2	Rectangle	Wall	$E_3, E_4, \ldots$
F_3	Circle	Bottom	$E_5, E_6, \ldots$
...	...	...	...

Edge

Edge ID	Beginning Point	Ending Point
E_1	P_1	P_2
E_2	P_3	P_4
...	...	...

Point

Point ID	Space Coordinate		
	X	Y	Z
P_1	X_1	Y_1	Z_1
P_2	X_2	Y_2	Z_2
...	...	...	...

FIGURE 12.7 The extraction of building edges from CSG model for relative controls: (a) the CSG model of a building; (b) the building in an aerial image; (c) the extracted building lines.

in the CSG model. From the attributes of face data structure, L_1 and L_3 are automatically recognized as the edges of roofs, and L_2 as the edge of a wall. On the other hand, the control points, P_1, P_3, and P_5, are extracted, and they can be automatically constructed lines, l_1, l_2, and l_3. When the L_1, L_2, and L_3 are back-projected onto the original aerial image, L_1, L_2, and L_3, whose topographic relationships have been described in CSG model, will be matched with the lines, l_1, l_2, and l_3, respectively (Figure 12.5c). Thus, the attributes (e.g., edges of roof, and edges of wall), and topographic relationships (i.e., perpendicularity) of l_1, l_2, and l_3, can be inherited from L_1, L_2, and L_3. Thus, l_1 and l_3 are edges in the roof, and l_2 is a vertical line in the wall. Also, l_2 is perpendicular to l_1 and l_3, and l_1 is also perpendicular to l_3. Thus, when $l_1 \perp l_3$, Equation 12.5 would be applied to construct a perpendicular relative control condition (equation) by replacing A, B, and C by P_5, P_3, and P_1 (see Figure 12.7).

Similarly, when point P_4 was measured on the line l_3, P_3, P_4, and P_5 should theoretically lie on a straight line of the building roof. Thus, a collinear constraint condition (Equation 12.14) can be constructed by replacing M, N, and K by P_5, P_4, and P_3 (see Figure 12.7).

In this experiment, 106 buildings are back-projected onto the original aerial image, while 76 well-distributed lines are selected as relative controls. Figure 12.8 depicts these selected geometric control lines, in which the projected model wire is rendered with transparent mode, so that the occluded parts can be seen.

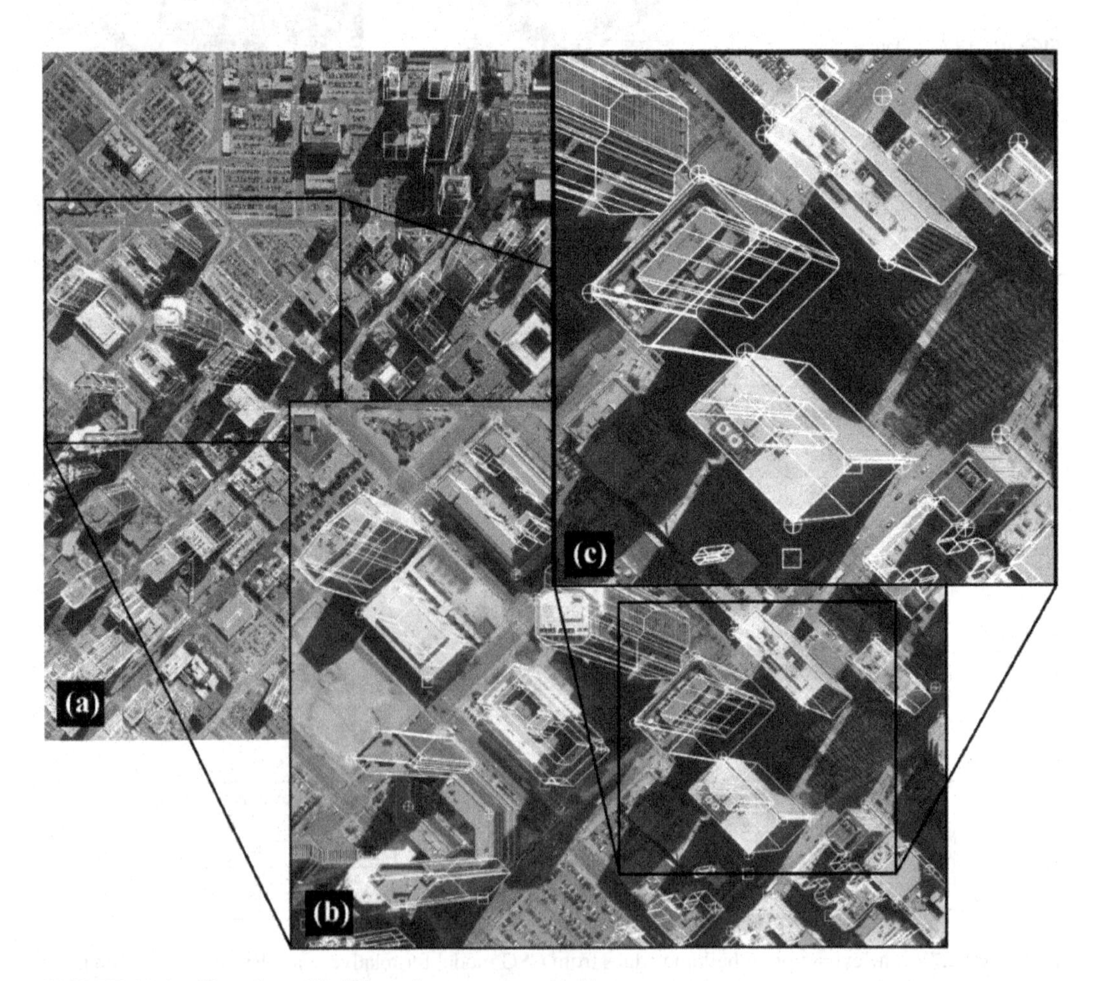

FIGURE 12.8 The selected building edges as constraint lines.

12.3.4 Accurate Comparison

Orthorectification was conducted through three methods. Method 1 only employed 8 GCPs; Method 2 used 232 GCPs; and Method 3, introduced in this chapter, employed 76 relative control lines and 56 GCPs. The IOPs were precisely calibrated and provided by the data vendor (Table 12.1). The exterior orientation parameters are determined by space intersection, in which the three types of control data are employed. The results and accuracy are listed in Table 12.2. As seen from Table 12.2, our method has the highest accuracy at standard deviation of 0.12 pixel.

After the above three groups of exterior orientation parameters are determined, the orthoimages were generated using the procedures as described by Zhou et al. (2002). The following two methods are employed to compare the accuracy of the three orthorectified images (orthoimages).

1. *Visual check.* The wire lines derived from the DBM model, which are taken as the "true" value, are superimposed onto the three orthorectified building roofs, respectively. Four orthorectified buildings with different heights and at different locations are selected for visually checking their accuracy achievable (Figure 12.9). As seen, there are significant offsets between the building wire lines and the building edges in the orthoimage generated by Method 1. The average deviations for four buildings are approximately 15 pixels and 10 pixels along the x and y directions. However, the average offsets from our method are only 1.0–1.5 pixels in both x and y directions. The results demonstrated that the accuracy of orthoimage generated by our method has been greatly increased.
2. *Checkpoint check.* To evaluate the absolute accuracy of the orthoimages orthorectified by three methods, 89 checkpoints are employed. The checkpoints are located in the corners of building and are considered as "true" values. The coordinates corresponding to the checkpoints in the three orthoimages are measured. The average deviations in both the x and y directions are listed in Table 12.3. As seen from Table 12.3, the offsets from Method 1 and Method 2 are approximately 3–5 feet and 2–4 feet, respectively, while they are 0.5–1.0 feet using our method. The results again demonstrated that our method can significantly improve the accuracy of the generated orthoimage.

TABLE 12.1
The Interior Orientation Parameter.

	Principal Point (mm)		Distortion Parameter		
Focal Length (mm)	x_0	y_0	k_1	p_1	p_2
153.022	0.002	−0.004	0.2076×10^{-4}	-0.1142×10^{-6}	0.1982×10^{-6}

TABLE 12.2
Accuracy Comparison of the Three Methods for the Exterior Orientation Parameter Determination.

Methods		Exterior Orientation Parameter						
	GCPs	ϕ (arc)	ω (arc)	κ (arc)	X_S (ft)	Y_S (ft)	Z_S (ft)	δ_o (pixel)
M 1	8 pts	0.0104	0.0158	−1.5503	3143007.3	1696340.4	9032.0	0.65
M 2	232 pts	−0.0025	−0.0405	−1.5546	3143041.4	1696562.6	9070.7	0.49
M 3	56 pts + 76 lines	−0.0016	−0.0298	−1.5538	3143040.6	1696520.9	9072.3	0.12

TABLE 12.3
Accuracy Comparison of Orthorectified Image.

	Method 1 (feet)		Method 2 (feet)		Method 3 (feet)	
	ΔX	ΔY	ΔX	ΔY	ΔX	ΔY
Mean	3.621	4.874	2.918	3.977	0.78	1.125

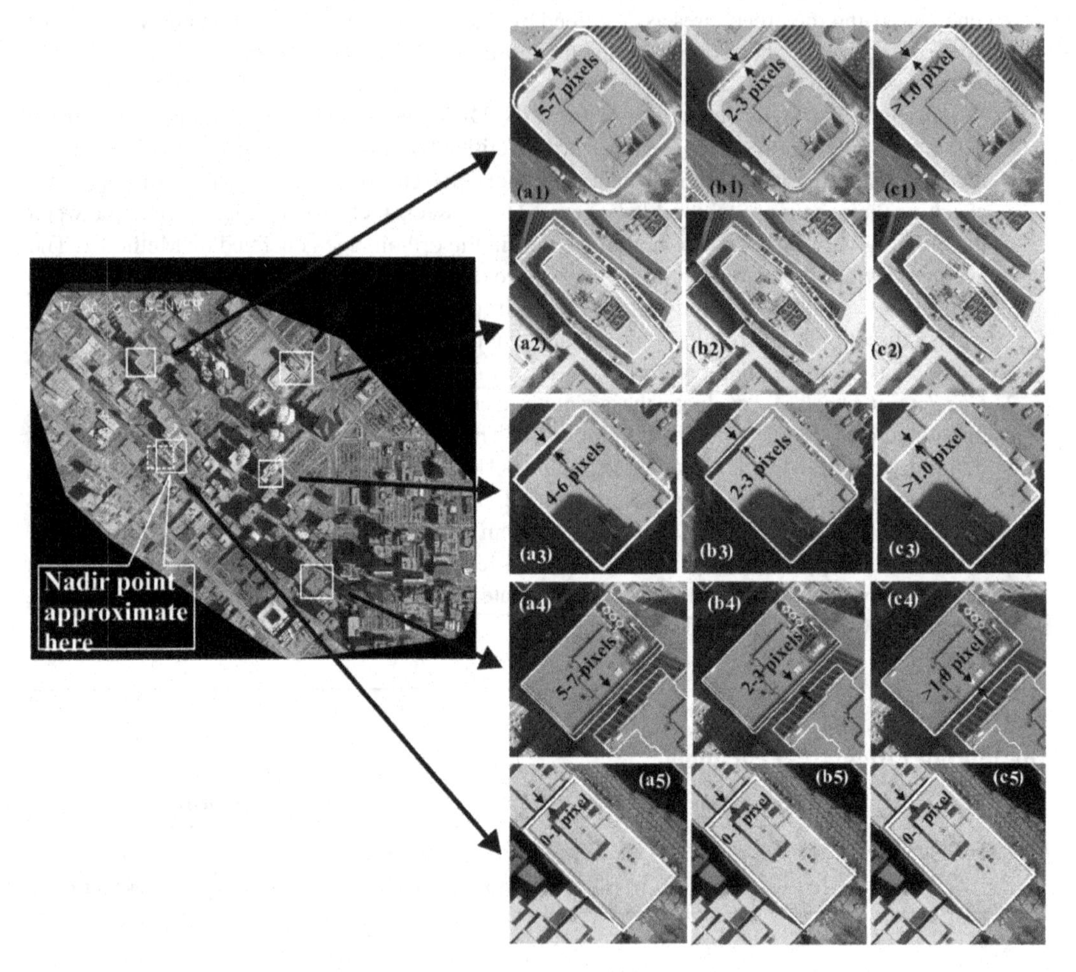

FIGURE 12.9 The comparison of the building's orthorectification using three control conditions: (a) 8 control points; (b) 232 GCPs; (c) 56 control points plus 76 control lines.

As mentioned previously, the relief displacements caused by high buildings occur along radial lines from the nadir point, which means that the relief displacements are zero for imaged objects at the nadir point and increase with increased radial distances from the nadir. In order to examine how much the proposed method could improve the accuracy for the marginal and central objects, the errors are assessed in the marginal and central orthoimage, and it has been found that the method introduced in this chapter could improve the accuracy of the orthoimage by approximately 5–7 feet for those objects in the image margin, and approximate 0–1 pixels for those high objects surrounding the nadir point (see Figure 12.9).

12.4 CONCLUSION

This chapter introduces a method for enhancing the accuracy of orthorectification for urban areas with extremely high buildings. The introduced method considers the building geometric features, including the perpendicularity and collinearity of building edges, and then models these geometric features as constraint conditions. These modeled geometric features are merged into the orthorectification model. A test field located in downtown Denver, Colorado has been used to evaluate the methods. The experimental results comparing the accuracy achieved by our method and other methods are conducted. The experimental results demonstrated that the proposed method could improve the accuracy by 2–5 feet for those buildings over 100 m high, and 5–7 feet for those buildings over 100 m high in the margin of imagery.

REFERENCES

Amhar, F., Josef, J., and Ries, C., The generation of true orthophotos using a 3D building model in conjunction with a conventional DTM, *The International Archives of the Photogrammetry and Remote Sensing*, part. 4, vol. 32, pp. 16–22, 1998.

Biason, A., Dequal, S., and Lingua, A., A new procedure for the automatic production of true orthophotos, *The International Archives of the Photogrammetry, Remote Sensing and Spatial Information Sciences*, edited by O. Altan, Vol. 35, July 12–23, 2004.

Habib, A., Kim, E., and Kim, C., New Methodologies for True Ortho-photo Generation, Photogrammetric Engineering & Remote Sensing (In press), 2006.

Jauregui, M., J. Vílchez, and L. Chacón, A procedure for map updating using digital mono-plotting. *Computers & Geosciences*, vol. 28, no. 4, pp. 513–523, 2002.

Joshua, G., Evaluating the accuracy of digital orthophotos quadrangles (DOQ) in the context of parcel-based GIS, *Photogrammetrlc Engineering & Remote Sensing,* vol. 67, no. 2, pp. 199–205, 2001.

Lawson, C. L., and Hanson, R. J., *Solving Least Squares Problems, Soc for Industrial & Applied Math*; New Ed edition, Philadelphia, PA: SIAM, September 1995.

Mayr, W., Bemerkungen zum Thema. True Ortho-image. *Photogrammetrie–Fernerkundung–Geoinformation*, vol. 4/2002, pp. 237–244, 2002.

Mikhail, E.M. et al. *Introduction to Model Photogrammetry*, John Wiley & Sons, Inc. ISBN: 0-471-30924-9, 2001.

Nielsen, M. Ø., True Orthophoto Generation, Informatics and Mathematical Modelling, Technical University of Denmark, 2004.

Palà, V., and Arbiol, R., True orthoimagery of urban areas, *GIM International*, vol. 16, p. 5051, December, 2002.

Passini, R. and K. Jacobsen, Accuracy Analysis of digital orthophotos from very height resolution imagery, *International Archives of the Photogrammery, Remote Sensing and Spatial Information Sciences*, edited by Orthan Altan, vol. XXXV, July 12–23 (DVD), 2004.

Rau, J. Y., Chen, N. Y., and Chen, L. C., True orthophoto generation of built-up areas using multi-view images, *Photogrammetry. Engm Remote Sensing*, vol. 68, no. 6, pp. 581–588, 2002.

Schickler, W., and A. Thorpe, Operational procedure for automatic true orthophoto generation, *International Archives of Photogrammetry and Remote Sensing,* vol. 32, Part 4, pp. 527–532, 1998.

Sheng, Y. W., Gong, P., and Biging, G. S., True orthoimage production for forested areas from large-scale aerial photographs. *Photogrammetry Engineering and Remote Sensing*, vol. 69, no. 3, pp. 259–266, March, 2003.

Siachalou, S., Urban orthoimage analysis generated from IKONOS data. *International archives of the photogrammetry, remote sensing and spatial information sciences*, edited by Orthan Altan, Vol. XXXV, July 2004 (DVD).

Skarlatos, D., Orthophotograph production in urban areas, *Photogrammetric Record,* vol. 16, no. 94, pp. 643–650, 1999.

Xie, W., and Zhou, G., Urban 3D building model applied to true orthoimage generation, *First Workshop of the EARSeL Special Interest Group on Urban Remote Sensing*, Berlin, Germany, March 23, 2006 (DVD).

Zhou, G., and Chen, W., Urban large-scale orthoimage standard for national orthophoto program, *IEEE Geoscience and Remote Sensing Annual Conference*, Seoul, Korea, July 25–29, 2005.

Zhou, G., Xie, W., and Cheng, P. Orthoimage creation of extremely high buildings. *IEEE Transactions on Geoscience & Remote Sensing,* vol. 46, no. 12, pp. 4132–4141, 2008.

Zhou, G. et al. Orthorectifying 1960's disclassified intelligence satellite photography (DISP) of Greenland, *IEEE Transactions on Geoscience and Remote Sensing,* vol. 40, no. 6, pp. 1247–1259, 2002.

Zhou, G. et al. A comprehensive study on urban true orthorectification. *IEEE Transactions on Geoscience and Remote sensing,* vol. 43, no. 9, pp. 2138–2147, 2005.

13 Near Real-Time Orthophotomap Generation from UAV Video

13.1 INTRODUCTION

The early development of small, long-endurance, and low-cost unmanned aerial vehicles (UAVs) with significant payload capabilities was initially for military application. For example, a small UAV initiative project was started by the Program Executive Officer of the Secretary of the Navy for cruise missiles in 1998 (Kaminer et al. 2004; Kelley 2000). Afterward, the Naval Research Laboratory tested their helicopter UAV system (Kahn and Foch 2003), and the Naval Surface Warfare Center developed a field compatible SWARM platform, which utilized a heavy fueled engine, an on-board power supply, and an autonomous command and control system (Furey 2004). The US Army has also deployed small UAV research and development (Kucera 2005).

The small UAVs have also generated significant interest for civil users. NASA Ames Research Center, NASA Dryden Research Center, and NASA Goddard Space Flight Center at Wallops Flight Facility have developed different types of UAV systems with different types of sensors on-board for a variety of civilian applications, such as homeland security demonstration (Herwitz et al. 2004), forestry fire monitoring (Casbeer et al. 2005), quick response measurements for emergency disaster (Postell and Thomas 2005), Earth science research (Bland et al. 2004), volcanic gas sampling (Caltabiano et al. 2005; Longo et al. 2004), humanitarian biological chemo-sensing demining tasks (Bermudez et al. 2007), and monitoring of gas pipelines (Hausamann et al. 2005). In particular, the civilian UAV users from private sector and local government agencies have a strong demand for a low-cost, moderately functional, small airborne platform, varying in size, computerization, and levels of autonomy (Moore et al. 2003). Therefore, applications of small UAV systems for small private sector businesses and non-military government agencies for small areas of interest are attracting many researchers. For example, Hruska et al. (2005) reported their small and low-cost UAVs to be primarily used for capturing and down-linking real-time video. A UAV-based still imagery workflow model, including initial UAV mission planning, sensor selection, UAV/sensor integration, and imagery collection, processing, and analysis, has been developed. To enhance the analysts' change detection ability, a UAV-specific, GIS-based change detection system called SADI for analyzing the differences was also developed. Dobrokhodov et al. (2006) also reported a small low-cost UAV system for autonomous target tracking, while simultaneously estimating the GPS coordinates of the target. A low-cost, primarily COTS system is utilized, with a modified RC aircraft airframe, gas engine, and servos. Tracking is enabled using a low-cost, miniature pan-tilt gimbal, driven by COTS servos and electronics. Oh (2004) and Narli and Oh (2006) presented their research results using micro-air-vehicle navigation for homeland security, disaster mitigation, and military operations in the environment of time-consuming, labor-intensive, and possibly dangerous tasks like bomb detection, search-and-rescue, and reconnaissance. Nelson et al. (2004) reported their initial experiments in cooperative control of a team of three small UAVs, and Barber et al. (2006) presented how they determined the GPS location of a ground-based object from a fixed-wing miniature air vehicle (MAV). Using the pixel location of the target in an image, measurements of MAV position and attitude, and camera pose angles, the target is localized in world coordinates. Logan et al. (2005) and Boer et al. (2006) have discussed the challenges of developing a UAV system, including the difficulties encountered, and proposed a list of technology shortfalls that need to be addressed. Noth et al. (2005), within the framework of an ESA program, presented an ultra-lightweight solar autonomous model airplane called Sky-Sailor. Johnson et al. (2006, 2004) described the design, development, and operation of UAVs developed at the Georgia Institute of Technology. Moreover, they tested autonomous fixed-wing UAVs with the ability to hover for applications in urban or other constrained environments where the combination of fast speed, endurance, and stable hovering flight can provide strategic advantages. Kang et al. (2006) presented a new method of building a probabilistic occupancy map for a UAV equipped with a laser scanning sensor. Ryan developed a decentralized hybrid controller for fixed-wing UAVs assisting a manned helicopter in a US Coast Guard search-and-rescue mission, during which two UAVs fly on either side of the helicopter, with constant velocity and maximum turn rate constraints. Sinopoli et al. (2001) developed a system for autonomous navigation of UAVs based on computer vision in combination with GPS/INS.

One of the most important forms of image data processing in UAV systems is real-time orthorectification and mosaicking so that the georeferenced UAV image can be merged with geospatial data for a fast response to time-critical events (Zhou et al. 2006). Many approaches to image orthorectification and mosaicking have been presented by researchers in the past decades. The previous methods have handled different operation platforms, such as spaceborne (e.g., Liu et al. 2007; Sheng and Alsdorf 2005) and airborne (e.g., Bielski et al. 2007; Shin et al. 1997; Zhou et al. 2005) and different sensors, such as radar (Simard et al. 1999), and visible and multispectral images (e.g., Du et al. 2001; Li et al. 2003; Zhou et al. 2002). The previously mathematical models, ranging from a simple affine transformation (which utilizes higher-order polynomials) to projective transformations, have also been discussed (Zhou et al. 2002). In summary, they can in general be divided into two types: (1) the first is called the non-parametric approach, a rigorous solution in which ground control points (GCPs) are generally used. The spatial relationships between an image pixel and its conjugate ground point are characterized by the imaging geometry, which is described by the collinearity condition of the central perspective images. (2) The second approach is called the parametric approach. This method does not need to recover the sensor orientation in advance of the processing. In this method, GCPs are collected at locations where identifiable points are coincident on both the image and a corresponding map. Once enough GCPs are collected, the image coordinates are modeled as functions of the map coordinates, using the least-square solution to fit the functions.

This chapter presents a method of real-time orthorectification and mosaicking of a UAV video stream in order to meet the need of UAV data processing for a fast response to time-critical events. In closely related work, Campbell and Wheeler (2006) presented a vision-based geolocation method based on a square root sigma point filter technology. However, both Dobrokhodov et al. (2006) and Campbell and Wheeler (2006) exhibited that the estimate biases caused by their method are sensitive to heavy wind conditions. Gibbins et al. (2004) reported their geolocation accuracy of over 20 m; Whang et al. (2005) described a geolocation solution, in which the range estimates were obtained using a terrain model, and a nonlinear filter was used to estimate the position and velocity of ground moving targets. Barber et al. (2006) proposed a method for georectification at localization errors of below 5 m.

13.2 MATHEMATICAL MODEL OF GEOREFERENCING

For a UAV system (see Zhou et al. 2006), the geometric configuration between the two navigation sensors and the video camera is depicted in Figure 13.1. The mathematical model can be expressed by

$$r_G^M = r_{GPS}^M(t) + R_{Att}^M(t)\cdot\left[s_G\cdot R_C^{Att}\cdot r_g^C(t) + r_{GPS}^C\right] \tag{13.1}$$

where r_G^M is a vector to be computed for any ground point G in the given mapping frame; $r_{GPS}^M(t)$ is a vector of the GPS antenna phase center in the given mapping frame, which is determined by the

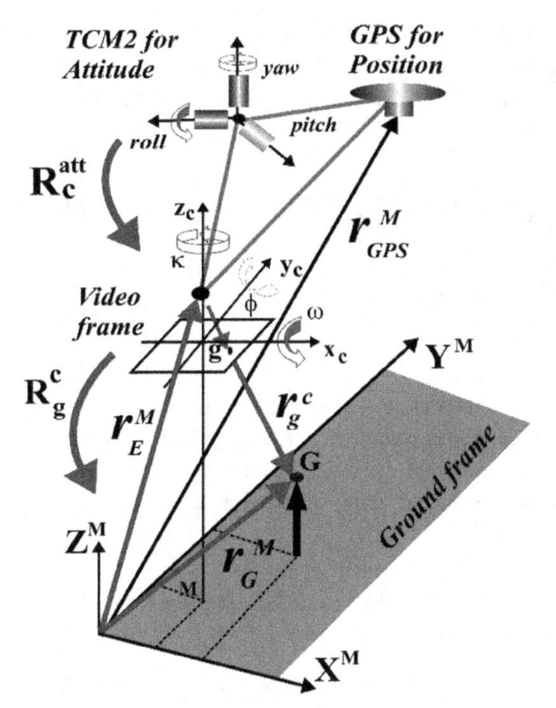

FIGURE 13.1 The geometric configuration for UAV-based multi-sensors, including video camera, GPS and attitude sensor.

on-board GPS at a certain epoch (t); s_G is a scale factor between the camera frame and the mapping frame; $r_g^C(t)$ is a vector observed in the image frame for point g, which is captured and synchronized with GPS epoch (t); R_C^{Att} is the so-called boresight matrix (orientation offset) between the camera frame and the attitude sensor body frame; and r_{GPS}^C is the vector of position offset between the GPS antenna geometric center and the camera lens center, which is usually determined by terrestrial measurements as part of the calibration process. $R_{Att}^M(t)$ is a rotation matrix from the UAV attitude sensor body frame to the given mapping frame, and is a function of the three attitude angles, that is,

$$R_{Att}^M = \begin{pmatrix} \cos\psi\cos\zeta & \cos\xi\sin k + \sin\xi\sin\psi\cos\zeta & \sin\xi\sin\zeta - \cos\xi\sin\psi\cos\zeta \\ -\cos\psi\sin\zeta & \cos\xi\cos k - \sin\xi\sin\psi\sin\zeta & \sin\xi\cos\zeta + \cos\xi\sin\psi\sin\zeta \\ \sin\psi & -\sin\xi\cos\psi & \cos\xi\cos\psi \end{pmatrix} \quad (13.2)$$

where ξ, ψ, and ζ represent roll, pitch, and yaw, respectively. So, the relationship between the two sensors is in fact a mathematically determinate matrix, R_C^{Att} through Equation 13.1. The determination of R_C^{Att} is usually solved by a least-squares adjustment on the basis of a number of well-distributed GCPs. Once this matrix is determined, its value is assumed to be a constant over the entire flight time in a traditional airborne mapping system (Cramer and Stallmann 2002; Grejner-Brzezinska 1999; Mostafa et al. 1998; Schwarz et al. 1993; Skaloud 1996; Yastikli and Jacobsen 2005). The basic procedures of UAV-based orthorectification and mosaicking are as follows.

13.2.1 Calibration of Video Camera

The calibration of the video camera includes the calibration of the parameters such as focal length, principal point coordinates, and lens distortion calibration, which is called interior orientation parameters (IOPs). The direct linear transformation (DLT) method, which was originally presented by Abdel-Aziz and Karara (1971), is used here. This method requires a set of GCPs whose object space and image coordinates are already known. In this step, the calibration process only considers the focal length and principal point coordinates because the solved IOPs and exterior orientation parameters (EOPs) will be employed as initial values in the later bundle adjustment model. The DLT model is

$$x_{g1} - x_0 + \rho_1 \left(x_{g1} - x_0 \right) r_1^{\ 2} = \frac{L_1 X_G + L_2 Y_G + L_3 Z_G + L_4}{L_9 X_G + L_{10} Y_G + L_{11} Z_G} = f_x^1 \tag{13.3a}$$

$$y_{g1} - y_0 + \rho_1 \left(y_{g1} - y_0 \right) r_1^{\ 2} = \frac{L_5 X_G + L_6 Y_G + L_7 Z_G + L_8}{L_9 X_G + L_{10} Y_G + L_{11} Z_G} = f_y^1 \tag{13.3b}$$

where $r_{(i)}^{\ 2} = (x_{g(i)} - x_0)^2 + (y_{g(i)} - y_0)^2$ $(i = 1,2)$; (x_{g1}, y_{g1}) and (x_{g2}, y_{g2}) are the coordinates of the image points g_1 and g_2 in the first and second image frames, respectively; (X_G, Y_G, Z_G) are the coordinates of the ground point, G; (x_0, y_0, f, ρ_1) are the IOPs; and $L_i (i = 1, \ldots, 9)$ are unknown parameters. Equation 13.3 is a nonlinear equation, and must be linearized using the Taylor series. The linearized equation is:

$$-\left[X_G L_1 + Y_G L_2 + Z_G L_3 + L_4 + x_{g1} X_G L_9 + x_{g1} Y_G L_{10} + x_{g1} Z_G L_{11} \right] / A + \left(x_{g1} - x_0 \right) r_1^{\ 2} \rho_1 + x_{g1} / A = v_x \tag{13.4a}$$

$$-\left[X_G L_5 + Y_G L_6 + Z_G L_7 + L_8 + y_{g1} X_G L_9 + y_{g1} Y_G L_{10} + y_{g1} Z_G L_{11} \right] / A + \left(y_{g1} - y_0 \right) r_1^{\ 2} \rho_1 + y_{g1} / A = v_y \tag{13.4b}$$

The matrix form of Equation 13.4 is

$$V = C\Delta + L \tag{13.5}$$

where

$$C = -\frac{1}{A} \begin{bmatrix} X_G & Y_G & Z_G & 1 & 0 & 0 & 0 & 0 & x_{g1} X_G & x_{g1} Y_G & x_{g1} Z_G & \left(x_{g1} - x_0 \right) r_1^{\ 2} \\ 0 & 0 & 0 & 0 & X_G & Y_G & Z_G & 1 & y_{g1} X_G & y_{g1} Y_G & y_{g1} Z_G & \left(y_{g1} - y_0 \right) r_1^{\ 2} \end{bmatrix}$$

$$\Delta = \left[L_1 \ L_1 \ L_1 \ L_1 \ L_1 \ L_1 \ L_1 \ L_1 \ L_1 \ L_1 \ L_1 \ \rho_1 \right]^T$$

$$V = \begin{bmatrix} v_x \\ v_y \end{bmatrix}$$

$$L = -\frac{1}{A} \begin{bmatrix} x \\ y \end{bmatrix}$$

With the iteration computation, the 11 parameters in Equation 13.5 can be solved. With the solved 11 parameters, the IOPs can be calculated by

$$x_0 = -\left(L_1 L_9 + L_2 L_{10} + L_3 L_{11} \right) / \left(L_9^2 + L_{10}^2 + L_{11}^2 \right) \tag{13.6}$$

$$y_0 = -\left(L_5L_9 + L_6L_{10} + L_7L_{11}\right) / \left(L_9^2 + L_{10}^2 + L_{11}^2\right) \tag{13.7}$$

$$f_x^2 = -x_0^2 + \left(L_1^2 + L_2^2 + L_3^2\right) / \left(L_9^2 + L_{10}^2 + L_{11}^2\right) \tag{13.8a}$$

$$f_y^2 = -y_0^2 + \left(L_5^2 + L_6^2 + L_7^2\right) / \left(L_9^2 + L_{10}^2 + L_{11}^2\right) \tag{13.8b}$$

$$f = \frac{f_x + f_y}{2} \tag{13.9}$$

The EOPs can be calculated by

$$a_3 = L_9 / \sqrt{\left(L_9^2 + L_{10}^2 + L_{11}^2\right)}$$

$$b_3 = L_{10} / \sqrt{\left(L_9^2 + L_{10}^2 + L_{11}^2\right)}$$

$$c_3 = L_{11} / \sqrt{\left(L_9^2 + L_{10}^2 + L_{11}^2\right)}$$

$$a_1 = \frac{1}{f_x}\left[L_1 / \sqrt{\left(L_9^2 + L_{10}^2 + L_{11}^2\right)} + a_3x_0\right]$$

$$b_1 = \frac{1}{f_x}\left[L_2 / \sqrt{\left(L_9^2 + L_{10}^2 + L_{11}^2\right)} + b_3x_0\right]$$

$$c_1 = \frac{1}{f_x}\left[L_3 / \sqrt{\left(L_9^2 + L_{10}^2 + L_{11}^2\right)} + c_3x_0\right]$$

$$a_2 = \frac{1}{f_y}\left[L_5 / \sqrt{\left(L_9^2 + L_{10}^2 + L_{11}^2\right)} + a_3y_0\right]$$

$$b_2 = \frac{1}{f_y}\left[L_6 / \sqrt{\left(L_9^2 + L_{10}^2 + L_{11}^2\right)} + b_3y_0\right]$$

$$c_2 = \frac{1}{f_y}\left[L_7 / \sqrt{\left(L_9^2 + L_{10}^2 + L_{11}^2\right)} + c_3y_0\right]$$

The rotation matrix can be expressed by

$$R_M^C = \begin{pmatrix} a_1 & a_2 & a_3 \\ b_1 & b_2 & b_3 \\ c_1 & c_2 & c_3 \end{pmatrix} \tag{13.10}$$

The exposure center coordinates (X_S, Y_S, Z_S) can be calculated by solving the following equations.

$$a_3X_S + b_3Y_S + c_3Z_S + L' = 0 \tag{13.11a}$$

$$x_0 + f_x\left(a_1X_S + b_1Y_S + c_1Z_S\right)/L' + L_4 = 0 \tag{13.11b}$$

$$y_0 + f_y\left(a_2X_S + b_2Y_S + c_2Z_S\right)/L' + L_8 = 0 \tag{13.11c}$$

13.2.2 Determination of the Offset Between GPS Antenna and Camera

The GPS antenna geometric center and the camera lens center cannot occupy an identical center. The offset $\left(r_{GPS}^{M}\right)$ between the two centers must precisely be measured so that the correction can be carried out in Equation 13.1. The precise measurement of the offset was conducted using a Topcon® GTS-2B Total Station, whose specification can be referenced at http://bujorel.com/_wsn/page16.html. The process is: (1) set up the Total Station 5–10 m away from the UAV aircraft; (2) take a shot to the GPS antenna and read the horizontal and vehicle distance and angles from the Total Station; (3) take a shot to the lens of camera, during which the vertical wire of the Total Station's telescope is aligned with the telescope axis, and the horizontal wire of the Total Station's telescope is aligned with the shut; (4) revise the of Total Station's telescope, and repeat Steps 2 and 3; (5) repeat Steps 2, 3, and 4 three times; (6) suppose that the origin of a presumed local coordinate is at the Total Station, and calculate the coordinates of the GPS antenna, (X_{GPS}, Y_{GPS}, Z_{GPS}), and the camera lens, (X_{lens}, Y_{lens}, Z_{lens}); and (7) calculate the offset between two centers by $D_{offset} = \sqrt{\left(X_{GPS} - X_{lens}\right)^2 + \left(Y_{GPS} - Y_{lens}\right)^2 + \left(Z_{GPS} - Z_{lens}\right)^2}$. The measurement accuracy will reach millimeter level, since the Total Station has a measurement capability of millimeter level.

13.2.3 Solution of Kinematic GPS Errors

For kinematic GPS errors, the baseline length away from the ground reference stations is limited for the on-board differential GPS survey. It has been demonstrated that a GPS receiver on-board an UAV can achieve an accuracy of a few centimeters using this limitation (Zhou et al. 2006). The other errors will be orthorectified in this chapter mathematically. Basically, the traditional differential rectification model is based on photogrammetric collinearity, in which the interior and exterior orientation elements and DEM (X, Y and Z coordinates) are known.

13.2.4 Estimation of Boresight Matrix

With the solved EOPs, estimation of an initial boresight matrix, R_C^{Att}, can be realized through multiplication of the attitude sensor orientation data derived from the on-board TCM2 sensor with the three angular elements of the EOPs solved by DLT. The formula is expressed by

$$R_C^{Att}(t) = \left[R_M^C(t) \cdot R_{Att}^M(t)\right]^T \tag{13.12}$$

where R_C^{Att} and R_{Att}^M are the same as in Equation 13.1; R_M^C is a rotation matrix, which is a function of three rotation angles (ω, ϕ, and κ) of a video frame, and is expressed by

$$R_M^C = \begin{pmatrix} a_1 & a_2 & a_3 \\ b_1 & b_2 & b_3 \\ c_1 & c_2 & c_3 \end{pmatrix} = \begin{pmatrix} \cos\phi\cos k & \cos\omega\sin k + \sin\omega\sin\phi\cos k & \sin\omega\sin k - \cos\omega\sin\phi\cos k \\ -\cos\phi\sin k & \cos\omega\cos k - \sin\omega\sin\phi\sin k & \sin\omega\cos k + \cos\omega\sin\phi\sin k \\ \sin\phi & -\sin\omega\cos\phi & \cos\omega\cos\phi \end{pmatrix} \tag{13.13}$$

With the initial values computed above, a rigorous mathematical model was established to simultaneously solve the camera's IOPs and EOPs of each video frame. In addition, because the stereo camera calibration method can increase the reliability and accuracy of the calibrated parameters due to coplanar constraints (Bethea et al. 1997), a stereo pair of images constructed by the first and the second video frames is selected. The mathematic model, for any ground point, G, can be expressed as

For the first video frame:

$$f_x^{g1} = -f\frac{r_{11}^1\left(X_G - X_S^1\right) + r_{12}^1\left(Y_G - Y_S^1\right) + r_{13}^1\left(Z_G - Z_S^1\right)}{r_{31}^1\left(X_G - X_S^1\right) + r_{32}^1\left(Y_G - Y_S^1\right) + r_{33}^1\left(Z_G - Z_S^1\right)} \quad (13.14a)$$

$$f_y^{g1} = -f\frac{r_{21}^1\left(X_G - X_S^1\right) + r_{22}^1\left(Y_G - Y_S^1\right) + r_{23}^1\left(Z_G - Z_S^1\right)}{r_{31}^1\left(X_G - X_S^1\right) + r_{32}^1\left(Y_G - Y_S^1\right) + r_{33}^1\left(Z_G - Z_S^1\right)} \quad (13.14b)$$

For the second video frame:

$$f_x^{g2} = -f\frac{r_{11}^2\left(X_G - X_S^2\right) + r_{12}^2\left(Y_G - Y_S^2\right) + r_{13}^2\left(Z_G - Z_S^2\right)}{r_{31}^2\left(X_G - X_S^2\right) + r_{32}^2\left(Y_G - Y_S^2\right) + r_{33}^2\left(Z_G - Z_S^2\right)} \quad (13.15a)$$

$$f_y^{g2} = -f\frac{r_{21}^2\left(X_G - X_S^2\right) + r_{22}^2\left(Y_G - Y_S^2\right) + r_{23}^2\left(Z_G - Z_S^2\right)}{r_{31}^2\left(X_G - X_S^2\right) + r_{32}^2\left(Y_G - Y_S^2\right) + r_{33}^2\left(Z_G - Z_S^2\right)} \quad (13.15b)$$

where $r_{(i)}{}^2 = (x_{g(i)} - x_0)^2 + (y_{g(i)} - y_0)^2$ $(i = 1,2)$; (x_{g1}, y_{g1}) and (x_{g2}, y_{g2}) are the coordinates of the image point g_1 and g_2 in the first and second video frames, respectively; (X_G, Y_G, Z_G) are the coordinates of the ground point, G; (x_0, y_0, f, ρ_1) are the IOPs; $r_{i,j}^m\left(i = 1,2,3; j = 1,2,3\right)$ are elements of the rotation matrix R for the first video frame (when m = 1) and the second video frame (when m = 2), which are a function of three rotation angles $(\omega_1, \phi_1, \kappa_1)$ and $(\omega_2, \phi_2, \kappa_2)$. The expression is described in Equation 13.13. In this model, the unknown parameters contain the camera's IOPs, (x_0, y_0, f, ρ_1), and the EOPs of the first and second video frames, $(X_S^1, Y_S^1, Z_S^1, \omega_1, \phi_1, \kappa_1)$ and $(X_S^2, Y_S^2, Z_S^2, \omega_2, \phi_2, \kappa_2)$, respectively. To solve these unknown parameters, Equations 13.14 and 13.15 must be linearized by using a Taylor series expansion including only the first-order terms. The vector form of the linearized equation is expressed by

$$v_1 = A_1X_1 + A_2X_2 - L \quad (13.16)$$

Where X_1 represents a vector of the EOPs of two video frames; X_2 denotes the vector of the camera IOPs; A_1, and A_2 are their coefficients; and v_1 is a vector containing the residual error. Their components can be referenced in Zhou et al. (2006).

13.3 GEORECTIFICATION OF VIDEO STREAM

After the orientation parameters of the individual video frame are determined by the model described in Section 13.2, each original video frame can be orthorectified. The procedures include: (1) the determination of the size of the orthorectified image; (2) the transformation of pixel locations from the original image to the resulting (rectified) image using Equation 13.1; and (3) resampling the original image pixels into the rectified image for assignment of gray values. The flowchart is illustrated in Figure 13.2.

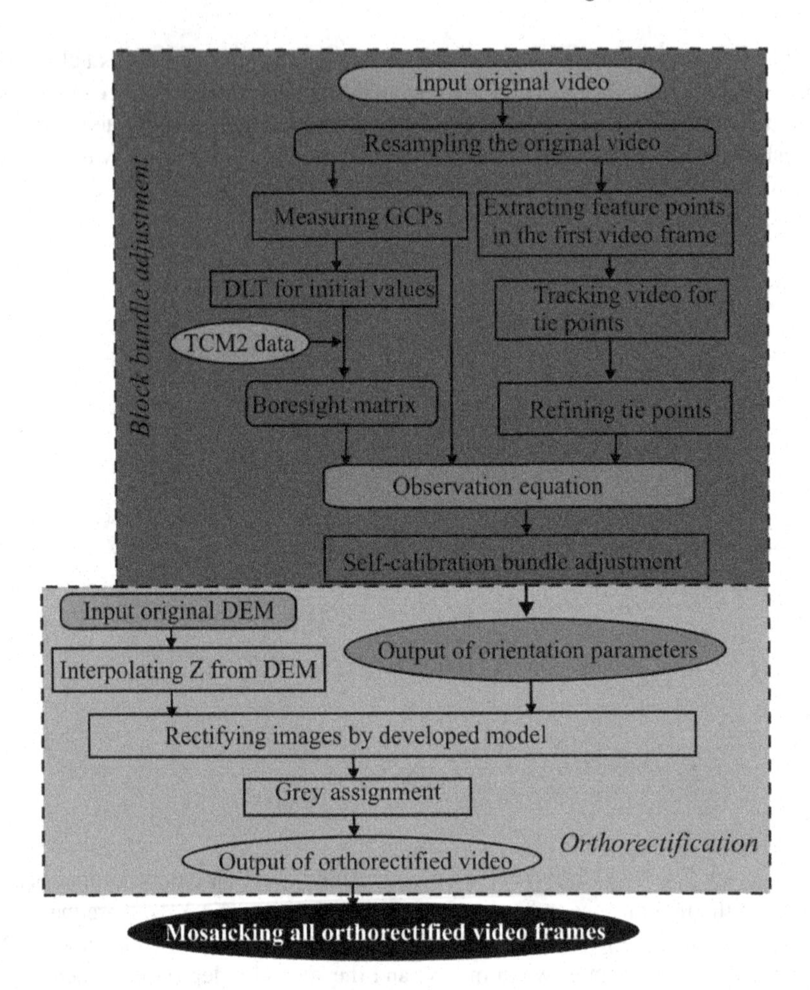

FIGURE 13.2 The flowchart of geometric rectification using the block bundle adjustment model.

13.3.1 Determination of Orthorectified Image Size

The orthorectification process registers the original image into a chosen map-based coordinate system, and invariably the size of the original image is changed. To properly set up the storage space requirements when programming, the size of the resulting image footprint (upper left, lower left, upper right, and lower right) has to be determined in advance. These procedures are as follows:

- ***The determination of four corner coordinates.*** For a given ground resolution of $\Delta_{Xsample}$ and $\Delta_{Ysample}$ along x and y directions in the original image, assume that the planimetric coordinates of any GCP are (X_{GCP}, Y_{GCP}), whose corresponding location in the original image plane is (row_{GCP}, col_{GCP}). The coordinates of four corner points can then be determined routinely. For example, Corner 1 can be calculated by:

$$X_1 = X_{GCP} - col_{GCP} \cdot \Delta_{Xsample}$$

$$Y_1 = Y_{GCP} - row_{GCP} \cdot \Delta_{Ysample}$$

The other corners can also be calculated accordingly.

- The determination of minimum and maximum coordinates from the above four corners. For example, the minimum x coordinate can be calculated by:

$$X_{\min} = min(X_1, X_3)$$

The maximum x, (X_{max}), and minimum and maximum y, (Y_{min}, Y_{max}), can be calculated accordingly.
- The determination of size of the resulting image can be calculated by

$$\text{N} = \text{col} = \frac{X_{\max} - X_{\min}}{\Delta X}, \quad \text{M} = \text{Row} = \frac{Y_{\max} - Y_{\min}}{\Delta Y}$$

where ΔX and ΔY are ground sampled distance (GSD) in the resulting image.

13.3.2 Orthorectification

The basic procedures of orthorectification are as follows:

1. For any point, P (I, J), in the resulting image, (I, J) are its image coordinates in the image plane;
2. Compute the planimetric coordinates of the point P (X_S, Y_S) with respect to the geodetic coordinate system by using the given cell size;
3. Interpolate the vertical coordinates, Z_S, from the given DEM using a bilinear interpolation algorithm;
4. Compute the photo coordinate (x, y) and the image coordinate (i, j) of the point P in the original image by using Equation 13.1, in which all of the parameters have been determined by the methods described in Section 13.2;
5. Calculate the gray value g_{orig} by a nearest neighbor resampling algorithm;
6. Assign the gray value g_{orig} as the brightness g_{result} of the resulting (rectified) image pixel.

The above procedure is then repeated for each pixel to be rectified. The details of the overall process of the orthorectification can be referenced in Zhou et al. (2002).

13.3.3 Mosaicking

The mathematical model for radiometric balancing and blending operations for scene-to-scene radiometric variations was developed for individual scenes to prevent a patchy or quilted appearance in the final mosaic. In this model, the weights for blending individual scenes along the specified buffer zone are calculated by the following cubic Hermite function:

$$W = 1 - 3d^2 + 2d^3 \tag{13.17}$$

$$G = W \cdot G_1 + (1 - W) \cdot G_2 \tag{13.18}$$

where W is the weighting function applied in the overlap area with values ranging from 0 to 1; d is the distance of a pixel to the buffer line which is normalized from 0 to 1; G_1 and G_2 are brightness of overlapping images, and G is the resulting brightness value. In the buffer zone, large-intensity values have lower weight, while lower brightness values have high weight.

13.4 EXPERIMENTS AND ANALYSIS

13.4.1 Experimental Field Establishment

An experimental field, located in Picayune, Mississippi, approximately 15 minutes north of the NASA John C. Stennis Space Center, was established. This test field is about four miles long along NW and three miles wide along SW. In this field, 21 non-traditional GCPs using DGPS (differential GPS) were collected. These GCPs are located in the corners of sidewalk, or parking lot, crossroad, and curb end (see Figure 13.3). Each point is observed for at least 30 minutes, and it was ensured that at least 4 GPS satellites are locked simultaneously. The height angles cutoff is 15 degrees. The planimetric and vertical accuracy of the GCPs is at decimeter level. This accuracy is enough for taking GCPs in the late processing of UAV-based georeferencing and 2D planimetric mapping because the accuracy evaluation of this system is carried out relative to the USGS DOQ (US Geological Survey, digital orthophoto quadrangle), whose cell size is 1 m. In addition to the 21 non-traditional GCPs, 1 m USGS DOQ imagery (see Figure 13.3) covering the control field was also downloaded from USGS website for the accuracy evaluation of UAV-based real-time video data georeferencing and 2D planimetric mapping.

13.4.2 UAV System

A cheap, small UAV system was developed by Zhou et al. (2006). The specifications of the UAV are listed in Table 13.1. This UAV system is specifically designed as an economical, moderately functional, small airborne platform intended to meet the requirement for a fast response to time-critical events in many small private sector or government agencies for the small areas of interest. Cheap materials, such as sturdy plywood, balsa wood, and fiberglass, are employed to feature a proven versatile high-wing design, tail dragger landing gear with excellent ground clearance that allows operation from semi-improved surfaces. Generous flaps enable short rolling take-offs and slow flight. The 1-½ hp, 2-stroke engine burns a commercial glow fuel mixed with gas (see Figure 13.4).

In addition, the UAV is constructed to break down into a few easy-to-handle components which quickly pack into a small size van, and are easily deployed, operated, and maintained by a crew of three. This UAV system, including hardware and software, is housed in a lightly converted (rear

FIGURE 13.3 USGS DOQ and the distribution of the measured 21 non-traditional GCPs.

TABLE 13.1
Specifications of a Low-Cost Civilian UAV Platform

Power Plant	2 stroke, 1½ hp	Endurance	45 Minutes at Cruise Speed	Fuel Capacity	0.46 kg
Length × Height	1.53 m × 1.53 m	Cruise Speed	56 km/h	Wingspan	2.44 m
Gross weight	10 kg	Max Speed	89 km/h	Payload	2.3 kg
Operating Altitudes	152 ~ 619 m	Operating Range	1.6–2.5 km		

FIGURE 13.4 UAV ground control station and field data collection.

seat removed and bench top installed) van (see Figure 13.4), on a mobile vehicle for providing command, control, and data recording to and from the UAV platform, and real-time data processing in order to meet the requirement of a fast response to time-critical events. The field control station houses the data stream monitoring and UAV position interface computer, radio downlinks, antenna array, and video terminal. All data (GPS data, UAV position and attitude data, and video data) are transmitted to the ground receiver station via wireless communication, with real-time data processing in the field for a fast response to rapidly evolving events. In this project, three on-board sensors, GPS, attitude sensor (TCM2), and video camera are integrated into a compact unit. The GPS is a Garmin eTrex Vista Personal Navigator Handheld GPS Receiver with 12 parallel channels, which continuously tracks and uses up to 12 satellites to compute and update the position. The eTrex Vista combines a basemap of North and South America with a barometric altimeter and electronic compass. The compass provides bearing information and the altimeter determines the UAV's precise altitude. An attitude navigation sensor (TCM2-20) is selected to provide the UAV's real-time attitude information. This sensor integrates a three-axis magneto-inductive magnetometer and a high-performance two-axis tilt sensor (inclinometer) in a package, and provides tilt-compensated compass headings (azimuth, yaw, or bearing angle) and precise tilt angles relative to Earth's gravity (pitch and roll angle) for precise three-axis orientation. The electronic gimbaling eliminates moving parts and provides information about the environment of pitch and roll angles, and three-dimensional magnetic field measurement. Data is output on a standard RS-232 serial interface with a simple text protocol that includes checksums. A video camera, *Topica Color TP 6001A CCD*, was used to acquire the video stream at a nominal focal length of 8.5 mm with auto and preset manual focus, and program and manual exposure. The camera was installed in the UAV payload bay at a nadir-looking direction. The video stream is recorded with a size of 720 (h) × 480 (v) $pixel^2$ and delivered in an MPEG-I format.

13.4.3 Data Collection

The data was collected over the established control field on April 3, 2005. The weather conditions were: mostly sunny; humidity: 53%; visibility: 10 mi; high/low temperature: near 70F/50F; and winds: NE at 5 to 10 mph. The UAV and all the other hardware, including computers, monitor, antennas, and the peripheral equipment (e.g., cable), and the software developed in this project were housed in, and transported to, the test field via the field control station (see Figure 13.3). After the UAV was assembled, all the instruments, such as antenna, computers, video recorder, battery, and so on, were set up, and the software system was tested, the test and validation being conducted at 10:10 am local time. Some of the tested results are described as follows.

1) **Navigation Data Collection.** The GPS antenna position and UAV attitude are also collected and transmitted to the ground control station. Figure 13.6 is part of the received navigation data string, which consists of two data stream lines in one text file. The first line is the GPS position line, followed by the attitude reference line. This text file can be opened with Microsoft Excel, and the details of two data stream lines are (see Figure 13.5):

   - ***Position – 1st data line.*** It contains a UTC time, latitude, longitude, and altitude (antenna height above MSL reference in meters). The data was taken from the $GPGGA NEMA-183 data string provided by the Garmin GPS and is updated approximately every two seconds.
   - ***Attitude Reference-2nd data line.*** It includes magnetic heading, roll, and pitch. The update or refresh rate is the same as the GPS update rate.

GPS
UTC 15:43:24 30 30.614N 089 11.116W Alt 49.3 Meters
HPR
Mag Heading 29.8 Pitch 0.6 Roll 10.4
GPS
UTC 15:43:25 30 30.614N 089 11.116W Alt 49.1 Meters
HPR
Mag Heading 32.6 Pitch 0.6 Roll 5.4
GPS
UTC 15:43:26 30 30.614N 089 11.116W Alt 48.8 Meters
HPR
Mag Heading 30.9 Pitch 0.9 Roll 0.7
……..

FIGURE 13.5 The sample of GPS and TCM2-20 attitude data string.

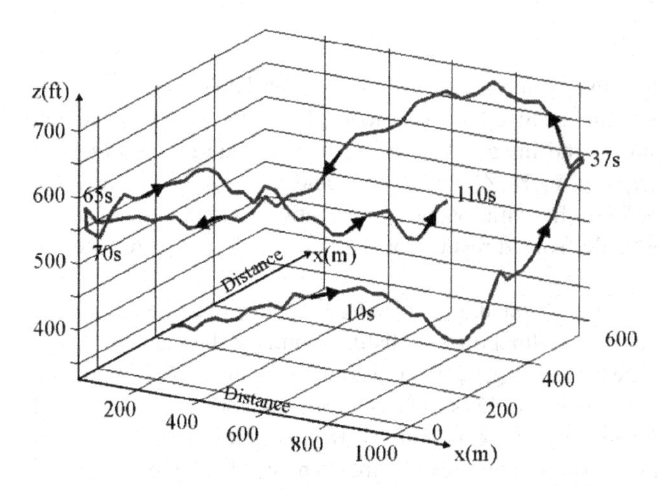

FIGURE 13.6 UAV 3D test flight trajectory for a 110-second flight duration.

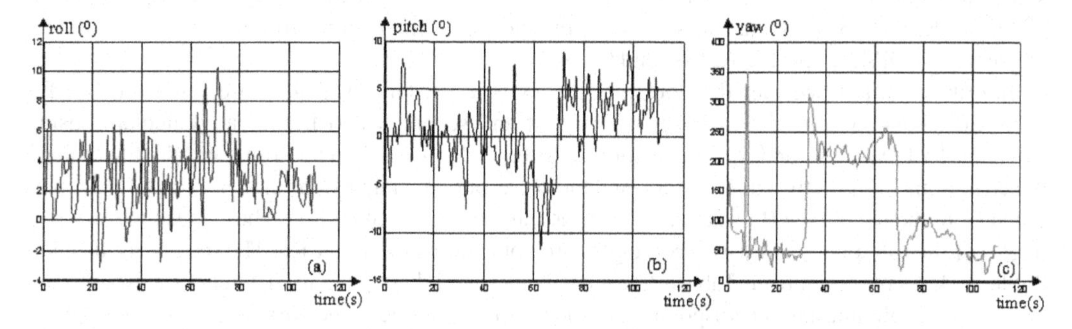

FIGURE 13.7 Changes of UAV attitude data: (a) roll, (b) pitch, and (c) yaw for a 110-second flight segment.

2) **Video Data Collection.** The video data stream was collected for approximately 30 minutes, and was transmitted to the field control station at real-time. The data collection process demonstrated that the received video is very clear (see Figure 13.3e). Moreover, the UTC time taken from the on-board GPS was overlaid onto the video in the lower right-hand corner (see Figure 13.3e). Meanwhile, the video is recorded on digital tape using the DHR-1000NP SONY DV Edit Recorder. The video was then converted from tape to MPEG-I format and delivered through CD or Internet.

In order to navigate the video data collection for reducing the complexity of post-processing, a UAV 3D trajectory and attitude navigation software was developed for real-time monitoring of the UAV flight trajectory, velocity, and attitude. The software was developed through retrieving on-board GPS and TCM2 data and plotting their 3D coordinates and attitudes (roll, pitch, and yaw) in a 3D/2D coordinate system at real-time (see Figures 13.6 and 13.7). Figures 13.6 and 13.7 are a 110-second segment of the UAV's test flight as recorded by the on-board GPS and TCM2 sensors. From Figures 13.6 and 13.7, the flight altitude is approximately 360–700 feet, and there is a dramatic change of heading (yaw) angle around 37 seconds, and two abrupt changes in the direction of flight. In addition, it can be concluded that this UAV is not capable of remaining at a stable altitude, navigation direction, and velocity. As a consequence, the image resolutions and overlaps of two neighbor video frames are probably not the same. Using the UAV's flight attitude and trajectory data, the video stream is resampled for the post-processing at a non-uniform rate depending on the changes of these parameters.

13.4.4 Bundle Adjustment of Video

With the measurement of a number of high-quality non-traditional GCPs described in Section 13.4.1, all unknown parameters in Equation 13.1 can be solved. In this model, 11 GCPs are employed and their imaged coordinates in the first and second images are also measured. The initial values of unknown parameters, including (x_0, y_0, f, ρ_1), $(X_S^1, Y_S^1, Z_S^1, \omega_1, \phi_1, \kappa_1)$ and $(X_S^2, Y_S^2, Z_S^2, \omega_2, \phi_2, \kappa_2)$, are provided by the above computation. With the initial values, an iterative computation updating the initial values is carried out, and the finally solved results for the first video frame are listed in Table 13.2.

The above computational processing can be extended into an entire strip, in which the interesting distinct points must be extracted and tracked. The final tracked distinct points in the video flow will be used as tie points to tie all overlap images together in the bundle adjustment model (i.e., Equation 13.16). Through the solution of Equation 13.16, the EOPs of each video frame can be obtained. A statistics analysis of EOPs for the video flow (correspondingly 18,200 video frames) is listed in the last column of Table 13.2. From our experimental results, the standard deviation (σ_0) of the six unknown parameters can reach 0.42 pixels. In addition, the maximum, minimum, and average standard deviations of the six EOPs are listed in Table 13.3. As seen, the average standard deviations of linear elements of EOPs are less than 1.5 m, and the average standard deviations of nonlinear elements of EOPs are less than 22 seconds.

In order to verify whether the boresight matrix is a constant over an entire flight mission for a low-cost UAV system, the matrixes of the first and second video frames are calculated using Equation 13.1, called R_1 and R_2. The results are listed in Table 13.4. In addition, in order to compare the error of matrixes between R_1 and R_2, and the corresponding angles, the error matrix (Re) is computed from the R_1 and the R_2, that is, $Re = R_1{*}R_2$, and the norm of (I–Re). The corresponding error angles of the roll, pitch, and yaw error angles are obtained from matrix Re. The results are listed in Table 13.4. As observed from Table 13.4, it has been found that apparent differences exist between the two boresight matrixes corresponding to the two video frames. The corresponding angles along the x, y, and z axis with respect to the ground coordinate system are 1.1955°, 2.7738°, and –1.7698°,

TABLE 13.2
Results of the Three Methods (σ_0 is Standard Deviation) for the First Video Frame.

	Roll (ω) (°)	*Pitch* (ϕ) (°)	*Yaw* (κ) (°)	x_0 (pixel)	y_0 (pixel)	f (pixel)	ρ_1	σ_0 (pixel)
On-board TCM2	0.07032	0.00245	1.08561	-	-	-	-	
DLT	–0.01039	0.00002	–1.06379	362.20	241.32	790.54	-	1.27
Our method	–0.01873	0.00032	–1.02943	361.15	239.96	804.09	–1.02e-7	0.42

TABLE 13.3
The Accuracy Statistics of Results of the Proposed Methods (σ_0 is Standard Deviation).

	X_S (m)	Y_S (m)	Z_S (m)	ω (sec)	ϕ (sec)	κ (sec)
Minimum σ_0	0.17	0.09	1.33	10.5	8.4	17.1
Maximum σ_0	2.20	1.94	1.21	30.8	24.4	13.3
Average σ_0	1.54	1.11	1.25	21.2	17.5	15.8

TABLE 13.4
Calculated Boresight Matrixes for the First and Second Video Frames (The Norm of (I-Re), Where Re is the Error Matrix from R_1 and R_2, and is Given by Re = R_1*R_2).

Video	Video Frame #1		Video Frame #2	
Element	**Boresight Matrix (R_1)**	**Angles (°)**	**Boresight Matrix (R_2)**	**Angles (°)**
Value	$\begin{pmatrix} -0.5470 & 0.8352 & 0.0575 \\ -0.8371 & -0.5465 & -0.0254 \\ 0.0102 & -0.0620 & 0.9980 \end{pmatrix}$	$\begin{pmatrix} -3.2955 \\ 1.4564 \\ 56.8605 \end{pmatrix}$	$\begin{pmatrix} -0.4581 & 0.8867 & 0.0621 \\ -0.8887 & -0.4554 & -0.0529 \\ -0.0187 & -0.0794 & 0.9967 \end{pmatrix}$	$\begin{pmatrix} -3.5634 \\ 3.0335 \\ 62.8683 \end{pmatrix}$
	Error matrixes		**Error angles**	
Error matrix (R_e=R_1*R_2)	$\begin{pmatrix} -0.492737 & -0.869941 & -0.020841 \\ 0.869625 & -0.491364 & -0.048390 \\ 0.031764 & -0.041962 & 0.998619 \end{pmatrix}$		$\begin{pmatrix} 1.1955 \\ 2.7738 \\ -1.7698 \end{pmatrix}$	
Norm of (I-Re)	$\begin{pmatrix} 1.492737 & 1.869941 & 1.020841 \\ 0.130375 & 1.491364 & 1.048390 \\ 0.968236 & 1.041962 & 0.001381 \end{pmatrix}$			

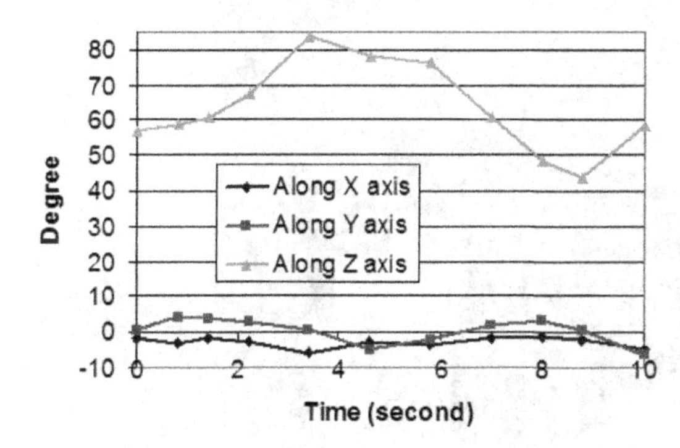

FIGURE 13.8 Boresight matrix changes over time in a low-cost UAV system.

respectively. Further investigation of boresight matrixes corresponding to other video frames in a strip yielded similar results. The angle changes relative to the first video frame are compared, and depicted in Figure 13.8 at the first 10 seconds. As found in Figure 13.8, the result reveals an important fact that the boresight matrix for a low-cost UAV system will not be able to remain a constant. This conclusion is inconsistent with the traditional one. This means that using a uniform boresight matrix derived from the boresight calibration for direct georeferencing on the basis of an on-board navigator is impracticable for a low-cost UAV mapping system. Therefore, the EOPs of each video frame in a low-cost UAV mapping system should be estimated individually in order to obtain a precise boresight matrix for high-accuracy planimetric mapping.

The causes were analyzed, it was discovered that slightly differential movement between the attitude sensor and video camera probably frequently exists in a low-cost UAV system. In particular, when the UAV suddenly changes its velocity (acceleration), roll, pitch, and yaw angles, the movement becomes obvious because the gimbals, which can effectively separate the GPS/TCM2 sensors from mechanical vibration, are ignored in our UAV system (Zhou et al. 2006). In other words, the camera and attitude sensor cannot be rigidly tightened to their chassis. As a result, the boresight matrix, as determined by the above method, will not remain a constant over the entire UAV flight mission. Thus, transforming the navigation data derived from on-board sensors into the camera coordinate system using a uniform boresight matrix is probably difficult for a small low-cost UAV system (Wu and Zhou 2006).

Based on the above fact, this chapter presents a method which can simultaneously determine R_C^{Att} and the camera's IOPs, that is, focal length *(f)*, principal point coordinates (x_0, y_0), and lens distortion (p_1). The details are as follows.

13.4.5 Orthorectification and Accuracy Analysis

With the above solved EOPs for each video frame, the generation of georeferencing video can be implemented using the proposed method described in Section 13.3. Further details of this method can be referenced in Zhou et al. (2002). With the method, each video frame is individually orthorectified and mosaicked together to create a 2D planimetric mapping covering the test area (see Figure 13.9). In order to quantitatively evaluate the accuracy (absolute accuracy) achieved by our method, 55 checkpoints are measured in both the mosaicked ortho-video and the USGS DOQ. The results are listed in Table 13.5. As seen from Table 13.5, the average accuracy can achieve 1.5–2.0 m (that is, 1–2 pixels) relative to USGS DOQ. Meanwhile, it is found that the lowest accuracy occurred in the middle area (Section 2), due to the paucity and poor distribution of GCPs used in the bundle

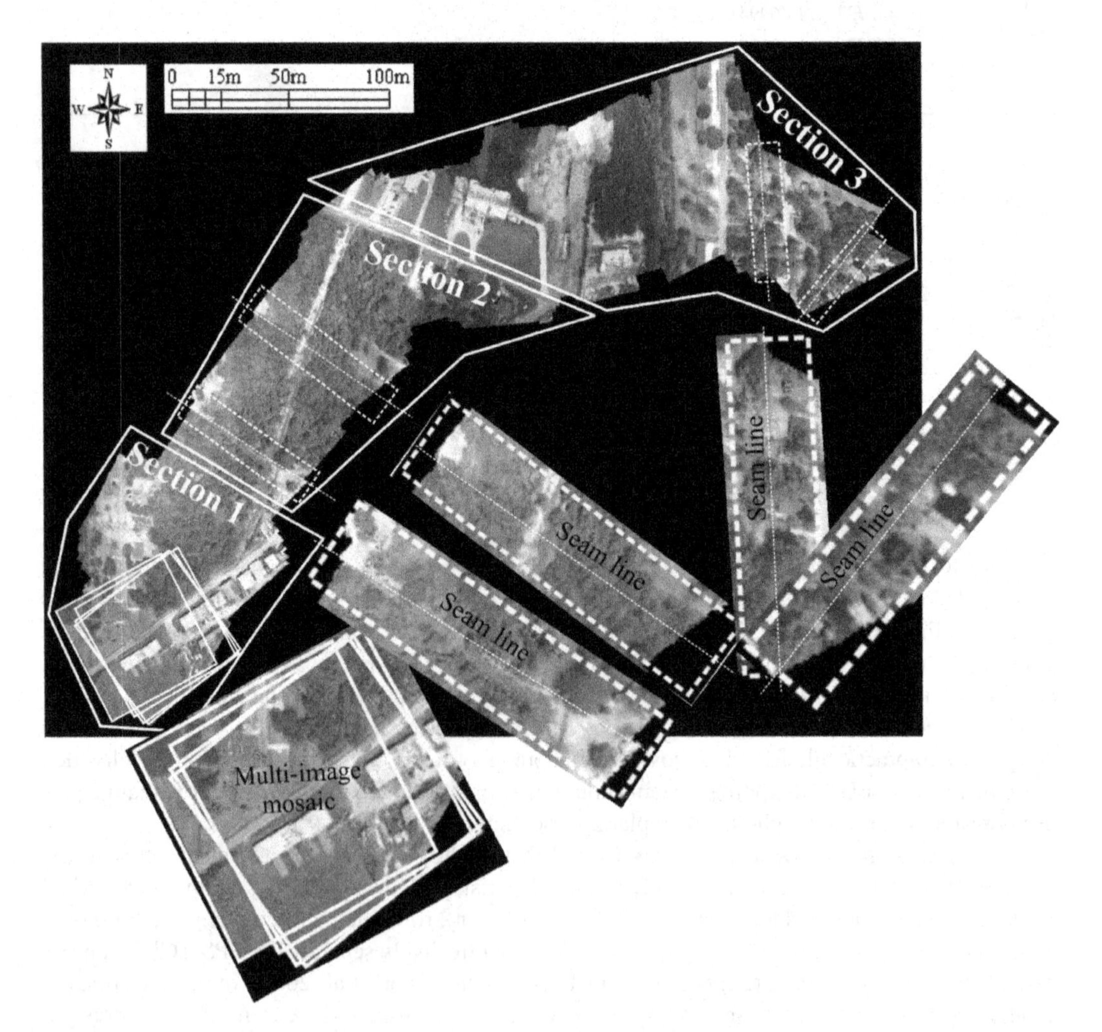

FIGURE 13.9 The mosaicked ortho-video, and the accuracy estimation of ground coordinates and seamlines in the different sections, in which 11, 9, and 15 GCPs are selected in Sections 1, 2, and 3, respectively.

TABLE 13.5
Accuracy Evaluation of the 2D Planimetric Mapping derived using Three Orientation Parameters, and $\delta X = \sqrt{(X - X')^2 / n}$ and $\delta Y = \sqrt{(Y - Y')^2 / n}$, where (X,Y) and (X',Y') are Coordinates in the 2D Planimetric Mapping and the USGS DOQ, Respectively.

Accuracy Relative to USGS DOQ	From Self-Calibration Bundle Adjustment	From Boresight Alignment	From GPS/TCM2
δX (m)	0.17	10.46	*44.04*
δY (m)	*0.25*	*10.33*	*56.26*

adjustment model. Sections 1 and 3 in Figure 13.9 have a relatively higher accuracy due to more and better-distributed GCPs. So, the experimental results demonstrated that the algorithms developed and the proposed method in this chapter can rapidly and correctly orthorectify a video image within acceptable accuracy limits.

In addition, the accuracy of seam lines of two overlapping mosaicked images are measured. The sub-windows of the magnified seamlines for the three sections are shown in Figure 13.9. The results showed that the accuracy of seamlines in the three sections can achieve less than 1.2 pixels.

13.5 CONCLUSIONS

This chapter presented a mathematical model for real-time orthorectification and mosaicking of video flow acquired by a small and low-cost UAV. The developed model is based on a photogrammetry bundle model, in which the DLT algorithm is used for calculating the initial values of unknown parameters. The developed model is able to simultaneously solve the video camera's IOPs and the EOPs of each video frame. A test field, located in Picayune, Mississippi, has been established and 60 minutes' video data were acquired by the UAV platform to test our method. The experimental results demonstrated that accuracy of the mosaicked video images (i.e., 2D planimetric map) is approximately 1–2 pixels, that is, 1–2 meters when compared with 55 checked points, which were measured by DGPS surveying. The accuracy of seamlines of two neighbor images is less than 1.2 pixels. The developed software and system are operated in the field ground station, which is housed in a van. Through the field test, the processing speed and achieved accuracy can meet the requirement for a UAV-based real-time response to time-critical events. The suggested method is specifically appropriate for an economical, moderately functional, small UAV platform intended to meet the requirement for a fast response to time-critical events in many small private sector or government agencies for small areas of interest, for example, forest fires.

REFERENCES

Abdel-Aziz, Y.I., and Karara, H.M., Direct linear transformation from comparator coordinates into object space coordinates in close-range photogrammetry. *Proceedings of the Symposium on Close-Range Photogrammetry*, Falls Church, VA, American Society of Photogrammetry, pp. 1–18, 1971.

Badia, S. B. I., Bernardet, U., Guanella, A., Pyk, P., Verschure, P. F., A biologically based chemo-sensing UAV for humanitarian demining, *International Journal of Advanced Robotic Systems*, vol. 4, no. 2, pp. 187–198, 2007.

Barber, D. et al. Vision-based target geo-location using a fixed-wing miniature air vehicle, *Journal of Intelligent and Robotic Systems: Theory and Applications*, vol. 47, no. 4, pp. 361–382, December, 2006.

Bethea, M.D. et al. Three-dimensional camera calibration technique for stereo imaging velocimetry experiments, *Optical Engineering*, vol. 36, no. 12, pp. 3445–3454, December, 1997.

Bielski, C., Grazzini, J., and Soille, P., Automated morphological image composition for mosaicing large image data sets, *IEEE International Geoscience and Remote Sensing Symposium, IGARSS, 2007*, Barcelona, Spain, pp. 4068–4071, 2007.

Bland, G. et al. "Sensors with wings" - Small UAVs for earth science, *Collection of Technical Papers - AIAA 3rd "Unmanned-Unlimited" Technical Conference, Workshop, and Exhibit*, vol. 2004, pp. 317–327, 2004.

Boer, J.F. et al. Specific aspects in the preliminary design process for small Rotorcraft UAVs, *Proceedings of International 62nd Annual Forum - Vertical Flight: Leading through Innovation*, Phoenix, AZ, pp. 887–898, 2006.

Caltabiano, D., Muscato, G., Orlando, A., et al., Architecture of a UAV for volcanic gas sampling, *IEEE Conference on Emerging Technologies & Factory Automation. IEEE*, Catania, Italy, vol. 12, pp. 739–744, 2005.

Campbell, M.E., and Wheeler, M., A vision based geolocation tracking system for UAVs. *Proceedings of the AIAA Guidance, Navigation, and Control Conference and Exhibit*, Keystone, CO. (Paper no. AIAA-2006-6246), 2006.

Casbeer, D. W., Beard, R. W., Mclain T. W., et al., Forest fire monitoring with multiple small UAVs, *American Control Conference. IEEE*, Portland, OR, vol. 5, pp. 3530–3535, 2005.

Cramer, M. and Stallmann, D., System calibration for direct georeferencing. *Proceedings of International Society for Photogrammetry and Remote Sensing*, Graz, Com. III, Part A, vol. 34, pp. 79–84, September 2002.

Dobrokhodov, V.N., Kaminer, I.I., and Jones, K.D., Vision-based tracking and motion estimation for moving targets using small UAVs. *Proceedings of the AIAA Guidance, Navigation, and Control Conference and Exhibit*, Keystone, CO. Paper no. AIAA-2006-6606, 2006.

Du, Y. et al. Radiometric normalization, compositing, and quality control for satellite high resolution image mosaics over large areas, *IEEE Transactions on Geoscience and Remote Sensing*, vol. 39, no. 3, pp. 623–634, March 2001.

Furey, D., SWARM UAV—Development of a small, very low cost, high endurance UAV, *Proceedings of AUVSI's Unmanned Systems North America*, pp. 1339–1349, 2004.

George, P., and Thomas, P., *Wallops Flight Facility Uninhabited Aerial Vehicle (UAV) User's Handbook, Suborbital and Special Orbital Projects Directorate, 840-HDBK-0002, NASA Wallops Flight Facility*, Documentation Web Site at http://www.wff.aasa.gov April 15, 2005.

Gibbins, D., Roberts, P., and Swierkowski, L., A video geo-location and image enhancement tool for small unmanned air vehicles (UAVs). *Proceedings of Intelligent Sensors, Sensor Networks and Information Processing Conference*, Melbourne, Australia, pp. 469–473, 2004.

Grejner-Brzezinska, A.G., Direct exterior orientation of airborne imagery with GPS/INS system-performance analysis. *Journal of the Institute of Navigation*, vol. 46, no. 4, pp. 261–270, 1999.

Hausamann, D. et al. Monitoring of gas pipelines - a civil UAV application, *Aircraft Engineering and Aerospace Technology*, vol. 77, no. 5, pp. 352–360, 2005.

Herwitz, S. et al. UAV homeland security demonstration, *AIAA 3rd "Unmanned-Unlimited" Technical Conference, Workshop, and Exhibit*, Chicago, IL, pp. 396–400, 2004.

Hruska, R. et al. Small UAV-acquired, high-resolution, georeferenced still imagery. *Proceedings, AUVSI's Unmanned Systems North America*, Las Vegas, Nevada, pp. 837–840, 2005.

Johnson, E.N., Schrage, D.P., and Prasad, J.V.R., UAV flight test programs at Georgia Tech, *Proceeding in AIAA 3rd "Unmanned-Unlimited" Technical Conference, Workshop, and Exhibit*, Chicago, IL, pp. 527–539, 2004.

Johnson, E.N. et al. Flight test results of autonomous fixed-wing UAV transitions to and from stationary hover, *Proceedings of AIAA Guidance, Navigation, and Control Conference 2006*, Keystone, CO, pp. 5144–5167, August 21–24 2006.

Kahn, A. D., Foch, R. J., Attitude command attitude hold and stability augmentation systems for a small-scale helicopter UAV, *Digital Avionics Systems Conference. IEEE*, Indianapolis, IN, vol. 2, pp.8.A.4/1–8.A.4/10, 2003.

Kaminer, I. I. et al., Cooperative control of small UAVs for naval applications, *Proceedings of IEEE Conference on Decision & Control*, Nassau, Bahamas, vol. 1, pp. 626–631, 2004.

Kang, Y., Caveney, D.S., and Hedrick, J.K., Probabilistic mapping for UAV using point-mass target detection, *Proceedings of AIAA Guidance, Navigation, and Control Conference*, Keystone, CO, pp. 1915–1926, August 21–24 2006.

Kelley, K. K., Small UAV initiative, *Proceedings of SPIE – The International Society for Optical Engineering*, San Diego, CA, vol. 4127, pp. 46–48, 2000.

Kucera, J., Army to deploy new small UAV, *Jane's Defence Weekly*, April, p. 1, 2005.

Li, M., Liew, S.C., and Kwoh, L.K., Producing cloud free and cloud-shadow free mosaic from cloudy IKONOS images, *IEEE International Geoscience and Remote Sensing Symposium. Proceedings*, vol. 6, pp. 3946–3948, 2003.

Liu, Y. et al. Real-time three-dimensional radar mosaic in CASA IP1 testbed, *IEEE International Geoscience and Remote Sensing Symposium, IGARSS 2007*, Barcelona, Spain, pp. 2754–2757, 2007.

Logan, M.J. et al. Technology challenges in small UAV development, *Collection of Technical Papers - InfoTech at Aerospace: Advancing Contemporary Aerospace Technologies and Their Integration*, vol. 3, pp. 1644–1648, 2005.

Longo, A. A., Pace, P., Marano, S., A system for monitoring volcanoes activities using high altitude platform stations, *55th International Astronautical Congress of the International Astronautical Federation, the International Academy of Astronautics, and the International Institute of Space Law*, Vancouver, British Columbia, Canada, vol. 2, pp. 1203–1210, 2004.

Moore, M., Rizos, C., Wang, J., Boyd, G., Mathews, K., Williams, W., Smith, R., Issues concerning the implementation of a low cost attitude solution for an unmanned airborne vehicle (UAV), *SatNav 2003 The 6th International Symposium on Satellite Navigation Technology Including Mobile Positioning & Location Services Melbourne*, Melbourne, Australia, 2003.

Mostafa, M.R., Schwarz, K.P., and Chapman, M.A., Development and testing of an airborne remote sensing multi-sensor system, *ISPRS Commission II, Symposium on Data Integration: Systems and Techniques*, Cambridge, UK, 1998.

Narli, V., and Oh, P.Y., Near-earth unmanned aerial vehicles: Sensor suite testing and evaluation, *Proceedings of 2006 ASME International Design Engineering Technical Conferences and Computers and Information In Engineering Conference, DETC2006*, Philadelphia, PA, 7p, 2006.

Nelson, D.R. et al. Initial experiments in cooperative control of unmanned air vehicles, *Collection of Technical Papers - AIAA 3rd "Unmanned-Unlimited" Technical Conference, Workshop, and Exhibit*, Chicago, IL, pp. 666–674, 2004.

Noth, A., Engel, W., Siegwart, R., Design of an ultra-lightweight autonomous solar airplane for continuous flight, In *Field and Service Robotics*. Springer, Berlin, Heidelberg, pp. 441–452, 2005.

Oh, P.Y, Flying insect inspired vision for micro-air-vehicle navigation, *AUVSI's Unmanned Systems North America 2004 - Proceedings*, San Francisco, CA, pp. 2201–2208, 2004.

Postell, G., Thomas, P., *Wallops flight facility uninhabited aerial vehicle (UAV) user's handbook.* Suborbital Special Orbital Projects Directorate, NASA Wallops Flight Facility, Wallops Island, Virginia, 2005 [Online]. Available: http://www.wff.aasa.gov.

Ryan, A.D., Nguyen, D.L., and Hedrick, J. K., Hybrid control for UAV-assisted search and rescue, *Proceedings of the ASME Dynamic Systems and Control Division*, Orlando, FL, pp. 187–195, November 5–11 2005.

Schwarz, K.P., Chapman, M.E., Cannon, E., and Gong, P., An integrated INS/GPS approach to the georeferencing of remotely sensed data. *Photogrammetric Engineering & Remote Sensing*, vol. 59, pp. 1667–1674, 1993.

Sheng, Y., and Alsdorf, D.E., Automated georeferencing and orthorectification of Amazon basin-wide SAR mosaics using SRTM DEM data. *IEEE Transactions on Geoscience and Remote Sensing*, vol. 43, no. 8, pp. 1929–1940, 2005.

Shin, D., Pollard, J.K., and Muller, J.P., Accuracy geometric correction of ATSR images. *IEEE Transactions on Geoscience and Remote Sensing*, vol. 35, no. 4, pp. 997–1006, 1997.

Simard, M., Saatchi, S., and DeGrandi, G., Classification of the Gabon SAR mosaic using a wavelet based rule classifier, *IEEE 1999 International Geoscience and Remote Sensing Symposium. IGARSS'99 (Cat. No.99CH36293)*, vol. 5, pp. 2768–2770, 1999.

Sinopoli, B., et al. Vision based navigation for an unmanned aerial vehicle, *Proceedings in IEEE International Conference on Robotics and Automation (ICRA)*, Seoul, Korea, pp. 1757–1764, May 21–26 2001.

Skaloud, J., Cramer, M., and Schwarz, K.P., Exterior orientation by direct measurement of camera and position, *The International Archives of the Photogrammetry and Remote Sensing*, Vienna, Austria, vol. XXXI, Part B3, pp. 125–130, 1996.

Skaloud, J., Cramer, M., Schwarz, K. P., Exterior orientation by direct measurement of camera position and attitude. *International Archives of Photogrammetry and Remote Sensing*, vol. 31, no. B3, pp. 125–130, 1996.

Whang, I.H., et al. On vision-based tracking and range estimation for small UAVs. *Proceedings of the AIAA Guidance, Navigation, and Control Conference and Exhibit*, San Francisco, CA, Paper no. AIAA-2005-6401, 2005.

Yastikli, N., and Jacobsen, K., Influence of system calibration on direct sensor orientation, *Photogrammetric Engineering & Remote Sensing*, vol. 71, no. 5, pp. 629, 2005.

Zhou, G., High-resolution UAV video data processing for forest fire surveillance, *Technical Report to National Science Foundation*, Old Dominion University, Norfolk, Virginia, USA, August, 82p, 2006.

Zhou, G. Near real-time orthorectification and mosaic of small uav video flow for time-critical event response. *IEEE Transactions on Geoscience and Remote Sensing,* vol. 47, no. 3, pp. 739–747, 2009.

Zhou, G., Jezek, K., and Wright, W., Orthorectifying 1960's disclassified intelligence satellite photography (DISP) of Greenland, *IEEE Transactions on Geoscience and Remote Sensing,* vol. 40, no. 6, pp. 1247–1259, 2002.

Zhou, G., Wu, J., Wright, S., Gao, J., High-resolution UAV video data processing for forest fire surveillance. Old Dominion Univ., Norfolk, VA, Tech. Rep. National Sci. Foundation, 82p, 2006.

Zhou, G., et al. A comprehensive study on urban true orthorectification. *IEEE Transactions on Geoscience and Remote Sensing*, vol. 43, no. 9, pp. 2138–2147, 2005.

14 Orthophotomap Generation from Satellite Imagery Without Camera Parameters

14.1 INTRODUCTION

Rocky karstification in karst areas (also called karst rocky desertification (KRD)) is considered one of the major factors that contribute to the global carbon balance as a global CO_2 sink (Xu and Jiang 1997; Jiang and Qin 2011; Zhou et al. 2015). With the increasing interest in global carbon emissions, studies and analyses through comparing historical data with current data may discover how rocky karstification contributes to long-term environmental changes over decadal spans.

Guangxi is located in the southwestern karst area in China, and the KRD area is approximately 23,790.80 km^2, accounting for 19.8% of the total KRD area in China in 2005. Fortunately, the declassified intelligence satellite photography (DISP) released to the public domain in February 1995 has provided researchers with a unique opportunity to investigate the KRD in Guangxi in the 1960s. The DISP was collected by the first generation of United States photoreconnaissance satellites between 1960 and 1972 through the systems named CORONA, ARGON, and LANYARD. More than 860,000 images of the Earth's surface were declassified with the issuance of this executive order and were contracted to the USGS for sale.

However, further processing and application of DISP has exposed various problems:

1. The USGS does not provide Chinese users with the parameters required to further process DISP. These parameters include satellite orbit parameters (e.g., inclination, flight height, descent time, etc.) and the camera's interior orientation parameters (IOPs) (e.g., focal length, principal point coordinates, fiducial marks, etc.). This implies that traditional bundle block adjustment based on the photogrammetric collinearity equation is not applicable (Zhou and Jezek 2002).
2. It is very difficult to obtain sufficient ground control points (GCPs) in the historical DISP imagery due to the time intervals of several decades and cloudy coverage in Southern China. Thus, it is almost impossible to rectify each DISP image on a frame-by-frame basis.

For the two reasons above, this chapter presents a second order polynomial equation-based rectification model for orthorectification of DISP images. The previous relevant studies on this topic are as follows. Kim et al. utilized a collinearity equation to rectify ARGON imagery from 1963 to study the seasonal variations of glaciers on the Queen Maud Land coast of Antarctica (Kim et al. 2001). Zhou and Jezek 2002 proposed a collinearity equation-based self-calibration block bundle adjustment method that integrates the bundle adjustment method and satellite orbital parameters, solving interior orientation parameters (IOPs) (including lens distortion) and exterior orientation parameters (EOPs) simultaneously to rectify ARGON images from 1962 and 1963 (Zhou and Jezek 2002). The rectified ARGON imagery was employed to mosaic Greenland ice sheets from the 1960s, which were then quantitatively compared to the ice sheet extent over a 30-year interval. Kim et al. (2007)

applied a state-of-the-art digital imaging technology based on an extended block adjustment to rectify ARGON imagery from 1963 that covered Antarctica. They assembled all images into a quality mosaic of coastal Antarctica to study glaciers. In addition, due to the imaging model limitations of high-resolution satellites such as IKONOS, rational polynomial-based block adjustment, also called the rational polynomial coefficient (RPC), was proposed by multiple authors. For example, Tao and Hu (2001, 2002) analyzed the accuracy of orthorectification of a SPOT image and an aerial image using the RPC model. Yang (2000) suggested that the RPC model can replace the rigorous sensor model orthorectification of SPOT images. Liu (2004) developed a stereotaxic method of IKONOS images based on the RPC model. Huang (2008) proposed a rational polynomial-based block adjustment for orthorectification of SAR images. Grodecki and Dial (2003) rectified IKONOS satellite imagery using the RPC method. The RPC model incorporates priori constraints into the images described by the RPC, and multiple independent images can be added in accordance with the needs of users. However, the RPC model requires a number of GCPs, and the computation is very time-consuming. Therefore, the RPC method is not applicable to the DISP images that the USGS provides because the imaging model of DISP was not provided by the USGS. Additionally, few GCPs are available in the study area. Thus, this chapter presents an effective and simple mathematical model for geometric rectification of DISP images, considerably improving the computational effectiveness.

14.2 THE SECOND ORDER POLYNOMIAL EQUATION-BASED RECTIFICATION MODEL METHOD

14.2.1 Polynomial Equation-Based Block Adjustment Model

The objective of polynomial equation-based block adjustment is to tie overlapping images together without the absolute need for GCPs in each image and obtain coordinates of TPs (TPs) and conversion parameters for rectification. Because the study area is a karst landform with a large wavy terrain and large elevation differences, the relief displacement is large. For correction of relief displacement, relief displacement is introduced into the block adjustment model shown below.

Figure 14.1 shows the imaging geometry of DISP from the CORONA mission, where $S - WVU(W'\ V'\ U')$ is a camera coordinate system, $o - xy$ is an image plane system, $O - XYZ$ is a geographic coordinate system, and Δh is relief displacement. First, distortion caused by elevation differences should be corrected. Then, other distortions should be corrected by utilizing a polynomial model.

Because the relief displacement only occurs in the direction of scanning, CORONA images are panoramic camera images, and the panoramic projection scan direction is the x-direction. Therefore, as shown in the imaging equation above, there is no relief displacement in the y-direction. Thereby, the relief displacement correction functions can be expressed by

$$\Delta h = Z \cdot h / M \tag{14.1}$$

$$\begin{cases} \Delta x = x \cdot Z / M \\ \Delta y = y \cdot Z / M = 0 \end{cases} \tag{14.2}$$

where x and y are image coordinates; Δx and Δy are image distortions in the x- and y-directions, respectively, caused by elevation differences; Z is elevation; h is the distance from the image point to the nadir point; and M is the satellite flight altitude. Because the relief displacement occurs in the direction of scanning and the KH-4A/B's images are panoramic camera images, the images can be rectified using the second order polynomial equation-based model.

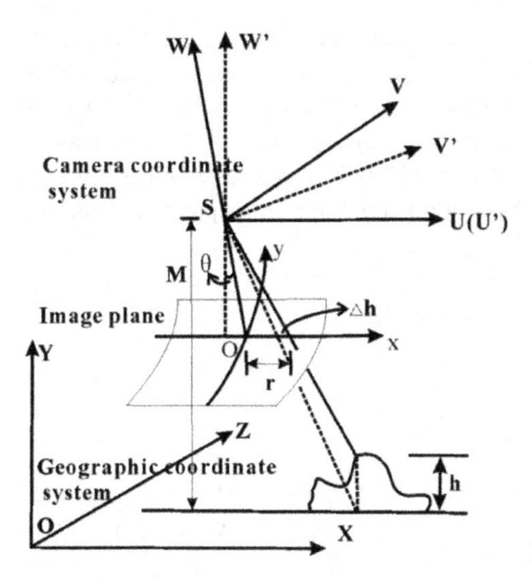

FIGURE 14.1 The imaging geometry of DISP from the CORONA mission.

- **Traditional Second Order Polynomial Equation**

The traditional second order polynomial model has been widely applied for image rectification. This chapter extends the traditional equation into a block situation by adding TPs, which tie overlapping images together. With the extended model, the 2D coordinates of TPs and the coefficients of second order polynomial equations are solved. Furthermore, these parameters are used for ortho-rectification of DISP imagery without the absolute requirement of at least six GCPs in each DISP image.

The traditional second order polynomial equations are expressed as follows (Valadan Zeoj et al. 2002):

$$\begin{cases} x+\Delta x = a_0 + a_1X + a_2Y + a_3XY + a_4X^2 + a_5Y^2 \\ y+\Delta y = b_0 + b_1X + b_2Y + b_3XY + b_4X^2 + b_5Y^2 \end{cases} \tag{14.3}$$

where $\alpha = (a_0, a_1, a_2, a_3, a_4, a_5)^T$ and $\beta = (b_0, b_1, b_2, b_3, b_4, b_5)^T$ are coefficients; x and y are image coordinates; Δx and Δy are image distortions in the x- and y-directions, respectively; and X and Y are 2D coordinates in a given map coordinate system.

For a given GCP, Equation 14.3 can be linearized using a Taylor series and is expressed as follows:

$$\begin{cases} v_x = \Delta a_0 + X\Delta a_1 + Y\Delta a_2 + XY\Delta a_3 + X^2\Delta a_4 + Y^2\Delta a_5 - l_x \\ v_y = \Delta b_0 + X\Delta b_1 + Y\Delta b_2 + XY\Delta b_3 + X^2\Delta b_4 + Y^2\Delta b_5 - l_y \end{cases} \tag{14.4}$$

where $\Delta a_i (i = 0, 1, \ldots, 5)$ and $\Delta b_i (i = 0, 1, \ldots, 5)$ are correction terms of coefficients; v_x, v_y are residuals; X and Y are 2D coordinates of GCPs; and l_x and l_y are constants expressed by Equation 14.5.

$$\begin{cases} l_x = x - \left(a_0 + a_1X + a_2Y + a_3XY + a_4X^2 + a_5Y^2\right) \\ l_y = y - \left(b_0 + b_1X + b_2Y + b_3XY + b_4X^2 + b_5Y^2\right) \end{cases} \tag{14.5}$$

As shown in Equation 14.4, one GCP only establishes two observations, but Equation 14.4 has 12 unknown parameters. Therefore, six GCPs, which establish 12 observation equations, are needed to solve the 12 coefficients that are used to rectify a single image. Generally, more than six GCPs are observed in each image to establish more than 12 observation equations. The least-squares estimation is employed to calculate the 12 coefficients. Mathematically, the solution can be described as follows.

Assuming that N GCPs (N≥6) are observed, the observation equations are expressed in matrix form as follows.

$$\boldsymbol{V} = \boldsymbol{A}\cdot\boldsymbol{\alpha} - \boldsymbol{L}, \tag{14.6}$$

where

$$V = \left[V_{x1}V_{y1}\cdots\cdots V_{xN}V_{yN}\right]^T,$$

$$A = \begin{pmatrix} 1 & X_1 & Y_1 & X_1Y_1 & X_1^2 & Y_1^2 & 0 & 0 & 0 & 0 & 0 & 0 \\ 0 & 0 & 0 & 0 & 0 & 0 & 1 & X_1 & Y_1 & X_1Y_1 & X_1^2 & Y_1^2 \\ \vdots & \vdots & \vdots & \vdots & \vdots & \vdots & \vdots & \vdots & \vdots & \vdots & \vdots & \vdots \\ \vdots & \vdots & \vdots & \vdots & \vdots & \vdots & \vdots & \vdots & \vdots & \vdots & \vdots & \vdots \\ 1 & X_N & Y_N & X_NY_N & X_N^2 & Y_N^2 & 0 & 0 & 0 & 0 & 0 & 0 \\ 0 & 0 & 0 & 0 & 0 & 0 & 1 & X_N & Y_N & X_NY_N & X_N^2 & Y_N^2 \end{pmatrix},$$

$$\alpha = \begin{pmatrix} a_0 & a_1 & a_2 & a_3 & a_4 & a_5 & b_0 & b_1 & b_2 & b_3 & b_4 & b_5 \end{pmatrix}^T$$

and

$$L = [l_{x1}\ l_{y1}\ \cdots\ \cdots\ l_{xN}\ l_{yN}]^T$$

The least-squares estimation, that is, $V^TPV = \min$, gives the solutions of the coefficients of the second order polynomial equation below.

$$\alpha = \left(A^TA\right)^{-1}A^TL \tag{14.7}$$

The following expressions can be further obtained from Equation 14.7:

$$a_i = a_i^0 + \sum_{j=1}^{N^{ite}} \Delta a_i^j \left(i = 1,\ldots,5; j = 1,\ldots,N^{ite}\right) \tag{14.8a}$$

$$b_i = b_i^0 + \sum_{j=1}^{N^{ite}} \Delta b_i^j \left(i = 1,\ldots,5; j = 1,\ldots,N^{ite}\right) \tag{14.8b}$$

where a_i^0, b_i^0 are initial values; $\Delta a_i^j, \Delta b_i^j$ are increases during each iteration; and N^{ite} is the number of iterations.

- **The Second Order Polynomial Equation-Based Rectification Model**

As mentioned above, due to the shortage of GCPs in each of the DISP images, the TPs must be identified to tie images with the same overlapping areas. Under this condition, the TPs whose XY-coordinates are unknown are introduced into the traditional second order polynomial equation. This extended model is called the second order polynomial equation-based rectification model

(2OPE-RM) in this chapter (see Figure 14.1). Equation 14.3 is extended by considering TPs as unknown parameters and linearized into the following form.

$$\begin{cases} v_x = \Delta a_0 + X\Delta a_1 + Y\Delta a_2 + XY\Delta a_3 + X^2\Delta a_4 + Y^2\Delta a_5 \\ \quad +\left(a_1 + a_3Y + 2a_4X\right)\Delta X + \left(a_2 + a_3X + 2a_5Y\right)\Delta Y - l_x \\ v_y = \Delta b_0 + X\Delta b_1 + Y\Delta b_2 + XY\Delta b_3 + X^2\Delta b_4 + Y^2\Delta b_5 \\ \quad +\left(b_1 + b_3Y + 2b_4X\right)\Delta X + \left(b_2 + b_3X + 2b_5Y\right)\Delta Y - l_y \end{cases} \tag{14.9}$$

Then, Equation 14.9 can be rewritten as follows:

$$\begin{cases} v_x = \Delta a_0 + X\Delta a_1 + Y\Delta a_2 + XY\Delta a_3 + X^2\Delta a_4 + Y^2\Delta a_5 + f_1\Delta X + f_2\Delta Y - l_x \\ v_y = \Delta b_0 + X\Delta b_1 + Y\Delta b_2 + XY\Delta b_3 + X^2\Delta b_4 + Y^2\Delta b_5 + g_1\Delta X + g_2\Delta Y - l_y \end{cases}, \tag{14.10}$$

where

$f_1 = a_1 + a_3Y + 2a_4X,$
$f_2 = a_2 + a_3X + 2a_5Y,$
$g_1 = b_1 + b_3Y + 2b_4X,$ and
$g_2 = b_2 + b_3X + 2b_5Y.$

The symbols above are the same as those in Equation 14.9. Additionally, assuming that there are N GCPs (N≥6), M TPs in t images are collected at the GCPs. Similarly, Equation 14.9 can be expressed in matrix form as follows:

$$V = A\cdot\alpha + B\cdot\beta - L \tag{14.11}$$

where

$$V = \left[v_{x_1}^{GCP}\ v_{y_1}^{GCP} \ldots \ldots v_{x_n}^{GCP}\ v_{x_n}^{GCP} | v_{x_1}^{TP}\ v_{y_1}^{TP} \ldots \ldots v_{x_M}^{TP}\ v_{x_M}^{TP}\right]^T,$$

$$\alpha = \left(\Delta\alpha_0^1, \Delta\alpha_1^1, \Delta\alpha_2^1, \Delta\alpha_3^1, \Delta\alpha_4^1, \Delta\alpha_5^1, \ldots\ \Delta\alpha_0^t, \Delta\alpha_1^t, \Delta\alpha_2^t, \Delta\alpha_3^t, \Delta\alpha_4^t, \Delta\alpha_5^t\right),$$

$$\beta = \left(\Delta X_1^1, \Delta Y_1^1, \ldots, \ldots, \Delta X_M^t, \Delta Y_M^t\right),$$

$$A = \begin{pmatrix} \left.\begin{matrix} 1 & X_1^1 & Y_1^1 & X_1^1Y_1^1 & X_1^{1^2} & Y_1^{1^2} & 0 & 0 & 0 & 0 & 0 & 0 \\ \vdots & \vdots & \vdots & \vdots & \vdots & \vdots & \vdots & \vdots & \vdots & \vdots & \vdots & \vdots \\ 0 & 0 & 0 & 0 & 0 & 0 & 1 & X_{N_1}^i & Y_{N_1}^i & X_{N_1}^iY_{N_1}^i & X_{N_1}^{1^2} & Y_{N_1}^{1^2} \end{matrix}\right\} \text{The } 1^{\text{st}} \text{ image with } \mathrm{N}_1 \text{ GCPs} \\ \left.\begin{matrix} 1 & X_1^i & Y_1^i & X_1^iY_1^i & X_1^{i^2} & Y_1^{i^2} & 0 & 0 & 0 & 0 & 0 & 0 \\ \vdots & \vdots & \vdots & \vdots & \vdots & \vdots & \vdots & \vdots & \vdots & \vdots & \vdots & \vdots \\ 0 & 0 & 0 & 0 & 0 & 0 & 1 & X_{N_i}^i & Y_{N_i}^i & X_{N_i}^iY_{N_i}^i & X_{N_i}^{i^2} & Y_{N_i}^{i^2} \end{matrix}\right\} \text{The } i^{\text{th}} \text{ image with } \mathrm{N}_i \text{ GCPs} \\ \left.\begin{matrix} 1 & X_1^N & Y_1^N & X_1^NY_1^N & X_1^{N^2} & Y_1^{N^2} & 0 & 0 & 0 & 0 & 0 & 0 \\ \vdots & \vdots & \vdots & \vdots & \vdots & \vdots & \vdots & \vdots & \vdots & \vdots & \vdots & \vdots \\ 0 & 0 & 0 & 0 & 0 & 0 & 1 & X_{Nn}^N & Y_{N_n}^N & X_{N_n}^NY_{N_n}^N & X_{N_n}^{N^2} & Y_{N_n}^{N^2} \end{matrix}\right\} \text{The } \mathrm{N}^{\text{th}} \text{ image with } \mathrm{N}_n \text{ GCPs} \end{pmatrix},$$

and

$$B = \begin{pmatrix} \left.\begin{matrix} f_1^1 & f_2^1 \\ \vdots & \vdots \\ f_2^{M_1} & f_2^{M_1} \end{matrix}\right\} \text{The } 1^{\text{st}} \text{ image with } \mathrm{M}_1 \text{ TPs} \\ \left.\begin{matrix} f_1^i & f_2^i \\ \vdots & \vdots \\ f_1^{M_i} & f_2^{M_i} \end{matrix}\right\} \text{The } i^{\text{th}} \text{ image with } \mathrm{M}_i \text{ TPs} \\ \left.\begin{matrix} f_1^M & f_2^M \\ \vdots & \vdots \\ f_1^{M_n} & f_2^{M_n} \end{matrix}\right\} \text{The } \mathrm{M}^{\text{th}} \text{ image with } \mathrm{M}_n \text{ TPs} \end{pmatrix}.$$

Equation 14.11 is the 2OPE-RM model derived in this chapter. Relative to the traditional model in Equation 14.6, this model introduces TPs as unknown parameters. Finally, Equation 14.11 is usually solved using least-squares estimation, which is expressed as follows.

$$\Phi = V^T V = min \tag{14.12}$$

With least-squares estimation, the normal equation matrix can be written as follows.

$$\begin{pmatrix} A^T A & A^T B \\ B^T A & B^T B \end{pmatrix} \begin{pmatrix} \delta\alpha \\ \delta\beta \end{pmatrix} = \begin{pmatrix} A^T L \\ B^T A \end{pmatrix} \tag{14.13}$$

Thus, the solution of the unknown parameters is given by Equation 14.14:

$$\begin{cases} \Delta a = -\left(Q_{\alpha\alpha} A^T L + Q_{\alpha\beta} B^T L\right) \\ \Delta b = -\left(Q_{\beta\alpha} A^T L + Q_{\beta\beta} B^T L\right) \end{cases} \tag{14.14}$$

where $\boldsymbol{Q}_{ij}(i,j = 1,2)$ gives the components of the covariance matrix, which is the inverse of the normal matrix, as shown in Equation 14.15.

$$Q_{ij} = \begin{pmatrix} A^T A & A^T B \\ B^T A & B^T B \end{pmatrix}^{-1} = \begin{pmatrix} Q_{\alpha\alpha} & Q_{\alpha\beta} \\ Q_{\beta\alpha} & Q_{\beta\beta} \end{pmatrix} (i, j = 1,2) \tag{14.15}$$

The coefficients of the 2OPE-RM in each image and the 2D coordinates (XY) of each TP are as follows:

$$a_i = a_i^0 + \sum_{j=1}^{N^{ite}} \Delta a_i^j \ \left(i = 1, \ldots, 5; j = 1, \ldots, N^{ite}\right) \tag{14.16a}$$

$$b_i = b_i^0 + \sum_{j=1}^{N^{ite}} \Delta b_i^j \ \left(i = 1, \ldots, 5; j = 1, \ldots, N^{ite}\right) \tag{14.16b}$$

$$X_i^{t_i} = X_i^0 + \sum_{j=1}^{N^{ite}} \Delta X_i^j \left(i = 1, \dots, 5; j = 1, \dots, N^{ite};\ t_i = 1, \dots, t\right) \tag{14.17a}$$

$$Y_i^{t_i} = Y_i^0 + \sum_{j=1}^{N^{ite}} \Delta Y_i^j \left(i = 1, \dots, 5;\ j = 1, \dots, N^{ite};\ t_i = 1, \dots, t\right) \tag{14.17b}$$

where X_i, Y_i are coordinates of the *i-th* TP in image t_i; ΔX_i, ΔY_i are increases in X_i and Y_i; a_i^0 and b_i^0 are initial values; and Δa_i and Δb_i are increases in the coefficients in each iteration.

As shown in Equation 14.11, each image has 12 unknown parameters ($a_i, b_i; i = 0, 1, \dots, 5$), and each TP has two unknown parameters (XY-coordinates). Two equations can be established for each GCP or TP. Moreover, the TPs and/or GCPs should be well distributed in each image. For example, there are 4 images, 12 GCPs, and 9 TPs in Figure 14.2. The four images imply that there are 48 unknown parameters. The 12 GCPs can be used to establish 42 observation equations (i.e., 7 GCPs in Image 1 can be used to establish 14 observations, 3 GCPs in Image 2 can be used to establish 6 observations, 6 GCPs in Image 3 can be used to establish 12 observations, and 5 GCPs in Image 4 can be used to establish 10 observations). The 9 TPs can be used to establish 34 observation equations (i.e., 3 TPs in Image 1 can be used to establish 6 observations, 3 TPs in Image 2 can be used to establish 6 observations, 6 TPs in Image 3 can be used to establish 12 observations, and 5 GCPs in Image 4 can be used to establish 10 observations). With this model, there are 76 (76 = 42 + 34) observations and 66 (66 = 48 + 18) unknown parameters. Thus, 2OPE-RM does not require each DISP image to have more than six GCPs.

The accuracy of the adjustment computation is evaluated using Equation 14.18:

$$\delta_o = \sqrt{\frac{V^T V}{r}} \tag{14.18}$$

where δ_o is the standard deviation of the unit weight, V is the matrix of residuals, and r is the number of redundant observations. Thus, the standard deviations of individual unknown parameters can be calculated as follows.

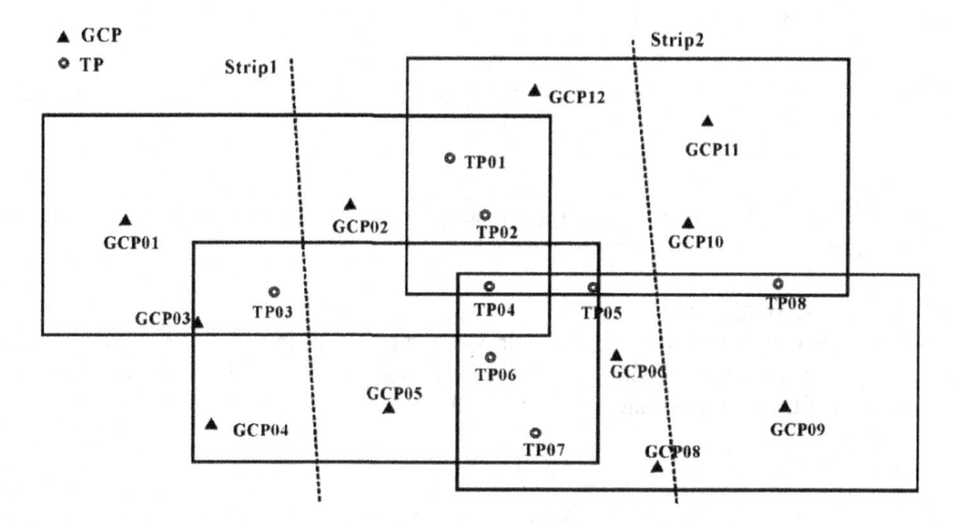

FIGURE 14.2 Illustration of the second order polynomial equation-based rectification model (2OPE-RM).

$$\delta_{X_i} = \delta_o \sqrt{Q_{X_i}} \tag{14.19}$$

To evaluate the accuracies of TPs, assuming that there are n TPs, the average of δ_{X_i} is as follows:

$$\mu_X = \frac{1}{n} \Sigma \delta_{X_i} \tag{14.20}$$

where n is the number of TPs.

14.2.2 Orthorectification of DISP Images

With the established model and the coefficients determined in Section 14.2.1, each original DISP image can be orthorectified. The steps are as follows.

Step 1: Determination of the Rectified Image Size
To properly establish the storage space of the orthorectified image, the size of the resulting image (upper left, lower left, upper right, and lower right) must be determined in advance. This procedure is proposed as follows.

- *Determination of the four corner coordinates.* The four corner coordinates of the original image are projected into the UTM coordinate system. Then, eight coordinates are obtained:

$$\left(X_{ul}, Y_{ul}\right), \left(X_{ll}, Y_{ll}\right), \left(X_{ur}, Y_{ur}\right), \left(X_{lr}, Y_{lr}\right).$$

The maximum and minimum values of X and Y (X_{min}, X_{max}, Y_{min}, and Y_{max}) are calculated from the eight coordinates above to constitute four coordinate pairs. These pairs are the map coordinates of the four boundaries of the resulting image's scope.

$$\begin{aligned} X_{\min} &= \min\left(X_{ul}, X_{ll}, X_{ur}, X_{lr}\right), X_{\max} = \min\left(X_{ul}, X_{ll}, X_{ur}, X_{lr}\right) \\ Y_{\min} &= \min\left(Y_{ul}, Y_{ll}, Y_{ur}, Y_{lr}\right), Y_{\max} = \min\left(Y_{ul}, Y_{ll}, Y_{ur}, Y_{lr}\right) \end{aligned} \tag{14.21}$$

- *Determination of the resulting image's size.* The size of the resulting image can be determined by M and N as follows:

$$M = \frac{Y_{max}}{Y_{GSD} \dfrac{X_{min}}{X_{GSD}}} \tag{14.22}$$

Where $M = row$, $N = col$, and Y_{GSD}, X_{GSD} are the ground-sampled distances (GSD) in the resulting image.

Step 2: Coordinate Transformation
Because the orthorectification model only expresses the relationship between the original coordinates (x_{ori}, y_{ori}) and ground coordinates (X_{gro}, Y_{gro}), the ground coordinates should be transformed into the coordinates of the resulting image (x_{re}, y_{re}) as follows:

$$x_{re} = \frac{Ygro_{max}}{Y_{GSDre} \dfrac{X_{gro} - X_{min}}{X_{GSD}}} \tag{14.23}$$

Where Y_{gro}, X_{gro} are the ground coordinates of the pixel after rectification.

Step 3: Orthorectification
The calculation of the geographic coordinates of individual pixels, resampling of the original image, and registration of the chosen map coordinates system are carried out as follows.

1. The process can be applied to any point $P(I,J)$ in the resulting image with image coordinates(I,J).
2. In accordance with image coordinates (I,J) and GSD, calculate the geographic coordinates (X,Y).
3. Compute the image coordinates (i,j) of point P in the original image using Equation 14.5.
4. Calculate the gray value g_{ori} via bilinear resampling interpolation.
5. Assign the gray value g_{ori} to point P as g_{res} in the resulting (rectified) image/pixel.

The above procedure is then repeated for each pixel that must be rectified until the entire image is completely rectified.

14.2.3 Data Set

- Study area

The study area is located in Guangxi, China, spanning from latitudes 20.54°N to 26.24°N and longitudes 104.26°E to 112.04°E (see Figure 14.3) and encompassing 23,790.8 km^2. The study area is in the south-central subtropics of China.

- Data Set

DISP imagery. In total, 444 DISP images from five orbits of different missions, including the CORONA 1035-1 Mission (24 images) on September 25, 1966, the CORONA 1102-2 Mission (48 images) on December 18, 1967, and the 1106-1/2 Mission (39 images) on February 7, 1969, were purchased from the USGS (see Figure 14.4).

Aerial photos. Five aerial photos with film formats of 18 × 18 cm^2 from 1961 were acquired at a photographic scale of 1:14000. Each photo covers approximately 6.35 km^2. Five aerial photos were purchased from the Guangxi Bureau of Geospatial Information, China.

Coordinate data of GCPs. The coordinate data associated with GCPs in the KRD area were collected from Google Earth.

14.3 RESULTS AND ACCURACY ANALYSIS

14.3.1 Image Preprocessing

The DISP film was scanned into digital images, producing film-grain noise and resulting in image quality degradation. Many noise filters have been used in the public domain. However, most of these approaches are either time-consuming because of complex modeling or they erroneously remove geophysical features because of noise in the overall image. The filter algorithm developed by Zhou et al. (2002) was used to remove noise in this study. One of the advantages of the algorithm is that it avoids the problems noted above because this approach performs statistical calculations within variable-size and variable-shape sub-windows (see Figure 14.5) that are determined individually for every pixel in the image, rather than modeling the noise in the overall image. The algorithm is briefly described as follows.

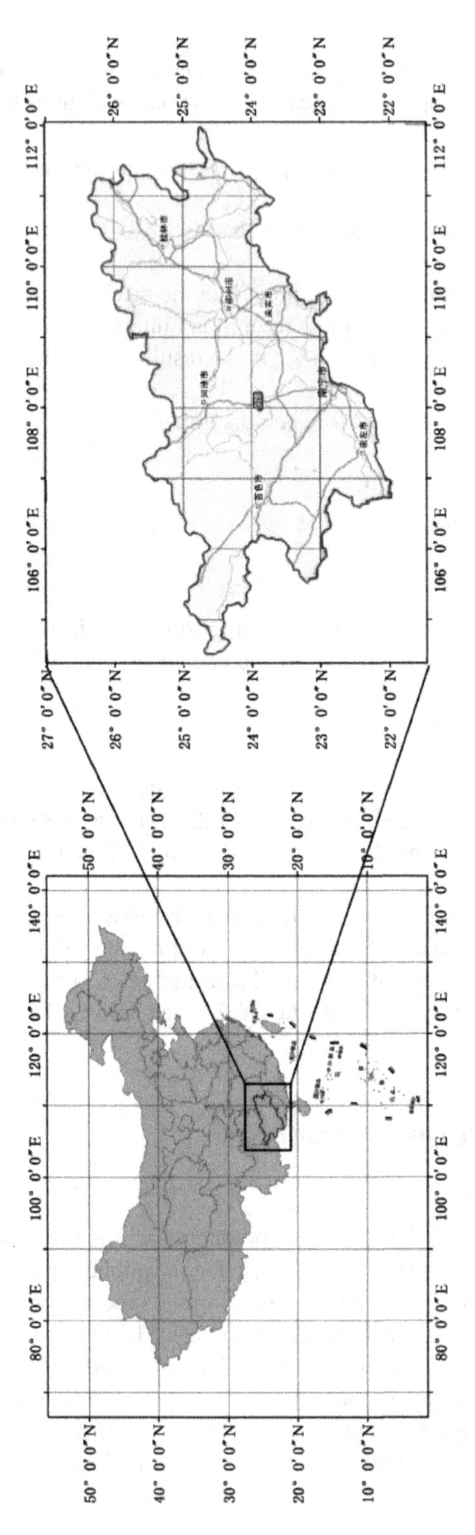

FIGURE 14.3 Study area.

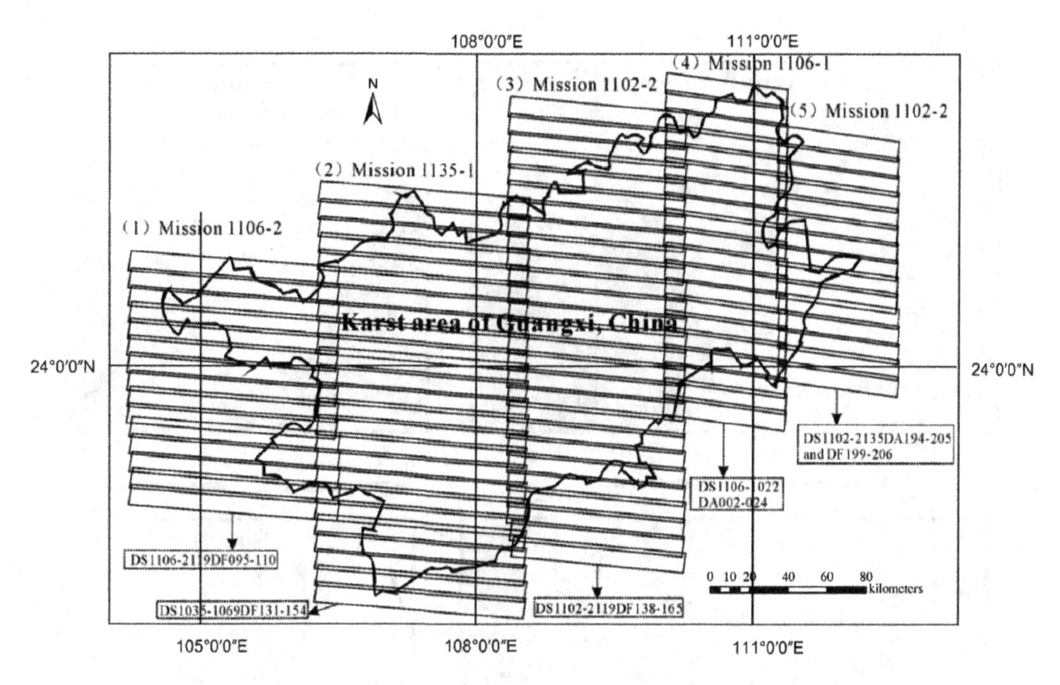

FIGURE 14.4 DISP image data set.

$$\begin{bmatrix} a_{11} & a_{12} & a_{13} & a_{14} & a_{15} \\ a_{21} & a_{22} & a_{23} & a_{24} & a_{25} \\ a_{31} & a_{32} & a_{33} & a_{34} & a_{35} \\ a_{41} & a_{42} & a_{43} & a_{44} & a_{45} \\ a_{51} & a_{52} & a_{53} & a_{54} & a_{55} \end{bmatrix} \begin{bmatrix} a_{11} & a_{12} & a_{13} & a_{14} & a_{15} \\ a_{21} & a_{22} & a_{23} & a_{24} & a_{25} \\ a_{31} & a_{32} & a_{33} & a_{34} & a_{35} \\ a_{41} & a_{42} & a_{43} & a_{44} & a_{45} \\ a_{51} & a_{52} & a_{53} & a_{54} & a_{55} \end{bmatrix} \begin{bmatrix} a_{11} & a_{12} & a_{13} & a_{14} & a_{15} \\ a_{21} & a_{22} & a_{23} & a_{24} & a_{25} \\ a_{31} & a_{32} & a_{33} & a_{34} & a_{35} \\ a_{41} & a_{42} & a_{43} & a_{44} & a_{45} \\ a_{51} & a_{52} & a_{53} & a_{54} & a_{55} \end{bmatrix} \begin{bmatrix} a_{11} & a_{12} & a_{13} & a_{14} & a_{15} \\ a_{21} & a_{22} & a_{23} & a_{24} & a_{25} \\ a_{31} & a_{32} & a_{33} & a_{34} & a_{35} \\ a_{41} & a_{42} & a_{43} & a_{44} & a_{45} \\ a_{51} & a_{52} & a_{53} & a_{54} & a_{55} \end{bmatrix} \begin{bmatrix} a_{11} & a_{12} & a_{13} & a_{14} & a_{15} \\ a_{21} & a_{22} & a_{23} & a_{24} & a_{25} \\ a_{31} & a_{32} & a_{33} & a_{34} & a_{35} \\ a_{41} & a_{42} & a_{43} & a_{44} & a_{45} \\ a_{51} & a_{52} & a_{53} & a_{54} & a_{55} \end{bmatrix}$$

$$\begin{bmatrix} a_{11} & a_{12} & a_{13} & a_{14} & a_{15} \\ a_{21} & a_{22} & a_{23} & a_{24} & a_{25} \\ a_{31} & a_{32} & a_{33} & a_{34} & a_{35} \\ a_{41} & a_{42} & a_{43} & a_{44} & a_{45} \\ a_{51} & a_{52} & a_{53} & a_{54} & a_{55} \end{bmatrix} \begin{bmatrix} a_{11} & a_{12} & a_{13} & a_{14} & a_{15} \\ a_{21} & a_{22} & a_{23} & a_{24} & a_{25} \\ a_{31} & a_{32} & a_{33} & a_{34} & a_{35} \\ a_{41} & a_{42} & a_{43} & a_{44} & a_{45} \\ a_{51} & a_{52} & a_{53} & a_{54} & a_{55} \end{bmatrix} \begin{bmatrix} a_{11} & a_{12} & a_{13} & a_{14} & a_{15} \\ a_{21} & a_{22} & a_{23} & a_{24} & a_{25} \\ a_{31} & a_{32} & a_{33} & a_{34} & a_{35} \\ a_{41} & a_{42} & a_{43} & a_{44} & a_{45} \\ a_{51} & a_{52} & a_{53} & a_{54} & a_{55} \end{bmatrix} \begin{bmatrix} a_{11} & a_{12} & a_{13} & a_{14} & a_{15} \\ a_{21} & a_{22} & a_{23} & a_{24} & a_{25} \\ a_{31} & a_{32} & a_{33} & a_{34} & a_{35} \\ a_{41} & a_{42} & a_{43} & a_{44} & a_{45} \\ a_{51} & a_{52} & a_{53} & a_{54} & a_{55} \end{bmatrix}$$

9 masks

FIGURE 14.5 The adaptive filter algorithm.

1. Select a window of 5 × 5 pixels.
2. Calculate the mean $n_i(i = 1,2\ldots9)$ and variance $\alpha_i(i = 1,2\ldots9)$ of nine masks.
3. Select one mask with the lowest variance α_k and mean n_i, and calculate the weights of every pixel within the kth mask using the following equation.

$$\omega_i = e^{|\Delta_i|},\ \Delta_i = gray_i - n_k \tag{14.24}$$

4. Calculate the output using Equation 14.25:

$$gray_{output} = \left(\sum_{i=1}^{M} W_i \cdot gray_i\right) / \left(\sum_{i=1}^{M} \omega_i\right) \tag{14.25}$$

where M is the number of pixels in the kth mask and $gray_i(i = 1,2\ldots9)$ is the intensity.

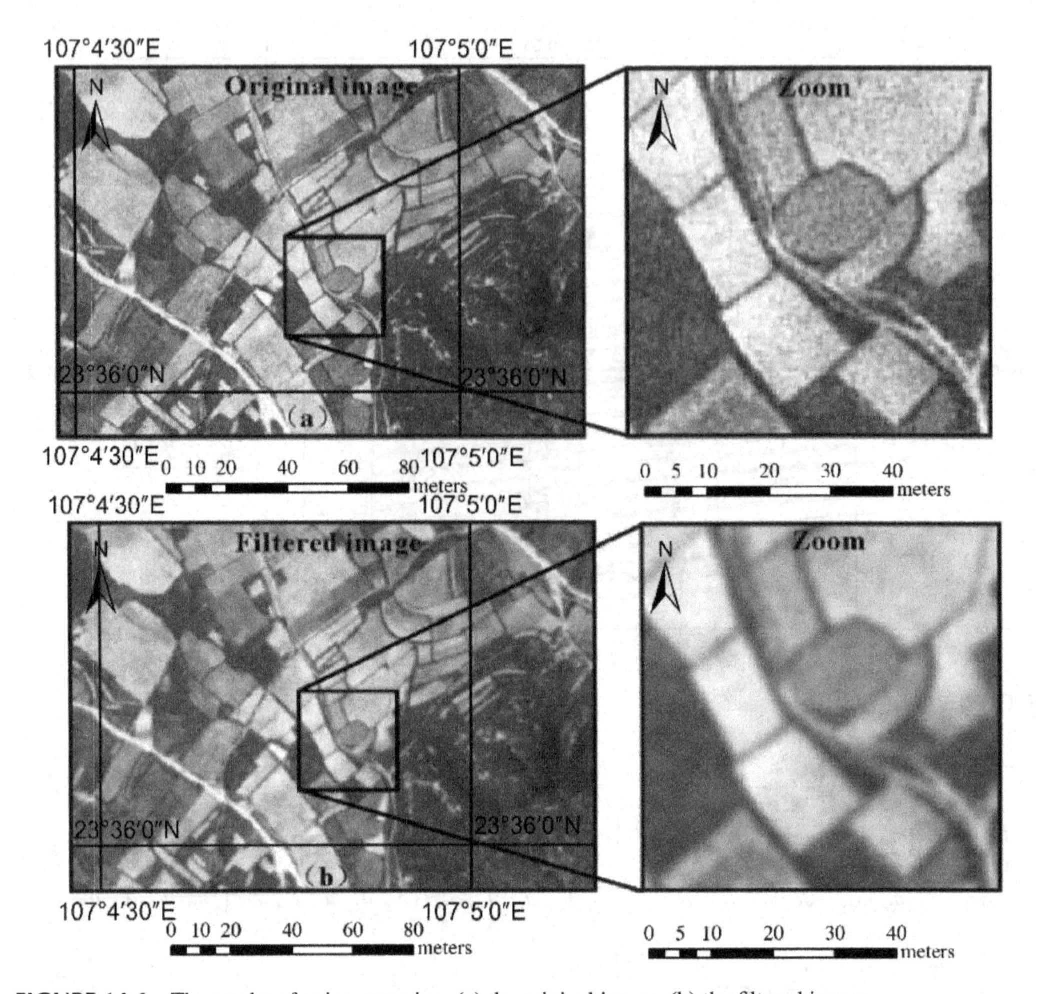

FIGURE 14.6 The results of noise removing: (a) the original image, (b) the filtered image.

With the filter algorithm above, the results of removing the DISP image noise are depicted in Figure 14.6, which demonstrates the effectiveness of the proposed approach.

14.3.2 DISP Image Orthorectification and Accuracy Analysis

14.3.2.1 DISP Image Orthorectification

Because sufficient numbers of GCPs are not observed in each DISP image, TPs are identified to tie overlapping images together and solve for the coefficients of the 2OPE-RM. The study area consists of 355 DISP images (there are 444 DISP images in total, but only 355 high-quality images are employed). Thus, it is impractical to construct a block in the entire study area and then solve for the orthorectification parameters of all DISP images simultaneously because such a huge block will produce a significantly large number of observation equations, resulting in a huge computational burden during matrix inversion. Therefore, this chapter divides the study area into 24 blocks consisting of various DISP images (see Figure 14.7a). Each block was rectified independently. For example, Block 1 consists of 9 images in Figure 14.7b, in which 20 GCPs and 29 TPs were identified and measured. The 20 GCPs are employed 36 times to establish 72 observation equations. The 29 TPs are employed 60 times to establish 120 observation equations. Thus, 192 observation equations (72 + 120 = 192) are established in Block 1. There are 166 unknown

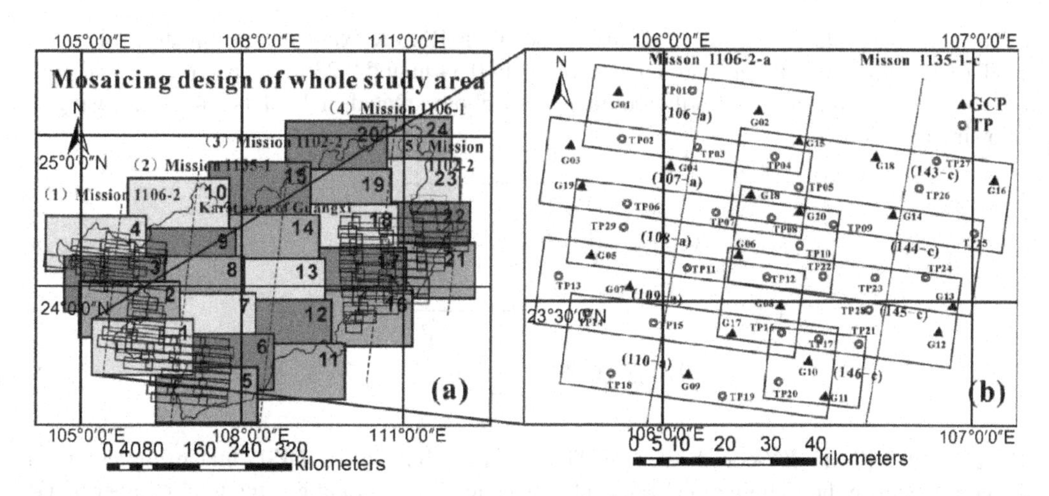

FIGURE 14.7 The polynomial block adjustment: (a) design of polynomial block adjustment in the entire area, which is divided into 24 blocks, and (b) one block, which is used to explain establishment of the observation equations.

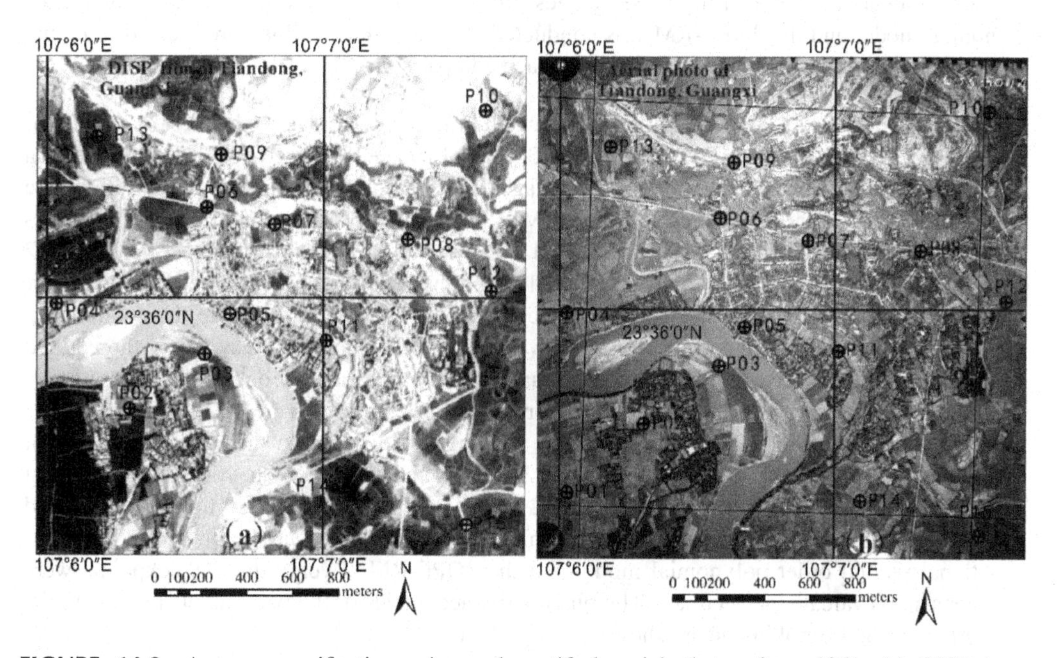

FIGURE 14.8 Accuracy verification using orthorectified aerial photos from 1961: (a) DISP image orthorectified using the proposed method, and (b) orthorectified aerial photo from 1961.

parameters (9 × 12 + 29 × 2 = 166). There are 26 redundant observations (i.e., 192 – 166 = 26), meeting the requirements of least-squares adjustment.

With the 192 observation equations established using Equation 14.11, the parameters used to rectify the 9 DISP images are solved simultaneously using Equation 14.14. The 2D coordinates of TPs are obtained using Equation 14.17. With the solved coefficients and TP coordinates in each image, orthorectification is performed for each DISP image at a GSD of 2.0 m. Figure 14.8a is part of one orthorectified DISP image.

The computational accuracies of TPs using the 2OPE-RM are evaluated by Equation 14.20. The standard deviations of TPs (μ_X and μ_Y) are averagely 0.34 m and 0.23 m, respectively. In addition, the "absolute" accuracy of the orthorectified aerial photo created in 1961 is calculated using the following equations:

$$\Delta X_{RMSE} = \sqrt{\frac{\sum_{k=1}^{n} \left(X_k - x_k\right)^2}{n-1}} \tag{14.26}$$

$$\Delta Y_{RMSE} = \sqrt{\frac{\sum_{k=1}^{n} \left(Y_k - y_k\right)^2}{n-1}} \tag{14.27}$$

where X_k and Y_k are XY-coordinates of TPs in the orthorectified DISP image, x_k and y_k are XY-coordinates in the orthorectified aerial photo created in 1961, and n is the total number of TPs. Using Equation 14.26 and Equation 14.27, ΔX_{RMSE} and ΔY_{RMSE} are 2.0 m and 1.6 m, respectively. These values are equivalent to approximately 2.0 pixels in the orthorectified DISP imagery.

14.3.2.2 Accuracy Comparison Analysis

Accuracy comparison between the DISP images orthorectified using the traditional second order polynomial model and the 2OPE-RM was conducted. Two test fields, which are located in mountainous and flat areas, were selected for the accuracy comparison.

1. The Bameng field is a mountainous area located in Bameng County to the west of the city of Baise, Guangxi, China, at 23.671°N to 24.135°N and 106.941°W to 107.698°W. This test area covers the entire DS1106-2119DF107a image. The maximum and minimum elevations are 1128 m and 790 m above mean sea level (MSL), respectively. Therefore, the relief displacement is significant. There are 12 GCPs and 7 TPs scattered throughout the test field. The 12 GCPs are used for second order polynomial equations to solve for the 12 rectification coefficients, and the 12 GCPs and 7 TPs are used in the 2OPE-RM to calculate the coefficients. Twenty-three checkpoints were chosen to evaluate the achievable accuracy. The orthorectified aerial photo provided by the Bureau of Guangxi Geomatics and Geographic Information is considered to represent the "true" values for validation. The results are listed in Table 14.1.
2. The Longzhou field is a flat area located in Longzhou County to the west of the city of Chongzuo, Guangxi, China, at 22.105°N to 22.469°N and 106.593°W to 106.878°W. This test field completely covers the entire DS1106-2119DF110a image. In this test field, 11 GCPs and 7 TPs are scattered throughout the DISP image. The same GCPs are employed in the traditional second order polynomial model and the 2OPE-RM. Twenty-three checkpoints were chosen to evaluate the accuracy. The planimetric accuracies of the two models relative to the orthorectified aerial image are shown in Table 14.1.

TABLE 14.1
Accuracy Comparison of Mountainous and Flat Areas.

Test area Models	Bameng Field (Located in a Mountainous Area)		Longzhou Field (Located in a Flat Area)	
	ΔX_{RMSE} (m)	ΔY_{RMSE} (m)	ΔX_{RMSE} (m)	ΔY_{RMSE} (m)
Second order polynomial model	1.96	1.84	1.85	1.57
Our model	1.85	1.69	1.67	1.49

14.3.3 Image Mosaicking

Based on the individual image orthorectification above, the next task is to mosaic the individual orthorectified DISP images into an image map. First, the characteristics of the study area and DISP images must be understood.

1. The study area covers 23,790.8 km^2 (between latitudes 20.54° and 26.24°N and longitudes 104.26° and 112.04°E), which consists of 355 DISP images that total 100 GB. A good mosaicking scheme may save computational time and computer storage.
2. The study area is located in a karst landscape, where mountainous and hilly terrain areas account for two-thirds of the total area.
3. The overlap between neighboring images must be less than 30%.
4. The study area is covered by five strips of DISP images from four missions (see Figure 14.9).

To minimize the influence of error propagation and avoid repeatedly sampling images, based on the characteristics above, the mosaicking is designed as follows (see Figure 14.9).

1. The 16 DISP images from Mission 1106 were first mosaicked, covering the western portion of the study area. The mosaicked map is depicted in Figure 14.10a. Twenty DISP images from Mission 1102-2 were mosaicked, and the mosaicked map is depicted in Figure 14.10b. Twenty-eight DISP images from Mission 1106 were mosaicked, and the mosaicked map is depicted in Figure 14.10c. Twenty-three DISP images from Mission 1106 were mosaicked, as the mosaicked map is depicted in Figure 14.10d. Finally, 18 DISP images from Mission 1106 were mosaicked, and the mosaicked map is depicted in Figure 14.10d, covering the eastern portion of the area.
2. With the five mosaicked maps above, a map image of the entire study area was assembled by merging the five mosaicked images. The order of mosaicking is from the east and west to the middle of the study area (see Figure 14.10).

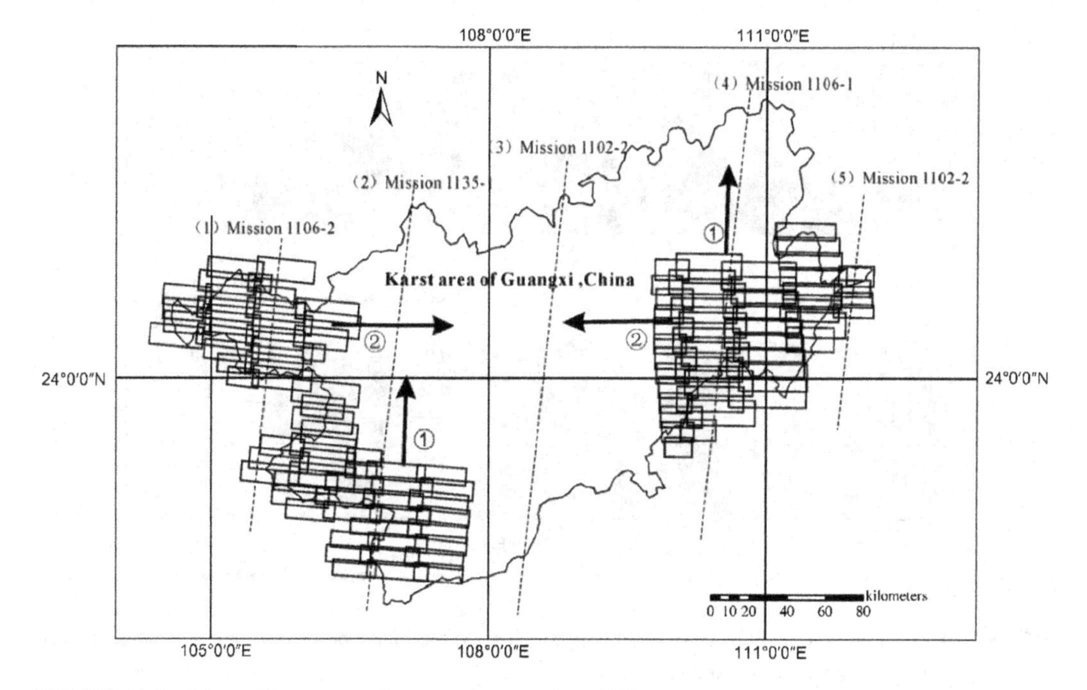

FIGURE 14.9 Mosaicking process from east/west to the middle.

14.3.4 Radiance Balance

Due to the differences in the imaging date/time and different imaging conditions during different missions, brightness differences between neighboring strips are unavoidable. In addition, patchwork lines are also unavoidable. To produce a seamless mosaic of the entire study area, this chapter used a histogram equalization method to adjust the brightness of two neighboring strips. The boundary line was chosen along the center image, and overlapping areas were feathered. Figure 14.11 shows the result of the radiance balance.

14.3.5 Mosaicking Result and Accuracy Evaluation

The entire study area has been mosaicked by 355 orthorectified DISP images (see Figure 14.12a). A mountainous area located in Du'an County (see Figure 14.12b) and a flat area located in Xingbin County (see Figure 14.12c) are selected as the samples for accurate validation. Seventy-eight GCPs,

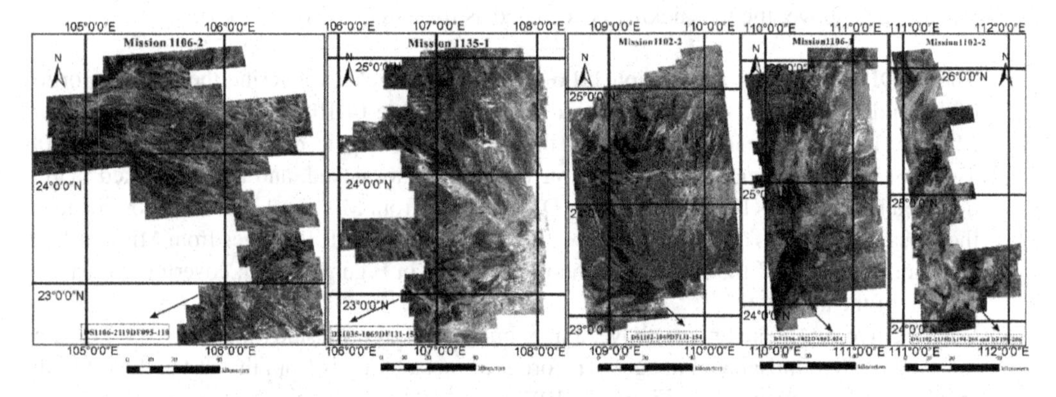

FIGURE 14.10 The mosaicked images of the various missions.

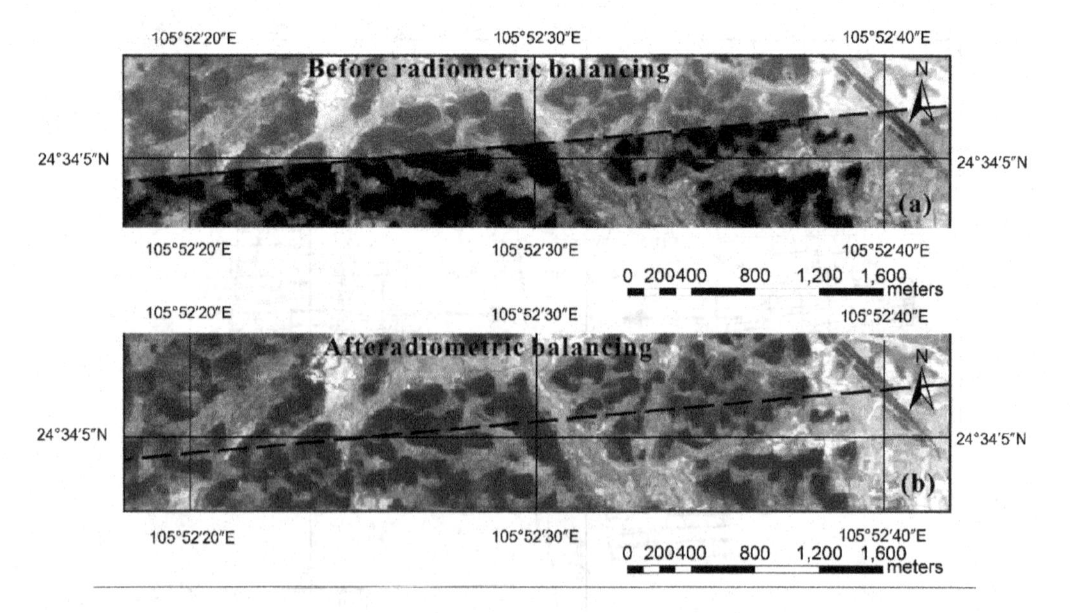

FIGURE 14.11 Radiometric balance of neighboring strips: (a) the image before radiometric balancing, and (b) the image after radiometric balancing.

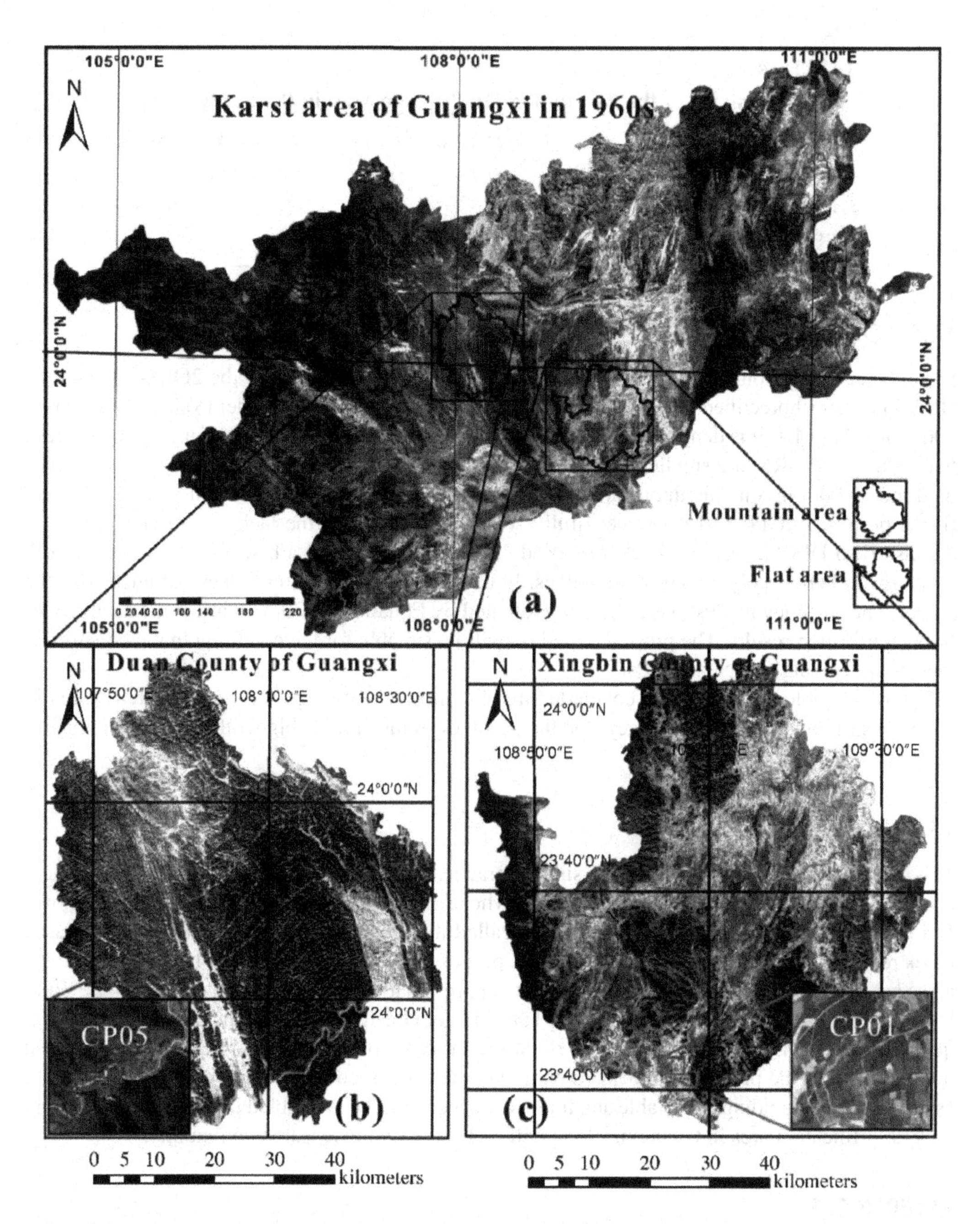

FIGURE 14.12 Mosaic results: (a) a completely assembled 1960s mosaic of the Guangxi karst area, (b) accuracy validation in a mountainous area, and (c) accuracy validation in a flat area.

which were measured by RTK GPS measurements, are uniformly distributed in other countries throughout the entire study area. These include 25 checkpoints (CPs) scattered throughout the two test fields. The equations ΔX_{RMSE} and ΔY_{RMSE} in Equations 14.26 and 14.27 are used to measure the accuracy. The results are listed in Table 14.2. As shown in Table 14.2, the accuracy in flat areas is better than that in mountainous areas, and the overall accuracies of the entire study area are 2.11 m and 1.74 m.

TABLE 14.2
Final Accuracies of the Assembled DISP Image Map in the Study Area.

Area	ΔX_{RMSE} (m)	ΔY_{RMSE} (m)
Mountainous area	2.07	1.60
Flat area	1.86	1.79
The entire study area	2.11	1.74

14.4 DISCUSSIONS

From the accuracy comparison between the DISP images orthorectified by the 2OPE-RM and the DISP images orthorectified by the traditional second order polynomial model (Valadan Zeoj et al. 2002) in Table 14.1, it is demonstrated that the RMSEs of XY-coordinates in the images orthorectified by the 2OPE-RM are smaller than those orthorectified by the traditional second order polynomial model both in a mountainous area and a flat area. This experiment result demonstrated that: (i) the proposed 2OPE-RM can successfully solve the problem of the lack of sufficient GCPs in the historical DISP imagery; (ii) the proposed 2OPE-RM can obtain a better result than the traditional second order polynomial method does. In the 2OPE-RM, TPs are the key parameters of the model. The accuracy of TPs, which can be evaluated by Equation 14.20, will affect the accuracy of orthorectification results. The proposed method will be suitable for the condition that the historical images are lacking sufficient GCPs.

From the Table 14.2, it can be concluded that the mosaicked images of the entire KRD area in Guangxi still remain high in accuracy, and the accuracy in flat areas is higher than that in mountainous areas.

14.5 CONCLUSIONS

This chapter presents a highly effective, simple, practical mathematical model for the orthorectification of CORONA DISP images from the 1960s, whose interior and exterior parameters are unknown and in which GCPs are lacking. The model is called the second order polynomial equation-based block rectification model (2OPE-RM). With the proposed model, all images can be orthorectified at an accuracy level of 2.0 pixels, corresponding to approximately 2.0–4.0 m with respect to the WGS 84 datum. All the images covering the entire karst area of Guangxi, China are assembled into a high-quality image map. The sampled distance of the assembled mosaicking map is 2.0 m. The proposed model can solve the problems associated with the traditional second order polynomial model, such as lack of GCPs, yielding acceptable and improved accuracy. The assembled image map of the entire rock desertification area in Guangxi, China will be for use by the research community.

REFERENCES

Grodecki, J., and Dial, G., Block adjustment of high-resolution satellite images described by rational polynomials, *Photogrammetric Engineering & Remote Sensing*, 2003, 69(1), pp. 59–68.

Huang, M., et al. Block adjustment with airborne SAR images based on polynomial ortho-rectification, *Geomatics and Information Science of Wuhan University*, 2008, 33(6), pp. 569–572.

Jiang, Z. C., and Qin, X. Q., Calculation of atmospheric CO_2 sink formed in karst progresses of the karst divided regions in China, *Geoscientia Sinica*, 2011, 30(4), pp. 363–367.

Kim, K., Jezek, K. C., and Liu, H., Orthorectified image mosaic of Antarctica from 1963 Argon satellite photography: image processing and glaciological applications. *International Journal of Remote Sensing*, 2007, 28(23), pp. 5357–5373.

Kim, K., Jezek, K.C., and Sohn, H., Ice shelf advance and retreat rates along the coast of Queen Maud Land, Antarctica, *Journal of Geophysical Research*, 2001, 106(C4), pp. 7097–7106.

Liu, J., Wang, D. H., and Mao, G. M., High accuracy stereo positioning of IKONOS satellite image based on RPC model, *Bulletin of Surveying and Mapping*, 2004, 0911(9), pp. 1–4.

Tao, C. V., and Hu, Y., A comprehensive study of the rational function model for photogrammetric processing, *Photogrammetric Engineering & Remote Sensing*, 2001, 67(12), pp. 1347–1357.

Tao, C. V., and Hu, Y., 3D reconstruction methods on the rational function model, *Photogrammetric Engineering& Remote Sensing*, 2002, 68(7), pp. 705–714.

Valadan Zoej, M., et al. 2D geometric correction of IKONOS imagery using genetic algorithm, *International Archives of Photogrammetry and Remote Sensing and Spatial Information Sciences*, Ottawa, Canada, 9–12, July, 2002.

Xu, S., and Jiang, Z., Preliminary estimate the source and sink relation of karstification and atmospheric greenhouse gas CO_2, *Chinese Science Bulletin*, 1997, 42, pp. 953–956.

Yang, X.H., Accuracy of rational function approximation in Photogrammetry, *Proceeding of ASPRS annual convention XXXIII (B3)*, Amsterdam, Netherlands, 16–22, July, 2000.

Zhou, F., et al., Second order polynomial equation-based block adjustment for orthorectification of DISP imagery. *Remote Sensing*, 2016, 8, p. 680.

Zhou, G., Huang, J., Tao, X., Luo, Q., Zhang, R., and Liu, Z., Overview of 30 years of research on solubility trapping in Chinese karst, *Earth-Science Reviews*, 2015, 146, pp. 183–194.

Zhou, G., and Jezek, K. C., Satellite photograph mosaics of Greenland from the 1960s era, *International Journal of Remote Sensing*, 2002, 23(6), pp. 1143–1159.

Zhou, G., Jezek, K.C., Wright, W., and Granger, J., Orthorectification of 1960s satellite photographs covering Greenland. *IEEE Transactions on Geoscience and Remote Sensing*, 2002, 40(6), pp. 1247–1259.

15 Building Occlusion Detection in an Urban True Orthophotomap

15.1 INTRODUCTION

In the previous chapter, the basic principles of high-resolution digital orthophotomap (DOM) and true orthophoto map (TOM) generation have been described. Briefly, the basic steps of high-resolution true orthophotomap generation in urban areas include both digital terrain model (DTM)-based orthophotomap generation and digital building model (DBM)-based orthophotomap generation as well as their merging (Zhou et al. 2005).

The traditional orthorectification methods for high-resolution remotely sensed images in urban areas suffer from two major problems.

1. *Ghost images.* Traditional methods apply a so-called "digital differential rectification model" to rectify the building's relief displacement on the basis of a digital surface model (DSM). Under this condition, a tall building can be rectified to its orthogonal position using a DSM, that is, a tall building in a DSM can find the corresponding gray pixels in an image plane. However, some surfaces (e.g., ground surfaces, low buildings) that are occluded by other tall buildings cannot effectively obtain their corresponding positions in the image plane, that is, only those visible building surfaces are orthorectified to their orthogonal positions, whereas the occluded surfaces still keep their original positions (pixels). As a result, the building's roof is copied, resulting in the same building's roof appearing twice in the orthorectified image plane. This phenomenon is called a "ghost image" (Skarlatos 1999; Zhou et al. 2002, 2005; Zhong et al. 2011) (see Figure 15.1(iii)).
2. *Occlusion detection.* One of the most important steps in the process of urban high-resolution true orthophotomap generation is detection of building occlusion. To effectively detect occlusion, a widely used method, called the Z-buffer algorithm, calculates the distance between the DSM surfaces and the projection center. This distance is called Z-distance, and the method is thereby called the Z-buffer algorithm (Amhar et al. 1998). In this algorithm, the DSM resolution (also known as the cell size) is theoretically required to be equal to the imaging ground sampling distance (GSD). This condition is very difficult to meet because the terrestrial surface undulates and because the central projection image contains various types of deformations. For example, as shown in Figure 15.1, there is a terrestrial profile, noted as $\overset{\frown}{AB}$, that is denoted a–b in the image plane. When the gradient of $\overset{\frown}{AB}$ is close to a projective ray ($\overrightarrow{SA}$) (see Figure 15.1), more than one ground cell in the DSM corresponds to one pixel in the image plane. For instance, the $\overset{\frown}{AB}$ has 12 ground cells in the DSM, but its imaged area a–b in the image plane contains only 3 pixels. In accordance with the principle of the Z-buffer method, one DSM cell corresponds to only one pixel. In other words, nine ground cells in $\overset{\frown}{AB}$ have no corresponding pixels, but they are still treated as an occlusion area. In fact, the surface of $\overset{\frown}{AB}$ is visible. Consequently, nine visible ground cells cannot be assigned to gray in orthorectification, resulting in false occlusion (see Figure 15.1).

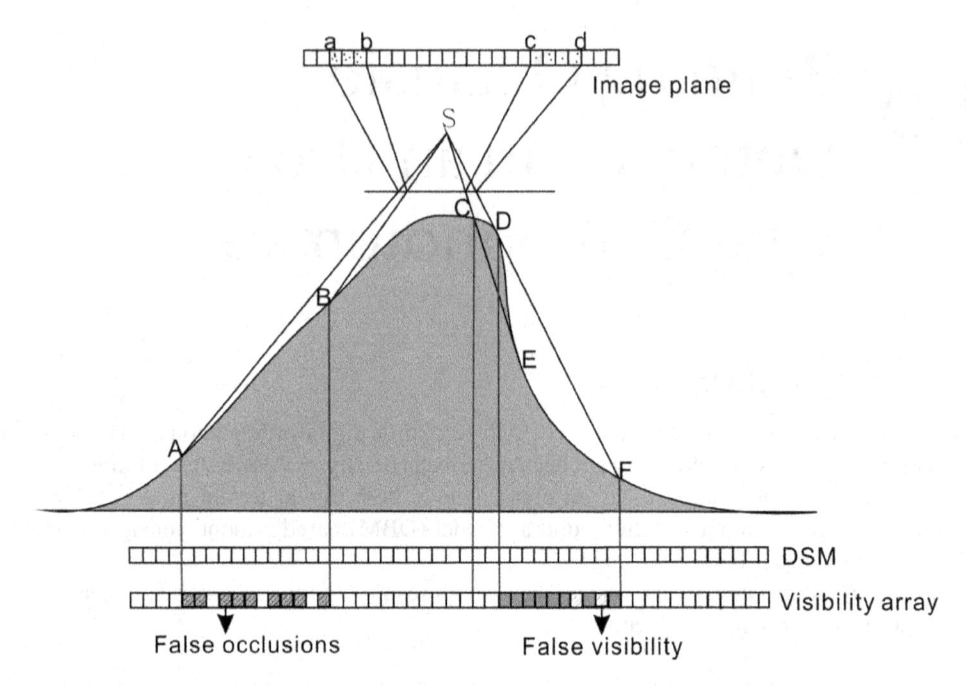

FIGURE 15.1 False occlusion and false visibility in the Z-buffer method.

Additionally, as shown in Figure 15.1, when the gradient of a terrestrial profile, noted $\overset{\frown}{DEF}$, is greater than that of a projection ray $\overrightarrow{SF}$, $\overset{\frown}{DEF}$ is occluded by the profile $\overset{\frown}{CD}$, which is imaged as c–d in the image plane. As can be observed in Figure 15.1, $\overset{\frown}{DEF}$ occupies 10 ground cells and $\overset{\frown}{CD}$ occupies 2 ground cells in the DSM, whereas the imaging area, c–d, occupies 4 pixels in the image plane. In accordance with the principle of the Z-buffer method, $\overset{\frown}{CD}$ should correspond to two pixels, that is, c–d. In other words, the remaining two pixels of gray need to be distributed through $\overset{\frown}{DEF}$. As a result, two ground cells in $\overset{\frown}{DEF}$ are wrongly detected as visible area, that is, false visibility occurs (see Figure 15.1) (Sheng 2007; Habib et al. 2007).

Considerable research has been performed to characterize false occlusion detection and false visibility problems. Developed methods can be cataloged into:

1. *The Z-buffer and improved methods.* For example, Rau (2002) introduced a method to determine the number of pseudo pixels, in which the building height difference and the projection length are estimated locally and the detection rules are applied again. However, it can be time-consuming if many pseudo points exist in the image. Bang (2007) presented the sorted DSM method. The DSM is sorted according to height values, and only visible points are assigned pixel values; consequently, computation speed is improved. Zhou et al. (2016) improved the Z-buffer algorithm by establishing the minimum bounding sector (MBS). The MBS can help to decrease the searching area of the Z-buffer. Although all of these Z-buffer methods improve detection accuracy relative to the traditional Z-buffer method, they do not completely eliminate the misdetection.
2. *Other methods.* In addition to the Z-buffer methods, Sheng (2004) analyzed the principles of ray-tracing (RT), iterative photogrammetric (IP), and iterative ray-tracing (IRT) and presented a comprehensive comparison report on their steps, parameter selection, divergence, occlusion-compliance, precision, robustness, and efficiency through tests on a variety of data sets. Wang (2008) concluded that these methods suffer from limitations that their iterations do not converge during the calculation process under the complicated topography. Nielsen

(2004) proposed an improved method by constructing a triangulated irregular network (TIN), which consists of a series of height measurements from the DSM, and then projecting the TIN onto the image, finally testing the intersections between the straight line formed by linking the camera center to the occluded point and the triangulation model of the DSM. Kuzmin et al. (2004) used a polygon-based algorithm to detect the occluded area. This method combined various available elevation data into a common polygonal surface and then projected the polygon onto an image plane to determine whether the surface polygons overlapped, and finally judged the distance of the overlapped region and the camera center to obtain the occluded area. The experimental results demonstrate that the occluded area can be effectively detected with a high accuracy. Habib (2007) proposed a method called the angle-based RT method on the basis of the off-nadir angles to the lines of sight connecting the perspective center to the DSM points along a radial direction starting from the object space nadir point to minimize the errors of false visibility and false occlusions. In this method, the projection angles of measured points change along a radial direction while moving away from the nadir point. This method establishes a search path between ground points and the camera center and uses both spiral sweep and adaptive radial sweep methods to traverse the entire image region. Sheng (2007) proposed the vector domain-based method. This method views the image pixel as a square patch and uses an RT method to calculate the location of square vertexes in the DSM. The new location of square vertexes forms a new polygon. The polygonal region that overlaps is identified as the occlusion area. This approach mitigates false occlusion and false visibility. However, misconvergence can occur in RT methods, and the process of image resampling is limited by round-off errors. Youn and Kim (2008) proposed a method called the height-based RT method, whose basic principle is that a certain object point has been successfully captured by the cameras and that no point along the search path will be higher than the projection ray. Zhong et al. (2011) modified the process of sweeping by analyzing the scan area, number of azimuth lines, and visibility judgment area. The major advantage of this approach is that it is capable of attaining high accuracy and eliminating the misdetection effects of false visibility and false occlusions. However, the method requires calculating the angles and heights of a potential point relative to all other points during the search, which is time-consuming.

Although much progress on occlusion detection and visibility analysis for urban true orthophotomap generation has been made in the past decades, false occlusion and false visibility problems have not been completely solved. For this reason, this chapter presents an innovative method, completely different from the traditional methods above, for occlusion detection from ghost images. The proposed method is expected to be more accurate than the previous ones for the following reasons:

1. The model developed in this chapter determines the sizes and boundaries of an occluded area through the building's roofs and visible facades. In other words, an enclosed polygon constructed by the building's roofs and facades is taken as a boundary of the occluded area over which the boundary of detected occlusion can be correctly determined. With the correct boundary, false occlusion and false visibility can be avoided. Therefore, the method proposed in this manuscript is different from Z-buffer methods (Kuzmin et al. 2004; Wang et al. 2008).
2. The traditional methods detect the occluded area by calculating the distances between DSM and the exposed center pixel by pixel, which requires iteration for solving the intersection point between the ray and DSM surface; the proposed method detects the occluded area on the ghost image without iterative computation. As a result, the proposed method is capable of better accuracy and reduced computational time in detecting occlusion.

15.2 THE PRINCIPLE OF BUILDING OCCLUSION DETECTION FROM GHOST IMAGES

15.2.1 The Relationship Between Building Occlusion and Ghost Images

Occlusion is caused by the fact that only the surface of an object closest to the camera center is imaged in perspective projection imaging. The occluded area's size and shape depend upon building height, camera center position, and the relative relationship between the camera centers and the building's position (see Figure 15.2).

The purpose of orthophotomap creation is to rectify a perspective projection aerial image into an orthogonal projection image through a differential rectification algorithm. The generated orthophotomap is thereby free of occlusion – theoretically. However, when one building is orthorectified on the basis of a DSM, the building roof appears twice in an orthophotomap. This phenomenon is called a ghost image (Zhou et al. 2002, 2005) (see Figure 15.2(iii)). The model of differential rectification is the collinearity equation, that is, a given point, such as A on the building roof in Figure 15.2(i), is expressed by

$$x_a = x_0 - f\frac{a_1(X_A - X_S) + b_1(Y_A - Y_S) + c_1(Z_A - Z_S)}{a_3(X_A - X_S) + b_3(Y_A - Y_S) + c_3(Z_A - Z_S)}$$
$$y_a = y_0 - f\frac{a_2(X_A - X_S) + b_2(Y_A - Y_S) + c_2(Z_A - Z_S)}{a_3(X_A - X_S) + b_3(Y_A - Y_S) + c_3(Z_A - Z_S)} \tag{15.1}$$

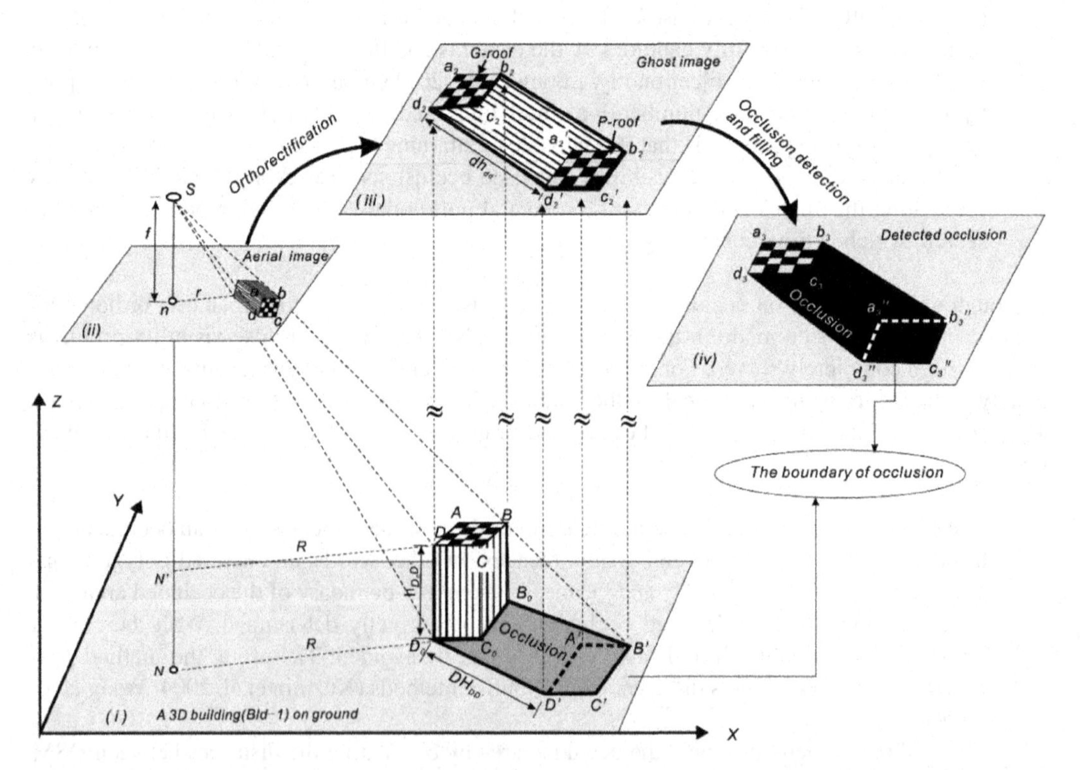

FIGURE 15.2 The relationship of the occluded region in both the image plane and the ghost image: (i) represents the DBM and the occluded area, (ii) represents the projected image on the image plane, (iii) represents the ghost image after orthorectification, and (iv) is the result of occlusion detection, whose boundary corresponds to the boundary of the occluded area in (ii) and (iii).

where x_0, y_0, f are the inner orientation parameters (IOPs) of the camera; (a_i, b_i, c_i; $i = 1, 2, 3$) are the elements of the rotation matrix, which is a function of three rotational angles; X_S, Y_S, Z_S are the exterior orientation parameters (EOPs); (X_A, Y_A, Y_A) are 3D coordinates of point A, and (x_a, y_a) are the imaged 2D coordinates in the image plane.

As shown in Figures 15.2(i) and (iii), S is the exposure center of a camera; f is the focal length of the camera; and N is the ground nadir point. The points $ABCD$ and $A'B'C'D'$ represent the corners of a building roof and the corresponding points in the occluded area, respectively. The points $abcd$ and $a'b'c'd'$ represent their projections in the ghost image, respectively. As shown in Figure 15.2(i), although points $A'(X_{A'}Y_{A'}Z_{A'})$ and $A(X_A, Y_A, Z_A)$ can be simultaneously imaged at the same image point $a(x_a, y_a)$, only point $A(X_A, Y_A, Z_A)$, which is closest to the camera center, is visibly imaged. This means that point A' is occluded. With the same imaging principle, other points (e.g., B, C, D) on the roof of the building are imaged, but B', C', and D' are not. Therefore, the polygon enclosed by $B_0C_0D_0D'C'B'$ in Figure 15.2(i) is an occluded area. This fact demonstrates that the occluded area has an inherent relationship with the ghost image of a building.

On the other hand, as shown in Figure 15.2(i), because the two points A' and A are imaged to the same points $a(a')$, the two points have the same gray value; that is, gray(a') = gray(a). Similarly, gray(b') = gray(b), gray(c') = gray(c), and gray(d') = gray(d). For convenience afterwards, two roofs of a building in the ghost image are named ***P-roof*** *(projected-roof)*, which is constructed by $a'_2b'_2c'_2d'_2$, and ***G-roof*** *(ghost-roof)*, which is constructed by $a_2b_2c_2d_2$, in Figure 15.2(iii). Therefore, building occlusion detection simply becomes the detection of the ghosted region in the ghost image. This means that the building occlusion caused by perspective projection in an original aerial image can be detected from its correspondingly orthorectified ghost image; that is, the occluded area, $B_0C_0D_0D'C'B'$ in Figure 15.2(i), corresponds to an enclosed polygon $b_2c_2d_2b'_2c'_2d'_2$ in the ghost image (see Figure 15.2(iii)). It should be emphasized that such a finding has not previously been applied to true orthophotomap generation.

This basic knowledge can be mathematically described and verified as follows.

1) *Determine the coordinates of point* $A'(X_{A'}, Y_{A'}, Z_{A'})$. Assuming that the imaging process is a pinhole model (i.e., that S, A and A' lie in a straight line in Figure 15.2), it has

$$\tan^{-1}\theta = \sqrt{\frac{(X_s - X_A)^2 + (Y_s - Y_A)^2}{(Z_S - Z_A)^2}} = \sqrt{\frac{(X_A - X_{A'})^2 + (Y_A - Y_{A'})^2}{Z_A^2}} \tag{15.2}$$

The slope of the line SA in the X–Y plane (noted as kSA) is calculated by

$$k_{SA} = \frac{Y_S - Y_A}{X_S - X_A} = \frac{Y_A - Y_{A'}}{X_A - X_{A'}} \tag{15.3}$$

With Equations 15.2 and 15.3, it obtains

$$X_{A'} = \frac{X_A - \tan^{-1}\theta \cdot Z_A}{\sqrt{1 + k_{SA}^2}} \tag{15.4a}$$

$$Y_{A'} = Y_A - k_{SA} \cdot (X_A - X_{A'}) \tag{15.4b}$$

$$Z_{A'} = Z_{ground} \tag{15.4c}$$

where Z_{ground} is the elevation, which can be obtained from the DSM by the corresponding planimetric coordinates of *X–Y*.

2. *Determine the coordinates of points* $a_2(x_{a2}, y_{a2})$, $a'_2(x_{a'_2}, y_{a'_2})$ *onto the ghost image*. For a given ground resolution of $\Delta_{Xsample}$, $\Delta_{Ysample}$ in the *x* and *y* directions, respectively, assume that the planimetric coordinates of any ground control point (GCP) are (X_{GCP}, Y_{GCP}), whose corresponding location in the original image plane is (row_{GCP}, col_{GCP}). The coordinates of the lower left boundary point can then be determined by (Zhou et al. 2002):

$$X_0 = X_{GCP} - \mathrm{col}_{GCP} \cdot \Delta_{Xsample} \tag{15.5a}$$

$$Y_0 = Y_{GCP} - \mathrm{row}_{GCP} \cdot \Delta_{Ysample} \tag{15.5b}$$

Thus, the coordinates of point a_2 and point a'_2 on the ghost image are

$$a_2{:}\begin{cases} x_{a2} = (X_A - X_0) \cdot M \\ y_{a2} = (Y_A - Y_0) \cdot M \end{cases} \tag{15.6a}$$

$$a'_2{:}\begin{cases} x_{a'_2} = (X_{A'} - X_0) \cdot M \\ y_{a'_2} = (Y_{A'} - Y_0) \cdot M \end{cases} \tag{15.6b}$$

where *M* is the scale of the aerial image.

3. *Verify that the gray value of a point in the ghost image is the same as that in the original image.* In differential orthorectification, the gray value of a pixel is sampled and interpolated from the original aerial image pixel by the neighborhood interpolation method, namely,

$$gray_{ghost}(x_{a2}, y_{a2}) = gray_{ghost}(x_{a'2}, y_{a'2}) = Neighbor\ interpolation(gray_{original}(x_a, y_a)) \tag{15.7}$$

where $Neighbor\ interpolation(gray_{original}(x_a, y_a))$ describes the neighbor interpolation method. The other points, *B, C*, and *D*, are the same.

On the basis of descriptions above, it can be concluded that the polygon $B_0C_0D_0D'C'B'$ in Figure 15.2(i) corresponds to the polygon $b_2c_2d_2b'_2c'_2d'_2$ in the ghost image in Figure 15.2(iii); that is, the role of occlusion detection in the aerial image is to detect the polygon $b_2c_2d_2b'_2c'_2d'_2$ in the ghost image. In other words, the occlusion's boundary can ideally be determined from a ghost image through the polygon's characteristics. In this chapter, it is assumed that the DBM is accurately co-registered with an aerial image, since the ghost image generated is based on the correctness of both the DBM and DTM. As a result, 3D coordinates of each corner of a building can correctly be back-projected onto their correspondingly imaged buildings. This strategy for occlusion detection is especially useful for true orthophotomap generation and can greatly simplify processing and increase the reliability of occlusion detection for true orthophotomap generation.

15.2.2 The Methodologies for Two Typical Cases

Building occlusion in an urban region can be divided into two typical cases: buildings occluding the ground and buildings occluding other buildings.

To describe how to employ the method above to carry out occlusion detection, two concepts are first introduced:

- *Maximum angle* (*MAX–AZ*). This is an angle between North and a line projected onto the X–Y plane that is formed by linking a point of the building roof boundary and the image nadir (see Figure 15.4). Because there are many points in a building roof boundary, the largest angle is defined as the *maximum angle* or *MAX–AZ*.
- *Minimum angle* (*MIN–AZ*). The smallest angle is defined as the *minimum angle* or *MIN–AZ*.

15.2.2.1 Buildings Occluding the Ground

In order to describe the methodology for one typical case – buildings occluding the ground – Figure 15.2(i) is taken as an example to describe the details of our methodology.

a) Determining the boundary of the occluded area

The occlusion area caused by a building should be "superimposed" by the same building imaged on the image plane. Thus, detecting an occluded area becomes determining the building boundary. As shown in Figure 15.3, the boundary of the occluded area consists of points $g_1, g_2, g_3, \ldots, g_9$, which are located on the G-roof, and the points $p_1, p_2, p_3, \ldots, p_9$, which are located on the P-roof. However, some of the corners of the roofs are not located on the boundary of the occluded area, such as points p10 and g10 in Figure 15.3. To determine the boundary points of the occluded area correctly, a line connecting the point of MAX–AZ (i.e., A(AZmax) in Figure 15.3) and the point of MIN–AZ (i.e., B (AZmin) in Figure 15.3) is chosen as a reference line. In accordance with the characteristics of a central projection, the projected building should be inclined along the direction of a projection ray. This means that, with the reference line, the corner points away from the exposure center should be chosen as the occluded boundary points in both the G-roof and the P-roof. Therefore, the occlusion detection is proposed in this chapter through determining these points. In order to explain the proposed method in detail, Point A in Figure 15.3 is taken as an example.

(1) Assuming that $\overrightarrow{AN}$ points to North in the ghost image (which has in fact been orthorectified) and $\overrightarrow{AS}$ is a vector from point A pointing to the camera center projected onto x–y plane on the ghost image, and is expressed by $\overrightarrow{AS} = (x_S - x_A, y_S - y_A)$ in Figure 15.4, the angle of the vector $\overrightarrow{AS}$ can be determined by

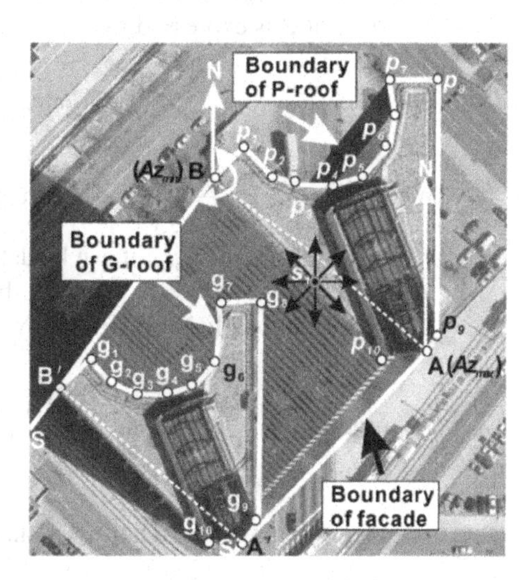

FIGURE 15.3 Determination of maximum and minimum angles and boundary points on both the P-roof and the G-roof points, respectively.

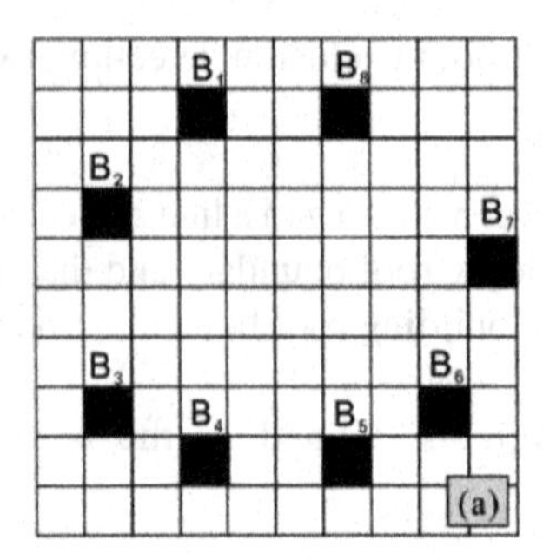

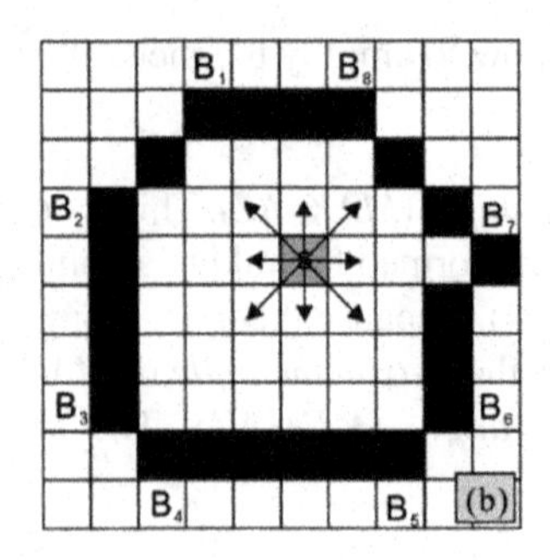

FIGURE 15.4 (a) Connection of two successive points to construct an enclosed boundary, and (b) eight-connection region growing method to determine the pixels inside the boundary.

$$AZ_{\overrightarrow{AS}} = \begin{cases} \arccos\left(\dfrac{\overrightarrow{AN} \times \overrightarrow{AS}}{\left|\overrightarrow{AN}\right| \times \left|\overrightarrow{AS}\right|}\right) & when \quad \Delta x < 0 \\ 2\pi - \arccos\left(\dfrac{\overrightarrow{AN} \times \overrightarrow{AS}}{\left|\overrightarrow{AN}\right| \times \left|\overrightarrow{AS}\right|}\right) & when \quad \Delta x > 0 \end{cases} \tag{15.8}$$

Where $\Delta x = x_s - x_A$. Similarly, for the other points, $g_1, g_2, g_3, \ldots, g_N$ on the G-roof and $p_1, p_2, p_3, \ldots, p_N$ on the P-roof, the angles of each point are calculated accordingly, noted as $\{AZ_{g1}, AZ_{g2}, \ldots, AZ_{gN}\}$ for those corresponding points on the G-roof and $\{AZ_{p1}, AZ_{p2}, \ldots, AZ_{pN}\}$ for those corresponding points on the P-roof. With these angles, the maximum and minimum angles are determined by

$$AZ_{max} = Max\left\{AZ_{p1}, AZ_{p2}, \ldots, AZ_{pN}\right\} \tag{15.9a}$$

$$AZ_{min} = Min\left\{AZ_{p1}, AZ_{p2}, \ldots, AZ_{pN}\right\} \tag{15.9b}$$

2. With the maximum angle, AZ_{max}, which is supposed as Point A on the P-roof, and the minimum angle, AZ_{min}, which is supposed as Point B on the P-roof, the straight line linking Point A and Point B is expressed by

$$y = Slope_{AB} \cdot x + Intercept_{AB} \tag{15.10}$$

Where $Slope_{AB} = \dfrac{x_B - x_A}{y_B - y_A}$ and $Intercept_{AB} = y_A - x_A \cdot \dfrac{x_B - x_A}{y_B - y_A}$.

With the AZ_{max}, and AZ_{min}, it can be determined whether the other points $g_1, g_2, g_3, \ldots, g_N$ on the G-roof and $p_1, p_2, p_3, \ldots, p_N$ on the P-roof are located on the boundary of the occluded area. For example, for Point p1 (x_{p1}, y_{p1}), it has

$$y = Slope_{AB} \cdot x_{p1} + Intercept_{AB} \measuredangle \begin{cases} y_{p1} < y & when\ Point\ p_1\ is\ on\ boundary \\ y_{p1} \geq y & when\ Point\ p_1\ is\ NOT\ on\ boundary \end{cases} \tag{15.11}$$

3. The other points can be determined accordingly.

b) Detecting the pixels inside the boundary of the occluded area

After determining the occluded boundary, which consists of $B_1, B_2, B_3, \ldots, B_N$ in Figure 15.4, the next step is to detect the pixels inside the boundary of the occluded area. To this end, the steps below are suggested:

1. Connect each successive boundary point (e.g., B_1, B_2), through which an enclosed boundary is constructed (see Figure 15.4).
2. Select a seed point by calculating the center coordinates of the occluded area by

$$\left(x_{Bc} = \frac{\sum_{i=1}^{N} x_{B_i}}{N}, \; y_{B_c} = \frac{\sum_{i=1}^{N} y_{B_i}}{N} \right) \tag{15.12}$$

3. Employ the eight-connection region growing method (Shariat et al. 2008) for detection of pixels inside the boundary of the occluded area (that is, the occluded area). The brightness of the detected pixels located within the occlusion is set as zero, *that is*, $G_{occlusion}$ ($I_{occlusion}$, $J_{occlusion}$) = 0.

15.2.2.2 Building Occluding Other Buildings

As shown in Figure 15.5, there are two buildings, labeled Bld-1 and Bld-2. Bld-1 is closer to the ground nadir, N, than is Bld-2, and Bld-1 occludes part of Bld-2. In this case, Bld-1 should first be handled, and then Bld-2. The steps are provided below.

a) Extraction of Bld-1's G-roof

The seed region growing (SRG) algorithm (Adams and Bischof 1994) is used to extract the G-roofs, and the steps are similar to the description in Section 15.2.1. (b). A slight modification is made below:

1. With the steps described in Section 15.2.1. (b), and Equation 15., an enclosed polygon on the G-roof is obtained.
2. Select a seed point by calculating the center coordinates of the G-roof by

$$\left(x_{g_c} = \frac{\sum_{i=1}^{N} x_{g_i}}{N}, \; y_{g_c} = \frac{\sum_{i=1}^{N} y_{g_i}}{N} \right) \tag{15.13}$$

   where x_{g_i}, y_{g_i} are the coordinates of Bld-1 G-roof's boundary points and N is the number of points.
3. Obtain all point coordinates of Bld-1 G-roofs by the eight-connection region growing method. The coordinates of extracted pixels within the Bld-1's G-roof are noted as $EXR\left(x_{G-roof_i}, y_{G-roof_i}\right)$ where i =1, 2, ..., n.
4. Extract the gray value from the ghost image by

$$gray\left(EXR\left(x_{G-roof-i}, y_{G-roof-i}\right)\right) = gray\left(Image\left(x_{G-roof-i}, y_{G-roof-i}\right)\right) where\ i = 1,2,3,\ldots,n \tag{15.14}$$

   Equation (15.) implies that the gray of each pixel on the building's G-roof image is assigned to the corresponding one in the extracted area; that is, $EXR\left(X_{G-roof-i}, Y_{G-roof-i}\right)$.

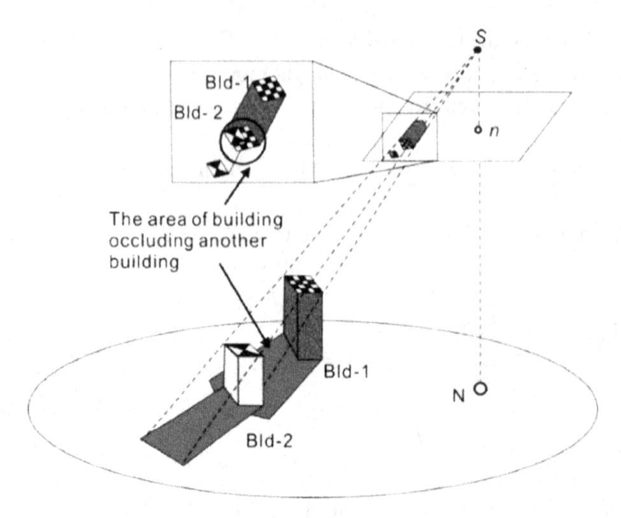

FIGURE 15.5 Example of building (Bld-1) occluding another building (Bld-2).

b) Determining the boundary of the occluded area

Using the steps above, the G-roof of the Bld-1can be extracted from the ghost image. To further detect the boundary of the occluded area of the building occluding other buildings, it can be considered that the Bld-1's G-roof and the area occluded by Bld-1 are a new large area and the boundary of this new large area can be detected through:

1. The boundary of the occluded area on the G-roof. In this case, it has

$$y = Slope_{A'B'} \cdot x_{g1} + Intercept_{A'B'} \nless \begin{cases} y_{g_1} \geq y & \text{when Point } g_1 \text{ is on boundary} \\ y_{g_1} < y & \text{when Point } g_1 \text{ is NOT on boundary} \end{cases} \quad (15.15)$$

Equation 15. indicates that when the line $A'B'$ is taken as a reference, the other points closer to the camera center on the G-roof are considered to lie on the boundary of the occluded area.

2. The boundary of the occluded area on the P-roof. In this case, it has

$$y = Slope_{AB} \cdot x_{p_1} + Intercept_{AB} \nless \begin{cases} y_{p_1} < y & \text{when Point } p_1 \text{ is on boundary} \\ y_{p_1} \geq y & \text{when Point } p_1 \text{ is NOT on boundary} \end{cases} \quad (15.16)$$

Equation 15.1 indicates that when the line AB is taken as a reference, the other points far from the camera center on the P-roof are considered to lie on the boundary of the occluded area.

3. Merging Steps (1) and (2) above, the boundary of the occluded area can be determined.

c) Assigning black to the boundary of the occluded area

With Step (b) above, the following step is to assign gray to the occluded area on the P-roof. Usually, the gray is assigned zero (black), that is

$$Gray\left(Image\left(x_{(occlusion+G-roof)-i}, y_{(occlusion+G-roof)-i}\right)\right) = 0 \text{ where } i = 1,2,3,\ldots,n. \quad (15.17)$$

where $\left(x_{(occlusion+G-roof)-i}, y_{(occlusion+G-roof)-i}\right)$ represents the coordinates within the occluded area, which are solved by the SRG algorithm, and n is the number of points on the boundary.

d) Combining the boundary on the G-roof image and the detected occlusion points on the P-roof image

To handle the issue of Bld-1 being partially occluded by Bld-2, the G-roof of Bld-1 extracted in Step (a) and the occluded area detected in Step (b) are merged. This step is repeated for all buildings until all occlusion detection is completed. Afterward, a Boolean additional operation is employed to obtain the entire occluded area.

15.3 EXPERIMENTAL RESULTS AND ANALYSIS

15.3.1 Experimental Data Set

The details of the experimental data can be found in Zhou et al. (2005), which includes image data and DBMs. A brief description is as follows:

- *Image data.* The experimental data, covering the City of Denver, CO, consist of high-resolution aerial images taken at a height of 1650 m above the mean ground elevation of the imaged region (see Figure 15.6). The endlap and sidelap of the images are approximately 65% and 30%, respectively. A DSM is provided by the City of Denver, CO. The ghost images are created from the original images of dv1118, dv1119 and dv1120 by orthorectification.
- *DBM data.* The DBM is provided by the City of Denver, CO. It contains only the three-dimensional information describing buildings, including the plane coordinates and elevation of roofs (see Figure 15.7).

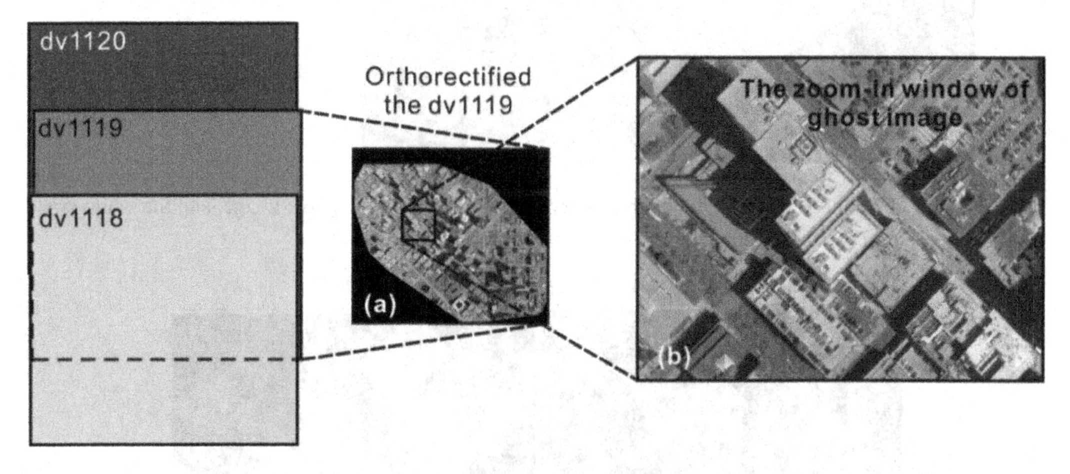

FIGURE 15.6 Ghost image generation dv1118, dv1119m and dv1120 created by orthorectification and the overlapping area is used to compensate for the detected occluded area.

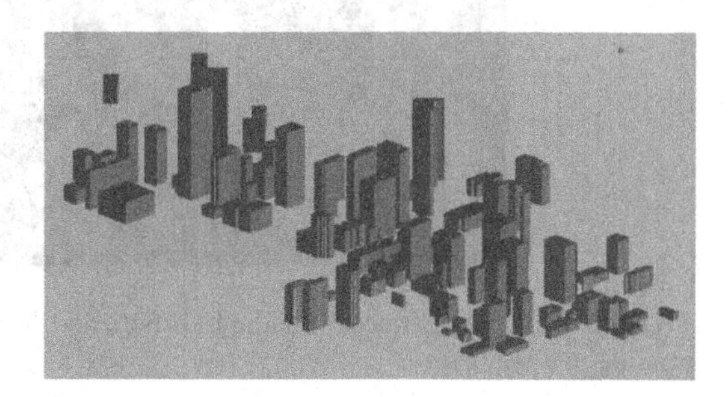

FIGURE 15.7 Digital building model (DBM).

15.3.1.1 Experimental Results and Analysis

Experimental results acquired using the steps described in Section 15.2.2 are summarized below.

1) G-roof extraction

Using the steps described in Section 15.2.2 and Equation 15.11, the G-roof boundaries for 121 buildings are extracted from the ghost image and the coordinates of the seeding points are calculated using Equation 15.; the results are depicted in Figure 15.8. With the calculated coordinates of the seeding points, the SRG algorithm (Adams and Bischof 1994) is employed to extract the full G-roof of each building, including the gray areas (see Figure 15.9).

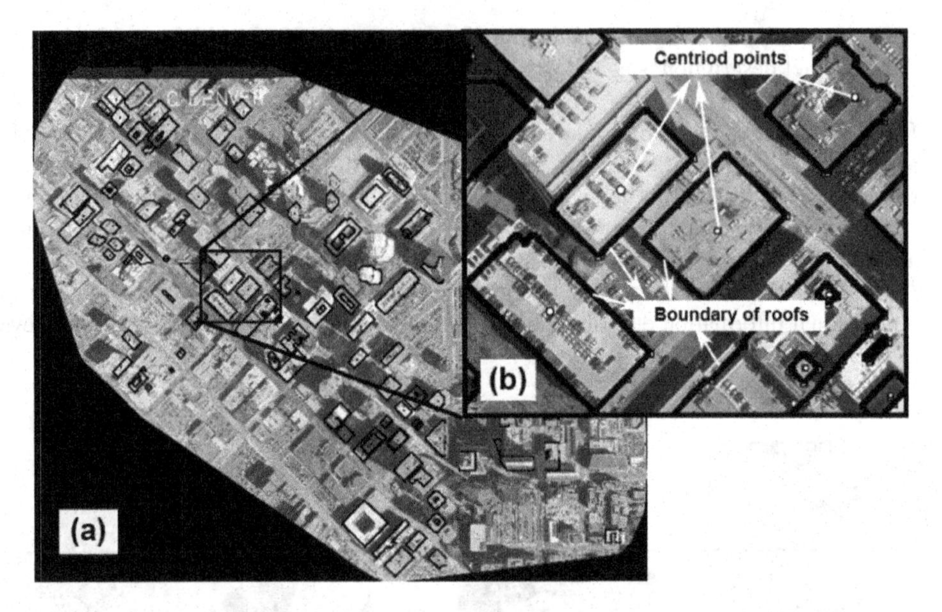

FIGURE 15.8 The black lines denote the boundary of the G-roofs on the ghost image, and the white circle denotes the centroid of seeding points on the ghost image.

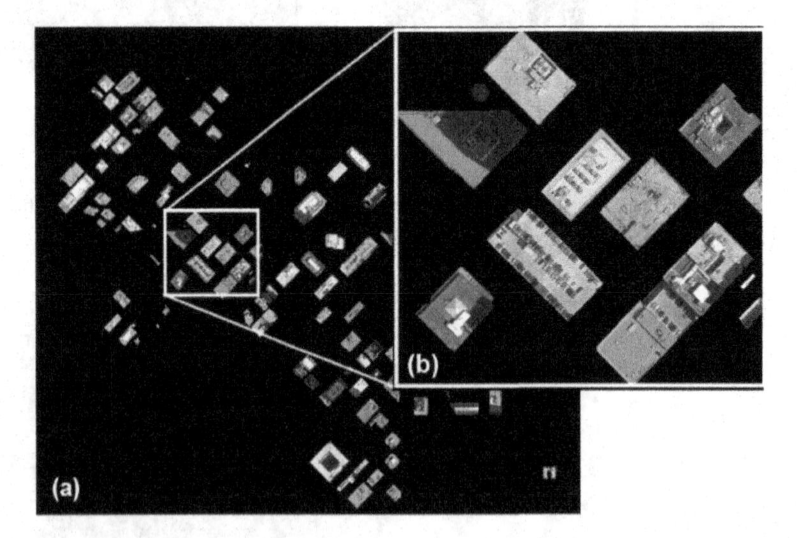

FIGURE 15.9 (a) Extracted building G-roofs and (b) a magnified window to demonstrate the extracted G-roofs.

2) Boundary detection of occluded area

Using Equations 15.9a and b, *MAX–AZ* and *MIN–AZ* are calculated and the reference line, *AB* in Figure 15.4, is determined. The boundary points on the G-roof and P-roof are detected and depicted in Figure 15.10. On the basis of boundary points on the building roofs and the *MAX–AZ* and *MIN–AZ*, the boundary on the G-roof and the boundary on the P-roof can effectively be determined. In practice, the two cases – that is, building occluding the ground and building occluding other buildings – are combined, when determining the boundary of occluded area using Equations 15. and 15.1. The results of the detected boundaries are depicted in Figure 15.11.

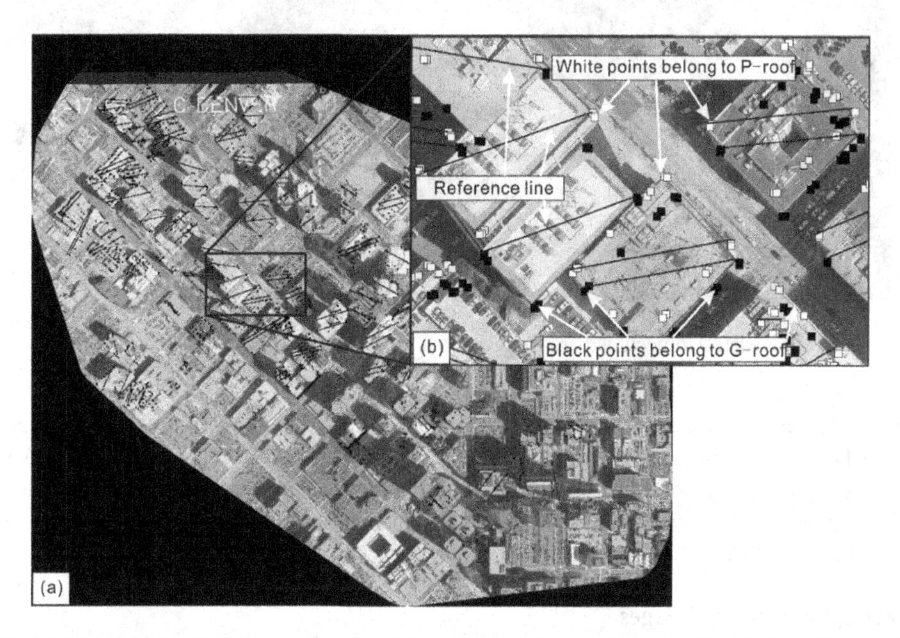

FIGURE 15.10 White points locate the P-roof, black points locate the G-roof, and black lines inside the P-roof and G-roof are the reference lines.

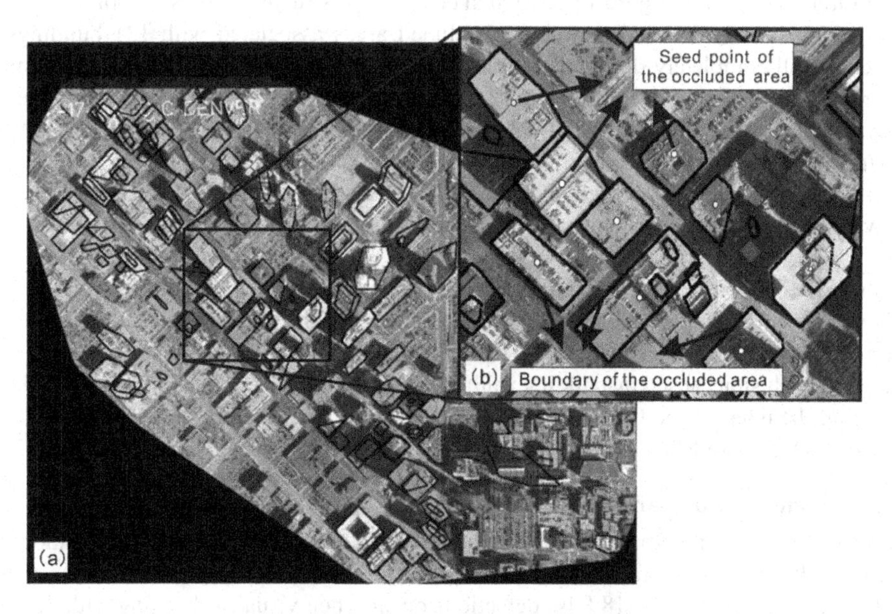

FIGURE 15.11 Black lines on the building represent the boundary of the occluded area, and the white points denote the centroid points.

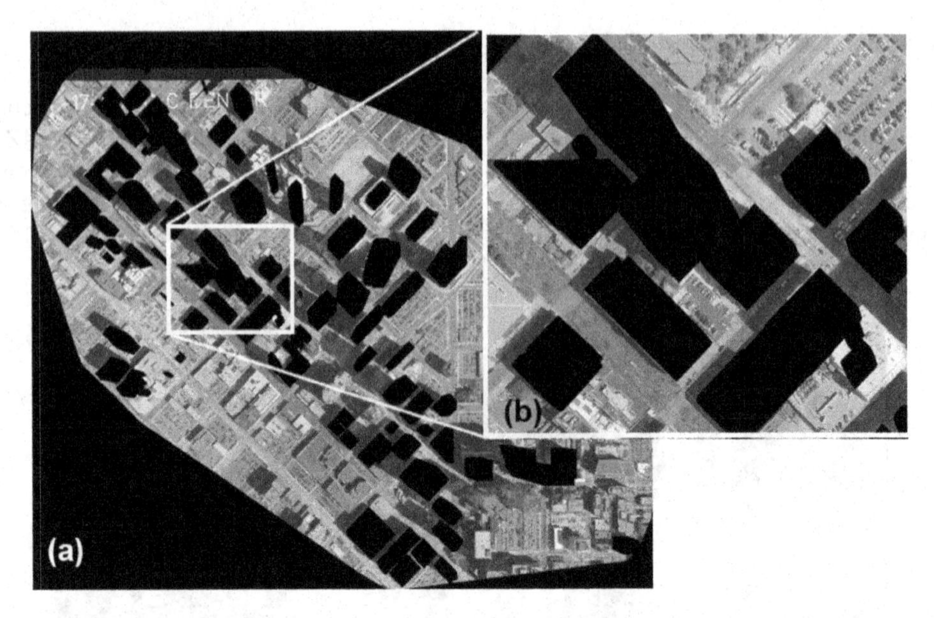

FIGURE 15.12 The detected occluded area whose grays are assigned as zero (i.e., black).

With Equation 15.1, the seeding point coordinates are calculated and the results are depicted in Figure 15.11(a), and a magnified window in Figure 15.11(b) is used to demonstrate the position of the seeding points at each building. With the obtained seeding points at each of the buildings, the occluded area is detected using the SRG algorithm and the gray area is assigned as zero (that is, black) (see Figure 15.12).

3) Merging two types of images

After the occlusion boundaries are detected on the G-roof image and the occluded area is detected on the P-roof image above, the next step is to merge the two types of images using a Boolean addition operation, as shown in Figure 15.13, which contains only the building's G-roof.

With the operations outlined above, the occluded areas associated with 121 buildings in the study area are all completely detected using the method proposed in this chapter. To demonstrate the accuracy achievable, the detected boundary is overlapped onto the original aerial image and the accuracy is statistically analyzed. It is found that the average offset is 1.0–2.5 pixels. Figure 15.14 demonstrates the accuracy achievable by showing four magnified buildings; it can be observed that Bld-1, Bld-2, Bld-3, and Bld-4 have offsets of 2–3 pixels, 2–3 pixels, 2–4 pixels, and 0–1 pixels, respectively. The different sizes of errors are caused by the following:

1) For the occlusion caused by a building roof: because of the different intervals of the selected boundary points in the building G-roof, the detected occluded area exposes different sizes of errors in practice, such as with Bld-2 and Bld-3 in Figure 15.14.
2) For the occlusion caused by building wall: the occluded area and boundary are formed through linking the lines on the G-roof and P-roof. Therefore, the accuracy of occlusion detection is high, such as with Bld-1 and Bld-4 in Figure 15.14.

15.3.1.2 Accuracy Comparison

The occlusion detection is compared between the proposed method and the Z-buffer method, and the statistical results are listed in Table 15.1. As observed from Table 15.1, both methods can effectively detect 121 buildings. However, 18 false detections occur when visually checking with the Z-buffer method, whereas they do not with our method. In addition, the visual comparisons are depicted

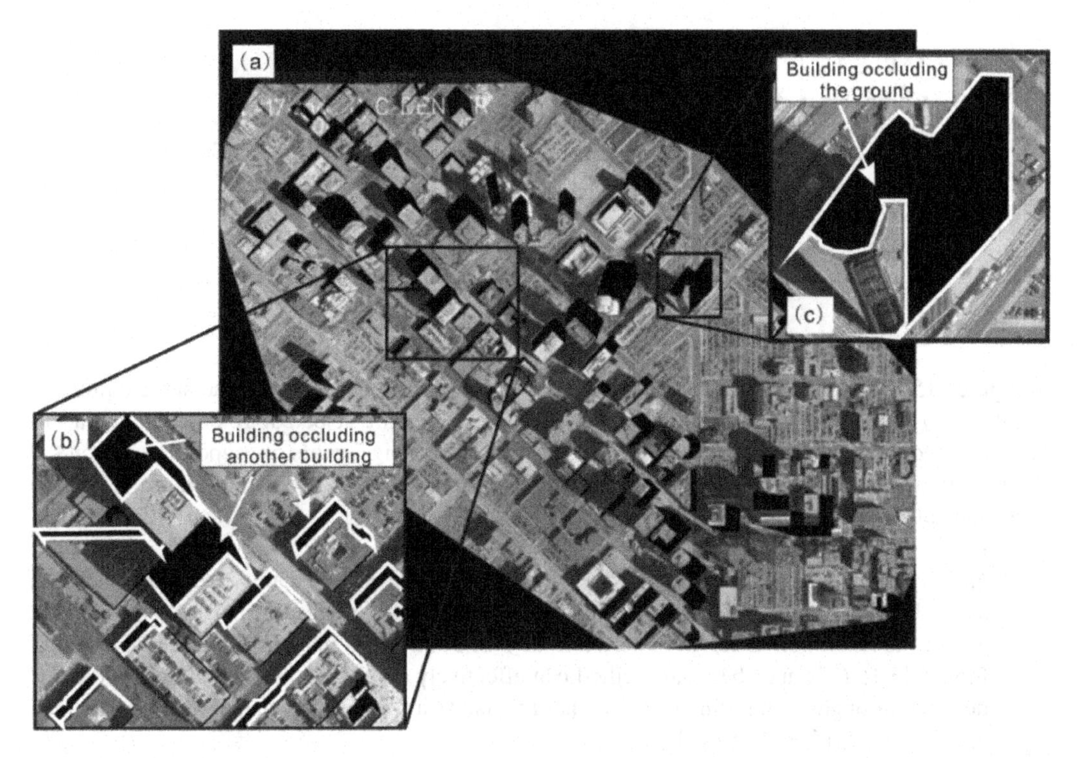

FIGURE 15.13 The result of occlusion detection for the entire study area in which the buildings contain the correct roofs.

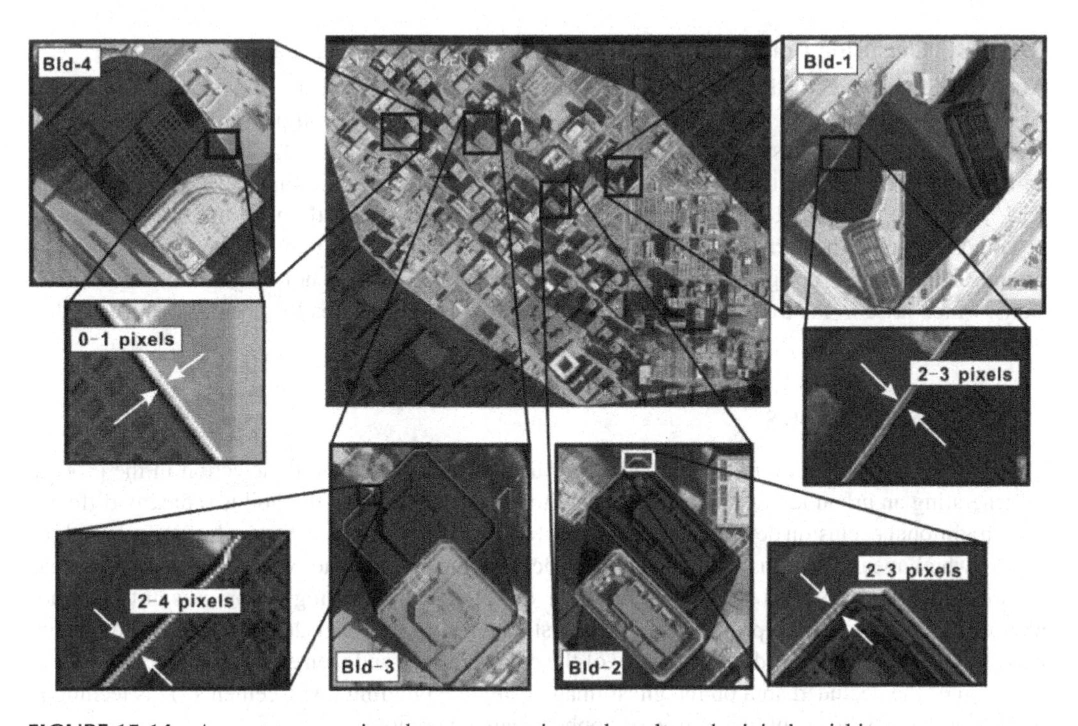

FIGURE 15.14 Accuracy comparison between experimental results and original aerial image.

TABLE 15.1
Comparison of the Two Methods

Approaches	Number of Detected Buildings	Number of False Visibility and False Occlusion Errors
Our method	121	0
Z-buffer	121	18

in Figures 15.15 (i) and (ii), respectively. In the two resulting images, six of the same buildings, noted *a1–a6* in Figure 15.15(i) and *b1–b6* in Figure 15.15(ii), are selected. To compare the accuracy achieved by the two methods, the boundaries of the buildings' roofs are superimposed on the resulting image by back-projection (see Figure 15.15). As can be observed from Figure 15.15, the following can be found:

a) *False visibility.* As shown in Figure 15.15 (a2) and (a4), the occluded areas are falsely detected as visible when applying the Z-buffer method. This may be caused by irregular structures and/or the resolution differences between the DSM and aerial image. As can be observed in Figure 15.15 (b2) and (b4), our method can effectively determine the occluded boundary as detected through the building roofs' corner points. As a result, false visibility can be avoided (see Figure 15.15 (b2) and (b4)).
b) *False occlusions.* As shown in Figure 15.15 (a3), (a5), and (a6), it is found that many false occlusions are detected when applying the Z-buffer method. For example, part of the visible area is wrongly detected as the occluded area in Figure 15.15 (a3), (a5), and (a6). However, our method can effectively avoid false occlusions (see Figure 15.15 (b3), (b5), and (b6)).

15.3.1.3 Occlusion Compensation for True Orthophotomap Generation

After an occluded area is detected, the occluded area remains blank. Thus, the occluded area must be filled using the neighboring overlap "slave" orthophotomap. The compensation method for the occluded area on the "master" orthophotomap is described by Zhou et al. (2005). With occlusion compensation, a complete true orthophotomap is created and depicted in Figure 15.16. From Figure 15.16, it can be concluded that: (1) the occluded area in the master orthophotomap can effectively be compensated by the corresponding patch of the "slave" orthophotomap; (2) the buildings can be orthorectified in their upright positions; and (3) the proposed method not only avoids false occlusions, which arise in the traditional Z-buffer method, but also improves the accuracy of true orthophotomap generation.

15.4 CONCLUSIONS

This chapter proposes an innovative method for accomplishing occlusion detection in the process of generating an urban large-scale true orthophotomap. The proposed method is radically different from traditional occlusion detection because the proposed method detects the occlusion from a ghost image, which means that this method separates occlusion detection and orthorectification, whereas traditional methods, such as the Z-buffer method, detect occlusion during the process of true orthophotomap generation. The proposed method first establishes a model describing the relationship between the ghost image and the boundary of the occluded area and then employs a model for identification of the occluded area on the ghost image using the building displacements. This technique has not previously been applied in true orthophotomap generation.

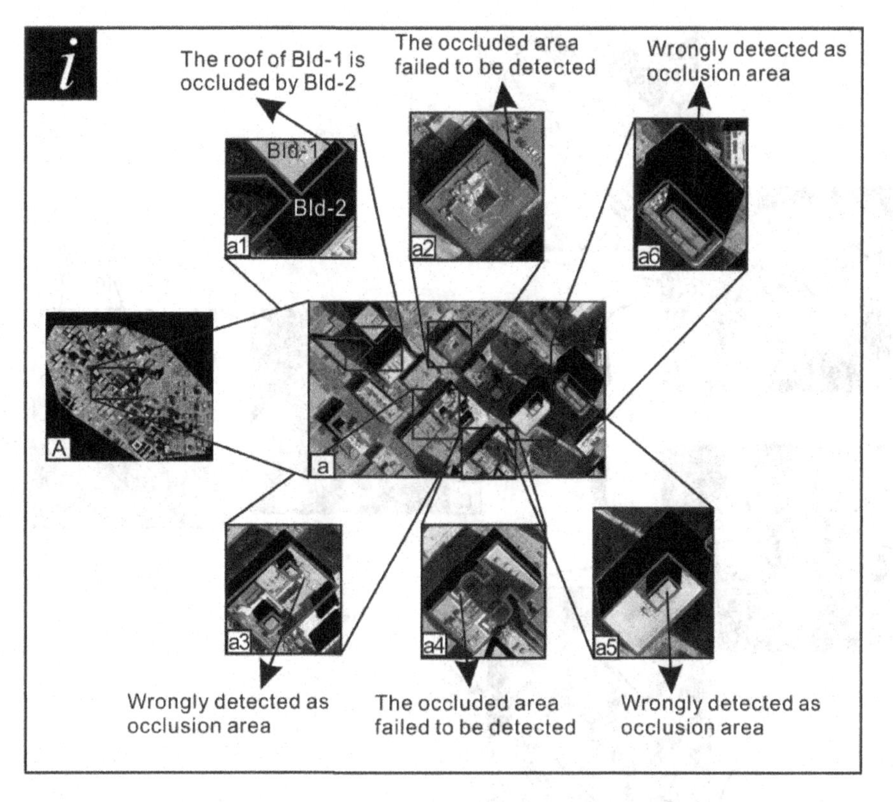

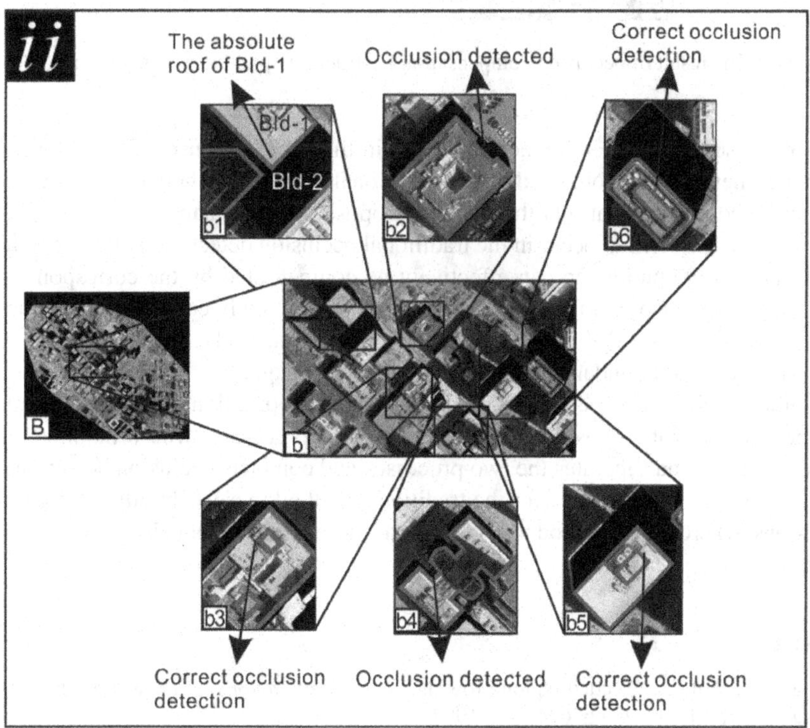

FIGURE 15.15 (i) is the result from the Z-buffer method, where (a1)–(a5) are the incorrect results; (ii) is the result from our method, where (b1)–(b5) are the correct results.

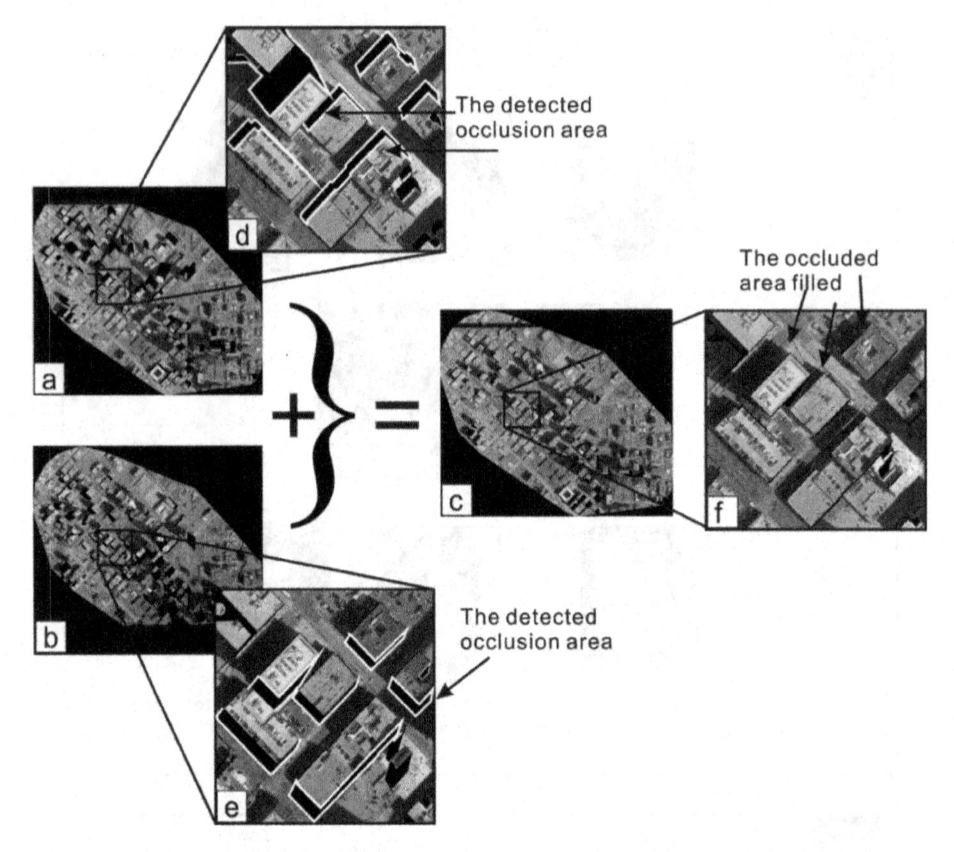

FIGURE 15.16 The result of occlusion compensation for true orthophotomap generation.

This chapter uses high-resolution aerial images in the City of Denver, CO as the experimental area. The experimental results obtained by the traditional Z-buffer method and our method are compared. It can be concluded that: (1) the method proposed in this chapter completely avoids false occlusion detection, which can occur in the traditional occlusion detection method; (2) the occluded area in the master orthophotomap can effectively be compensated by the corresponding patch of the "slave" orthophotomap; (3) the proposed method is completely capable of detecting all building occlusion occurrences; (4) the accuracy of the proposed method can reach 1.0–2.5 pixels when overlapping the detected boundary onto the original aerial image.

The significant difference between our method and the traditional method is that the traditional method carries out occlusion detection during the process of true orthophotomap generation, whereas our method separates the two processes and conducts occlusion detection from ghost images. Therefore, it is unavoidable for the traditional method to cause building false visibility and false occlusions, whereas our method not only avoids these problems but also improves the accuracy of true orthophotomap generation.

REFERENCES

Adams, R., and Bischof, L., "Seeded region growing," *IEEE Transactions on Pattern Analysis and Machine Intelligence*, vol. 16, no. 6, pp. 614–647, 1994.

Amhar, F., Josef, J., and Ries, C., "The generation of true orthophotomaps using a 3D building model in conjunction with a conventional DTM," *The International Archives of the Photogrammetry and Remote Sensing,* vol. 32, no. 4, pp. 16–22, 1998.

Bang, K.I., et al. "Comprehensive analysis of alternative methodologies for true orthophotomap generation from high resolution satellite and aerial imagery," *American Society for Photogrammetry and Remote Sensing, Annual Conference*, Tampa, Florida, USA,May 7–11, 2007.

Habib, A., Kim, E., and Kim, C., "New methodologies for true orthophotomap generation," *Photogrammetric Engineering and Remote Sensing,* vol. 73, pp. 25–36, 2007.

Kuzmin, Y.P., Korytnik, S.A., and Long, O., "Polygon-based true orthophotomap generation," *XXth ISPRS Congress Proceedings–23*, Istanbul, 2004.

Nielsen, M.Ø., "True orthophotomap generation," Informatics and Mathematical Modelling, Lyngby, Technical University of Denmark, Master, 125, 2004.

Rau, J.Y., Chen, N.Y., and Chen, L.C., "True orthophotomap generation of built-up areas using multi-view images," *Photogrammetric Engineering & Remote Sensing*, vol. 68, no. 6, pp. 581–588, 2002.

Shariat, M., Azizi, A., and Saadatseresht, M., "Analysis and the solutions for generating a true digital orthophotomap in close range photogrammetry," *The International Archives of the Photogrammetry, Remote Sensing and Spatial Information Sciences*, vol. XXXVII, Beijing, China, 2008.

Sheng, Y.W., "Comparative evaluation of iterative and non-iterative methods to ground coordinate determination from single aerial images," *Computers and Geosciences*, vol. 30, pp. 267–279, 2004.

Sheng, Y.W., "Minimising algorithm induced artefacts in true orthophotomap generation: a direct method implemented in the vector domain," *Photogrammetric Record*, vol. 22, no. 118, pp. 151–163, 2007.

Skarlatos, D., "Orthophotomap graph production in urban areas," *Photogrammetric Record*, vol. 16, no. 94, pp. 643–650, 1999.

Wang. X, Jiang, W., and Bian, F., "Occlusion detection analysis based on two different DSM models in true orthophotomap generation," *The 16th International Conference on Geoinformatics*, Guangzhou, China, 2008.

Youn, J., and Kim, G.H., "Visible height based occlusion area detection in true orthophotomap generation," *Journal of The Korean Society of Civil Engineers*, vol. 28, pp. 417–422, 2008.

Zhong, C., Li, H., and Huang, X., "A fast and effective approach to generate true orthophotomap in built-up area," *Sensor Review*, vol. 31, no. 4, pp. 341–348, 2011.

Zhou, G., et al. "Orthorectification of 1960s satellite photographs covering Greenland," *IEEE Transactions on Geoscience and Remote Sensing*, vol. 40, no. 6, pp. 1247–1259, 2002.

Zhou, G., et al. "A comprehensive study on urban true orthorectification," *IEEE Transactions on Geoscience & Remote Sensing,* vol. 43, pp. 2138–2147, 2005.

Zhou, G., et al. Building occlusion detection from ghost images. *IEEE Transactions on Geoscience & Remote Sensing,* vol. 55, no. 2, pp. 1074–1084, 2016.

Section IV

Advanced Algorithms Urban Remote Sensing Application

16 Hierarchical Spatial Features Learning for Image Classification

16.1 INTRODUCTION

Very high resolution (VHR) remotely sensed (RS) images are widely available because of the fast development of advanced remote sensing technology. As the VHR RS images contain abundant ground information at pixel-level, they have therefore attracted end users' attention to their application in urban mapping, precision agriculture, forest monitoring, and so on. However, the traditional RS image classification methods depending only on spectral information have been demonstrated to be inadequate (Dell'Acqua et al. 2004). On the one hand, the VHR RS image classification suffers from uncertainty of spectral information due to the low intra-class variance and the high inter-class variance, which causes a decrease in the separability of different kinds of objects in the spectral domain, particularly for the spectrally similar classes (Huang and Zhang 2013). How to effectively implement VHR RS image classification has become a research focus. On the other hand, spatial correlation is more important and spatial properties should be considered for the VHR images processing.

An effective method for solving this problem is to combine spatial information with spectral information (Haralick et al. 1973; Acharyya et al. 2003; Dell'Acqua et al. 2004; Chanussot et al. 2006; Huang et al. 2007, 2014; Valero et al. 2010; Mura et al. 2010; Pesaresi and Gerhardinger 2011; Rizvi and Mohan 2011; Huang and Zhang 2013; Raja et al. 2013). It has been found that combining spectral and spatial features can achieve better results than using either pure spectral or spatial analysis of VHR images (Dell'Acqua et al. 2004). Many methods for extraction of spatial features were proposed for image classification, such as the gray-level co-occurrence matrix (GLCM) (Haralick et al. 1973; Huang et al. 2014), the wavelet-based spatial features (Acharyya et al. 2003; Raja et al. 2013), and the morphological profiles (Valero et al. 2010; Mura et al. 2010). These methods need to artificially pre-design a kernel or a template based on users' expert knowledge. Additionally, they mainly extract low-level features, such as shapes, edges, and gradients. It is critical to extract higher-level data-driven features to fully characterize VHR data (Castelluccio et al. 2015).

Deep neural networks (DNNs) trained in an end-to-end manner have been found to be more powerful in high-level features extraction than the existing feature extractors depending on carefully engineered representations, such as the histogram of oriented gradient (HOG), morphological profiles, and scale-invariant feature transform (SIFT) features. This is because of the invariance properties of DNNs to the transformation of local imagery (Chen et al. 2014). Convolutional neural networks (CNNs) are effective DNNs, given that they can be adjusted by setting adaptive breadth and depth. Moreover, CNNs have fewer parameters and connections compared to other DNNs (LeCun et al. 1998; Krizhevsky et al. 2012). In recent years, CNNs have been widely studied in object recognition/detection (LeCun et al. 2015; Lenz et al. 2015), scene retrieval (Krizhevsky et al. 2012; Razavian et al. 2014), and semantic segmentation (Chen et al. 2014; Long et al. 2015).

This method has provided outstanding results because of their ability in complex/high-level features extraction. CNNs have also been utilized in the RS community recently (Mnih and Hinton 2012; Penatti et al. 2015; Chen et al. 2016; Ma et al. 2016; Zhang et al. 2016a, 2016b), such as for object detection (Mnih and Hinton 2012; Chen et al. 2016; Zhang et al. 2016b), or image retrieval/whole-image classification (Castelluccio et al. 2015; Penatti et al. 2015; Zhang et al. 2016a) for VHR RS images. For RS image labeling, several works have been carried out in the last few years using fully convolutional networks (Marmanis et al. 2016; Audebert et al. 2016; Maggiori et al. 2016a, 2016b, 2017). The fully convolutional networks (Long et al. 2015) can realize the end-to-end learning for pixelwise classification and at the same time keep the boundary information by applying the low-layer features during the upsampling.

A large convolutional window, covering abundant contextual information, is necessary for capturing long-distance correlation and correct decision making. However, to design a large kernel window, a CNNs model would be unmanageably large (Farabet et al. 2013). To deal with this problem, the multiscale method can be considered. The multiscale method is popular in RS image analysis (Santos et al. 2012, 2013; Tilton et al. 2012; Valero et al. 2013). Many researchers integrated the multiscale analysis with the classification method to make the training procedure simple and increase the classification accuracy. Farabet et al. (2013) proposed multiscale CNNs, which utilized the Laplacian pyramid to transfer the raw input image into three scales followed by upsampling. Although the scene parsing and pixelwise prediction results of the multiscale CNNs were well designed, the multiscale CNNs did not effectively use the original spectral information of images. As is well known, spectral information is very important for RS image classification, as it is associated with the spectral absorption features of various materials (Zhang et al. 2018).

This chapter introduces a hierarchical multiscale spatial feature extraction approach (HCNNs) based on three pixel-level CNNs (PCNNs), namely PCNNs-α, PCNNs-β, and PCNNs-γ. The PCNNs-α contains three three-layer stages, one full connection layer, and a final softmaxloss layer, while the PCNNs-β and PCNNs-γ only have three three-layer stages. Each of them is applied separately to the pyramid images obtained by the Gaussian pyramid. All parameters are learned in PCNNs-α, and then shared with PCNNs-β and PCNNs-γ. After the hierarchical spatial features (HSFs) are generated, they are upsampled to the size of the raw image. As a consequent, the size of feature maps and labeled output images are consistent. To making use of the OSFs of land covers, they are concatenated with the upsampled HSFs as a combined feature data set, used as the input of a support vector machine (SVM) classifier. Through combining OSFs and HSFs of land covers, the classification accuracy of the VHR RS image is improved significantly.

16.2 THE BASIC PRINCIPLE OF HCNNs

The flowchart of the proposed method is depicted in Figure 16.1, in which three main components are included. (1) An HSFs extractor is presented. To obtain large-scale spatial features for land cover types, the raw VHR RS images are preprocessed by the Gaussian pyramid method to produce an image pyramid. (2) These different scale images are fed to three PCNNs to produce a series of dense and multiscale spatial features. Then, all spatial features are upsampled to the size of the raw image. (3) A back-propagating algorithm is used to finetune the parameters of the PCNN. In the third component, the original spectrum and HSFs are combined to form a single data set as the input of an SVM classifier for image classification.

16.2.1 Image Pyramid Product

To obtain large-scale spatial features and limit the size of CNNs at the same time, the Gaussian pyramid method is used to produce different scale images as the inputs of PCNNs. The reasons

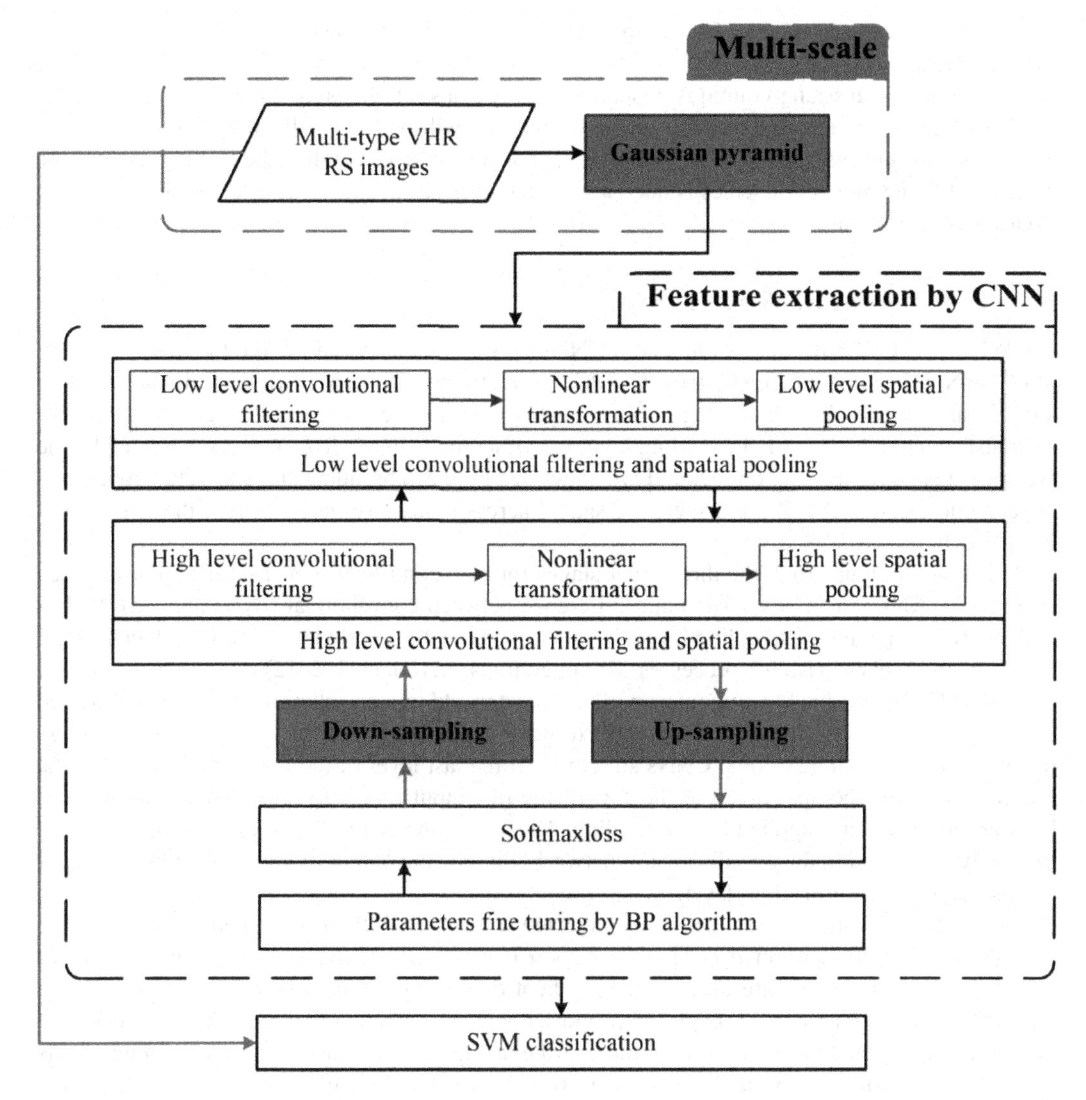

FIGURE 16.1 The flowchart of the proposed method.

for using the Gaussian pyramid method in this study are that it is a simpler method compared to other methods (for example, Wavelet transformation), and can keep all structures in the image. The Gaussian pyramid is a popular method in multiscale vision processing. Thus, images are represented as three-dimensional arrays $\mathbf{I} \in \mathrm{R}^{M \times N \times P}$, where M represents the row of images, N is the column, and P is the number of images or channels. Let the function of the Gaussian pyramid transformation be represented as $G(\mathbf{I})$. It can be mathematically expressed by

$$G_{n+1}\left(I\left(i,j\right)\right)=\sum_{r=-2}^{2}\sum_{s=-2}^{2}W\left(r,s\right)G_n\left(I\left(2i-r,2j-s\right)\right),\ i=1,\ \ldots,M;\ j=1,\ \ldots,N \quad (16.1)$$

where $\mathbf{W}(r, s) = \mathbf{W}(r) \times \mathbf{W}(s)$ is a Gaussian kernel with the length 5, $\mathbf{W}(\bullet) = (1/4\text{-}a/2, 1/4, a, 1/4, 1/4\text{-}a/2)$ is a vector, and i and j are the row and column indexes, respectively. After the processing of the $G(\mathbf{I})$ function, the raw VHR RS image $\mathbf{I}_0$ is transformed to multiscale images $\mathbf{I}_{m0}$, where $\forall m \in \{1, 2, \ldots, n\}$ and the size of $\mathbf{I}_{10}$ is the same with $\mathbf{I}_0$. The number of images in the pyramid decides the number of PCNNs in the HCNNs.

A three-layer Gaussian pyramid is generated. The reasons for choosing three scales is that if the number of scales is increased, the top image of pyramid will be too small, so that the last feature map of the last stage will be too small, which is not conducive to upsampling the feature map. On the other hand, if the number of scales is decreased, a large kernel window is needed to obtain enough contextual information.

16.2.2 Pixel-Level HSFS Extraction

All PCNNs (PCNNs-α, PCNNs-β, and PCNNs-γ) structures are based on the framework of MatConvNet (Vedaldi and Lenc 2015). The PCNNs-α structure, used to learn the trainable parameters, consists of three three-layer stages. Each stage consists of one convolution layer, one nonlinearity transformation layer, and one pooling layer to extract the features. The features extracted by the previous stage are used as the input and are combined by the subsequent stage to detect high-level or semantic features. As all parameters are shared across scales, we only describe the structure of PCNNs-α in this section.

As shown in Figure 16.2, the three-layer stages for extracting feature maps are similar between traditional CNNs and PCNNs. The main difference between conventional CNNs and PCNNs lies in the upsampling and subsampling layers in order to realize the pixelwise training for feedforward pass (FP) and backpropagation processes (BP) operations (LeCun et al. 2012) (see Figure 16.3). For traditional CNNs, the final feature maps will be concatenated into a vector and then fully connected to a classifier. However, in order to extract pixel-level HSFs, PCNNs have to apply a new strategy for FP compared with traditional CNNs structure. At the last layer of the last stage in PCNNs, the final features must be upsampled to the size of the raw input image for end-to-end training. The bilinear interpolation is applied for upsampling. The complete vector of each pixel is taken out and then fully connected to the one-dimension network. Finally, the output of the fully connected layer passes through the softmaxloss layer.

As a special FP strategy for the fully connected layer of PCNNs-α is applied, a particular BP strategy for the fully connected layer of PCNNs-α is necessary. In this study, sensitivity maps of each layer (Bouvrie 2006) are used by the gradient decent algorithm, where we can observe the errors propagating backward through the proposed network. When completing the BP processing of the fully connected layer, sensitivity maps of the last layer of the last stage (upsampling results of pooling and nonlinearity layers in Figure 16.3(b)) will be obtained. However, the size of the sensitivity maps of $\mathbf{P}^*_{\alpha,8}/\mathbf{N}^*_{\alpha,9}$ is the same as the raw image $\mathbf{I}_0$, which will mean that the sensitivities

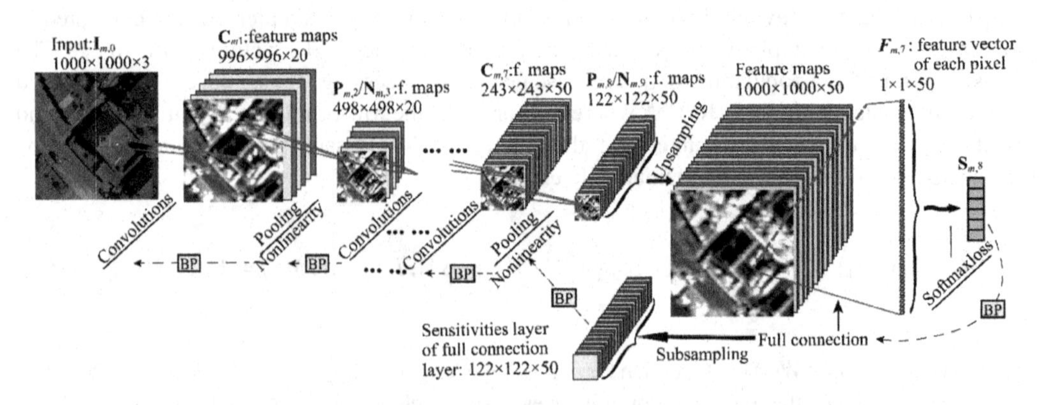

FIGURE 16.2 Structure of PCNNs-α in HCNNs (Note: a three-layer stage is ignored).

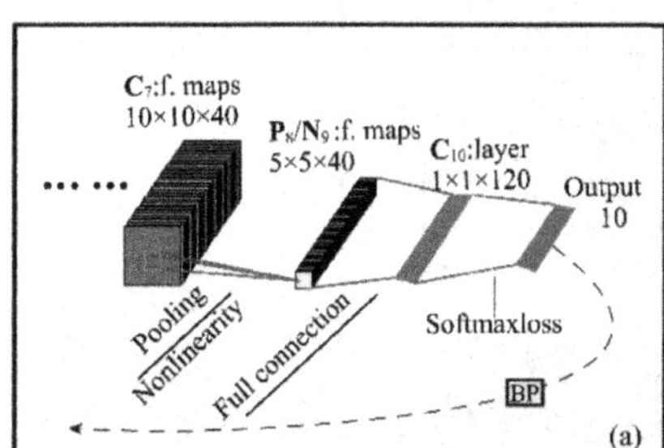

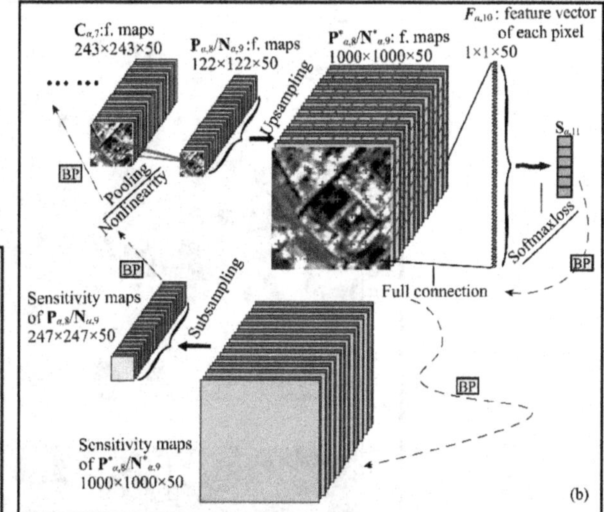

FIGURE 16.3 The difference between conventional CNNs and PCNNs in FP and BP processes of full connection layer. (a) The FP and BP processes of full connection layer in traditional CNNs, and (b) the FP and BP processes of full connection layer in PCNNs-α.

cannot be propagated back to the convolution layer of the last stage because of different sizes ($\mathbf{C}_{\alpha,7}$ in Figure 16.3(b)). Thereby, the sensitivity maps of $\mathbf{P}^*_{\alpha,8}/\mathbf{N}^*_{\alpha,9}$ must be subsampled to the size of $P_{\alpha,8}/N_{\alpha,9}$. Through iteratively implementing the FP and BP processes for PCNNs-α, the parameters will be learned. After obtaining all parameters, the parameters of three three-layer stages in PCNNs-α are shared with other PCNNs to obtain hierarchically multiscale spatial features.

16.3 EXPERIMENTS AND ANALYSIS

In this section, the experimental data set and environment are introduced, and the results of the VHR RS image classification using the method proposed above are presented.

16.3.1 Environmental Variables

To evaluate the proposed method, the subsets of aerial images (see Figure 16.4) in the center of the city of Vaihingen in Germany are used for testing. The ground resolution of the data set is 9 cm. The aerial images are pan-sharpened color infrared images, which include three spectral bands: near-infrared band, green band, and blue band. In these areas, there are dense historic buildings having rather complex shapes and being surrounded by some trees in Area 1, a few high-rising residential buildings in Area 2, and purely residential buildings with small, detached houses in Area 3. The land covers in the data set can be generally classified into five classes (Blue=Building, Yellow=Car, Cyan=Grass, Thistle=Road, Green=Tree) according to the ground truth provided by the International Society for Photogrammetry and Remote Sensing (ISPRS).

A part of the raw VHR RS image of size 1000 × 1000 pixels with its ground truth was selected from the whole image (Figure 16.5) to train the HCNNs model. The whole image (3007 × 2006 pixels) was used as the testing data. Nine hundred samples from each land cover were selected randomly from the ground truth as training samples to train the SVM classifier. The reasons for using Area 3 only are: (1) this area contains all ground cover types (except water) present in the whole image; (2) the proposed network is designed to cope with a small number of training samples

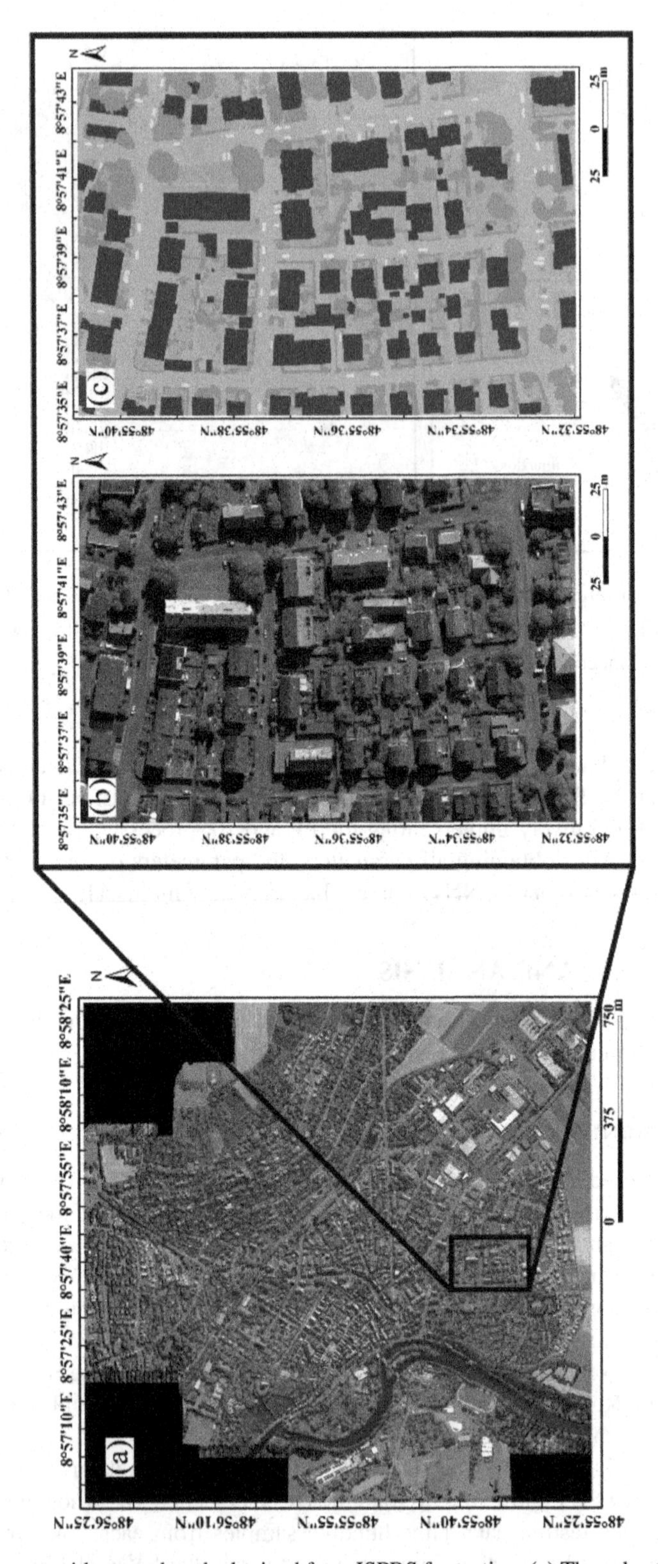

FIGURE 16.4 Images with ground truth obtained from ISPRS for testing. (a) The subsets of aerial images in the center of the city of Vaihingen in Germany, (b) the whole 3007 × 2006 image of Area 3, and (c) the ground truth of Area 3.

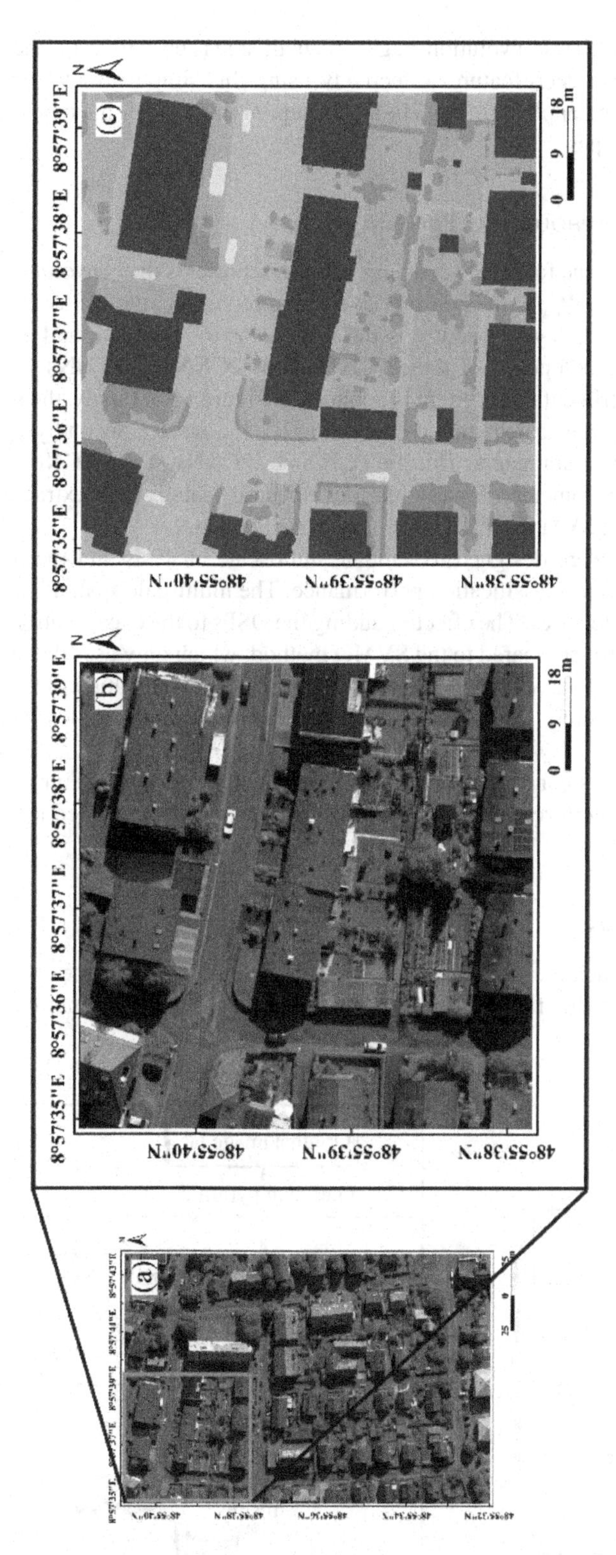

FIGURE 16.5 Images with ground truth obtained from ISPRS for training. (a) The whole 3007 × 2006 image of Area 3, (b) 1000 × 1000 part of the whole image, and (c) ground truth of 1000 × 1000 image.

because of the use of three convolution stages. A small area is used to test if the proposed multiscale framework can extract deep features effectively using this simple structure network. The trained ConvNet model can be applied to classify the remaining images of this data set if they contain the same ground cover types.

16.3.2 Tract Morphological Pattern

In this chapter, we use the following conventions. $\mathbf{C}_{p,x}$ represents the convolution layers, $\mathbf{N}_{p,x}$ denotes the nonlinearity layers, $\mathbf{P}_{p,x}$ is the pooling layers, $\mathbf{F}_{p,x}$ represents the full connection layer, and $\mathbf{S}_{p,x}$ denotes the softmaxloss layer, where p is the layer index of pyramid, and x is the layer index of PCNNs. HCNNs are composed of three PCNNs (i.e., PCNNs-α, PCNNs-β, and PCNNs-γ). The flowchart of the experimental procedure is depicted in Figure 16.6. Firstly, the raw VHR RS image $\mathbf{I}_0$ is transferred into a three-scale pyramid containing $\mathbf{I}_{10}$, $\mathbf{I}_{20}$, and $\mathbf{I}_{30}$. Secondly, $\mathbf{I}_{10}$ is fed to PCNNs-α for learning all shared parameters. Thirdly, PCNNs-α, PCNNs-β, and PCNNs-γ are used to extract HSFs using shared parameters. Finally, the raw spectral features and extracted HSFs are concatenated as the input of SVM for VHR RS image classification.

The experiments were designed to validate multiscale analysis and the utilization of spectral features can improve the classification performance. The multiscale spatial features were compared with the single scale features. The effect of adding the OSFs to the classification was tested. The proposed method was also compared to the SVMO method, which only uses the raw VHR RS image to classify the image, and the SVMOS method, which uses the combination of the raw VHR RS image and other spatial features, for example, extended morphological profiles (EMPs) (Benediktsson et al. 2005).

The experiment was conducted using a computer with an Intel (R) eight-core 3.6GHz i7-4790 CPU (with 8 MB three-level cache) and 8 GB of RAM. We implemented the proposed method in Matlab 2015a.

16.3.3 Parameters

For the ConvNet model, the Gaussian kernel parameter a is set to 0.4; the convolutional kernel $w_{a,x}$ is selected randomly; the trainable bias $b_{a,x}$ is initialized as 0; and the number of the iteration for

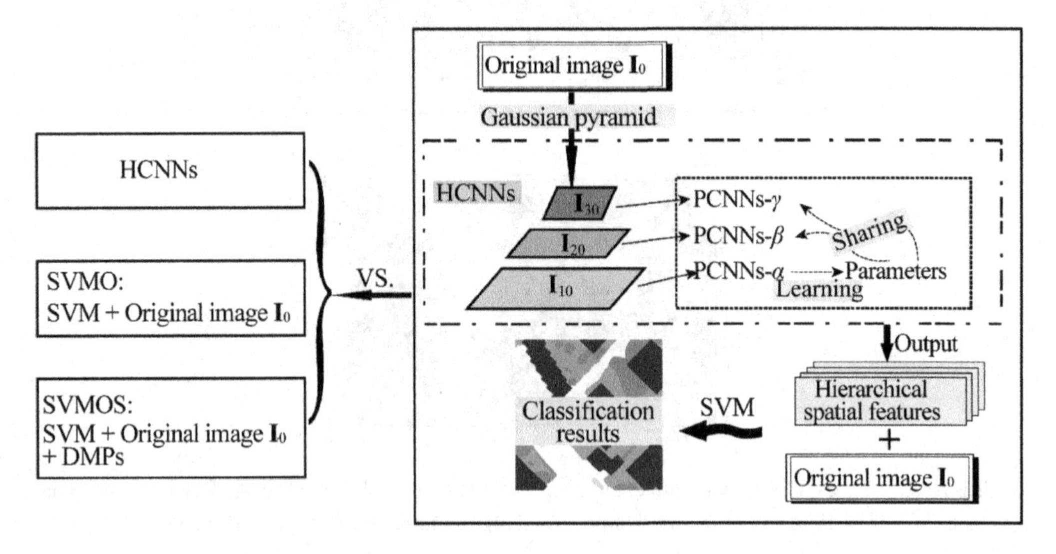

FIGURE 16.6 Flowchart of experimental procedure.

training the HCNNs model is set to 5000. For the SVM classifier, the kernel type is Radial Basis Function, and the penalty parameter is 100.0.

16.3.4 Pyramid Images Production for Training the HCNNs

Part of the raw VHR RS image $\mathbf{I}_0$ of size 1000 × 1000 is used to train the HCNNs model. Based on Equation 16.1 and a = 0.4, the three-layer Gaussian pyramid can be generated (see Figure 16.7). Their sizes are [1000, 1000, 3], [500, 500, 3], and [250, 250, 3] for $\mathbf{I}_{10}$, $\mathbf{I}_{20}$, and $\mathbf{I}_{30}$, respectively.

16.3.5 Structure Details of HCNNs

Given the training image $\mathbf{I}_{10}$ of size 1000 × 1000, three convolutional stages and two fully connected layers for PCNNs-α are described as follows. The parameters about the CNNs are set according to the LeNet model.

- **Convolutional stage 1:**

Layer $\mathbf{C}_1$ is a convolutional layer, whose size is 996 × 996 × 20, containing 20 feature maps obtained by using twenty 5 × 5 convolution kernels to interact with $\mathbf{I}_{10}$, respectively. Each 5 × 5 local receptive field in $\mathbf{I}_{10}$ is connected to a pixel in each feature map (with different convolution kernels). Through moving convolution kernels in $\mathbf{I}_{10}$, 20 feature maps of size 996 × 996 are produced in $\mathbf{C}_1$.

Layer $\mathbf{P}_2$ is a max pooling layer (i.e., subsampling layer), whose size is 498 × 498 × 20, with 20 feature maps of size 498 × 498 for each $\mathbf{P}_2$, respectively. Each 2 × 2 neighborhood in the feature maps in $\mathbf{C}_1$ is connected to a pixel in the corresponding feature map in $\mathbf{P}_2$. The 2 × 2 receptive fields are non-overlapping, thus the size of the feature maps in $\mathbf{P}_2$ is one-fourth of the size of the feature maps in $\mathbf{C}_1$.

Layer $\mathbf{N}_3$ is a nonlinearity layer, whose size is 498 × 498 × 20, with 20 feature maps whose size is the same as $\mathbf{P}_2$. The feature maps in $\mathbf{P}_2$ are passed through a ReLU function to produce the feature maps in $\mathbf{N}_3$.

- **Convolutional stage 2:**

Layer $\mathbf{C}_4$ is a convolutional layer, whose size is 494 × 494 × 100, with 100 feature maps of size 494 × 494. Each pixel in each feature map is connected to several 5 × 5 neighborhoods at an identical location in a subset of $\mathbf{N}_3$ feature maps.

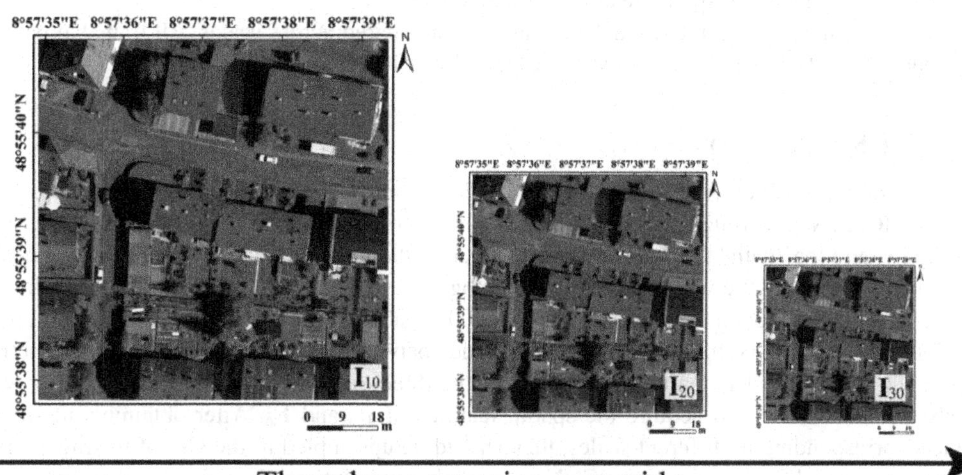

FIGURE 16.7 Three-layer Gaussian pyramid of VHR RS image $\mathbf{I}_0$.

Layer $\mathbf{P}_5$ is a max pooling layer, whose size is 247 × 247 × 100. This layer contains 100 feature maps for each $\mathbf{P}_5$ of size 247 × 247. Each 2 × 2 neighborhood of each feature map in $\mathbf{C}_4$ is connected to a pixel in the corresponding feature map in $\mathbf{P}_5$, in a similar way as $\mathbf{C}_1$ and $\mathbf{P}_2$.

Layer $\mathbf{N}_6$ is a nonlinearity layer, whose size is 247 × 247 × 100, with 100 feature maps whose size is the same as $\mathbf{P}_5$. The feature maps in $\mathbf{P}_5$ are passed through a ReLU function to produce the feature maps in $\mathbf{N}_6$.

- **Convolutional stage 3:**

Layer $\mathbf{C}_7$ is a convolutional layer, whose size is 243 × 243 × 50, with 50 feature maps for each $\mathbf{C}_7$ of size 243 × 243. Each pixel in each feature map is connected to several 5 × 5 neighborhoods at identical locations in a subset of $\mathbf{N}_6$ feature maps.

Layer $\mathbf{P}_8$ is a max pooling layer, whose size is 122 × 122 × 50. This layer contains 50 feature maps for each $\mathbf{P}_8$ of size 122 × 122. Each 2 × 2 neighborhood of each feature map in $\mathbf{C}_7$ is connected to a pixel in the corresponding feature map in $\mathbf{P}_8$, in a similar way as $\mathbf{C}_1$ and $\mathbf{P}_2$.

Layer $\mathbf{N}_9$ is a nonlinearity layer, whose size is 122 × 122 × 50, with 50 feature maps whose size is the same as $\mathbf{P}_8$. The feature maps in $\mathbf{P}_8$ are passed through a ReLU function to produce the feature maps in $\mathbf{N}_9$. Before the next stage of processing, each feature map in $\mathbf{N}_9$ is upsampled to the size 1000 × 1000 using the bilinear interpolation method, given that the spatial resolution of the image is high, and the linear model is reasonable to adopt. We denote the upsampled feature maps as $\mathbf{N}^*_9$, whose size is 1000 × 1000 × 50.

- **Two fully connected layers:**

Layer $\mathbf{F}_{10}$ is a fully connected layer only contained in PCNNs-α. In this layer, a vector of feature maps for each pixel is extracted from the upsampled feature maps. Thus, there are one million vectors. The length of each vector is 50. Based on these vectors, a matrix $\mathbf{M}_{10}$ with one million rows and 50 columns can be formed. Then the matrix $\mathbf{M}_{10}$ multiplies a trainable weight matrix $\mathbf{W}_{10}$ with 50 rows and 5 columns to produce the input of the softmax classifier. In fact, the matrix $\mathbf{W}_{10}$ is the weight matrix of the softmax classifier in $\mathbf{S}_{11}$.

Layer $\mathbf{S}_{11}$ is a composited layer combining the softmax classifier and logistic loss only contained in PCNNs-α. As the inputs of $\mathbf{S}_{11}$ are the results of the product between $\mathbf{M}_{10}$ and $\mathbf{W}_{10}$, the softmax and loss functions can be directly used to classify a class for each pixel and calculate the loss. However, we do not use softmax to make the final classification for each pixel, because the original spectral information should be introduced when classifying a class for each pixel. Thus the HCNNs are supervised hierarchical spatial feature extractors.

It should be noticed that the trainable parameters are shared across scales. All trainable parameters learned by PCNNs-α are shared with PCNNs-β, and PCNNs-γ.

16.3.6 HSFS Extraction

The gradient decent algorithm is used to optimize the parameters which are the same as those used in the MatConvNet. Through 5000 iterations, the trainable parameters of PCNNs-α are learned completely, and the spatial features of the first layer of the image pyramid, $\mathbf{I}_{10}$, are obtained. The time taken for training the whole model for an image with 1000 × 1000 pixels is about 24 hours, and the labeling time for an image size of 3007 × 2006 is about 30 seconds. The number of training iterations was determined to make sure the difference between two iterations for the loss function is as small as defined. The learned parameters of the three three-layer stages will be shared with PCNNs-β and PCNNs-γ to acquire the spatial features of $\mathbf{I}_{20}$ and $\mathbf{I}_{30}$. After obtaining all spatial features corresponding to different scales, they should be upsampled to the size of the raw image. There are 50 spatial features for each PCNNs. Part of the convolutional kernels and HSFs are shown in Figures 16.8 and 16.9, respectively.

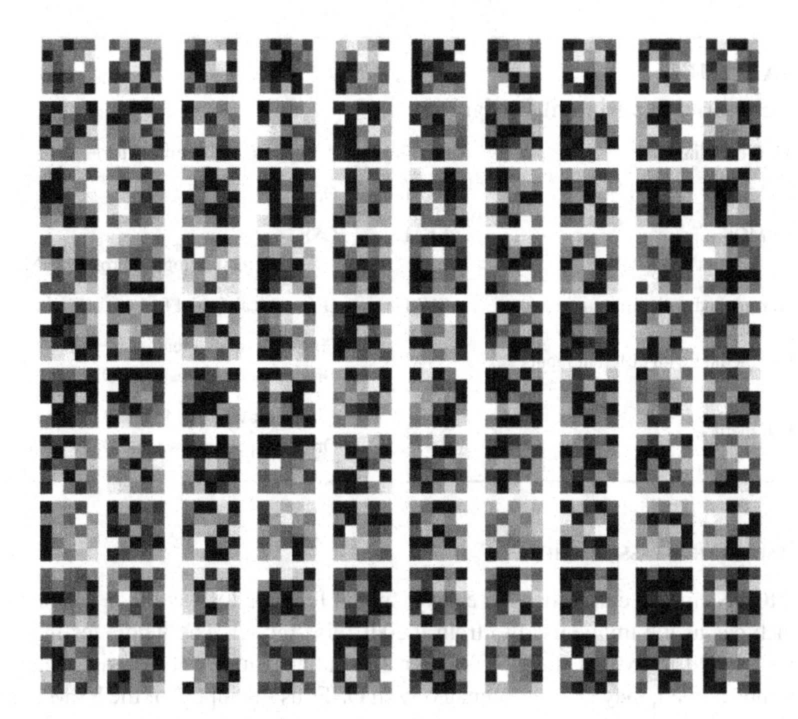

FIGURE 16.8 Part of convolutional kernels in $\mathbf{C}_{\alpha 7}$.

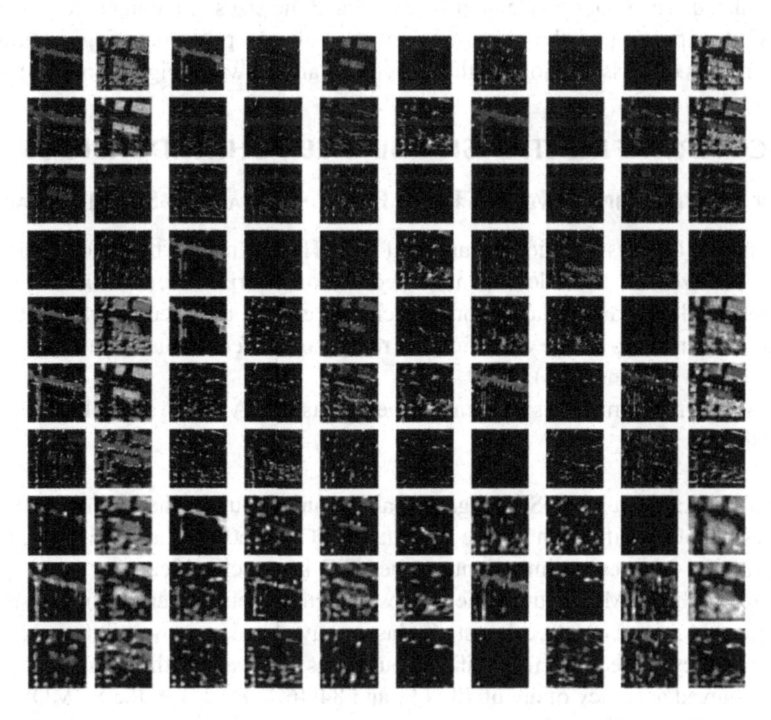

FIGURE 16.9 Part of hierarchically multiscale spatial features by HCNNs (Note: Rows 1–4 rows are obtained from $\mathbf{I}_{10}$, rows 5–7 rows are obtained from $\mathbf{I}_{20}$, and rows 8–10 are obtained from $\mathbf{I}_{30}$).

TABLE 16.1
Compared Experiment Design

Method Alias	Classifier	Combined Features
SVMO	SVM	OSF
SVMOS	SVM	OSF Extended morphological profiles (EMP)
Our method with different inputs	SVM	Hierarchical spatial features (HSF)
Our method with different inputs	SVM	First scale spatial features (FSF) Original spectral features (OSF)
Our method	SVM	Hierarchical spatial features (HSF) Original spectral features (OSF)

16.3.7 Results of Classification

The trained HCNNs model can be used to extract spatial features for any other image with the same ground cover types as the image used to train the HCNNs model. The whole testing 3007 × 2006 image is here applied to extract the HSFs. After all spatial features at different scales have been extracted and upsampled, they are concatenated with OSFs as the inputs of the SVM classifier. Nine hundred sample pixels for each class (i.e., building, car, grass, road, and tree) are randomly selected ten times from the ground truth reference to train the SVM model. For the SVM classifier, the kernel type is the Radial Basis Function, and the penalty parameter is 100.0.

To validate the proposed method, the popular spectral-spatial method (see Table 16.1), using EMPs, is completed. The structure element is "disc," and the sizes of element are 2, 4, 6, 8, and 10. The results of classification by the proposed method with non-postprocessing is shown in Figure 16.10(a). The results of classification for these methods are shown in Figure 16.10(b)–(e).

16.4 DISCOVERY OF RELATIONSHIP BETWEEN UHI AND VARIOUS VARIABLES

16.4.1 The Relationship Between LST and Environmental and Social Variables

In order to compare the classification accuracy of the VHR RS image, the whole ground truth (see Figure 16.4) is utilized for the calculation of a confusion matrix. After calculating the confusion matrix, the mean and standard variation for each class accuracy, the over accuracy (OA), and kappa coefficient (κ) for ten times of the classification results by different classification methods can be obtained (see Tables 16.2 and 16.3).

By comparing different methods in terms of the results of OA and κ, several comments are summarized as follows:

1) Original spectral features (OSFs) regarded as the only input for the SVM classifier obtain the worst results of classification for the test data set (OA = 60.85% and $\kappa = 0.51$), which means that the spatial features are important for the VHR image classification.
2) Compared to the SVMO method, the proposed method gets higher classification with almost 13% accuracy improvement. Almost all classes have been improved in terms of the classification accuracy, especially for artificial buildings and cars with strong spatial structures, which achieved accuracy of about 78.16% and 84.46%. However, the SVMOS method with the EMP features cannot extract such features, which can be seen from the classification results with 61.75% and 57.66% for the SVMOS method compared to 64.91% and 67.59% for the SVMO method.

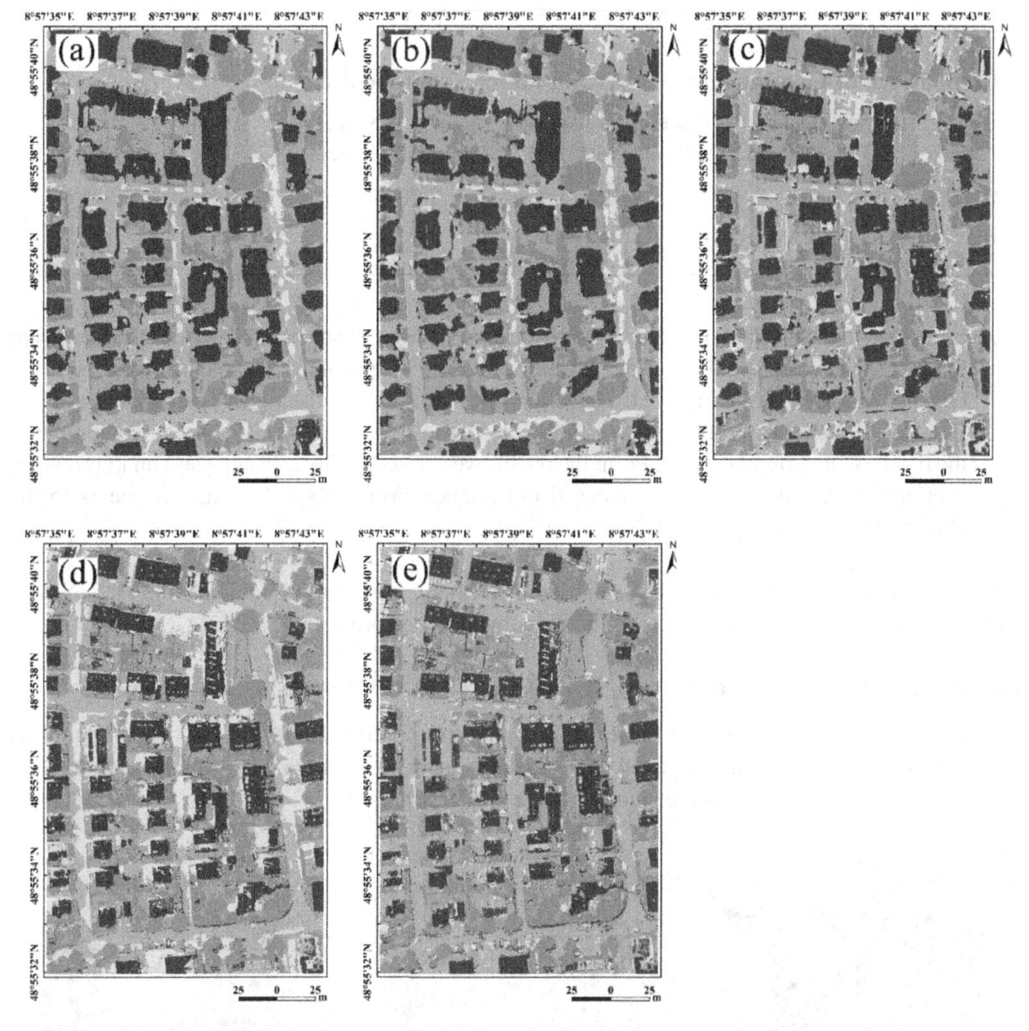

FIGURE 16.10 The results of classification in the whole test area by: (a) our method, (b) the proposed method with only FSF, (c) the proposed method with only one scale SF and spectral features, (d) SVMO, and (e) SVMOS.

TABLE 16.2
Confusion Matrix of the Proposed Method

	The Proposed Method	The Proposed Method with only FSF	One Scale SF and Spectral Features	SVMO	SVMOS
Building (%)	78.16 ± 0.83	76.90 ± 0.25	71.50 ± 0.67	64.91 ± 0.74	61.75 ± 0.89
Car (%)	84.46 ± 0.95	82.87 ± 0.786	68.58 ± 1.14	67.59 ± 0.95	57.66 ± 2.36
Grass (%)	69.78 ± 1.04	65.66 ± 1.48	68.24 ± 0.73	57.58 ± 1.41	58.86 ± 2.12
Road (%)	68.84 ± 1.14	66.82 ± 1.76	68.49 ± 0.89	51.83 ± 0.47	75.33 ± 2.65
Tree (%)	77.66 ± 0.93	75.77 ± 0.73	77.15 ± 0.21	74.47 ± 0.77	75.42 ± 0.56

TABLE 16.3
Comparison of Over Accuracy (OA) and Kappa Coefficient (κ)

	Our Method	The Proposed Method with only FSF	One Scale SF and Spectral Features	SVMO	SVMOS
OA (%)	73.33	71.15	70.88	60.85	68.35
κ	0.65	0.61	0.61	0.51	0.58

3) Compared to the SVMOS method with EMP features, the proposed method gets 5% improvement for the OA accuracy and 0.07 for the κ. Compared to other spatial features (i.e., EMPs), the HSFs obtained by HCNNs can improve the accuracy of classification of the higher-level spatial features.
4) In order to demonstrate the roles of multiscale analysis and spectral features, we compared the results with different inputs. From the results we can see, multiscale analysis can improve the accuracy by about 3%, and the spectral information brings about 2% improvements to the classification accuracy.

From the comments above, it is clear that the combination of spatial features from HCNNs and spectral features can effectively improve the classification performance.

16.4.2 The Relationship Between LST and Urban Landscape Metrics

To further validate the results of classification by the proposed method, a series of other subset images is extracted to compare the results of classification by all methods for a visual inspection (see Figure 16.11). A few comments are summarized as follows:

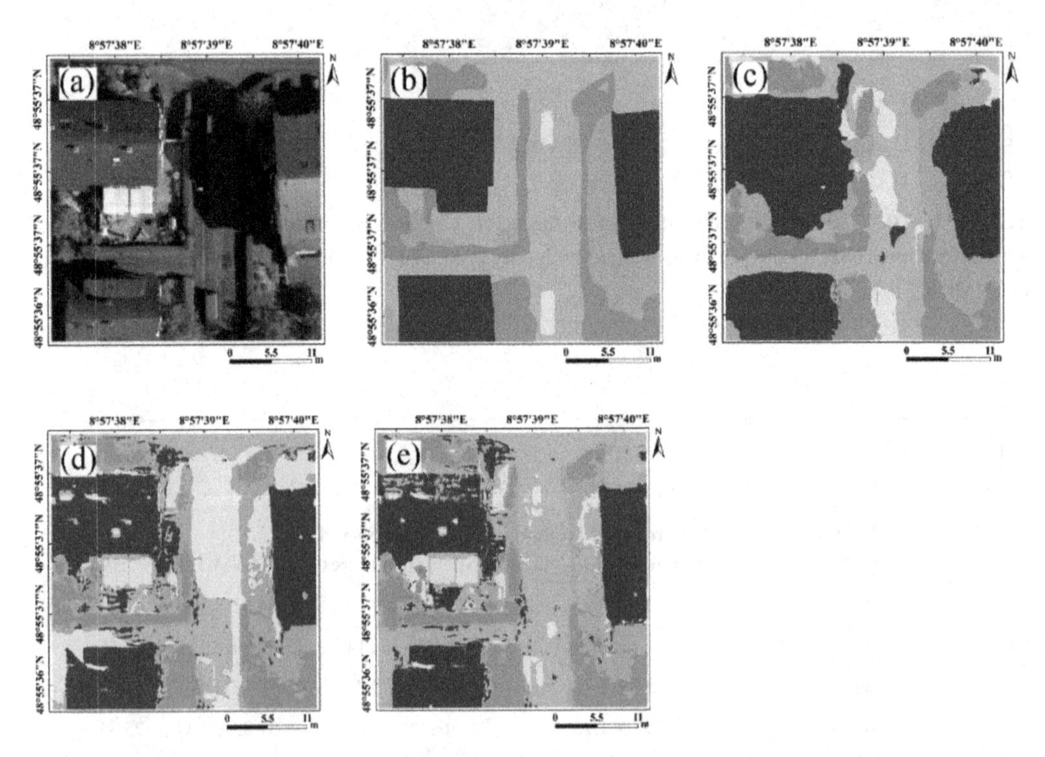

FIGURE 16.11 Subset image of classification images by different methods for visual inspection (Blue=Building, Yellow=Car, Cyan=Grass, Thistle=Road, Green=Tree). (a) Raw image (spectral features), (b) ground truth, (c) the proposed method, (d) SVMO, and (e) SVMOS.

1) Concerning the subset image of classification map at Location 1, the land covers in the image are complex, especially, for example, two buildings with very large differences in the roof materials which represent different colors. The SVMO and SVMOS methods misclassify the right-hand building, which has a similar color to the road and car (in shadow); however, the proposed method can effectively distinguish buildings from other land covers due to the introduction of HSFs.
2) As observed from the results of classification by different methods, the proposed method's ability to deal with shadow outperforms the other methods, because the method combines the original spectral features and HSFs. Although different land covers (in the shadows) are similar in gray level, the proposed method can effectively discriminate between them based on their high-level HSFs.

16.5 CONCLUSIONS

The purpose of this chapter is to propose a method to extract HMSFs for the classification of RS images with VHR. The study is inspired by the fact that CNNs are capable of extracting high-level spatial features without expert knowledge about the data. In this chapter, a novel architecture of CNNs, namely, hierarchical CNNs (HCNNs), containing several pixel-level CNNs (PCNNs), called PCNNs-α, PCNNs-β, and PCNNs-γ, are proposed to extract HMSFs from pyramid images obtained by the Gaussian pyramid method. In order to cope with the drawbacks for CNNs without considering the original spectral information, the combined features, containing all scale HMSFs, upsampled to the size of the raw image, and OSFs, are fed into a SVM classifier for the classification of the VHR RS image.

The proposed method has been applied to classify the VHR RS images in Vaihingen, Germany. When compared to the SVMO method, where the SVM classifier only uses the OSFs, and the SVMOS method, where the SVM classifier applied OSFs and EMPs, it can hereby be demonstrated that: (1) the OA for the proposed method reaches 73.33% and the κ is 0.65 in the test data set with an improvement of 13% and 5% compared to the other two methods; (2) the classification accuracy of almost all classes has been improved, especially for the artificial classes with strong structure information, such as buildings and cars with about 20% improvement compared to other methods. With the experimental results and comparison analysis in Sections 16.3 and 16.4, it can be concluded that the proposed method is capable of improving classification accuracy significantly for the VHR RS image with complex land covers types.

A limitation of the proposed method is training time compared to other methods. In future work, we plan to study an unsupervised preprocessing method for pre-training the parameters in CNNs. Another limitation is the blurring introduced by the convolutional manipulation, and how to cope with this problem is another research direction.

REFERENCES

Acharyya, M., R. K. De, and M. K. Kundu. 2003. "Segmentation of remotely sensed images using wavelet features and their evaluation in soft computing framework." *IEEE Transactions on Geoscience and Remote Sensing* 41: 2900–2905. doi:10.1109/TGRS.2003.815398.

Audebert, N., B. Le Saux, and S. Lefèvre. 2016. "Semantic segmentation of earth observation data using multimodal and multi-scale deep networks." *arXiv preprint arXiv:1609.06846*. doi:10.1007/978-3-319-54181-5_12.

Benediktsson, J. A., J. A. Palmason, and J. R. Sveinsson. 2005. "Classification of hyperspectral data from urban areas based on extended morphological profiles." *IEEE Transactions on Geoscience and Remote Sensing* 43: 480–491. doi:10.1109/TGRS.2004.842478.

Bouvrie, J. 2006. "Notes on convolutional neural networks." (unpublished). http://cogprints.org/5869/.

Castelluccio, M., G. Poggi, C. Sansone, and L. Verdoliva. 2015. "Land use classification in remote sensing images by convolutional neural networks." *arXiv:1508.00092*.

Chanussot, J., J. A. Benediktsson, and M. Fauvel. 2006. "Classification of remote sensing images from urban areas using a fuzzy possibilistic model." *IEEE Geoscience and Remote Sensing Letters* 3: 40–44. doi:10.1109/LGRS.2005.856117.

Chen, L. C., G. Papandreou, I. Kokkinos, K. Murphy, and A. L. Yuille. 2014. "Semantic image segmentation with deep convolutional nets and fully connected crfs." *arXiv preprint arXiv*:1412.7062.

Chen, Y., H. Jiang, C. Li, X. Jia, and P. Ghamisi. 2016. "Deep feature extraction and classification of hyperspectral images based on convolutional neural networks." *IEEE Transactions on Geoscience and Remote Sensing* 54: 6232–6251. doi:10.1109/TGRS.2016.2584107.

Dell'Acqua, F., P. Gamba, A. Ferrari, J. A. Palmason, J. A. Benediktsson, and K. Arnason. 2004. "Exploiting spectral and spatial information in hyperspectral urban data with high resolution." *IEEE Geoscience and Remote Sensing Letters* 1: 322–326. doi:10.1109/LGRS.2004.837009.

Farabet, C., C. Couprie, L. Najman, and Y. LeCun. 2013. "Learning hierarchical features for scene labeling." *IEEE Transactions on Pattern Analysis and Machine Intelligence* 35: 1915–1929. doi:10.1109/TPAMI.2012.231.

Haralick, R. M., K. Shanmugam, and I. Dinstein. 1973. "Textural features for image classification." *IEEE Transactions on Systems, Man, and Cybernetics* SMC-3: 610–621. doi:10.1109/TSMC.1973.4309314.

Huang, X., X. Liu, and L. Zhang. 2014. "A multichannel gray level co-occurrence matrix for multi/hyperspectral image texture representation." *Remote Sensing* 6(9): 8424–8445. doi:10.3390/rs6098424.

Huang, X., and L. Zhang. 2013. "An SVM ensemble approach combining spectral, structural, and semantic features for the classification of high-resolution remotely sensed imagery." *IEEE Transactions on Geoscience and Remote Sensing* 51: 257–272. doi:10.1109/TGRS.2012.2202912.

Huang, X., L. Zhang, and P. Li. 2007. "An adaptive multiscale information fusion approach for feature extraction and classification of IKONOS multispectral imagery over urban areas." *IEEE Geoscience and Remote Sensing Letters* 4: 654–658. doi:10.1109/LGRS.2007.905121.

Krizhevsky, A., I. Sutskever, and G. E. Hinton. 2012. Imagenet classification with deep convolutional neural networks. *Paper presented at advances in neural information processing systems*, Lake Tahoe, NV, 1097–1105.

LeCun, Y., Y. Bengio, and G. Hinton. 2015. "Deep learning." *Nature* 521(7553): 436–444. doi:10.1038/nature14539.

LeCun, Y., L. Bottou, Y. Bengio, and P. Haffner. 1998. "Gradient-based learning applied to document recognition." *Proceedings of the IEEE* 86(11): 2278–2324. doi:10.1109/5.726791.

LeCun, Y. A., L. Bottou, G. B. Orr, and K.-R. Müller. 2012. "Efficient backprop," in *Neural Networks: Tricks of the Trade: Second Edition*, G. Montavon, G. B. Orr, and K.-R. Müller, Eds., Berlin, Heidelberg: Springer Berlin Heidelberg, 9–48.

Lenz, I., H. Lee, and A. Saxena. 2015. "Deep learning for detecting robotic grasps." *The International Journal of Robotics Research* 34(4–5): 705–724. doi:10.1177/0278364914549607.

Long, J., E. Shelhamer, and T. Darrell. 2015. Fully convolutional networks for semantic segmentation. *Paper presented at Proceedings of the IEEE Conference on Computer Vision and Pattern Recognition* 3431–3440. doi:10.1109/CVPR.2015.7298965.

Ma, X., H. Wang, and J. Geng. 2016. "Spectral-spatial classification of hyperspectral image based on deep auto-encoder." *IEEE Journal of Selected Topics in Applied Earth Observations & Remote Sensing* 9(9): 4073–4085. doi:10.1109/JSTARS.2016.2517204.

Maggiori, E., Y. Tarabalka, G. Charpiat and P. Alliez. 2016a. Fully convolutional neural networks for remote sensing image classification. *Paper presented at IEEE International Geoscience and Remote Sensing Symposium (IGARSS)*, 5071–5074. doi:10.1109/IGARSS.2016.7730322.

Maggiori, E., Y. Tarabalka, G. Charpiat, and P. Alliez. 2016b. "High-resolution semantic labeling with convolutional neural networks." *arXiv preprint arXiv:1611.01962.*

Maggiori, E., Y. Tarabalka, G. Charpiat, and P. Alliez. 2017. "Convolutional neural networks for large-scale remote-sensing image classification." *IEEE Transactions on Geoscience and Remote Sensing* 55(2): 645–657. doi:10.1109/TGRS.2016.2612821.

Marmanis, D., J. Wegner, S. Galliani, and U. Stilla. 2016. "Semantic segmentation of aerial images with an ensemble of CNSS." *ISPRS Annals of the Photogrammetry, Remote Sensing and Spatial Information Sciences,* 473–480. doi:10.5194/isprs-annals-III-3-473-2016.

Mnih, V., and G. E. Hinton. 2012. "Learning to label aerial images from noisy data." *Paper presented at Proceedings of the 29th International Conference on Machine Learning (ICML-12)*, Edinburgh, Scotland, UK, 567–574.

Mura, M. D., J. A. Benediktsson, B. Waske, and L. Bruzzone. 2010. "Morphological attribute profiles for the analysis of very high resolution images." *IEEE Transactions on Geoscience and Remote Sensing* 48(10): 3747–3762. doi:10.1109/TGRS.2010.2048116.

Penatti, O. A. B., K. Nogueira, and J. A. dos Santos. 2015. "Do deep features generalize from everyday objects to remote sensing and aerial scenes domains?" *Presented at The IEEE Conference on Computer Vision and Pattern Recognition (CVPR) Workshops*, 44–51. doi:10.1109/CVPRW.2015.7301382.

Pesaresi, M., and A. Gerhardinger. 2011. "Improved textural built-up presence index for automatic recognition of human settlements in arid regions with scattered vegetation." *IEEE Journal of Selected Topics in Applied Earth Observations and Remote Sensing* 4(1): 16–26. doi:10.1109/JSTARS.2010.2049478.

Raja, R. A. A., V. Anand, A. S. Kumar, S. Maithani, and V. A. Kumar. 2013. "Wavelet based post classification change detection technique for urban growth monitoring." *Journal of the Indian Society of Remote Sensing* 41: 35–43. doi:10.1007/s12524-011-0199-7.

Razavian, A. S., H. Azizpour, J. Sullivan, and S. Carlsson. 2014. "CNN features off-the-shelf: an astounding baseline for recognition." *Paper presented at IEEE Conference on Computer Vision and Pattern Recognition Workshops (CVPRW)*, 512–519. doi:10.1109/CVPRW.2014.131.

Rizvi, I. A., and B. K. Mohan. 2011. "Object-based image analysis of high-resolution satellite images using modified cloud basis function neural network and probabilistic relaxation labeling process." *IEEE Transactions on Geoscience and Remote Sensing* 49(12): 4815–4820. doi:10.1109/TGRS.2011.2171695.

Santos, J. A., P. H. Gosselin, S. Philipp-Foliguet, R. S. Torres, and A. X. Falao. 2012. "Multiscale classification of remote sensing images." *IEEE Transactions on Geoscience and Remote Sensing* 50(10): 3764–3775. doi:10.1109/TGRS.2012.2186582.

Santos, J. A., P. H. Gosselin, S. Philipp-Foliguet, R. S. Torres, and A. X. Falcão. 2013. "Interactive Multiscale Classification of High-Resolution Remote Sensing Images." *IEEE Journal of Selected Topics in Applied Earth Observations and Remote Sensing* 6 (4): 2020–2034. doi:10.1109/JSTARS.2012.2237013.

Tilton, J. C., Y. Tarabalka, P. M. Montesano, and E. Gofman. 2012. "Best merge region-growing segmentation with integrated nonadjacent region object aggregation." *IEEE Transactions on Geoscience and Remote Sensing* 50(11): 4454–4467. doi:10.1109/TGRS.2012.2190079.

Valero, S., J. Chanussot, J. A. Benediktsson, H. Talbot, and B. Waske. 2010. "Advanced directional mathematical morphology for the detection of the road network in very high resolution remote sensing images." *Pattern Recognition Letters* 31: 1120–1127. doi:10.1016/j.patrec.2009.12.018.

Valero, S., P. Salembier, and J. Chanussot. 2013. "Hyperspectral image representation and processing with binary partition trees." *IEEE Transactions on Image Processing* 22(4): 1430–1443. doi:10.1109/TIP.2012.2231687.

Vedaldi, A., and K. Lenc. 2015. "MatConvNet: Convolutional neural networks for MATLAB." *Paper presented at the Proceedings of the 23rd ACM international conference on Multimedia*, Brisbane, Australia.

Zhang, F., B. Du, and L. Zhang. 2016a. "Scene classification via a gradient boosting random convolutional network framework." *IEEE Transactions on Geoscience and Remote Sensing* 54(3): 1793–1802. doi:10.1109/TGRS.2015.2488681.

Zhang F., B. Du, L. Zhang, and M. Xu. 2016b. "Weakly supervised learning based on coupled convolutional neural networks for aircraft detection." *IEEE Transactions on Geoscience and Remote Sensing* 54: 5553–5563. doi:10.1109/TGRS.2016.2569141.

Zhang, G., R. Zhang, G. Zhou, and X. Jia. 2018. "Hierarchical spatial features learning with deep CNNs for very high resolution remote sensing image classification." *International Journal of Remote Sensing* 39(18): 5978–5996.

17 Surface Soil Moisture Retrieval from CBERS-02B Satellite Imagery

17.1 INTRODUCTION

Surface soil moisture (SSM) plays an important role in hydrological, agricultural, and ecological applications. With the rapid development of spaceborne sensor technologies, many algorithms have been developed to estimate the SSM at regional and global scales in the last three decades. These algorithms can be categorized as spectrum method, thermal inertia method, the crop water stress index (CWSI) method, microwave remote sensing method, and SAR imagery method. For example, for an optical satellite imagery, the optical vegetation coverage method has widely been applied to retrieve SSM by bands 2, 3, and 4 of Landsat TM satellite imagery with a retrieval accuracy of 90% approximately (Wang et al. 2004). The thermal inertia method is an indicator for soil moisture (Lu et al. 2009). Donincka et al. (2011) developed a flexible multitemporal approach to derive an approximation of thermal inertia (ATI) from daily Aqua and Terra MODIS observations and aimed at retrieving high accuracy of the SSMs. Surface temperature and albedo as the parameters of the CWSI method are also used for retrieving SSM. For the passive microwave remote sensing method, the most current studies are Mladenova et al. (2014), who discussed the theoretical basis and overviewed the selected algorithms of soil moisture using passive microwave-based techniques, and Goodberlet and Mead (2014), who developed a model of surface roughness using passive microwave remote sensing of bare soil moisture. In addition, Nagarajan et al. (2012) analyzed the impact of assimilating passive microwave observations on root-zone soil moisture under dynamic vegetation conditions, while Parinussa (2012) discussed the soil moisture retrievals from the WindSat spaceborne polarimetric microwave radiometer. Champagne et al. (2010) evaluated the soil moisture derived from passive microwave remote sensing over agricultural sites using ground-based soil moisture monitoring networks, and Jackson et al. (2010) carried out the validation of the advanced microwave scanning radiometer soil moisture products. On the other hand, the studies on retrieval soil moisture from active microwave sensors have also developed. For example, Baup et al. (2007) retrieved soil moisture in the African Sahel region by using ENVISAT/ASAR data, and the results show that the RMSE can be 0.028 when ASAR is at a low incidence angle when taking into account vegetation effects by using multi-angular radar data. Other researchers, such as Velde et al. (2012) and Brocca et al. (2011, 2010), also made significant efforts in the method development for SSM retrieval. High accuracy of the SSM retrieval is restricted by the sensors' condition when only using active/passive sensed imagery. For the active SAR imager method, a few investigators, such as Pierdicca et al. (2013), Gherboudj et al. (2011), Merzouki et al. (2011), De Zan et al. (2014), and Zhou et al. (2015), have developed different methods. Baghdadi et al. (2012) and Doubková et al. (2012) used the C-band polarimetric SAR parameters to characterize the soil moistures, Lievens and Verhoest (2011) used L-band SAR based on water cloud modeling, the IEM, and effective roughness parameters for the SSM retrieval in wheat fields, Balenzano et al. (2011) used a dense temporal series of C- and L-band SAR data, and Balenzano et al. (2013) used a temporal series of L- and X-band SAR data for SSM retrieval over agricultural crops.

Due to the complement of different data, a few investigators have presented improvements by integrating active and passive sensor data. Lee and Anagnostou (2004) combined TMI channel in TRMM satellite (passive) with 13.8 GHz precipitation radar (PR) observation (active) to retrieve and evaluate near-surface (5 cm) soil moisture for three consecutive years in Oklahoma State. Liu et al. (2012) presented an approach for combining four passive microwave and two active microwave products, and readjusted the integrated data sets to the common soil moisture parameters to estimate the soil moisture at global scale by surface model (GLDAS-1-Noah). Panciera et al. (2014) used active passive experiments (SMAPEx) for soil moisture retrieval in the SMAP mission; Piepmeier et al. (2014) used the radio-frequency interference mitigation for the soil moisture active passive microwave radiometer; and Prakash et al. (2012) developed a fusion approach to retrieve soil moisture with SAR and optical data.

This chapter presents a new model to calculate the SSM from CBERS-02B imagery. The advantages of the CBERS 02B satellite are the relatively short revisit period and the high spatial resolution. In particular, this satellite is a collaborative environmental satellite between China and Brazil. The development of a new model for SSM retrieval will largely be helpful in promotion of CBERS 02B satellite data in the global remote sensing community.

17.2 MODEL FOR SSM RETRIEVAL FROM CBERS-02B IMAGERY

17.2.1 SSM Retrieval Model of Landsat TM Image

The spectral reflection factor decreases exponentially with increasing soil moisture content. It enables the retrieval of soil moisture from an optical image. The relationship between the soil moisture content and spectral reflectance can be expressed by

$$R = me^{nP} \tag{17.1}$$

where R and P are the soil spectral reflectance and soil water content, respectively; m and n are undetermined coefficients.

Pixels of remotely sensed images usually contain soil and non-soil information (such as vegetation). For this reason, it must theoretically be bare soil reflectance used as a retrieval parameter (i.e., excluding the interference of non-soil) to achieve a high accuracy of retrieved SSM. The method, called optical vegetation coverage, was suggested to obtain the bare soil spectral radiance, and further obtains the bare soil spectral reflectance in accordance with the relationship between the spectral radiance and reflectance.

Compared with the visible bands (bands 2 and 3), soil spectral reflectance of the near-infrared band (Band 4) has the highest correlation to the measured data from the spectrophotometer. Therefore, Liu et al. (1997) established a soil moisture retrieval model from Landsat TM. After a lot of verification and experiments in field, this model was expressed by

$$P_{TM4} = 220 - 42.91\, lg\left[\frac{0.6968B_2 + 0.5228B_3 - 0.2237B_4 + 20.26}{0.003308B_2 + 0.002482B_3 - 0.00579B_4 + 1.089} - 18.0\right] \tag{17.2}$$

where P_{TM4} represents the retrieved bare soil moisture content; B_2, B_3, and B_4 represent spectral radiance of Landsat TM Band 2, Band 3, and Band 4, respectively.

17.2.2 Spectral Radiance Relationships Between Landsat TM and CBERS-02B Images

The relationship between the average spectral radiance of each band and its average spectral reflectance can be expressed as

$$B_i = R_i \tau_i B_{si} \tag{17.3}$$

where i is band number; B_i represents average spectral radiance for band i; R_i is average spectral reflectance of the soil; τ_i represents average atmospheric transmittance; B_{si} is average solar radiance.

Landsat TM images applied in estimation of the SSM content have similar bands of CBERS-02B CCD spectrum range (see Table 17.1) (Pan et al. 2008). For example, Band 2 is green, Band 3 is red, and Band 4 is near-infrared.

With Equation 17.3, the average spectral radiance of each band in Landsat TM and CBERS-02B image can be expressed by

$$\frac{B'_i}{B''_i} = \frac{R'_i \tau'_i B'_{si}}{R''_i \tau''_i B''_{si}} = \frac{R'_i}{R''_i} \cdot \frac{\tau'_i}{\tau''_i} \cdot \frac{B'_{si}}{B''_{si}} \tag{17.4}$$

where and B_i'' stands for average spectral radiance of Landsat TM and CBERS-02B image per band, respectively; the other symbols have similar meanings. The values for each sensor can be obtained by sample statistics, ideal atmospheric transmittance curve analysis, and the solar irradiance analysis, which are discussed below, individually.

17.2.3 Average Spectral Reflectance

The average spectral reflectance for each band can be obtained via a sample statistical method, but the simplified model for calculating the spectral reflectance in the visible and near-infrared bands is through calculating the surface reflectance, that is,

$$R_i = \frac{pi * L_i}{ESUNI * cos(SZ)} \tag{17.5}$$

where R_i is surface spectral reflectance; L_i denotes the radiance value in sensor entrance; *ESUNI* is the atmosphere solar irradiance, which *pi* is provided by official data; *SZ* represents the solar zenith angle, which can be obtained from the image header file. For the purpose of accuracy comparison, the selected Landsat TM and CBERS-02B CCD imagery should ideally capture the same area in the same time epoch. Therefore, the ratio of average spectral reflectance of the two sensors per band can be obtained by Equation 17.5, that is,

TABLE 17.1
Spectral Range of Landsat TM and CBERS-02B Images

Sensor	Landsat TM Spectrum Range (μm)	CBERS-02B
Band 2	0.52–0.60	0.52–0.59
Band 3	0.63–0.69	0.63–0.69
Band 4	0.76–0.90	0.77–0.89

$$\begin{cases} \dfrac{R'_2}{R''_2} = 0.9875 \\ \dfrac{R'_3}{R''_3} = 1.0033 \\ \dfrac{R'_4}{R''_4} = 1.0151 \end{cases} \tag{17.6}$$

17.2.4 Average Atmospheric Transmittance

The average atmospheric transmittance of each band can be solved via the ideal atmospheric transmittance curve analysis, since the light through the atmosphere, solar radiation, would be absorbed and scattered by a variety of gasses and aerosols (see Figure 17.1).

As observed in Figure 17.1, different wavelengths and solar zenith angles correspond to different atmospheric transmittance rates, individually. In other words, we can estimate the ideal mean atmospheric transmittance of each band using Figure 17.1 if the spectral range of the band and solar zenith angle are determined. This means that solar zenith angle (*SZ*) can be calculated by.

$$SZ = 90° - SE \tag{17.7}$$

where *SE* represents solar elevation angle, which can be obtained from the image header file. For example, the solar elevation angles of Landsat TM and CBERS-02B CCD image are 41° and 43°, respectively, in the selected image of the first test area. Therefore, the solar zenith angles were 49° and 47°. According to the two sensors' spectral ranges in Table 17.1, the mean atmospheric transmittance for each band can be estimated via Figure 17.1, and the estimated results are listed in Table 17.2.

With Table 17.2 and Equation 17.4, we have

$$\begin{cases} \dfrac{\tau'_2}{\tau''_2} = 0.9965 \\ \dfrac{\tau'_3}{\tau''_3} = 0.9989 \\ \dfrac{\tau'_4}{\tau''_4} = 0.9968 \end{cases} \tag{17.8}$$

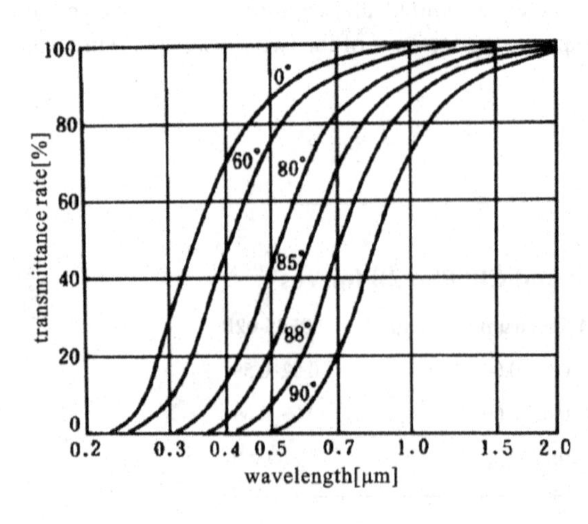

FIGURE 17.1 The ideal atmospheric transmittance diagram (Qian et al. 1995).

TABLE 17.2
Average Atmospheric Transmittance of Landsat TM and CBERS-02B Images.

Sensor	Band 2	Band 3	Band 4
Landsat TM	84.8%	92.2%	95.2%
CBERS-02B CCD	85.1%	92.3%	95.5%

TABLE 17.3
Average Solar Irradiance of Landsat TM and CBERS-02B (Unit: W/(m²·sr·μm)).

Sensor	Band 2	Band 3	Band 4
Landsat TM	1829	1557	1047
CBERS-02B CCD	1782.83	1546.52	1065.62

17.2.5 Average Solar Radiance

The average soil radiance ratio of Landsat TM and CBERS-02B CCD images can be estimated by the solar irradiance at the top of the atmosphere. Solar irradiance of Landsat TM is obtained from the published document, which is determined by satellite authority regularly. The CBERS-02B's data, which can be obtained by SBDART, are shown in Table 17.3.

With Table 17.3 and Equation 17.4, we obtain

$$\begin{cases} \dfrac{B'_{s2}}{B''_{s2}} = 1.0259 \\ \dfrac{B'_{s3}}{B''_{s3}} = 1.0068 \\ \dfrac{B'_{s4}}{B''_{s4}} = 0.9825 \end{cases} \tag{17.9}$$

17.2.6 SSM Model for Retrieval from CBERS-02B Image

When substituting the ratio of average spectral reflectance, average atmospheric transmittance, and average solar radiance above into Equation 17.4, it yields

$$\begin{cases} \dfrac{B'_{2}}{B''_{2}} = 1.0095 \\ \dfrac{B'_{3}}{B''_{3}} = 1.0009 \\ \dfrac{B'_{4}}{B''_{4}} = 0.9942 \end{cases} \tag{17.10}$$

Rewriting Equation 17.10, it yields

$$\begin{cases} B'_2 = 1.0095 * B''_2 \\ B'_3 = 1.0009 * B''_3 \\ B'_4 = 0.9942 * B''_4 \end{cases} \tag{17.11}$$

Substituting Equation 17.11 into Equation 17.2, the soil moisture retrieval model from CBERS-02B image can be expressed by

$$P_{CBERS4} = 209 - 42.91 lg\left[\frac{0.7034B_2 + 0.5233B_3 - 0.2224B_4 + 12.0}{0.003339B_2 + 0.002484B_3 - 0.005746B_4 + 0.5} - 18.0\right] \tag{17.12}$$

Where, B_2, B_3 and B_4 represent the spectral radiance of CBERS-02B images at band 2, 3, and 4, respectively.

17.2.7 Accuracy Evaluation Method

Previous literature demonstrated that the SSM retrieval model from Landsat TM images has an accuracy of higher than 90%. This accuracy index can be used to verify the retrieval results using the model proposed in this chapter. The measures include correlation coefficient (R), RMSE, and bias, which are expressed as follows.

$$R_{XY} = \frac{\sum_{i=1}^{N}\left[(X_i - \bar{X})(Y_i - \bar{Y})\right]}{\sqrt{\sum_{i=1}^{N}(X_i - \bar{X})^2} \cdot \sqrt{\sum_{i=1}^{N}(Y_i - \bar{Y})^2}} \tag{17.13}$$

$$RMSE = \sqrt{\frac{1}{N}\sum_{i=1}^{N}(Y_i - X_i)^2} \tag{17.14}$$

$$Bias = \frac{1}{N}\sum_{i=1}^{N} Y_i - \frac{1}{N}\sum_{i=1}^{N} X_i \tag{17.15}$$

Where X represents the SSM retrieved from Landsat TM image, which are taken as true values; Y is the SSM retrieved by the proposed model from the CBERS-02B image. The correlation coefficient is used to measure the linear correlation between the two variables. The greater the absolute value of R, the higher the linear correlation between the variables. The RMSE is used to measure the deviation between two retrieved SSMs. The smaller the RMSE, the greater the reliability of the retrieved SSMs. The bias reflects the degree of deviation of the observed values from the true value overall.

17.3 EXPERIMENT AND ANALYSIS

17.3.1 Test Area and Data

- **The First Test Area and Data Set**

The first test area is located in Jili Village of Laibin county, Guangxi Province, China, at latitude 23°32′20″ N through 23°37′8″ N and longitude 109°07′50″ E through 109°14′ E (see Figure 17.2) and it is nearly 93 km^2. The test area is a typical karst plain landform in which there are many

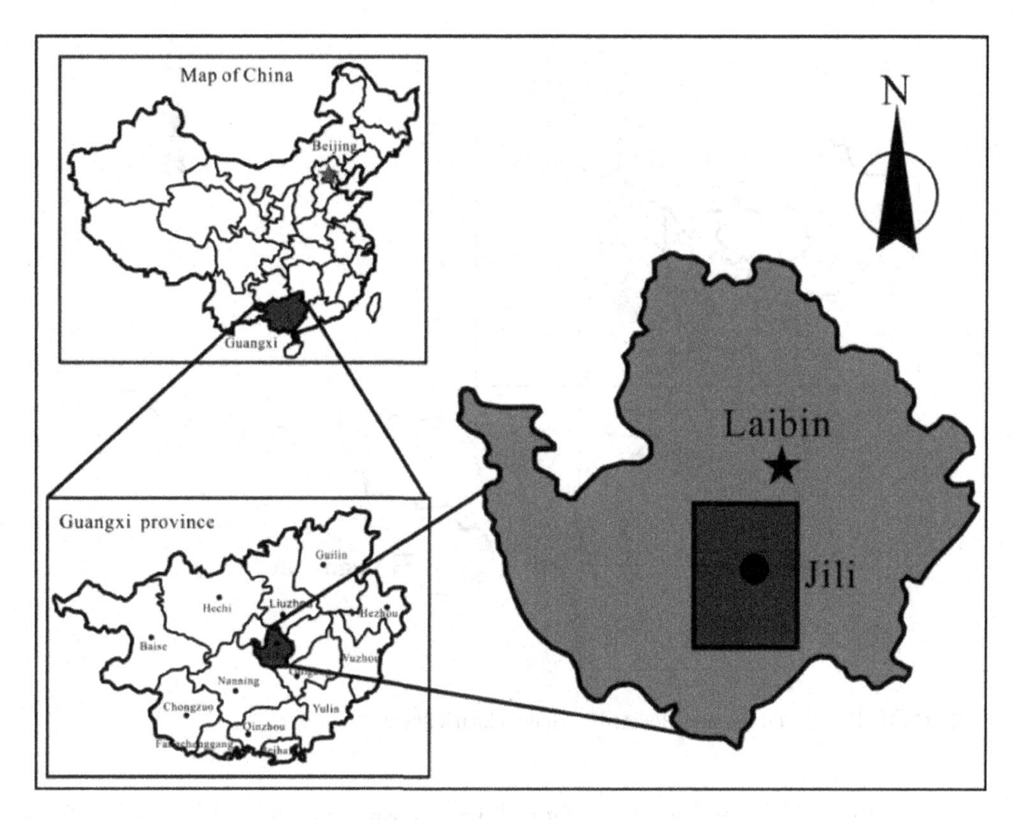

FIGURE 17.2 The test area of Jili Village.

desertified rocks, and the main crops are sugarcane and rice. The image was captured at the end of November, which is a harvest season.

For the purpose of comparison analysis, the selected Landsat TM and CBERS-02B images should ideally be captured at an exact time epoch in the same area. However, the Landsat and CBERS-02B satellites rarely simultaneously meet the two conditions in practice.

The CBERS-02B CCD image data (PATH73 ROW4), acquired at 03:33:48 on November 25, 2008, covered the Jili test area and were provided by the Chinese Ground Station and China Resources Satellite Application Center (http://www.cresda.com/n16/index.html). CBERS-02B satellite was successfully launched in September 2007 and carried CCD, HR, and WFI sensors. The CBERS-02B CCD sensor contains visible, infrared, and panchromatic bands, with 19.5 m spatial resolution and a width of 113 km, and a revisit time of 26 days.

The Landsat 5 TM data with international frame number of 125–43 was acquired at 02:53:41 on November 20, 2008 and covered the Jili area. The test data was downloaded from the website at http://www.nasa.gov/. The spatial resolution is 30 m. The Landsat-5, with the main detector TM, is operated on a sun-synchronous, 705 km orbit height, and a 16 days' revisit cycle.

Although the two data sets have a five-day interval, no rain, significant climate change, or ground surface vegetation changes occurred during this time frame. It can therefore be used for comparison analysis of two data sets.

- **The Second Test Area and Data Set**

The second test area is located in Yuanjiaduan Village of Jiujiang county, Jiangxi Province, China, at latitude 28°51′26″ N through 28°54′17″ N, and longitude 114°28′34″ E through 114°32′22″ E (see

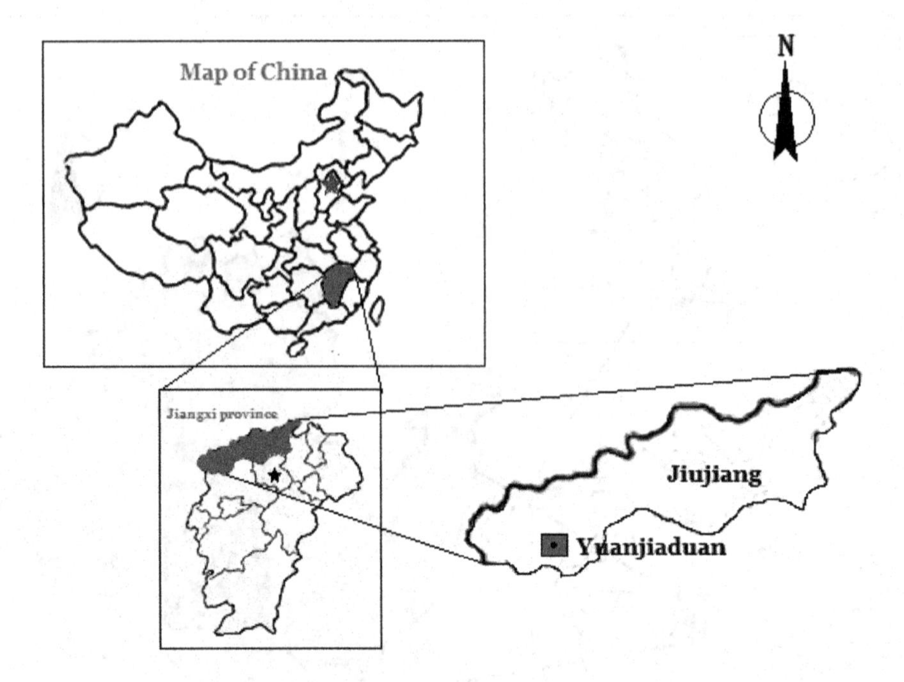

FIGURE 17.3 The second test area, located in Yuanjiaduan Village.

Figure 17.3). The second test area covers nearly 33 km^2, consisting of desertified rock, vegetables, and rice.

In the second test area, we found a pair of images captured by Landsat 5 TM at 02:35:14.619 on January 02, 2009 and captured by CBERS-02B at 03:17:11 on January 02, 2009, respectively. The two data sets only have a 41-minute time interval, but no rain, significant climate change, or ground surface vegetation changes occurred within 41 minutes. It therefore meets the requirement for comparison analysis of two data sets. The details of the data sets are: the CBERS-02B CCD image (PATH373 ROW67) capturing Yuanjiaduan test area was obtained from the Chinese ground receiving station and the Landsat 5 image international frame number 122–40 was downloaded from http://glovis.usgs.gov/.

17.3.2 Image Preprocessing

Data preprocessing, finished using ENVI 5.0, contains stripe noise removal, atmospheric correction, and geometric registration. A 3 × 3 low-pass filtering method is applied for noise removal, and the band control method is applied for the atmospheric radiation correction. The CBERS-02B and Landsat TM images are co-registered in the Beijing 54 coordinate system, which employs a Gauss-Kruger projection at 6° zonation. The registration model is a quadratic polynomial corrective model, and the resampling method is nearest-neighbor. To ensure Landsat TM data has a consistent resolution with the CBERS-02B image, the Landsat TM image is resampled at a resolution of 19.5 m.

17.3.3 SSM Retrieval and Analysis

The soil moisture retrieval algorithm from Landsat TM spectral radiance is calculated by Chander and Markham (2003).

$$L_{\lambda} = DN_{\lambda} \times Gain + Bias \tag{17.16}$$

where *Gain* and *Bias* can be obtained from the image header file. The CBERS-02B CCD images' spectral radiance can be obtained by

$$L_k = \frac{DN_k}{A_k} \tag{17.17}$$

where L_k indicates radiation of band k in sensor entrance; k is band number; DN_k represents the gray value of band k; A_k is the gain of band k, which is usually from the image header file.

(1) For the First Test Area

With Equation 17.2 and Equation 17.14, combining Equation 17.16 and Equation 17.17, the SSMs in the Jili experimental area are calculated and depicted in Figure 17.4 and Figure 17.5. We take the SSMs retrieved from the Landsat TM image as true values, and accumulate the total SSMs by adding up the SSM of each pixel in the CBERS-02B image. This result reveals that the relative difference between the two SSMs is 8.74%. This means that the SSM retrieved by the model proposed in this chapter can achieve 91.26%, relative to the SSMs retrieved from Landsat 5 TM.

To facilitate the deep comparison analysis, the retrieved SSM content is divided into six levels: <5%, 5–10%, 10–15%, 15–20%, 20–25%, and > 25% (see Figures 17.4 and 17.5). The SSM less than 5% (blue areas) presents generally artificial facilities, such as roads, bare areas, and rocky desertified areas, while higher than 25% (red areas) stands for water and forest areas with high vegetation coverage. The others may be the shrub land, dry land, paddy field, sugarcane land, and so on. To further analyze the differences in SSMs retrieved by the two satellite images, three magnified windows, which stand for rocky desertified land, sugarcane land, and shrub land, respectively, are employed to visually check the accuracy of the retrievals of SSM from Landsat TM and CBERS-02B images. As observed from the enlarged windows in Figure 17.4a, 17.4b and 17.4c and Figure 17.5a, 17.5b, and 17.5c, the SSMs retrieved by the two satellites are very close.

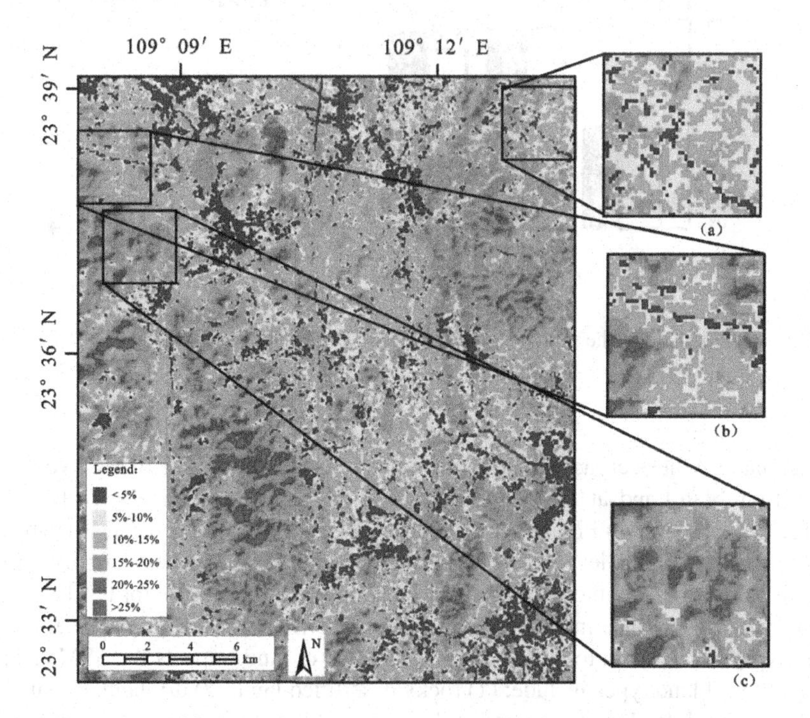

FIGURE 17.4 The retrieved SSMs level diagram from Landsat TM satellite imagery.

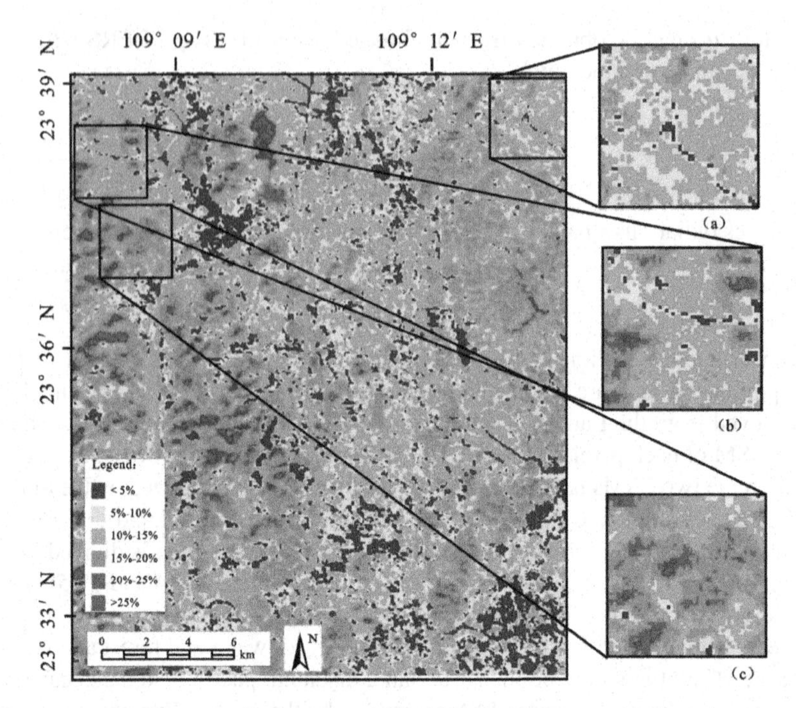

FIGURE 17.5 The retrieved SSMs level diagram from CBERS-02B satellite imagery.

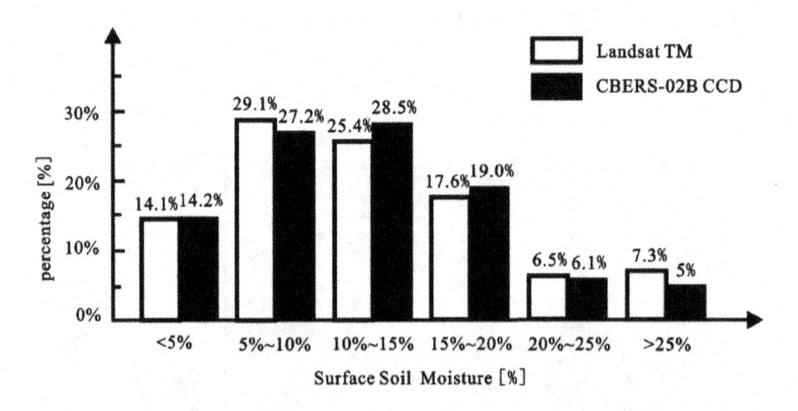

FIGURE 17.6 Comparison of the proportion of retrieved SSMs from Landsat TM and CBERS-02B satellite imagery.

To further analyze the accuracy of the retrieved SSMs, the proportions of six levels of soil moisture content from both Landsat TM and CBERS-02B images are statistically analyzed and shown in Figure 17.6. As observed in Figure 17.6, the proportions of the retrieved SSMs from Landsat TM and CEBRS imagery for each level are very close. For example, the smallest difference in two SSMs is about 0.1% for level 1, and the biggest difference in two SSMs is 2.9% for level 3.

Different types of land use impact the accuracy of the SSM retrieval. For the Jili test area, we selected six typical land types to evaluate how each type of land impacts the SSMs from CBERS-02B. The six typical land types include: (1) rocky desertified land, (2) dry land, (3) sugarcane land, (4) rice field, (5) shrub land, and (6) woodland.

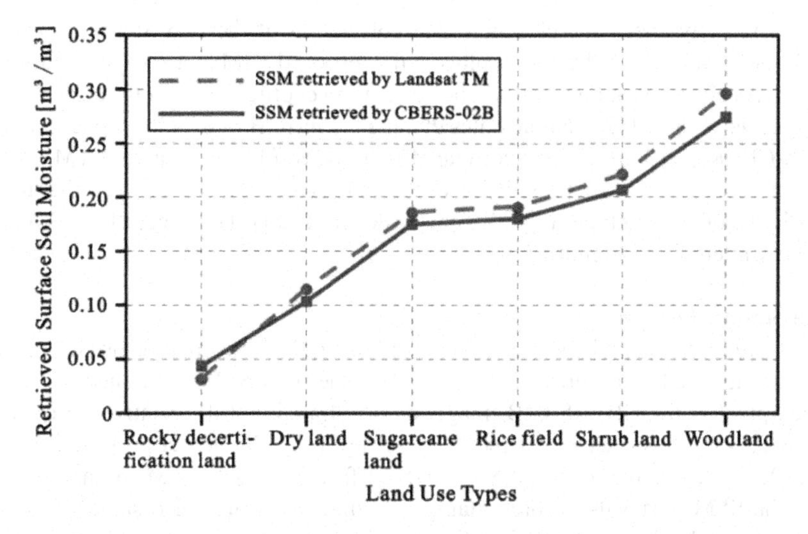

FIGURE 17.7 The contrast diagram of mean soil moisture retrieved by Landsat TM and CBERS-02B CCD data for six difference land use categories, in which the horizontal axis represents the retrieved SSMs from the Landsat TM image, and the vertical axis represents the retrieved SSMs from the CBERS-02B image.

For the six types of land, we randomly select 100 sample points from each type of land for statistical analysis. The planar scatter plots in Figure 17.7 illustrate the relationship between the average SSMs retrieved from the Landsat TM and CBERS-02B imagery. As observed in Figure 17.7, it can be noted that the retrieved SSMs from Landsat TM are slightly higher than those from CBERS-02B in five types of lands, excluding rocky desertified land, where the conclusion is opposite. This phenomenon may be caused by the mix pixels, whose percentage of vegetation is bigger than that of rocky desertified land. This means that the retrieved SSMs are affected by vegetation spectral reflectivity.

In addition, the correlation coefficients calculated using Equation 17.15, the RMSE calculated using Equation 17.14, and bias calculated using Equation 17.15 are used for evaluation of the relative accuracy of SSMs retrieved from CBERS-02B imagery relative to the Landsat TM. The results are shown in Table 17.4. It can be concluded from Table 17.4 that the correlation between the retrieved SSMs from Landsat TM and CBERS in karst rocky land is highest, and the corresponding RMSE and bias are smallest; meanwhile, the correlation in woodland is lowest, and the corresponding RMSE and bias are biggest. The phenomena may be caused by the vegetation coverage in the woodland, which significantly impacts the SSM retrieval. In contrast, the rocky desertified land has little vegetation coverage. The phenomena also reveal that although the optical vegetation coverage model is used for SSM retrieval, it does not eliminate the influence of vegetation completely.

TABLE 17.4
The Statistical Analysis of Accuracy of SSM for six Types of Land

Feature Types	R	RMSE [m³/m³]	Bias [m³/m³]
Rocky desertified land	0.939	0.005	0.003
Dry land	0.901	0.011	–0.005
Sugarcane land	0.891	0.018	–0.012
Rice field	0.880	0.015	–0.007
Shrub land	0.821	0.024	–0.016
Woodland	0.808	0.039	–0.022

In order to visually check the closeness and correlation of the two SSMs retrieved from the CBERS-02B and Landsat TM, the scatter plots of the retrieved SSMs for six types of land are shown in Figure 17.8. As observed in Figure 17.8, the correlation coefficients in both rocky desertified land and dry land achieve 0.9, which demonstrates that the SSMs retrieved by the model proposed in this chapter from CBERS-02B are closely correlated with the SSMs from Landsat TM. Moreover, the SSM points in the two land types (see Figure 17.8a, and Figure 17.8b) gather together more than the SSM points in the other land types (e.g., Figure 17.8e and 17.8f). This means the SSM points in the woodland are scattered (see Figure 17.8f).

(2) For the Second Test Area

The soil moisture in the second test area is retrieved using the same method as at the first test area. The results are depicted in Figures 17.9 and 17.10. The total SSMs, obtained by adding up the SSM of each pixel in the CBERS-02B image, demonstrated that the relative difference between two SSMs is 8.91%. This means that the SSM retrieved by the model proposed in this chapter can achieve 91.09%, relative to the SSMs retrieved from Landsat 5 TM. As a visual check, the accuracy of the SSM retrievals in three magnified windows, which also stand for rocky desertified land, sugarcane land, and shrub land (see Figure 17.9a, 17.9b and 17.9c and Figure 17.10a, 17.10b, 17.10c), respectively, are very close, when comparing the retrieved SSM contents, which are divided into six levels: <5%, 5–10%, 10–15%, 15–20%, 20–25%, and >25% (see Figure 17.9 and Figure 17.10).

In the second test area, six typical land types – (1) rocky desertified land, (2) dry land, (3) vegetable land, (4) rice field, (5) shrub land, and (6) woodland – are chosen to evaluate how each type of land impacts the SSMs from CBERS-02B. The correlation coefficients, RMSE, and bias are also used for evaluation of the relative accuracy of SSMs retrieved from CBERS-02B imagery relative to the Landsat TM. The results are shown in Table 17.5. It can be concluded from Table 17.5 that the correlation between the SSMs retrieved from Landsat TM and from CBERS-02B in karst rocky land is highest, and the corresponding RMSE and bias are smallest; meanwhile, the correlation in woodland is lowest, and the corresponding RMSE and bias are bigger. This conclusion is consistent with that in the first test area. The phenomena are mainly caused by the vegetation coverage in the woodland, which significantly impacts the SSM retrieval. In addition, since the images in the second test area were taken in January, when there were no crops in the rice fields, the correlations for rice fields are higher than the correlations for vegetable land.

Similarly, the scatter plots of the retrieved SSMs for six types of land are shown in Figure 17.11. As observed in Figure 17.11, the correlation coefficients for both rocky land and dry land achieve 0.90, which demonstrates that the SSMs retrieved by the model proposed in this chapter from CBERS-02B are closely correlated with the SSMs from Landsat TM. Moreover, the SSM points in the two land types (see Figure 17.11a) gather together more than the SSM points in the other land types (e.g., Figure 17.11e and 17.11f), and the SSM points in the woodland are scattered (see Figure 17.11f).

Discussions for Two Test Areas

To compare the impact of two test areas to the model developed in this chapter, the correlation coefficients in both the first and second test areas are depicted in Figure 17.12. The data sets for Landsat TM and CBERS-02B have an time interval of 5 days in the first test area, but only 41 minutes in the second test area, which should have less impact on the retrievals of SSMs than that in the first test area. As observed in Figure 17.12, it can be seen that the correlations of rocky desertified land, rice land, and shrub land are higher in the second test area than in the first test area. In contrast, the dry land and woodland have almost the same correlation coefficient, which may be because of the fact that the images in the first test area were captured at the end of November, which is a harvest season. This means that the vegetation coverage of sugarcane land was less in the first test area than in the second test area. Therefore, it can be concluded that the shorter the two image acquisition time intervals, the higher the correlation of SSMs.

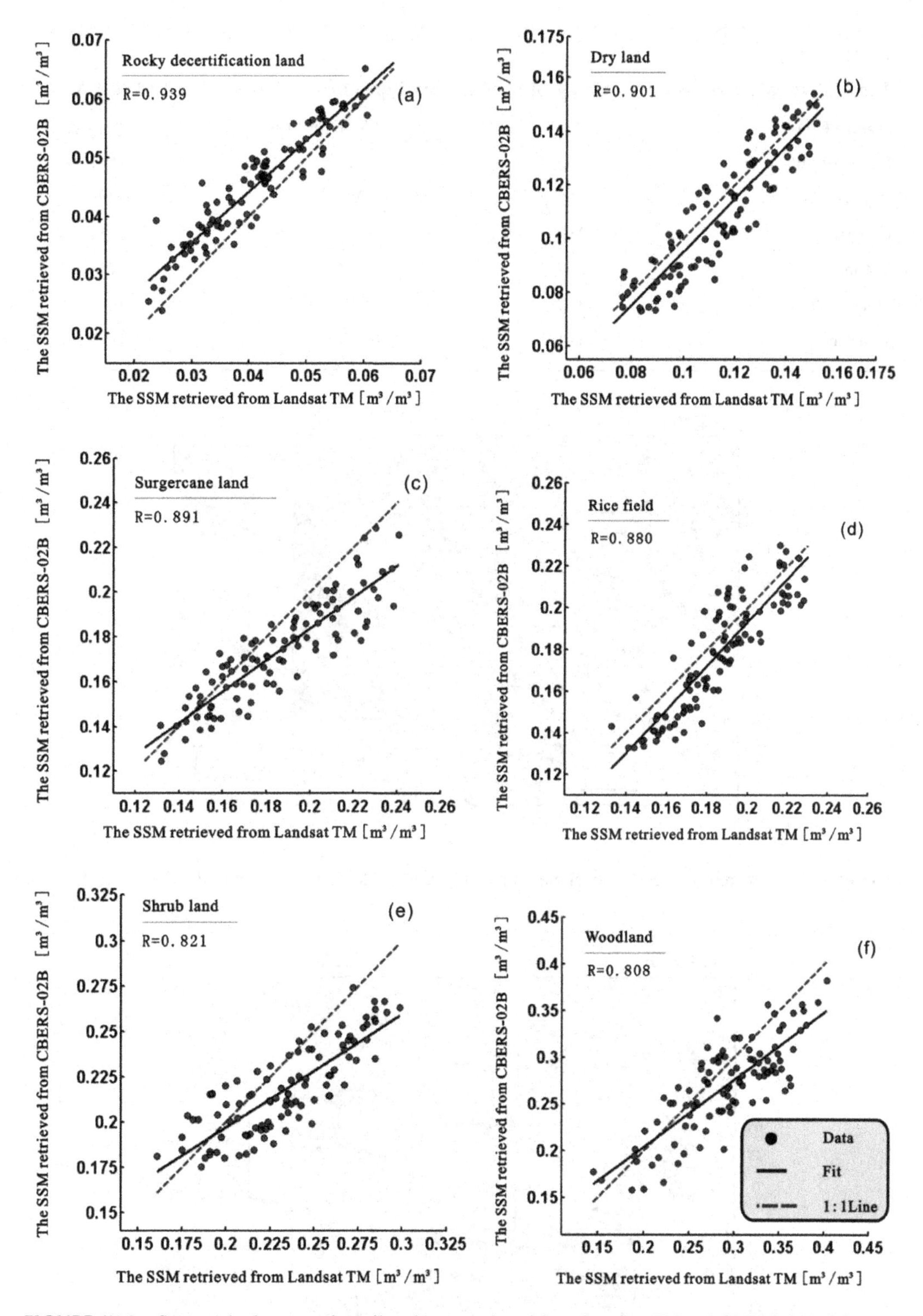

FIGURE 17.8 Scatter plot between the soil moisture retrieved from Landsat TM and CBERS-02B CCD data for six different feature types.

TABLE 17.5
The Statistical Analysis of Accuracy of SSM for Six Types of Land in the Second Test Area

Feature Types	R	RMSE [m³/m³]	Bias [m³/m³]
Rocky desertified land	0.95	0.045	0.001
Dry land	0.90	0.057	0.099
Rice field	0.90	0.070	–0.011
Vegetable land	0.87	0.099	–0.011
Shrub land	0.82	0.099	0.013
Woodland	0.81	0.097	0.016

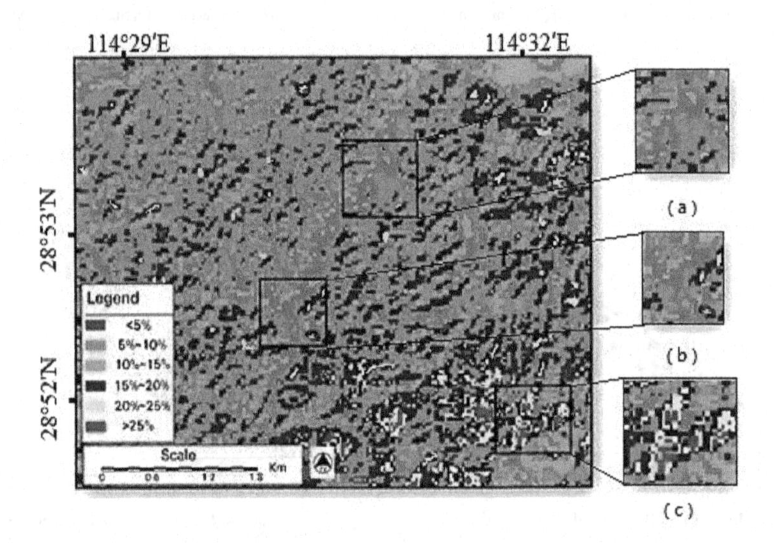

FIGURE 17.9 The retrieved SSMs level diagram from Landsat TM satellite imagery in the second test area.

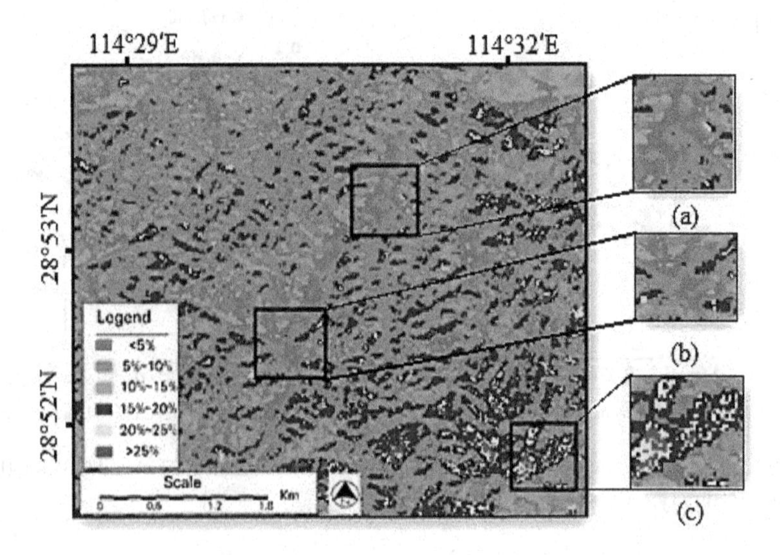

FIGURE 17.10 The retrieved SSMs level diagram from CBERS-02B satellite imagery in the second test area.

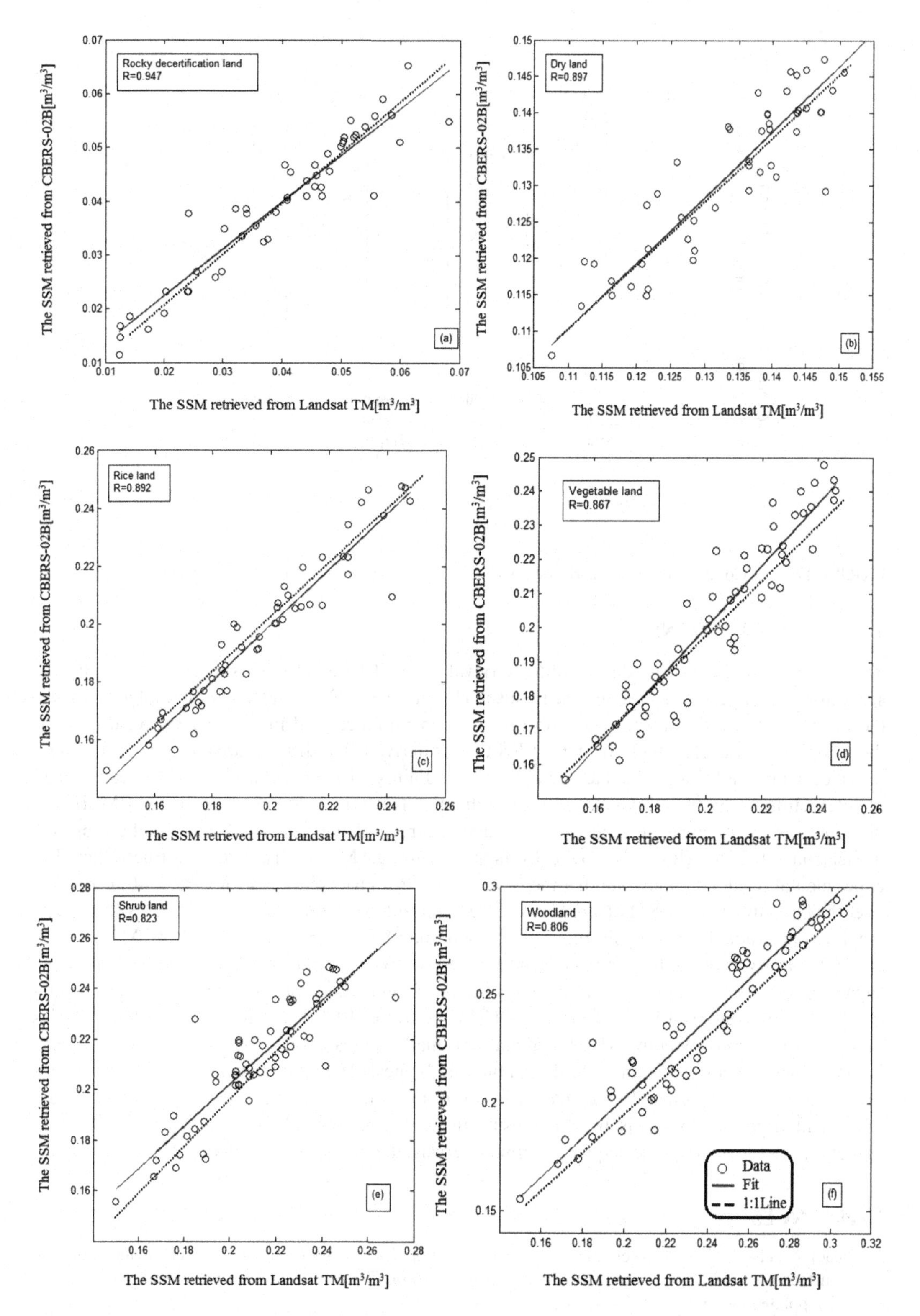

FIGURE 17.11 Scatter plot between the soil moisture retrieved from Landsat TM and CBERS-02B CCD data for six different feature types in the second test area.

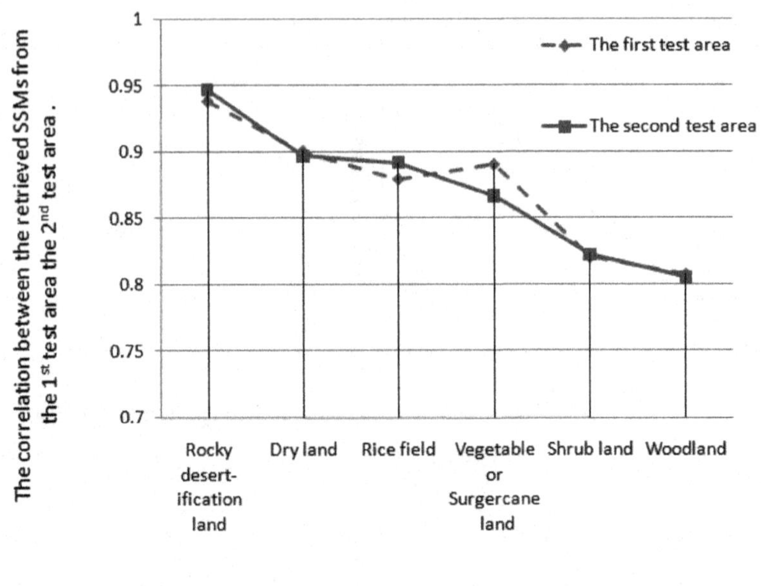

FIGURE 17.12 Comparison analysis of the retrieved SSMs from the first and the second test areas.

17.4 CONCLUSIONS

This chapter introduced a model for SSM retrieval from CBERS-02B. This model first obtains the spectral radiance relationship between Landsat TM and CBERS-02B imagery, and then calculates the mean spectral reflectance, mean atmospheric transmittance, and mean solar radiance. Finally, the model for retrieval of SSM from CBERS-02B is derived. The Jili test area, located in Guangxi province, China and Yuanjiaduan test area, located in Jiangxi province, China are employed to verify the correction of the model developed in this chapter. The correlation coefficient, the RMSE, and bias are used as the measure indexes to evaluate accuracy of the SSMs retrieved by the proposed model from CBERS-02B relative to the SSMs from Landsat TM, which are taken as true values. The experimental results demonstrate that the developed SSM retrieval model for CBERS-02B reaches over 90% relative to the SSMs from Landsat TM, and the SSM retrieved from CBERS-02B using the developed model has a consistent accuracy with the SSM retrieved from Landsat TM.

Further analysis through six types of land in both test areas (karst rocky desertified land, dry land, sugarcane land or vegetable land, rice field, shrub land, and woodland) discovered that the correlation between the retrieved SSMs from Landsat TM and from CBERS-02B in rocky desertified land is highest, and the corresponding RMSE and bias are smallest; meanwhile, the correlation in woodland is lowest, and the corresponding RMSE and bias are biggest. The phenomena reveal that mix pixels involving both vegetation and rocky have a lower accuracy than the pixels only involving pure bare rocky land. In addition, with discussion of results in the first and second test areas, it can be concluded that the shorter the two image acquisition time intervals, the higher the correlation of the SSMs.

REFERENCES

Baghdadi, N., et al. (2012). A potential use for the C-band polarimetric SAR parameters to characterize the soil surface over bare agriculture fields, *IEEE Transactions on Geoscience and Remote Sensing*, vol. 50, no. 10, pp. 3844–3858.

Balenzano, A., et al. Dense temporal series of C- and L-band SAR data for soil moisture retrieval over agricultural crops, *IEEE Journal of Selected Topics in Applied Earth Observations and Remote Sensing*, vol. 4, no. 2, pp. 439–450, 2011.

Balenzano, A., et al. 2013 On the use of temporal series of L- and X-band SAR data for soil moisture retrieval, Capitanata plain case study, *European Journal of Remote Sensing*, vol. 46, no. 1, pp. 721–737.

Baup, F., et al. Surface soil moisture estimation over the AMMA Sahelian site in Mali using ENVISAT/ASAR data, *Remote Sensing of Environment*, vol. 109, pp. 473–481, 2007.

Brocca, L., et al. ASCAT soil wetness index validation through in situ and modeled soil moisture data in central Italy, *Remote Sensing of Environment*, vol.114, pp. 2745–2755, 2010.

Brocca, L., et al. Soil moisture estimation through ASCAT and AMSR-E sensors: An inter-comparison and validation study across Europe, *Remote Sensing of Environment*, vol. 115, pp. 3390–3408, 2011.

Champagne, C., et al., Evaluation of soil moisture derived from passive microwave remote sensing over agricultural sites in Canada using ground-based soil moisture monitoring networks, *International Journal of Remote Sensing*, vol. 31, no. 14, pp. 3669–3690, 2010.

Chander, G., and Markham, B.L., Revised Landsat-5 TM radiometric calibration procedures, and post-calibration dynamic ranges. *IEEE Transactions on Geoscience and Remote Sensing*, vol. 41, pp. 2674–2677, 2003.

Donincka, J.V., et al. The potential of multitemporal aqua and terra MODIS apparent thermal inertia as a soil moisture indicator, *International Journal of Applied Earth Observation and Geoinformation*, vol.13, pp. 934–941, 2011.

Doubková, M., et al. Evaluation of the predicted error of the soil moisture retrieval from C-band SAR by comparison against modeled soil moisture estimates over Australia, *Remote Sensing of Environment*, vol. 120, pp. 188–196, 2012.

Gherboudj, I., et al. Soil moisture retrieval over agricultural fields from multi-polarized and multi-angular RADARSAT-2 SAR data, *Remote Sensing of Environment*, vol. 115, no. 1, pp. 33–43, 2011.

Goodberlet, Mark and Mead, James. A model of surface roughness for use in passive remote sensing of bare soil moisture, *IEEE Transactions on Geoscience and Remote Sensing*, vol. 52, no. 9, pp. 5498–5505, 2014.

Jackson, T., et al. (2010). Validation of advanced microwave scanning radiometer soil moisture products, *IEEE Transactions on Geoscience and Remote Sensing*, vol. 48, no. 12, pp. 4256–4272.

Lee, K.H., and Anagnostou E.N., A combined passive/active microwave remote sensing approach for surface variable retrieval using tropical rainfall measuring mission observations, *Remote Sensing of Environment*, vol. 92, pp.112–125, 2004.

Lievens, H., and Verhoest, N., On the retrieval of soil moisture in wheat fields from L-band SAR based on water cloud modeling, the IEM, and effective roughness parameters, *IEEE Geoscience and Remote Sensing Letters*, vol. 8, no. 4, pp. 740–744, 2011.

Liu, P., et al. A method for monitoring soil water contents using satellite remote sensing, *Journal of Remote Sensing (Chinese)*, vol. 1, no. 2, pp. 135–138, 1997.

Liu, Y., et al. Trend-preserving blending of passive and active microwave soil moisture retrievals, *Remote Sensing of Environment*, vol. 123, pp. 280–297, 2012.

Lu, S., et al. A general approach to estimate soil water content from thermal inertia, *Agricultural and Forest Meteorology*, vol. 149, pp. 1693–1698, 2009.

Merzouki, A., McNairn, H., and Pacheco, A., Mapping soil moisture using RADARSAT-2 data and local autocorrelation statistics, *IEEE Journal of Selected Topics in Applied Earth Observations and Remote Sensing*, vol. 4, no. 1, pp. 128–137, 2011.

Mladenova, E., et al. Remote monitoring of soil moisture using passive microwave-based techniques—Theoretical basis and overview of selected algorithms for AMSR-E, *Remote Sensing of Environment*, vol. 144, pp. 197–213, 2014.

Nagarajan, K., et al. Impact of assimilating passive microwave observations on root-zone soil moisture under dynamic vegetation conditions, *IEEE Transactions on Geoscience and Remote Sensing*, vol. 50, no. 11 PART 1, pp. 4279–4291, 2012.

Pan, Zhiqiang, Qiaoyans Fu, and Haoping Zhang Retrieval and application of band mean solar irradiance of CBERS-02 CCD, *Geo-information science*, vol.10, no. 1, pp.109–113, 2008.

Panciera, R., et al. The soil moisture active passive experiments (SMAPEx): toward soil moisture retrieval from the SMAP mission, *IEEE Transactions on Geoscience and Remote Sensing*, vol. 52, no. 1, pt.2, pp. 490–507, 2014.

Parinussa, R., Holmes, T., and Jeu, R., Soil moisture retrievals from the windSat spaceborne polarimetric microwave radiometer, *IEEE Transactions on Geoscience and Remote Sensing*, vol. 50, no. 7 PART 2, pp. 2683–2694, 2012.

Piepmeier, J.R., et al. Radio-frequency interference mitigation for the Soil Moisture Active Passive microwave radiometer, *IEEE Transactions on Geoscience and Remote Sensing*, vol. 52, no. 1, pt.2, pp. 761–775, 2014.

Pierdicca, N., et al. Monitoring soil moisture in an agricultural test site using SAR data: Design and test of a pre-operational procedure, *IEEE Journal of Selected Topics in Applied Earth Observations and Remote Sensing*, vol. 6, no. 3, pp. 1199–1210, 2013.

Prakash, R., Singh, D., and Pathak, N.P., A fusion approach to retrieve soil moisture with SAR and optical data, *IEEE Journal of Selected Topics in Applied Earth Observations and Remote Sensing*, vol. 5, no. 1, pp. 196–206, 2012.

Qian, Cangui, et al. Conversion pattern between MSS four channels radiometer and the channels associated with NOAA-AVHRR, and applying, *Application and Technology of Remote Sensing*, vol. 10, no.1, pp. 10–16, 1995.

Velde, V.D., et al. Soil moisture mapping over the central part of the Tibetan Plateau using a series of ASAR WS images, *Remote Sensing of Environment*, vol. 120, pp.175–187, 2012.

Wang, Cuizhen, et al. Soil moisture estimation in a semiarid rangeland using ERS-2 and TM imagery, *Remote Sensing of Environment*, vol. 90, pp. 178–189, 2004.

Zan, F.D., et al. A SAR interferometric model for soil moisture, *IEEE Transactions on Geoscience and Remote Sensing*, vol. 52, no. 1, pp. 418–425, 2014.

Zhou, G., et al. A new model for surface soil moisture retrieval from cbers-02b satellite imagery. *IEEE Journal of Selected Topics in Applied Earth Observations & Remote Sensing,* 8(2), 628–637, 2015.

18 Measuring Control Delay at Signalized Intersections Using GPS and Video Flow

18.1 INTRODUCTION

Delay at signalized intersections is considered as one of the primary elements of evaluating the level of service (LOS) (Bared 2005). A number of efforts have been made, such as those by Colyar and Rouphail (2003), Fu and Hellinga (2000), Hereth et al. (2006), Ji and Prevedouros (2005), Jiang et al. (2005), Kokkalis and Lakakis (2004), Koller et al. (1993), Li et al. (2002), Lin and Thomas (2005), Liu et al. (2006), Oh and Ritchie (2002), Van Leeuwaarden (2006), Washburn and Larson (2002), and Wolshon and Taylor (1999), to exactly estimate the actual control delay. The methods for collecting travel time have been presented by many authors in the past decades, such as Buehler et al. (1976) using ground-based time-lapse photography, Benekohal (1991) using aerial time-lapse photography and video, Quiroga et al. (2002) using a test car with GPS, Zhang et al. (2007) using the floating car technique, and Olszewski (1993) and Mousa (2002) tracing vehicle trajectories, known as path tracing. Other methods include a loop detector, manual counter, electronic distance measuring instrument (DMI), license plate matching, cellular phone tracking, and automatic vehicle identification (AVI). The manual method is labor-intensive and susceptible to human errors. Signal and dual loop detector methods cannot measure deceleration/acceleration rates. Video imaging processing, license plate matching, and AVI methods cannot exactly extract the traffic signal delay information because of the inherent image mismatch property. The DMI method is capable of providing accurate and reliable travel times, but it costs much in terms of equipment investment and maintenance (Hunter et al. 2006; Pan et al. 2007). Thus, the GPS method in recent years has increasingly been interesting to researchers for collecting traffic time data. The advantages of the GPS method can be summarized as follows:

1. It locates and tracks objects with an accuracy of centimeter level in real-time.
2. Stopped-time delay can be directly measured by GPS, and approach delay, and time in queue delay can indirectly be derived from GPS data.
3. Its measurement mode is flexible for any car.
4. GPS can directly provide positioning and timing data. The positioning data can be immediately fed into a geospatial database for integrating with other geospatial data, such as orthoimage, for temporal–spatial analysis.

However, the travel time measured by a test/probe vehicle with GPS is in fact a representative of the stream at an intersection. That is, the GPS data set is a point data mode, and cannot correctly represent area data describing other vehicles' travel time in a stream. For example, if a vehicle crosses the lane when approaching an intersection, or if a heavy vehicle decelerates/accelerates as a leading vehicle at an intersection, a single GPS test vehicle is obviously not able to represent other vehicles' speed in this steam. For this reason, an integration of the real-time kinematic (RTK) GPS technology and video image processing is proposed in this section to enhance the current existing method of calculating the control delay at an intersection. The RTK GPS can provide users with real-time positioning at an accurate level of decimeters, and the video can provide area data continuously recording all vehicles. Their integration attempts to increase the accuracy and simplifies the computation procedures.

18.2 MODEL FOR CALCULATING THE CONTROL DELAY

It has been widely accepted that the control delay components consist of the following (Quiroga et al. 2002; Olszewski 1993; Mousa 2002) (see Figure 18.1):

- *Total delay* includes deceleration, stopped, and acceleration delays. It is calculated as the difference between crossing times at accelerating to and decelerating from the normal travel speed.
- *Approach delay* includes stopped-time delay and the time lost when a vehicle decelerates from its normal speed to a stop and accelerates from the stop to its normal speed.
- *Stopped delay* implies the waiting time when the vehicle is completely stopped in a stationary position.
- *Deceleration delay* is the time difference that the vehicle experiences from the normal speed to the complete stop upstream of the stop line.
- *Acceleration delay* is the time difference that vehicle experiences from the complete stop to the normal velocity. Thus, it consists of two components: the delay upstream of the stop line; and the delay downstream of the stop line.

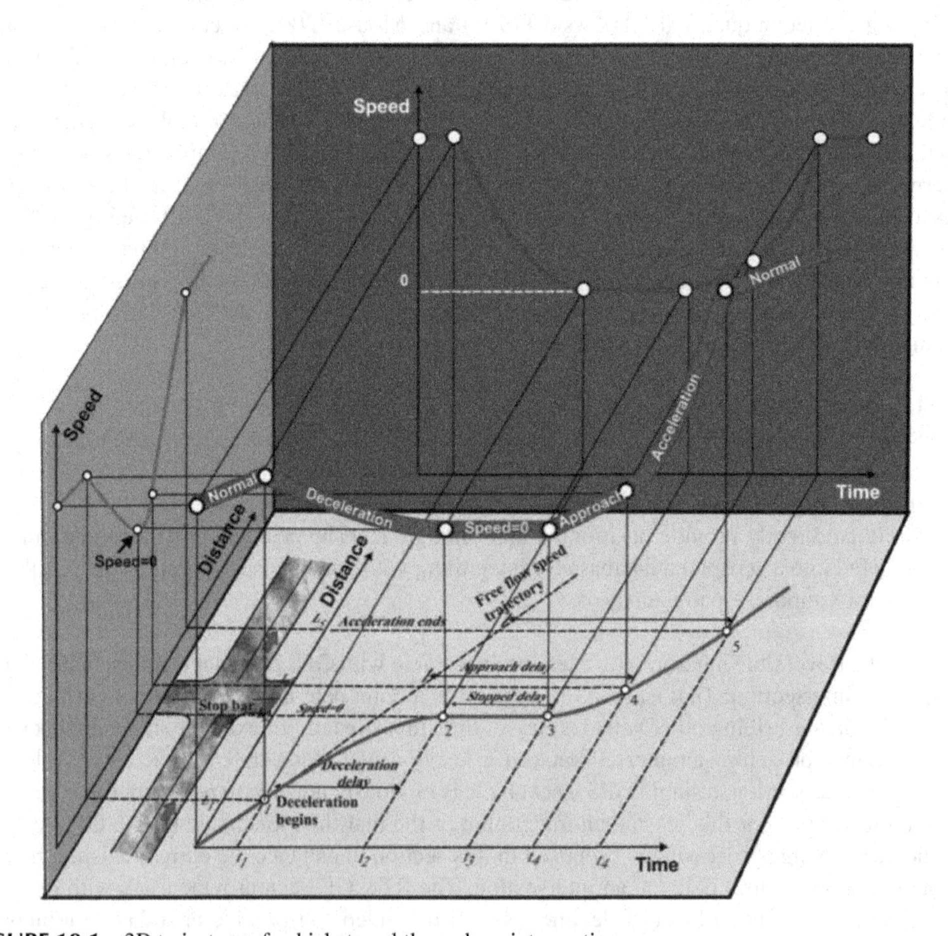

FIGURE 18.1 3D trajectory of vehicle travel through an intersection.

On the other hand, the control delay at an intersection can also be understood in three-dimensional (3D) space, which is constructed by time, distance, and speed (see Figure 18.1). The time axis and position axis lie on a horizontal plane, and the speed axis lies on a vertical plane. So, the spatial "trajectory" of a vehicle traveling at an intersection can be illustrated in this 3D space. If the 3D "trajectory" is projected down onto a 2D time–distance plane, the projected diagram is a well-known time–distance diagram (see Figure 18.1a). If the 3D "trajectory" is projected onto the 2D plane constructed by time and speed, a so-called time–speed diagram is obtained (see Figure 18.1b). If the 3D "trajectory" is projected onto the 2D plane constructed by distance and speed, a so-called distance–speed diagram is obtained.

Researchers used to apply the 2D time–distance diagram to calculate the control delay. They were in fact analyzing the change rates of speed, that is, acceleration and deceleration, in order to recognize the different types of delays. The mathematical model can be expressed by

$$d_t = d_d + d_s + d_{ac} \tag{18.1}$$

where

d_t is total delay;
d_d is deceleration delay, which is incurred by a vehicle's acceleration;
d_s is stopped delay, which occurs when a vehicle is fully immobilized; and
d_{ac} is acceleration delay, which is incurred by a vehicle's acceleration.

In fact, the 2D speed–time diagram is used to calculate the total control delay, that is, the speed change rate from the GPS data can be directly determined. The total control delay can be written by

$$\Delta t_i = T_i - t_i \tag{18.2}$$

where

t_i is the travel time of a GPS test vehicle that is affected by the controlled intersection;
T_i is the travel time that is unaffected by the controlled intersection;
Δt_i is total time delay, that is, time difference under the two conditions; and
i is *i-th* vehicle, where the i represents the GPS test vehicle.

For measurement of the T_i, when a GPS test vehicle passes through an intersection, the actual path is only being recorded, that is, the desired (ideal) path cannot be recorded directly using GPS. Thus, the typical vehicle free-flow speed (i.e., the desired speed) is measured where the upstream intersection is adequately far away from the intersection of interest and when the traffic volume is relatively low.

For measurement of the t_i, it used to be calculated by fact that the GPS test vehicle passes through two points – entrance and exit – where the speed change rates reach maximization, that is, the beginning of deceleration and ending of acceleration. Researchers used to express this change using the slope of curve in the 2D time–distance diagram (see Figure 18.1a). This implies that only the beginning and ending points, where the vehicle has a maximum deceleration and acceleration rates, can be determined from the GPS data, and the actual travel time, t_i, can be measured from the GPS data. Thus, the computation of the GPS test vehicle's speed and its change rate is critical.

The RTK GPS can provide users with exact instantaneous position (X, Y, and Z coordinates) and instantaneous time, t, at a real-time mode, thus speed at unit time frame (e.g., one second) can be calculated by

$$u_i = \frac{D_i}{\lambda} \tag{18.3}$$

where

u_i is speed of GPS test vehicle;
λ is a preset time interval (e.g., one second); and
D_i is distance at point p1, and p2.

It is calculated by

$$D_i = \sqrt{(X_i^{p_1} - X_i^{p_2})^2 + (Y_i^{p_1} - Y_i^{p_2})^2 + (Z_i^{p_1} - Z_i^{p_2})^2}.$$

where

p1 and p2 represent instantaneous position of the two points; and
(X, Y, Z) are coordinates at instantaneous position, p1, and p2.

So, the deceleration and acceleration can be calculated by

$$\left.\begin{aligned} a_i &= \frac{\partial u_i}{\partial \lambda} \\ a'_i &= \frac{\partial u_i}{\partial \lambda} \end{aligned}\right\} = \max \tag{18.4}$$

a_i, a'_i are maximum deceleration at upstream intersection and acceleration rate and downstream intersection, respectively.

The above formula is for computation of the control delay using one single GPS test vehicle. In combination with the video data, interpolation of the other vehicles' travel time (see Figure 18.2) can be conducted (see Section 18.4), and an average travel time of all vehicles is given by

$$\bar{t} = \frac{1}{N}\sum_{i=1}^{N} t_i \tag{18.5}$$

where

t_i $(i = 1, \ldots, N)$ is the travel time of *i-th* vehicle that is affected by the controlled intersection;
N is number of observed vehicles; and
$\bar{t}$ is average control time.

So, the average total time delay is

$$\overline{\Delta t} = \bar{t} - T \tag{18.6}$$

In summary, the steps of computing total delay using the GPS test vehicle in combination with video are:

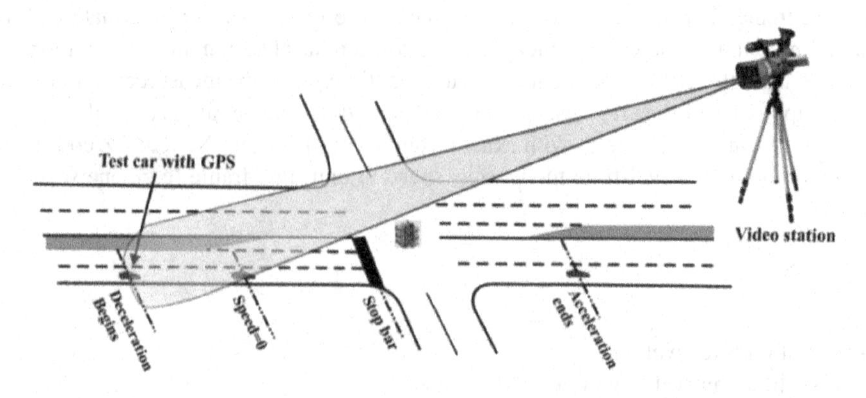

FIGURE 18.2 A test vehicle with GPS and video recorder for data collection at the intersection.

1. Interpolate the instantaneous position and time of the video recorded GPS test vehicle, which has a sampling rate of 20 frames per second, using the GPS test vehicle.
2. Obtain the instantaneous position and time of video recorded other vehicles using the video recorded GPS test vehicle.
3. Compute the travel speed of all vehicles during approaching/departing an intersection.
4. Determine the maximum deceleration/acceleration rates for all vehicles when approaching an intersection using a one second time interval.
5. Separately compute each individual vehicle's total travel time delay, acceleration/acceleration time delay, approach time delay, and time delay in the queue.
6. Compute average values of different time delay.

18.3 TRAVEL TIME DATA COLLECTION

18.3.1 RTK GPS Technology

RTK GPS measurement is a GPS surveying technology, which requires two GPS receivers. One is located at a reference (base) station, whose coordinates are known; the other receiver may be free to travel from point to point, and is called the roving receiver. Both receivers observe at least four GPS satellites simultaneously, and then the reference station GPS receiver broadcasts the correction to the roving receiver using a radio link, and the position of the rover station is solved by a double differencing algorithm (Zhou and Wang 2012; Zhou 2010). The RTK GPS surveying can provide real-time positional and time data at an accuracy of centimeters.

18.3.2 GPS Test Car Configuration

The GPS test car in this study consists of different hardware components. They are:

1. **Trimble 5700 GPS:** A Trimble® 5700 GPS receiver is a 24-channel dual-frequency RTK GPS receiver with options of 1 Hz, 2 Hz, 5 Hz, and 10 Hz positioning and data logging and 1 pulse per second output capability. A UHF radio modem is used as the receiver for RTK communications. The porTABLE 19.ZephyrTM antenna is used for RTK roving, and the Zephyr GeodeticTM antenna is used for the base station. The code DGPS can reach ± (0.25 m + 1 ppm) and ± (0.5 m + 1 ppm) for horizontal and vertical positioning, respectively.
2. **UHF Radio Modem:** The TRIMMARK 3 radio modem is used as the base station transmitter for correction communication. The UHF radio modem is narrowband with three UHF bands (410–420 MHz, 430–450 MHz, 450–470 MHz) with a wireless data rate of 4800, 9600, or 19,200 bps.
3. **Camcorder:** Sony's DCR-SR100 is a CCD camcorder with 30 GB non-removable hard drive storage space. It can record seven hours of high-quality video at dimension of 720 × 480 in MPEG form. The lens is a 30 mm Carl-Zeiss Vario Sonnar T with 10× optical zoom and 120× digital zoom capability.
4. **Test Car Configuration:** A Toyota Corolla is used as a test vehicle. A Trimble 5700 RTK GPS receiver is housed inside the car. An antenna is placed outside the car in order to minimize multipath error (see Figure 18.3). The extra power is provided through the vehicle cigarette lighter jack.

18.3.3 Field Data Collection

Test Sites: Data collection was conducted at State Route (SR)-337, near Old Dominion University and four miles away from the downtown area of the City of Norfolk. The test intersection SR-337 is a major north–south route that serves as a main arterial of primary state highway in the South

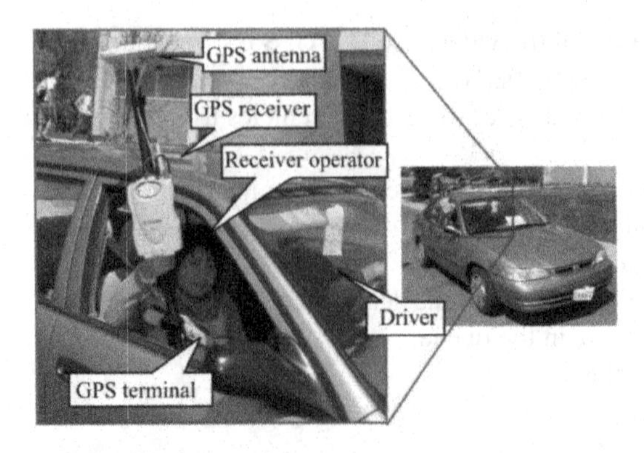

FIGURE 18.3 The configuration of the GPS test vehicle.

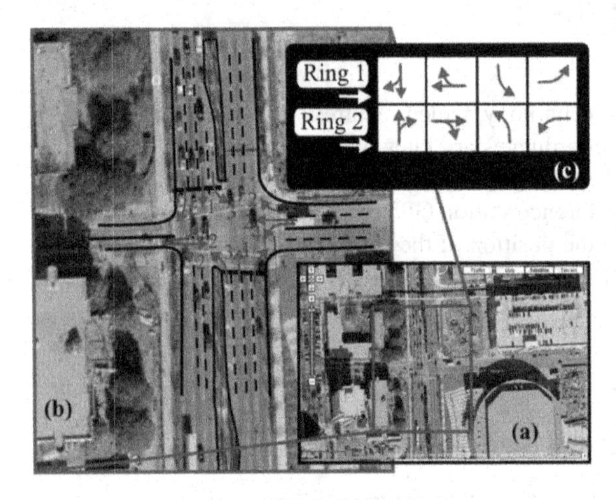

FIGURE 18.4 The roadway segments at the first intersections.

Hampton Roads area of the Commonwealth of Virginia, USA. On the SR-337 section, the southbound approach of the intersection with 43th Street was chosen to measure traffic delays. At this location, SR-337 was a three-lane highway with a post speed limit of 35 miles per hour (mph), or 56.3 kilometers per hour (km/h). This intersection had a directional traffic volume of 1200 vehicles per hour (vph) on SR-337. The traffic signal was operating in a fixed-timed mode with a total cycle length of 112 seconds and signal phasing for the through traffic movement consisting of 52, 4, and 56 seconds for the green, yellow, and red indications, respectively (see Figure 18.4c). No delay was caused to the through traffic due to left turns, as these vehicles are stored in a pocket lane of adequate length and are discharged from the intersection in a protected left turn phase (see Figure 18.4b). Parking was not permitted on the selected approach. The distance between the two intersections was about 0.7 km (0.45 miles).

Reference Station Setup: The reference station was set up on USGS Bench Mark (BM), named ODU-3. The coordinates and datum information were downloaded from the National Geodetic Survey (NGS) website. A Trimble® 5700 GPS receiver and Zephyr antenna were installed and set up at this reference station (see Figure 18.5).

Field Data Collections: The field GPS data collections were conducted at 11:00 am (off-peak hour). One student took care of the base station, one student was responsible for the operation of the GPS rover receiver, and one driver was in the test car. The cell phones were used for crew communication. The base station was turned on about 30–45 minutes earlier than the rover receiver. After about 1–5 minutes' initialization of the rover receiver, the test car entered SR-337 from several upstream intersections in order to merge naturally with the traffic stream. In addition, a video camera was set up at the test intersection for simultaneously collecting the video data. The

FIGURE 18.5 The reference station setup at USGS BM, ODU-3.

GPS
UTC 15:43:24 36 53 11.7328N 076 18.5061423W
GPS
UTC 15:43:25 36 53 11.7332N 076 13.5051723W
GPS
UTC 15:43:26 36 53 11.7344N 076 13.5051722W
……..

FIGURE 18.6 The sample of GPS data string.

same operations were repeated three times. The GPS recorded the time, latitude, and longitude at a sampling rate of one point per second (see Figure 18.6), and the video camera recorded the GPS test car's path at a sampling rate of 30 frames per second. In addition, a log sheet was required for the driver to record start time, end time, route ID, and any comments for the reference during post-processing.

18.3.4 Data Preprocessing

GPS and Video Data Preprocessing: After the field data were collected, they were downloaded onto computer for post-processing using Trimble Geomatics Office v. 1.6 software. Meanwhile, the video data were imported into the computer. For each route's starting/ending location, a unique ID was assigned to GPS and video data.

GPS Coordinate Transportation: The geographical coordinates (latitude and longitude) collected by GPS were based on the WGS 84 datum. The waypoint coordinates were converted into Universal Transverse Mercator (UTM) coordinates with NAD 83 using the software, UTMS—Version 2.1.

Accuracy Evaluation of GPS Data: In order to eliminate possible GPS blunders caused by issues such as unlocked GPS satellites and ambiguity solution, the GPS waypoints were superimposed onto the orthoimage. After the GPS data were preprocessed, data were archived in an MS Excel database (see Table 18.1).

TABLE 18.1
The archived GPS data.

Vehicle ID	Vehicle Type	Accumulative Distance (m)	Number of Lane	Distance Traveled (m)	Speed (km/h)
1	1	2	1	2	100
1	1	26	1	26	90
1	1	49	1	49	86
1	1	73	1	73	85
1	1	97	1	97	84
1	1	236	1	236	83

18.4 DELAY MEASURES WITH GPS AT SIGNALIZED INTERSECTIONS

18.4.1 Video Data Resampling

Because the original sampling rate of the video stream is 30 frames per second, the data redundancy is huge. In order to speed up the processing of the video data, resampling of the video stream at a uniform resampling rate (e.g., 20 frames/second) for an entire video record at the test intersection was conducted.

18.4.2 Interpolation for Video Recorded GPS Test Car

The collected GPS data are sequentially discrete points at a sampling rate of one point per second, while the recorded video data are an image sequence at a resampling rate of 20 frames per second. A second polynomial equation interpolation was used to interpolate the instantaneous position and time (epoch) for each video frame of the video recorded GPS test car. The core technology of this work is to seek for the same time (epoch) between GPS time and the stamped time on the video, since the GPS applies the UTC time, while the video image applies the local time. To do so, Adobe® Premiere® 6.5 software is used to exactly indicate the time of each video frame and GPS test vehicle's position, and then compare the two values. With the comparison, a corresponding relationship between the two time systems can be determined.

After two time systems were identified, the second polynomial equation interpolation was implemented for X-coordinates and Y-coordinates. The results are depicted in Figure 18.7a and Figure 18.7b. The accuracy of RSM achieves 0.13 m. Using the interpolated X- and Y-coordinates, the position of the GPS test vehicle is calculated, and relationship between the position vs time is plotted in Figure 18.8. As seen, the interpolation error of position and time in the duration of the test vehicle stopping at stop bar is equal to zero, and approximately 0.09–0.14 m in the duration of acceleration and deceleration.

18.4.3 Interpolation for Other Video Recorded Vehicles

18.4.3.1 Multi-Vehicle Tracking

The interpolation of other video recorded vehicles using the video recorded GPS test vehicle was implemented using a multi-object tracking algorithm, which was developed in the computer vision community, but much of the background work has been in non-transportation applications. A detailed review may be beyond this project. This study presents our vehicle tracking method, which includes feature detection, and feature/vehicle tracking.

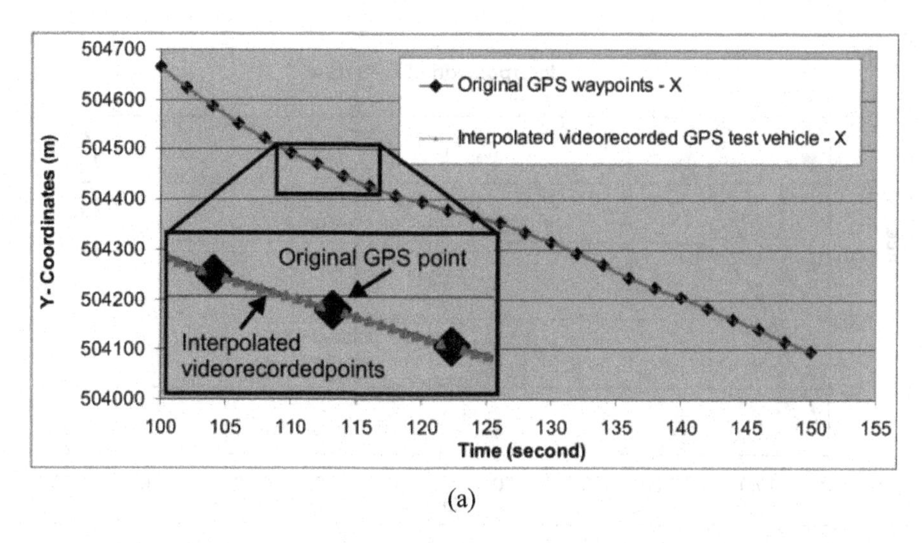

(a)

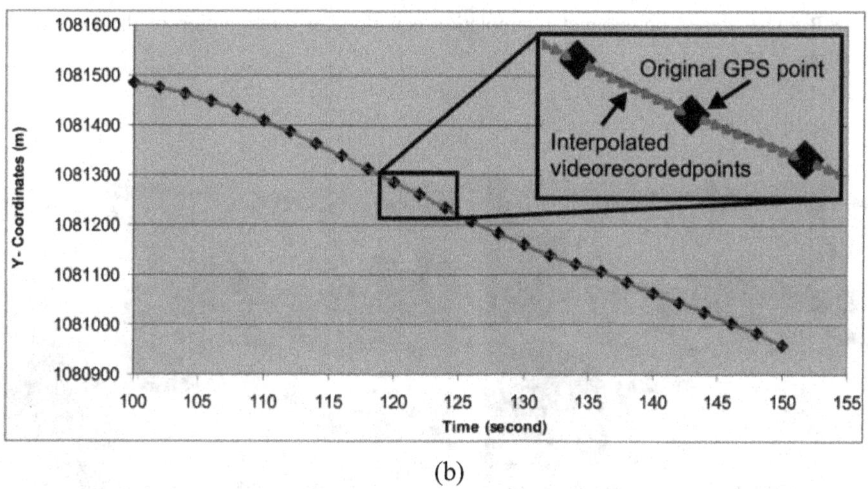

(b)

FIGURE 18.7 The interpolated (a) X-coordinates and (b) Y-coordinates using GPS waypoints.

A) Interest Point Feature Extraction

The Moravec operator (Moravec 1977) was employed to extract distinct points from the first frame (approximately at the beginning of entry into the test intersection). Details of this operator may be referred to in Moravec (1977). For our video frames, the size of the image sub-window in the Moravec operator is 7 × 7 pixel^2, and the assigned threshold magnitude of threshold (M_T) is 200. As a result, 412 interest points were totally extracted in the first video frame (see Figure 18.9a). The Moravec operator was tested using different sizes of sub-window and different magnitudes of threshold, M_T, and it was found that the Moravec operator has the following shortcoming as follows:

(1) *Too many distinct points.* These consume a lot of computational time.
(2) *Clustering.* Some of the distinct points are clustered together in a few areas. Such a distribution pattern directly results in mismatching due to the repeated texture.

For the above reasons, the Moravec operator is slightly improved by constraining the image processing area, that is, only processing the area along the road (see Figure 18.9) (Bared 2005), and using

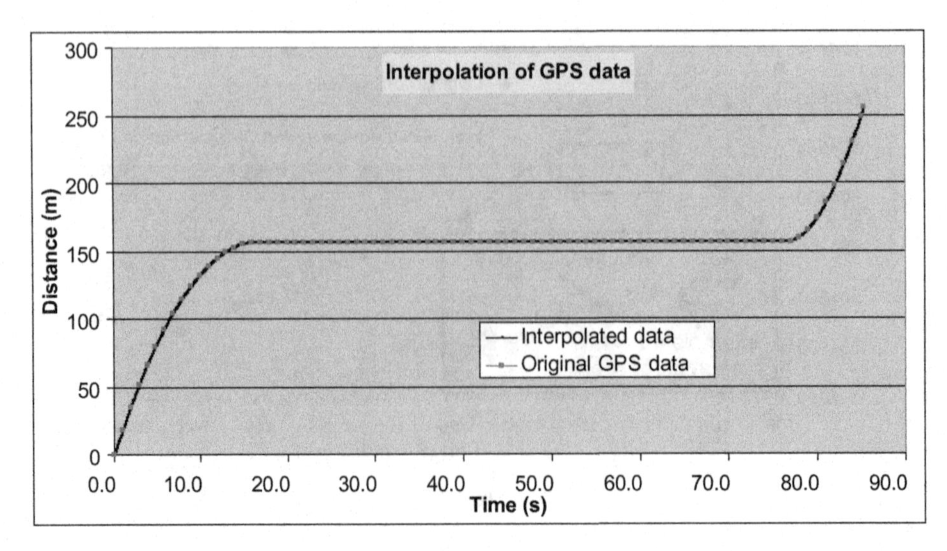

FIGURE 18.8 The second polynomial interpolation equation for interpolation of the GPS test vehicle's position vs time.

FIGURE 18.9 A virtual grid method is employed for the Moravec operator to constrain the distribution and number of the extracted distinct points.

a virtual grid method (see Figure 18.9b) to constrain the distribution and number of the extracted points (Benekohal 1991). In other words, only 1–4 points were finally allowed to be extracted at each cell. With the improvement above, 75 distinct points at the same video frame were extracted (see Figure 18.9b). Compared with Figure 18.9a, it can be seen that the extracted distinct points are not clustered together and are distributed in a reasonable distribution pattern.

B) Feature Point-Based Matching for Tracking Vehicles

Based on the extracted distinct points in the first frame, the conjugate points in consecutive frames for tracking vehicles must be found. The feature-based matching technology was used in this section. In order to decrease the search space (i.e., pull-in range) and save the computation time during the image matching, a pyramid transform-based method was developed to predict the locations of all the conjugate points in the following frame. The predication of the conjugate points in the following video frame was conducted as follows:

(1) Based on analysis of the imaging condition, the coarse x and y parallax were set at 3 and 3 pixels, respectively;
(2) For each feature point in the first video frame, a 27 × 27 pixel^2 sub-window was retrieved from the original video frame, and was then formed into a three-level pyramid;
(3) The location and gray information of each feature point in the highest level of the pyramid were recorded and then transformed to the bottom level of the following video frame.

Based on the predicted conjugate points, cross-correlation between the "target" sub-window centered at the extracted feature point in the first frame and the "matching" sub-window centered on the predicted points was employed to exactly locate the conjugate points. The criteria of cross-correlation was a correlation coefficient maximum, namely,

$$\rho = S_{xy} / \sqrt{S_{xx} S_{yy}} = max \tag{18.7}$$

where ρ is a correlation coefficient; S_{xy} is a covariance of the "target" sub-window and the "matching" sub-window; and S_{xx} and S_{yy} are variances of the "target" and the "matching" sub-windows, respectively. The size of the sub-windows in this project was 11 × 11 pixels^2, and the threshold was 0.86.

Figure 18.10 shows the results of the predicted conjugate feature points on the basis of the distinct point extracted in the previous video frame. With the suggested method, the predicted position of the feature points are 5 pixels left along the *column* direction and 6 pixels down along the *row* direction.

As seen from Figure 18.11, a few extracted feature points are in fact the features describing ground, such as signal pole's shadow, crosswalk, and curb. These feature points are meaningless for tracking vehicles. For this reason, two thresholds were set up to delete these meaningless points, as follows:

- The correlation coefficient is close to 0.95–1.0 when conducting matching;
- The feature point's (pixel) position does not change when conducting matching.

With the above two criteria, most of the meaningless ground feature points were deleted. Figure 18.11 depicts the results after the cross-correlation matching, in which 34 points were refined as the matched feature points.

The above description only explains the tracking vehicles using two images matching technology. This procedure needs to be repeated one by one until the last video frame. After all processes are finished, vehicles in a stream should be tracked and identified.

FIGURE 18.10 The predicted conjugate tie points in the consecutive video frame.

FIGURE 18.11 Tracking vehicles using the point feature-based cross-correlation matching technology.

18.4.3.2 Interpolation of Vehicle Position

The video camera was not calibrated in this study, which implies that the pixel size cannot directly be transformed into the ground size (called ground sample distance (GSD)). In this study, each pixel represents different GSDs due to the oblique video imaging mode. The nearer the vehicle to the video camera, the bigger the GSD (see Figure 18.12). For this reason, a simple method was developed to determine the GSD. The basic idea was to find each vehicle's pixel position (row and column) first, and then compare its pixel position to the GPS test vehicle's pixel position (row and column). As the GPS test vehicle's position and time were recorded by GPS, that is, the GSD of the video recorded GPS test vehicle is known, the GSDs of other video recorded vehicles were determined through the relative pixel position (see Figure 18.12). With the interpolated positions of all vehicles and their time, 2D distance–time diagram for all vehicles are depicted in Figure 18.13.

18.4.3.3 Extraction of Control Time Delay

18.4.3.3.1 Extraction of Vehicles' Speed

The above computation only obtains the instantaneous positions and time (epoch) of each vehicle. The speed and speed change rate of all vehicles are computed using Equation 18.4 under a given time period (one second in this study) in order to determine starting and ending points of each vehicle's passing at the test intersection, so that the length of each type of time delay can be calculated. To this end, two basic steps were implemented as follows:

(1) The GPS test vehicle driver was required to remember the approximate geographic position and travel speed when he/she decelerated and accelerated at the test intersection, so that the approximate starting point and ending points of all vehicles can be obtained. From the driver's information, the length of entry to the intersection was estimated as approximately 200 m, and the length of exit from the intersection was estimated as approximately 150 m; the speed of

FIGURE 18.12 Interpolation of other vehicles' positions using the GPS test vehicle.

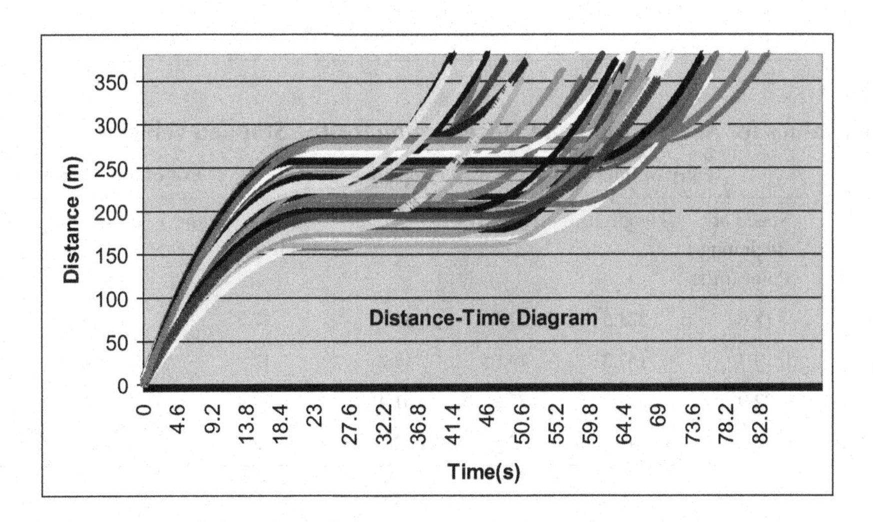

FIGURE 18.13 Traditional 2D distance–time for all vehicles.

entry to the intersection was estimated as approximately 18 m/s, and the speed of exit from the intersection was estimated as approximately 20 m/s.

(2) Exact estimation of each vehicle's speed was completed using Equation 18.4 under the time interval of one second. The approximate values for the speed and the length of entry and exit of the intersection were used as reference parameters in order to avoid multiple solutions, that is, several maximum deceleration/acceleration rates.

With the above calculation, Table 18.2 lists the results of average value, minimum value, maximum value, and standard deviation for speed at beginning point, length and time of entry to the intersection, and speed at ending point, length and time of exit from the intersection. As seen in Table 18.2, the minimum speed is 13.7 m/s, and an average of speed is 18.6 m/s for entry to the intersection. The length of entry to the intersection ranges from 157 m to 291 m. For entry to the intersection, the test vehicle started to decelerate at the twentieth second and the distance point of 280.8 meters. For the exit from the intersection, the speed at the ending point ranged from 12.9 m/s to 16.2 m/s.

18.4.3.3.2 Extraction of Average Control Time Delay

As mentioned above, the total delay at an intersection most frequently is quantified by: stopped-time delay, approach delay, travel time delay, and time in queue delay. Once the instantaneous location and time of all vehicles at maximum deceleration and acceleration rate are determined, these components of delay can be calculated. The stopped-time delay for all vehicles is relatively easy because all vehicles are completely stopped while waiting to pass an intersection, while other types of delays can be derived from the GPS data.

Table 18.3 lists the results of different types of delay for stopped and non-stopped vehicles. As can be seen, the average delay for non-stopped vehicles is about 6.9 seconds, which is probably caused by pedestrians because this test intersection is near the ODU campus, and many students crossed the streets along the pedestrian crosswalk. For the stopped vehicles, the average total delay is 43.2 seconds, the maximum total delay is 64 seconds, and the minimum total delay is 7.7 seconds. It can be found that the minimum total delay of the stopped vehicles is one second more than the average total delay for non-stopped vehicles. This means that a few vehicles probably accelerate when meeting a yellow light.

TABLE 18.2
Measured Values for Acceleration and Deceleration for the Stopped Vehicles.

	Entry of Intersection			Stop Time for Waiting to Pass (s)	Exit of Intersection		
	Speed at Beginning Point (m/s)	Length (m)	Time (s)		Speed at Ending Point (m/s)	Length (m)	Time (s)
Ave. values	18.6	221.5	22.5	35.6	17.2	143.8	17.5
Min. value	13.7	157.2	14.05	8.2	12.9	118.4	6.2
Max. value	22.9	291.3	25.9	61.4	22.4	153.5	33.4
Stand. deviation	2.6	40.7	2.5	12.1	2.3	24.4	4.7

TABLE 18.3
Delays of Stopped and Non-Stopped Vehicles at Traffic Intersection.

	Non-Stopped Vehicles	Stopped Vehicles (second)			
		d_d	d_s	d_a	d_t
Average values	6.9	10.62693	23.09875	9.439874	43.16556
Minimum values	0.0	2.147388	2.304664	1.62601	7.678062
Maximum value	13.3	13.94739	50.45466	16.12601	64.62806
Standard deviation	2.2	2.510381	12.16	3.840407	

TABLE 18.4
The comparison between HCM 200 model computed and GPS-measured.

	GPS-Measured (d_t)	HCM200 Manually Measured (d_t)	Difference of Two Methods (Δd_t)
Average values	43.17	29.34	13.82
Minimum values	7.68	8.99	–1.31
Maximum value	64.63	38.85	25.77

18.4.4 Comparison of GPS-Measured and Manually Measured Stopped-Time Delays

The HCM 2000 presented the following model to calculate the average control delay per vehicle.

$$d_t = \frac{0.5C(1-g/C)^2}{1-[\min(1,X)g/C]} + 900T\left[(X+1)+\sqrt{(X-1)^2+\frac{4X}{cT}}\right] \tag{18.8}$$

where
C is cycle length (second);
G is effective green time (second);
X is v/c ratio, or degree of saturation;
c is capacity of the intersection approach (vehicles per hour); and.
T is analysis period (hour).

The parameters, C, G, and c are obtained from the video data. Fifteen minutes of field video data at the test intersection were recorded, and then computation for the cycle length (C), the effective green time (G), and the capacity of the intersection approach are computed. With these parameters, the average control delay can be obtained, and is listed in Table 18.4. As seen, the GPS-measured average delay and HCM 200-measured average delay are different. The maximum difference is 25.8 seconds, and the minimum difference is approximately 1.3 seconds.

18.5 CONCLUSIONS

This chapter introduces a 3D understanding of computing the total control delay by the time–space–speed method, which demonstrated that traditional time–distance diagram is not unique in illustrating each component of the control delay. This chapter suggests a model for computing the total control delay on the basis of GPS test vehicle and video data. The details of the computation procedures for control delay are presented, which include the video recorded GPS test vehicle resampling and interpolation, other video recorded vehicle resampling and interpolation for positioning and time, the computation of each vehicle's speed and speed change rate, and the determination of maximum deceleration and acceleration. With the proposed method, the computation of each delay component, including deceleration/acceleration delay, stopped delay, and approach delay, becomes simple. The proposed method is verified by the real data collected by the GPS test vehicle in combination with video data. A comparison was also conducted with the HCM2000 model. The results demonstrated that the proposed methods can effectively measure the control delay in field with low cost.

The major advantages of the presented method in this study are that the GPS test vehicle in combination with video data can better represent the observations of a stream at an intersection (Bared 2005); with the method presented in this study, all of these delays can be directly determined or indirectly derived (Benekohal 1991); and although the GPS test vehicle only recorded the actual path of passing through an intersection, that is, the desired path cannot be recorded directly, the typical vehicle free-flow speed can be measured by driving the GPS test vehicle following the main traffic flow on a fairly distant upstream roadway section of the intersection (Buehler et al. 1976).

REFERENCES

Bared, Joe, A review of the signalized intersections: Informational guide, *ITE Journal (Institute of Transportation Engineers)*, vol. 75, no 7, July, 2005, p 37–44.

Benekohal, R. F. (1991). Procedures for validation of microscopic traffic flow simulation models. *Transportation Research Record.* vol. 1320, Transportation Research Board, Washington, D.C., 190–202.

Buehler, M. G., Hicks, T. J., and Berry, D.S. (1976). Measuring delay by sampling queue backup. *Transportation Research Record,* vol. 615, Transportation Research Board, Washington, D.C., p. 30–36.

Colyar, James D. Rouphail, Nagui M., Measured distributions of control delay on signalized arterials, *Transportation Research Record*, vol. 1852, 2003, p 1–9.

Fu, L.; Hellinga, B., Delay variability at signalized intersections, *Transportation Research Record*, vol. 1710, 2000, p 215–221.

Hereth, William R.; Zundel, Alan; Saito, Mitsuru, Automated estimation of average stopped delay at signalized intersections using digitized still-image analysis of actual traffic flow, *Journal of Computing in Civil Engineering*, vol. 20, no. 2, March/April, 2006, p 132–140.

Hunter, Michael P.; Seung, Kook Wu; Hoe, Kyoung Kim, Practical procedure to collect arterial travel time data using GPS-Instrumented test vehicles, *Transportation Research Record*, vol. 1978, *Traffic Signal Systems and Regional Systems Management*, 2006, p 160–168.

Ji, Xiaojin; Prevedouros, Panos D., Effects of parameter distributions and correlations on uncertainty analysis of highway capacity manual delay model for signalized intersections, *Transportation Research Record*, vol. 1920, 2005, p 118–124.

Jiang, Yi; Li, Shou; Zhu, Karen Qin, Traffic delay studies at signalized intersections with global positioning system devices, *ITE Journal (Institute of Transportation Engineers)*, vol. 75, no. 8, August, 2005, p 30–39.

Kokkalis, A.; Lakakis, K., *A Vehicle Guidance System Using GPS-GIS Integration in Highway Transportation and Administration, Proceedings of the International Conference on Applications of Advanced Technologies in Transportation Engineering*, Beijing, China, 2004, p 131–136.

Koller, D, Daniilidis, K, Nagel, H, (1993) Model-based object Tracking in monocular image sequences of road traffic scenes, *International Journal of Computer Vision*, vol. 10: 257–281.

Li, Shuo; Zhu, Karen; Nagle, John; Tuttle, Carl; Van Gelder, B.H.W., Reconsideration of sample size requirements for field traffic data collection with global positioning system devices, *Transportation Research Record*, vol. 1804, 2002, p 17–22.

Lin, Feng-Bor; Thomas, Daniel R., Headway compression during queue discharge at signalized intersections, *Transportation Research Record*, vol. 1920, 2005, p 81–85.

Liu, Kai; Yamamoto, Toshiyuki; Morikawa, Taka, *Estimating Delay Time at Signalized Intersections by Probe Vehicles, Proceedings of the Fifth International Conference on Traffic and Transportation Studies, ICTTS 2006*, Xi'an, China, 2006, p 644–655.

Moravec, H.P., 1977. *Towards Automatic Visual Obstacle Avoidance, Proceeding of 5th International Joint Conference on Artificial Intelligent*, Cambridge, MA, pp. 584.

Mousa, R.M. (2002). Measuring and modeling of delay components at signalized intersections, *Journal of Engineering and Applied Science*, vol. 49, no. 2, April, 2002, p 225–240.

Oh, Cheol; Ritchie, Stephen G., Real-time inductive-signature-based level of service for signalized intersections, *Transportation Research Record*, vol. 1802, 2002, p 97–104.

Olszewski, P. (1993). Overall delay, stopped delay, and stops at signalized intersection. *Journal of Transportation Engineering,* vol.. 119, no.. 6, p. 835–852.

Pan, Changxuan; Lu, Jiangang; Wang, Dawei; Ran, Bin, Data collection based on global positioning system for travel time and delay for arterial roadway network, *Transportation Research Record*, vol. 2024, *Information Technology, Geographic Information Systems, and Artificial Intelligence*, 2007, p 35–43.

Quiroga, Cesar; Perez, Michael; Venglar, Steve, *Tool for Measuring Travel Time and Delay on Arterial Corridors, Proceedings of the International Conference on Applications of Advanced Technologies in Transportation Engineering*, Boston Marriot, Cambridge, MA, USA, 2002, p 600–607.

Van Leeuwaarden, J.S.H., Delay analysis for the fixed-cycle traffic-light queue, *Transportation Science*, vol. 40, no. 2, May, 2006, p 189–199.

Washburn, Scott S.; Larson, Nate, Signalized intersection delay estimation: Case study comparison of TRANSYT-7F, synchro and HCS, *ITE Journal (Institute of Transportation Engineers)*, vol. 72, no. 3, March, 2002, p 30–35.

Wolshon, Brian; Taylor, William C., Analysis of intersection delay under real-time adaptive signal control, *Transportation Research Part C: Emerging Technologies*, vol. 7, no. 1, Feb, 1999, p 53–72.

Zhang, C.; Yang, X.; Yan, X. (2007): Methods for Floating Car Sampling Period Optimization, *Journal of Transportation Systems Engineering and Information Technology*, vol. 7, no. 3, June, 2007, p 100–104

Zhou, G. (2010). *Co-location Decision Tree for Enhancing Decision-Making of Pavement Maintenance and Rehabilitation*, Ph.D. dissertation, Virginia Tech., Blacksburg, Virginia, USA, May 2010.

Zhou, G. and Wang, L., 2012. Co-location decision tree for enhancing decision-making of pavement maintenance and rehabilitation. *Transportation Research Part C*, vol. 21, 287–305.

19 Measurement of Dry Asphalt Road Surface Friction Using Hyperspectral Images

19.1 INTRODUCTION

A vehicle which is able to drive on a road must have adequate surface friction in the contact area between the tires and the road surface, so that loss of driving control does not occur in situations normally expected to be safe and comfortable (Anderson and Henry 1979; Heaton et al. 1990; Liu et al. 2003a, 2003b). Thereby, the Skid Accident Reduction Program issued by the Federal Highway Administration (FHWA 1980) encouraged each state highway agency to minimize skidding accidents (especially under wet weather) by identifying and correcting sections of roadway with high or potentially high skid accident incidence and ensuring that new surfaces have adequate, durable skid-resistant properties. The skid resistance is usually measured using surface friction. The surface friction, which is defined using a friction coefficient measured according to a standardized method, can be computed by

$$F = \mu \times W \tag{19.1}$$

where F is the tractive force applied to the tire at the tire–pavement contact, μ is the coefficient of friction, and W is the dynamic vertical load on the tire. Many variable factors influence the surface friction, including:

1. The texture of the road aggregate;
2. The covering of the road surface, for example, ice, water;
3. Tire design and condition; and
4. The vehicle dynamics, for example, the load and vehicle speed.

It has been demonstrated that the value of the skid resistance is sensitive to changes in the above factors. In particular, the road surface texture in dry road conditions makes a significant contribution to friction (Ong and Fwa 2005; Asi 2007). There are many methods to measure friction. One method is to measure the skid number (SN), which is defined by the coefficient of friction multiplied by 100, that is, $SN = 100 \times \mu = 100\ (F/W)$, as specified in ASTM E 274 "Standard Test Method for Skid Resistance of Paved Surfaces Using a Full-Scale Tire." The ASTM E 274 method has been used by more than 40 states. Another method, which measures the coefficient of friction between tire and pavement in the yaw mode, is used by at least four states (Kyriakopoulos and Skounakis 2003). A simple trailer called Mu Meter is commercially available, which measures the side force developed by two yawed wheels with smooth tires (Bazlamit and Reza 2005). The most natural method for determining the skid resistance is to drive an automobile on a pavement, lock the wheels after the desired speed is reached, and measure how far the vehicle slides until it comes to a full stop (Lee et al. 2004).

Recent advances in hyperspectral remote sensing technology have shown capabilities to derive physical and chemical material properties on a very detailed level (Clark 1999; Gomez 2002; Usher and Truax 2008). In particular, the National Center for Remote Sensing in Transportation (NCRST) at UCSB (http://www.ncgia.ucsb.edu/ncrst/) (Herold et al. 2004, 2008) has explored the hyperspectral capabilities in mapping road conditions. They have demonstrated that the spectral signal from different types of road materials, such as concrete and asphalt, are different and hyperspectral sensing can easily distinguish between them (Herold et al. 2004, 2008). They especially studied how the variability within asphalt road surface's aging, deterioration, and the different road surface textures can be reflected in spectral characteristics. The results demonstrated that the size of aggregate and deep texture seals produce decreasing brightness. On the basis of the previous studies, this study attempts to explore how the hyperspectral image analysis technologies are applied to estimate the friction (skid resistance).

19.2 STUDY AREA AND DATA SET

19.2.1 Study Area

The study area is located in the Goleta urban area, located 170 km northwest of Los Angeles in the foothills of the California Coast Range. The main roads of interest are Fairview Avenue and Cathedral Oaks and in particular the area near the intersection of both. Both roads have four lanes, two in each traffic direction, and represent major urban roads. The Santa Barbara (SB) County Pavement Condition Index (PCI) values for these roads show that the roads reflect a large variety of conditions. The eastern part of Cathedral Oaks has a PCI of nearly 100. This road was resurfaced just before the study started. Fairview and the western part of Cathedral Oaks have fair/poor conditions with PCI values in the order of 40–60. Fairview's pavement is in particularly poor condition. The central divider of Fairview was the boundary between two traffic management zones.

19.2.2 Data Set

The data set is provided by the University of California at Santa Barbara (UCSB). The data set contains:

19.2.2.1 Road Condition Data

The PCI scaled from 0 and 100 was surveyed in June and July 2003 (Fairview and Cathedral Oaks) by two expert groups: the Center for Transportation Research and Education from the Iowa State University at Ames, and representatives from Independent Seals, Western Paving, and Vulcan Paving (Californian firms associated with CALTRANS). The experts categorized the road condition in five categories (excellent, good, fair, poor, and very poor) for road management action (do nothing, maintenance, minor rehabilitation, major rehabilitation, and replacement). Another road data set was provided by the SB County road database at the MicroPaver Pavement Management System (PMS), which is linked to ArcGIS. The third set of road condition data was obtained by the Automatic Road Analyzer (ARAN) of the Roadware corporation (www.roadware.com) in December 2002. This data about road distresses provides geocoded information for over 30 individual parameters aggregated for 10 m road or lane segments.

19.2.2.2 Spectral Library

Spectral libraries, which were pure spectral samples of surfaces, were acquired in the study area in May 2001 and February 2004 using an Analytical Spectral Devices (ASD) full range (FR) spectrometer (Analytical Spectral Devices, Boulder, CO, USA). The FR spectrometer uses three detectors spanning the visible and near infrared (VNIR) and shortwave infrared (SWIR1 and SWIR2) at a spectral range of 350–2400 nm, and a sampling interval of 1.4 nm for the VNIR detector and 2.0 nm for the SWIR detectors. The spectral library includes a wide range of materials (asphalt, concrete, gravel) over a continuous wavelength range with high spectral detail, and additional information and documentation about surface characteristics and the quality of the spectra (i.e. metadata).

19.2.2.3 Remote Sensing Data

Two hyperspectral data sets were provided in the study area. The first data were acquired by the AVIRIS sensor on June 9, 2000 at a spatial resolution of approximately 4 m, and 224 individual bands with a nominal bandwidth of 8–11 nm, covering a spectral range from 370 to 2510 nm (Nie et al. 2001). The Jet Propulsion Laboratory (JPL) and the UCSB preprocessed the data for motion compensation and reduction of geometric distortions due to topography.

A second set of hyperspectral remote sensing observations was provided by Spectir Inc. located in Goleta, CA (www.spectir.com). The company developed a new sensor "HyperSpecTIR" (HST) with 227 bands over a range of 450–2450 nm. This is quite an advantage over AVIRIS for detailed mapping of road conditions at the narrow swath of only 40 m.

19.3 ROAD TEXTURE AND SPECTRAL IMAGE PROPERTIES

19.3.1 Road Texture and Friction

In order to analyze the relationship between hyperspectral reflectance and dry road surface friction (texture), it should first be discovered which texture descriptors correlate to road friction. Liu et al. (2005), Augustin (1990), and Do et al. (2000) have given the following parameters:

1) *Microtexture.* Road surface microtexture, which is defined as fine asperities of coarse aggregates and those of the mortar matrix less than 0.2 mm high and 0.5 mm long (Do et al. 2000), plays a significant role in the contact between wet roads and tires (Raisanen et al. 2003). On dry roads, a high friction coefficient is associated with a harsh aggregate microtexture. In other words, the harshness of the microtexture is determined by particle shape because the sharp and jagged edges produce high pressures and so a high friction coefficient, whereas rounded particles result in lower pressures and reduced traction (Mate 2002).
2) *Macrotexture.* Macrotexture refers to the "roughness" of the pavement surface created by the "hills and valleys" between aggregate particles (Raisanen et al. 2003). The aggregate "hills" above thin water films make contacted with the tire. The aggregate "valleys" provide an escape route for water. Macrotexture displaces water from the tire/pavement interface, and is visible and can be measured.
3) *Aggregate particle size.* An aggregate size is usually measured as one of the parameters affecting friction. It has been recognized that the range between 6 mm and 12 mm would be the most efficient surface for bulk water drainage while maintaining the maximum possible contact between the tire and pavement surface for a texture depth of 2 mm (Neuman et al. 1983).
4) *Texture depth.* The texture depth is usually evaluated as one of the parameters affecting fracture. Depending on particle size, a texture depth in the range of between 1.0 mm and 4.0 mm is recommended.
5) *Inter-particle spacing.* The optimal inter-particle spacing for a high friction coefficient in the dry state is zero. However, the spacing between particles is required, since it is bulk water removal which is the important action of the macrotexture. It is recommended that the spacing is in the region of between 1.0 mm and 4.0 mm.

Many methods have been presented for measuring surface texture, such as tracing by stylus, measuring the outflow rate, and using putty to fill surface irregularities. One method that has been used frequently is the sand patch method, specified as ASTM E 965 *"Standard Test Method for*

Measuring Surface Macrotexture Depth Using a Sand Volumetric Technique." One non-contact sensor for measuring texture parameters is the Road Surface Analyzer (ROSAN), which was developed jointly by FHWA and private industry to provide measurements of pavement surface textures (Lees and Williams 1974; Purushothaman et al. 1990; Woodward et al. 2004; Crocker et al. 2004). It utilizes high-speed lasers to collect surface profile data, which are then converted to surface textural measurements by use of custom analysis software. Data acquisition and storage are accomplished by use of an on-board computer.

19.3.2 Hyperspectral Images and Road Friction

19.3.2.1 Spectral Properties of Asphalt Aging

With the given data, the spectral properties of asphalt aging and road surface friction are first analyzed for increasing aging, the loss of oily components by volatility or absorption, changes of composition by oxidation, and molecular structuring that influences the viscosity of the asphalt mix (Herold et al. 2004). The results of these processes result in the different road surface friction (textures) which are exposed in spectral properties. Figure 19.1 depicts three spectral samples of road asphalt with varying ages of the pavement. Spectrum C reflects a paved road, whose surface is completely sealed with asphalt mix at an age of one year. As seen from Figure 19.1, the younger the paved road, the lower the spectral reflectance, and the minimum reflectance is near 350 nm with a

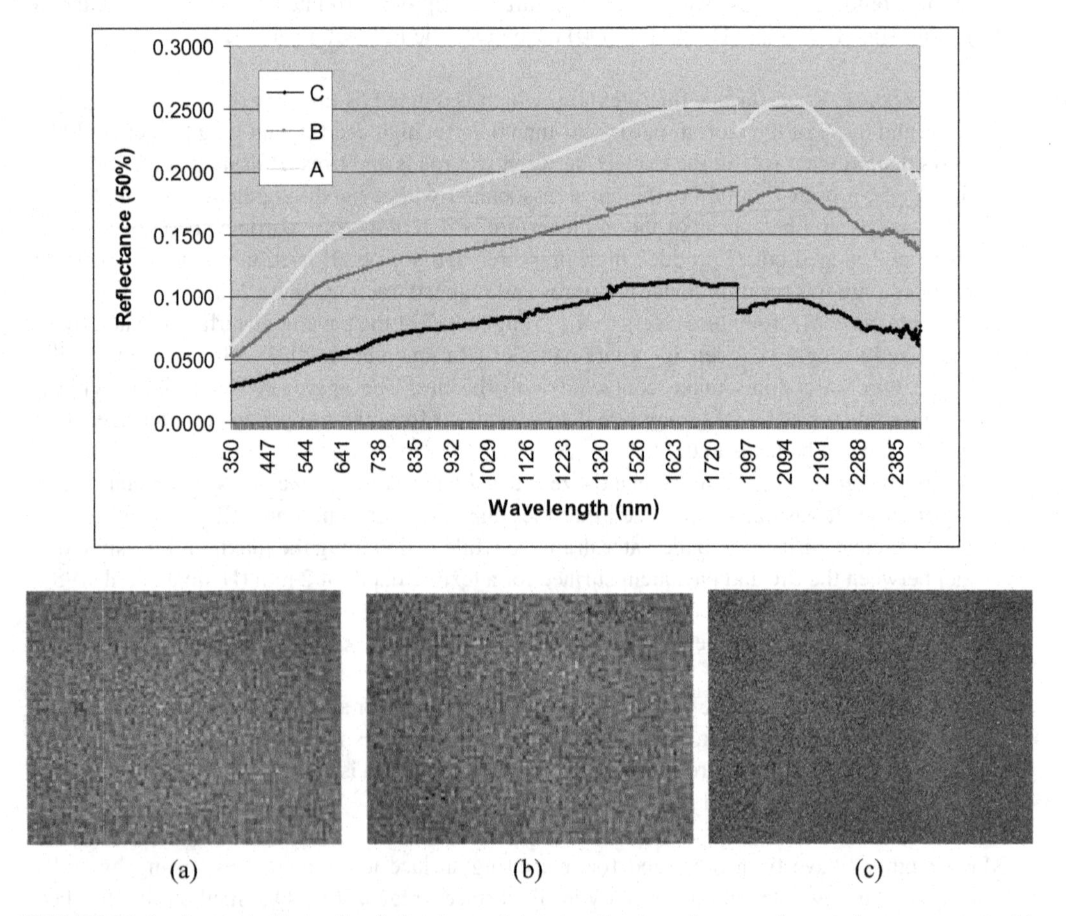

FIGURE 19.1 Spectral effects of asphalt aging and deterioration from the ground spectral measurements: (a) age greater than 10 years, (b) age less than 3 years, and (c) age less than one year.

linear rise towards longer wavelengths. Spectrum A represents a reflectance for an old, deteriorated road surface (see Figure 19.1a). With the erosion and aging of the asphalt mix, the road surface is less viscous and more prone to structural damage like cracking, resulting in high reflectance. In addition, as seen from Figure 19.1, for longer wavelengths, Spectrum C exhibits some obvious absorption features in the SWIR. The low overall reflectance suppresses most of the distinct features except the most prominent ones near 1750 nm and from 2200 to 2500 nm. This causes the strong reflectance decrease beyond 2200 nm.

19.3.2.2 Spectral Properties of Typical Asphalt Road Distresses

Usually, different types of asphalt road distress cause different frictions, thus the spectral properties of asphalt road distresses are first analyzed, and then their relationship with friction is further explored. Figure 19.2 shows the results of the spectral reflectance in response to structural damage or alligator cracks at different severity. As seen from Figure 19.2, the cracking distress has a significant impact on spectral reflectance in all spectra, and it exposes deeper layers of the pavement with higher contents of the original asphalt mix that is then manifested in increased hydrocarbon absorption features. This fact highlights the contrary spectral signal between road deterioration of the pavement itself (see Figure 19.1) and the severity of structural damage (see Figure 19.2). The concave shape in the VIS/NIR is more obvious for brighter, non-cracked road pavements.

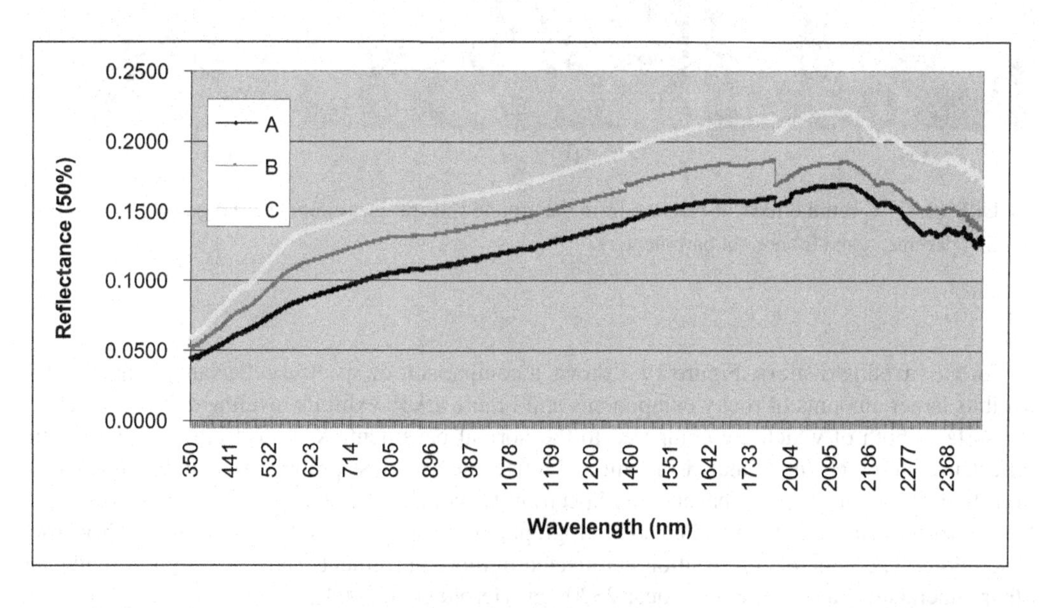

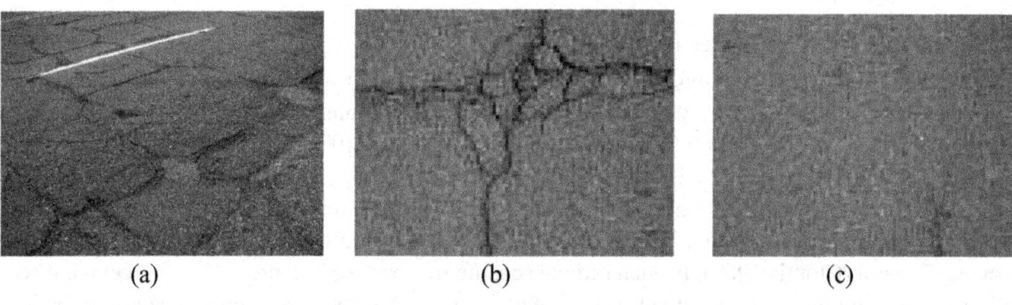

FIGURE 19.2 Spectral effects of severity of structural road damage from the ground spectral measurements: (a) high-severity cracking, (b) low-severity cracking, and (c) no cracking.

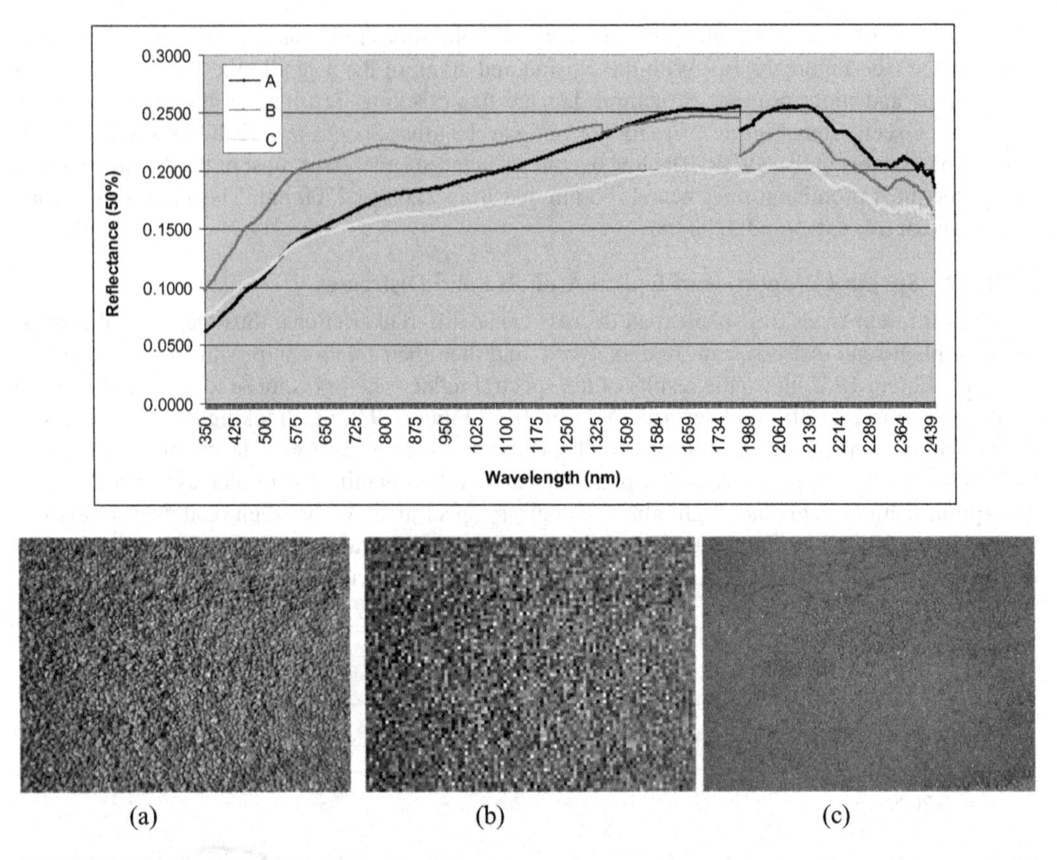

FIGURE 19.3 Spectral effects of raveling from the ground spectral measurements: (a) gravel pavement, (b) raveling pavement, and (c) normal pavement.

For the raveling distress, Figure 19.3 shows a comparison of spectral reflectance. Figure 19.3a exhibits larger amounts of rocky components and Figure 19.3b exhibits raveling debris (gravel) on the surface, both of which are compared to the normal pavement. As seen from Figure 19.3, the reflectance increases due to increasing mineral reflectance and less prominent hydrocarbon absorption. In comparison with the pavements, Spectrum C, which reflects a gravel parking lot surface, has higher reflectance in the visible and photographic near infrared due to the missing hydrocarbon absorptions. The mineral composition is reflected in more prominent features from iron oxide and other minerals like a calcite feature near 2320 nm (Herold et al. 2004).

19.3.2.3 Spectral Properties of Asphalt Paint

Herold et al. (2004) have demonstrated that street paint has a significant impact on pavement spectral analysis because it has highly reflective properties. For example, Spectra A and B in Figure 19.4 show reflectance values up to 60 % in the VIS/NIR region. The difference between Spectrum A and Spectrum B in the visible region is due to different colors, since Spectrum B represents yellow street paint and blue wavelengths are absorbed, while Spectrum A represents a white "turn" symbol. In addition, the street paint, with a typical asymmetric hydrocarbon doublet, has a strongest absorption at 1720 nm (Cloutis 1989). From a remote sensing perspective, the presence of street paint will increase the brightness of a road surface, especially in the VIS/NIR ,and emphasize hydrocarbon absorption in the SWIR.

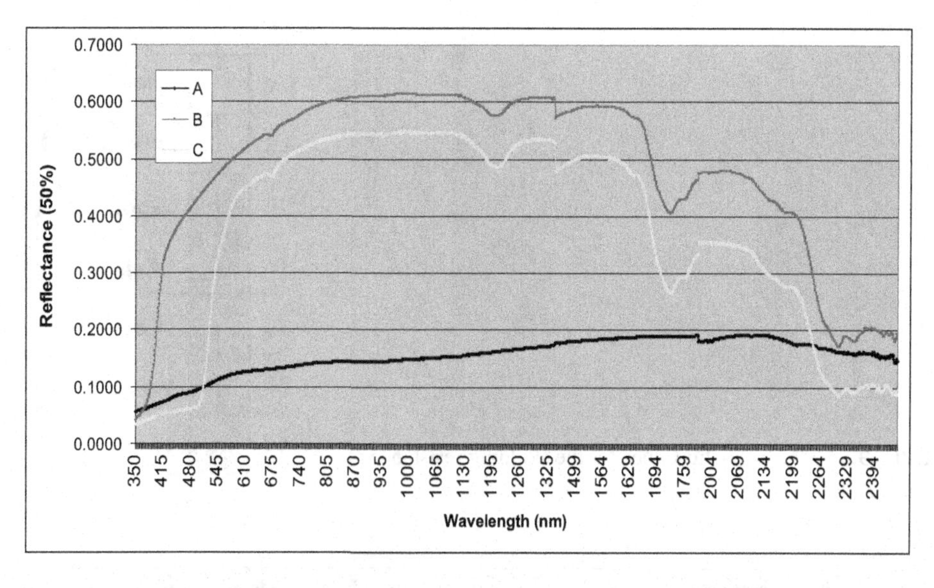

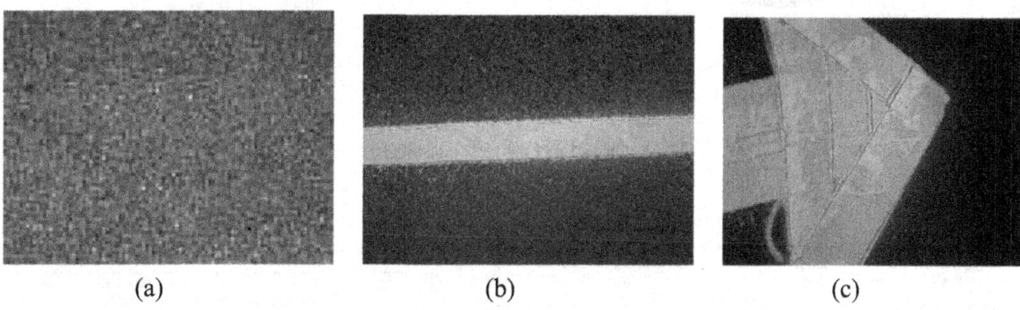

FIGURE 19.4 Spectral characteristics of street paint in different colors versus asphalt pavement from the ground spectral measurements: (1) normal pavement, (b) yellow street paint, and (c) white street paint.

19.4 MEASUREMENT AND MAPPING OF SKID RESISTANCE

19.4.1 Relationship Between PCI and Friction Coefficient

There are no skid resistance and/or friction coefficient data in the study area, but the PCI data are provided by a vendor. In order to deploy the study, the relationship between PCI and friction coefficient is first established. As described above, the PCI, scaled from 0 and 100, is surveyed by two expert groups, who categorized the road condition in five categories (excellent, good, fair, poor, and very poor) for road management action. On the other hand, for reasons of traffic safety there exist minimum requirements for the skidding resistance in the longitudinal direction for different roads. Thus, the relationship between the PCI and skidding resistance is established, and is depicted in Figure 19.5.

19.4.2 Modeling the Relationship Between Spectral Reflectance and Skid Resistance

In order to model the relationship between spectral reflectance and friction as precisely as possible, the surface road is divided into three categories: new asphalt road, asphalt road 1–3 years old, and asphalt road over 10 years old. For each category, a polynomial equation is applied to fit the reflectance curve. The results are depicted in Figures 19.6 to 19.8. For each regression analysis below, the values for reflectance are expressed as fractions of one (0.5 = 50 % reflectance). The

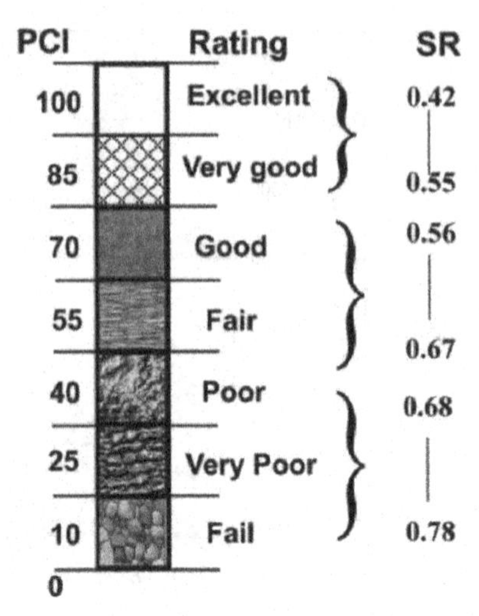

FIGURE 19.5 Characteristic values for the skidding resistance of roads and PCI.

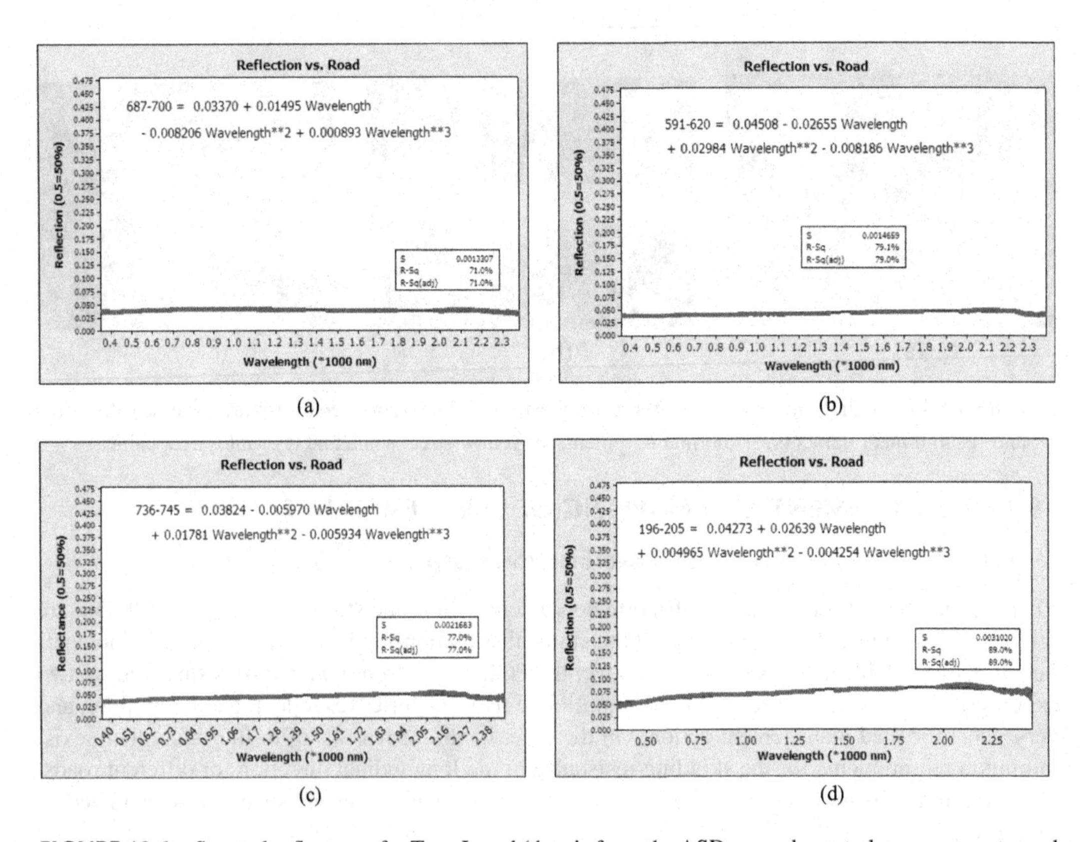

FIGURE 19.6 Spectral reflectance for Type I road (data is from the ASD ground spectral measurements, and the major water vapor absorption bands are interpolated).

spectral measurements range from 350 to 2450 nm (bandwidth 1 nm). The major water vapor absorption bands (1340–1450 nm and 1790–1970 nm) have been deleted. No noise removal or any related modifications have been performed and every user of the library should consider the following definitions:

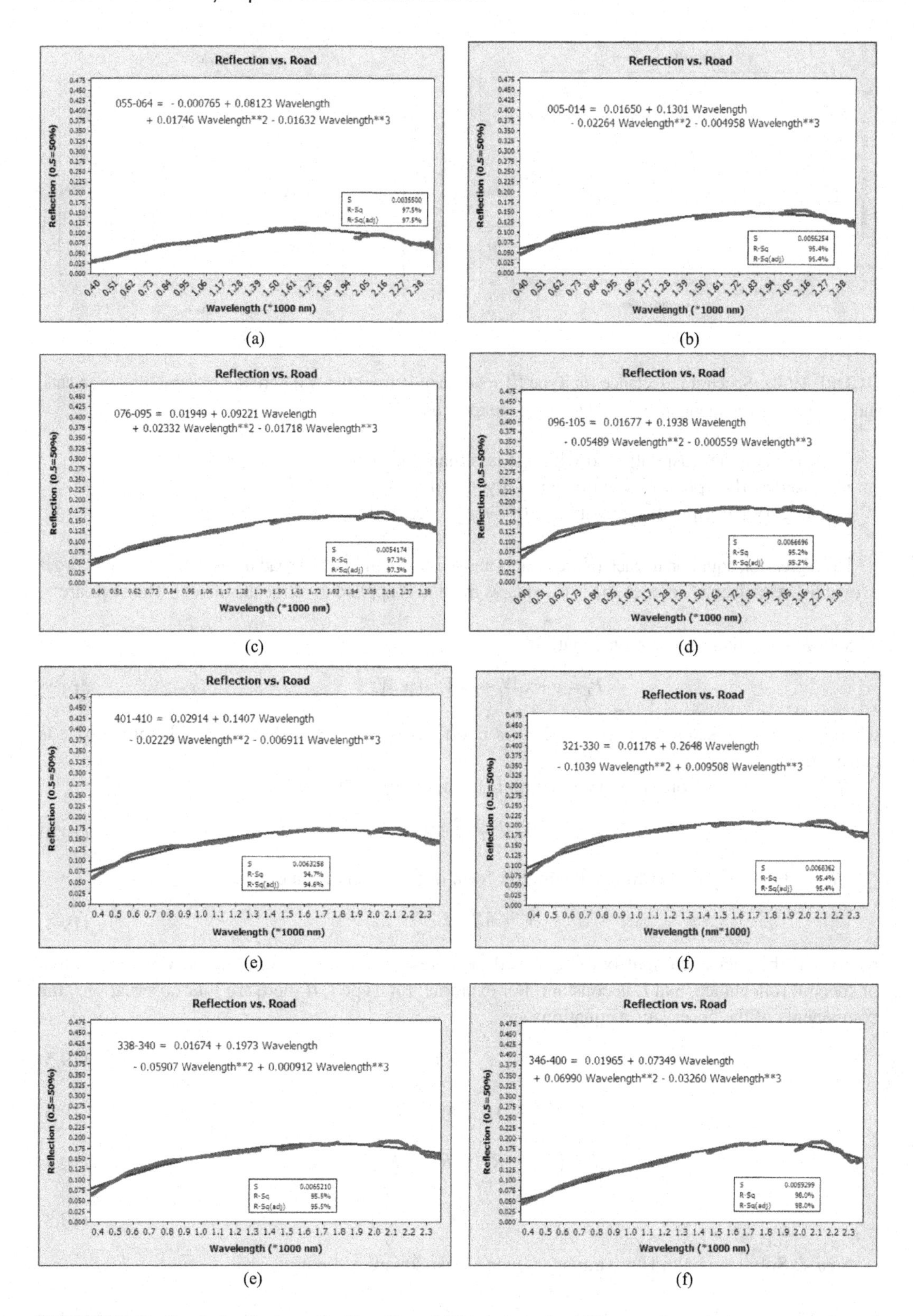

FIGURE 19.7 Spectral reflectance for Type II road (data is from the ASD ground spectral measurements, and the major water vapor absorption bands are interpolated).

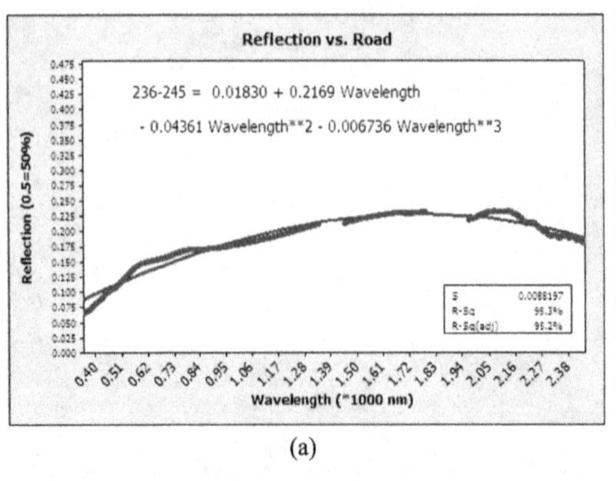

(a)

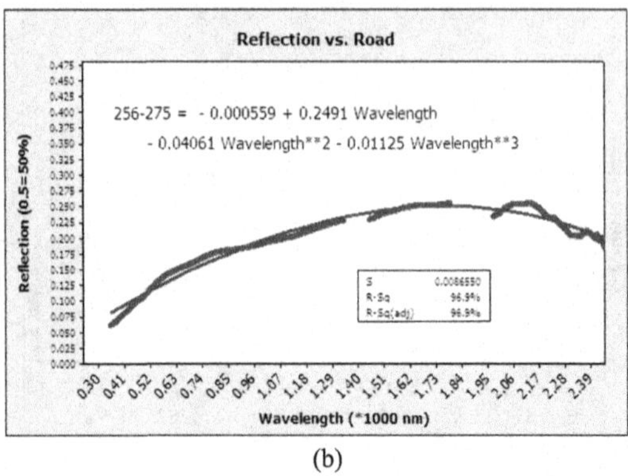

(b)

FIGURE 19.8 Spectral reflectance for Type III road (data is from the ASD ground spectral measurements, and the major water vapor absorption bands are interpolated).

- Category I: new asphalt road with age less than one year;
- Category II: asphalt road with age of 3 years old;
- Category III: asphalt road with age of over 10 years old.

The regression equation in each of the figures above is from individual road measurement data. In order to establish a "universal" model, the least-squares method is applied to refine this model. The steps are

Step 1: Establish observation equation

$$R_i = c + x_1 W_i + x_2 W_i^2 + x_3 W_i^3 \tag{19.2}$$

where R_i is skid resistance, x_i $(i = 1,2,3)$ is coefficients, W_i $(i = 1,2,3)$ is wavelength, and c is constant.

Step 2: Rewrite the observation equation for residual value by

$$v_i = c + x_1 W_i + x_2 W_i^2 + x_3 W_i^3 - R_i \tag{19.3}$$

Step 3: For many observations, the matrix form of the observation equations is

$$V = AX - L \tag{19.4}$$

where X is the coefficient matrix to be solved, $\boldsymbol{A}$ represents the coefficient matrix, $\boldsymbol{V}$ is the residual of spectral reflectance, and $\boldsymbol{L}$ is constant. For example, for Type I, if there are four observations, the components of the observation equations are

$$V = \begin{pmatrix} v_1 & v_2 & v_3 & v_4 \end{pmatrix}^T \tag{19.5}$$

$$X = \begin{pmatrix} c & x_1 & x_2 & x_3 \end{pmatrix} \tag{19.6}$$

$$L = \begin{pmatrix} R_1 & R_2 & R_3 & R_4 \end{pmatrix} \tag{19.7}$$

$$A = \begin{pmatrix} 1 & W & W^2 & W^3 \end{pmatrix} \tag{19.8}$$

Step 4: Solve the unknown parameters using least-square estimation, that is,

$$X = N^{-1} A^T L \tag{19.9}$$

where $N^{-1} = (A^T A)^{-1}$.

Through solving the four types of equation, the following equations are obtained:

$$\text{Type I}: R = 0.0399375 + 0.002205W + 0.01110225W^2 - 0.00437025W^3 \tag{19.10}$$

$$\text{Type II}: R = 0.02211143 + 0.1461657W - 0.01348714W^2 - 0.01022W^3 \tag{19.11}$$

$$\text{Type III}: R = 0.01931275 + 0.242425W - 0.0649375W^2 - 0.002317W^3 \tag{19.12}$$

With the above computational procedures, the three types of equations were established to represent three types of skidding resistance values: 0.42–0.55, 0.56–0.67, and 0.68–0.88, respectively.

19.4.3 Validation of Established Model

In this section, an algorithm, called spectral reflectance curve matching, is established to validate the established model. The basic idea is that the established model for describing the spectral reflectance of different types of road surface is taken as reference data, that is, the corresponding parameters (c, X1, X2, and X3) of the model will be used as reference data. For a specific type of road, its spectral reflectance curve can be obtained, that is, can be modeled (c¢, X1¢, X2¢, and X3¢). The corresponding coefficients of two models are matched. If the differences of all corresponding parameters are less than a given threshold, the model is accepted, and the corresponding skid resistance will be assigned to this type of road.

Figure 19.9 illustrates an example of how the established model is validated. The white curve is drawn by the above established model, and the green curves are drawn up by the given threshold for each parameter. With the given threshold, the spectral reflectance in the image is researched until the difference in spectral reflectance between the researched and modeled values is less a given threshold. For skid resistance of different regions of the road, the spectral reflectance curves are different (see Figure 19.9a and b for Regions 1 and 2).

When a spectral reflectance is accepted, the corresponding skidding resistance is assigned to this road. For example, if skid resistance of 0.56–0.67 has been assigned to a road, it implies that this road age is about 1–3 years old, and the corresponding spectral curve (see Figure 19.10a) and the imaged road surface can be pulled out in Figure 19.10b.

With the above the step, the AOI (area of interest) is found for mapping the skid resistance. ENVI software is utilized to map the skid resistance. The results are depicted in Figure 19.11, in which all of the red points indicate that their skid resistance is about 0.56–0.67. Similarly, the road is tested at an age of less than one year, which is associated with a skid resistance of 0.56–0.67. The results are depicted in Figure 19.12.

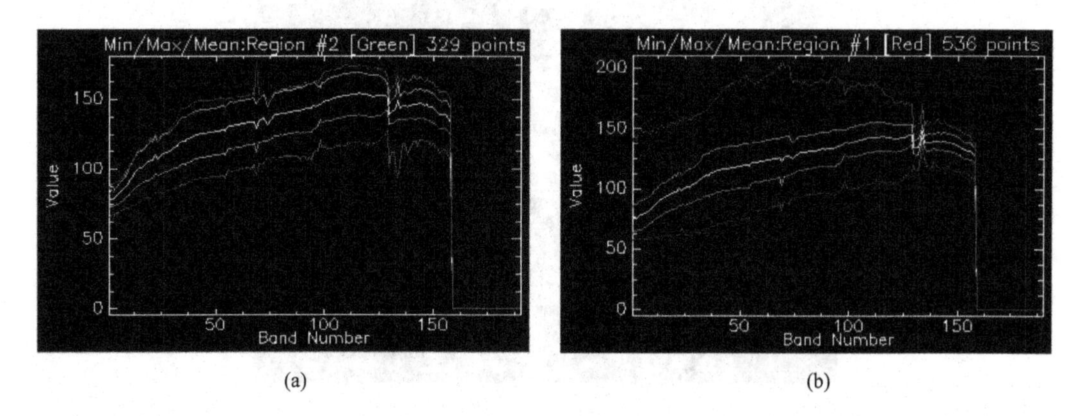

FIGURE 19.9 Spectral reflectance curve matching: (a) good matching, and (b) incorrect matching.

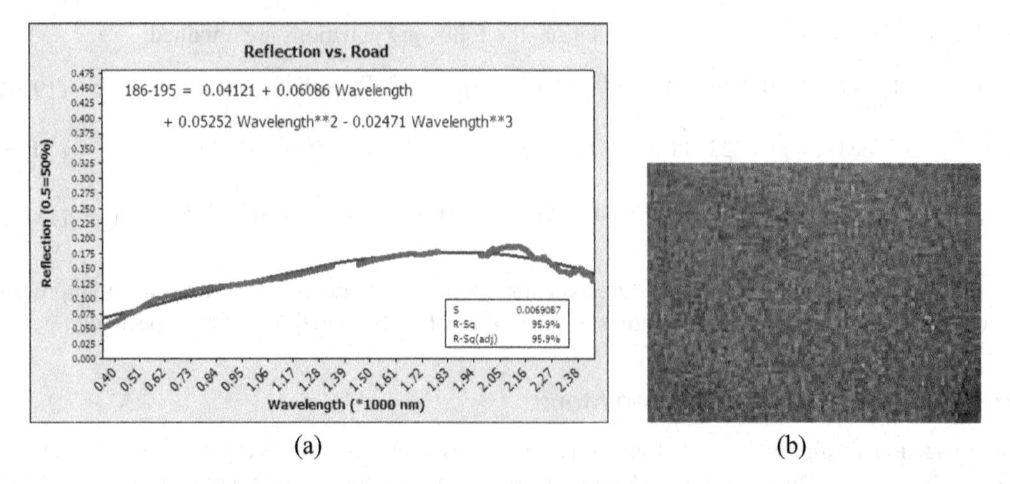

FIGURE 19.10 The recognized road associated with skid resistance at 0.56–0.67 through the spectral reflectance matching.

FIGURE 19.11 Mapping skid resistance at 0.56–0.67 using established model.

FIGURE 19.12 Mapping skid resistance at 0.42–0.55 using established model.

19.5 CONCLUSIONS

This chapter combined field spectrometry, *in situ* road surveys, and hyperspectral remote sensing to explore the potential in measuring skidding resistance from hyperspectral remote sensing imagery. The basic principle is that the magnitude of road surface skidding resistance is directly related to the road surface textures, while the road surface textures expose differences in spectral reflectance. Thus, the main contribution of this research is to develop a method to measure the skid resistance using hyperspectral images. Traditionally, the skid resistance measurements were made using a specific machine. With the presented technology in this chapter, the skid resistance measurements can be made in an office. The research results demonstrated that the aging and degradation of road surfaces, which have different skidding resistances, are represented by distinct spectral characteristics. Moreover, road skidding resistance predictions can be made using mapping from hyperspectral remote sensing imagery. This is the first time of presenting this idea and further studies are needed in the future.

With the study conducted in this chapter, successful applications of hyperspectral imaging in transportation will be able to facilitate the construction of better transportation facilities, lead to better transportation policies, and help the development of better ways of responding to natural and accidental disasters affecting transportation. Hyperspectral imagery has become widely used for a variety of applications in many fields. Even though the technology has not been fully exploited in the transportation field, the technology provides the opportunity to improve on existing procedures of skidding resistance measurement today in the transportation field, which is labor-intensive and time-consuming. Hyperspectral remote sensing imagery would be valuable in assessing the condition of the entire road network with less labor and less time. With the use of this technology, monitoring the skidding resistance of road networks is also important, and would save money and time. On the other hand, a few limitations to this study are: (1) this study only considers one factor, skid resistance. In fact, four major elements – driver, roadway, vehicle, and weather – have a significant impact on the skidding resistance, and they interact with others to cause skid accidents. (2) The investigation in this study only focused on a small study area and on asphalt road surfaces.

REFERENCES

Anderson, D.A., and Henry, J.J., Synthetic aggregates for skid-resistant surface courses, *Transportation Research Record*, vol. N712, pp. 61–68, 1979.

Asi, I.M., Evaluating skid resistance of different asphalt concrete mixes, *Building and Environment*, vol. 42, no. 1, pp. 325–329, January 2007.

Augustin, H., Skid resistance and road-surface texture, *Astm Special Technical Publication*, vol. N1031, pp. 5–13, June 1990.

Bazlamit, S.M., and Reza, F., Changes in asphalt pavement friction components and adjustment of skid number for temperature, *Journal of Transportation Engineering*, vol. 131, no. 6, pp. 470–476, June 2005.

Clark, R.N., Spectroscopy of Rocks and Minerals and Principles of Spectroscopy, In: A.N. Rencz (ed.). *Manual of Remote Sensing, Chapter 1*, John Wiley and Sons, New York, pp. 3–58, 1999.

Cloutis, A.E., Spectral reflectance properties of hydrocarbons: remote-sensing implications, *Science*, 4914, pp. 165–168, 1989.

Crocker, M., et al. Measurement of acoustical and mechanical properties of porous road surfaces and tire and road noise, *Transportation Research Record*, vol. 1891, pp. 16–22, 2004.

Do, M.T., Zahouani, H., and Vargiolu, R., Angular parameter for characterizing road surface microtexture, *Transportation Research Record*, vol. 1723, pp. 66–72, 2000.

FHWA, *Development and testing of INTRANS: A microscopic freeway simulation model*, Technical Report FHWA/RD-80/106-108, Federal Highway Administration, US-DOT, McLean, Virginia, USA, 1980.

Gomez, R.B., Hyperspectral imaging: a useful technology for transportation analysis, *Optical Engineering*, vol. 41, no. 9, pp. 2137–2143, 2002.

Heaton, B.S., Henry, J.J., and Wambold, J.C., Texture Measuring Equipment Vs Skid Testing Equipment, *Proceedings of The Australian Road Research Board, N Pt 2, Pavements and Materials*, pp. 53–64, 1990.

Herold, M., et al. Spectrometry for urban area remote sensing - Development and analysis of a spectral library from 350 to 2400 nm, *Remote Sensing of Environment*, vol. 91, no. 3-4, pp. 304–319, 2004.

Herold, M., et al. Spectrometry and Hyperspectral Remote Sensing of Urban Road Infrastructure, Online Journal of Space Communications, 3, http://satjournal.tcom.ohiou.edu/issue03/applications.html (access: March 2008).

Kyriakopoulos, K.J., and Skounakis, N., Moving obstacle detection for a skid-steered vehicle endowed with a single 2-D laser scanner, *IEEE International Conference on Robotics and Automation*, vol. 1, pp. 7–12, 2003.

Lee, Y.P.K., et al. Skid resistance prediction by computer simulation, *Proceedings of the Eighth International Conference on Applications of Advanced Technologies in Transportation Engineering*, Beijing, China, pp. 465–469, 2004.

Lees, G., and Williams, A.R., Machine for friction and wear testing of pavement surfacing materials and tyre tread compounds. *Rubber Industry*, vol. 8, no. 3, pp. 114–120, June 1974.

Liu, Y., Fwa, T.F., and Choo, Y.S., Effect of aggregate spacing on skid resistance of asphalt pavement, *Journal of Transportation Engineering*, vol. 129, no. 4, pp. 420–426, July/August 2003a.

Liu, Y., Fwa, T.F., and Choo, Y.S., Finite-element modeling of skid resistance test, *Journal of Transportation Engineering*, vol. 129, no. 3, pp. 316–321, May/June 2003b.

Liu, Y., Fwa, T.F., and Choo, Y.S., Effect of surface macrotexture on skid resistance measurements by the British pendulum test, *Journal of Testing and Evaluation*, vol. 33, no. 4, pp. 304–309, July 2005.

Mate, C.M., On the road to an atomic- and molecular-level understanding of friction, *MRS Bulletin*, vol. 27, no. 12, pp. 967–971, December 2002.

Neuman, R.F., Urban, J.A., and Mcninch, J.H., Performance characterization of dry friction materials, *I Mech E Conference Publications (Institution of Mechanical Engineers)*, London: Mechanical Engineering Publications Ltd, pp. 233–238, 1983.

Nie, Y., et al. 2001. Hyperspectral imaging in earth road construction planning, *Processing of SPIE*, vol. 4383, pp. 12–22, 2001.

Ong, G.P., and Fwa, T.F., Prediction of wet-pavement skid resistance and hydroplaning potential, *Transportation Research Record, no. 2005, Pavement Rehabilitation, Strength and Deformation Characteristics, and Surface Properties-Vehicle Interaction 2007*, pp. 160–171, 2005.

Purushothaman, N., Heaton, B.S., and Moore, ID., Numerical analysis of the friction mechanism of grooved road surfaces, *ASTM Special Technical Publication*, vol. N1031, pp. 127–137, June 1990.

Raisanen, M., Kupiainen, K., and Tervahattu, H., The effect of mineralogy, texture and Mechanical properties of anti-skid and asphalt aggregates on urban dust, *Bulletin of Engineering Geology and The Environment*, vol. 62, no. 4, pp. 359–368, November 2003.

Usher, J., and Truax, D., Exploration of Remote Sensing Applicability within Transportation. Remote Sensing Technologies Center final projects report, http://www.rstc.msstate.edu/publications/99-01/rstcofr01-005b.pdf (March 2008).

Woodward, D., Woodside, A., and Jellie, J., Improved prediction of aggregate skid resistance using modified PSV tests, *Proceedings of the Eighth International Conference on Applications of Advanced Technologies in Transportation Engineering*, Beijing, China, pp. 460–464, 2004.

Index

Page numbers in *italics* refer to content in *figures*; page numbers in **bold** refer to content in **tables**.

I

J

K

L

M

Printed in the United States
By Bookmasters